Landscape Series

Volume 36

Benjamin Kofi Nyarko, Dept of Geography & Regional Planning, University of Cape Coast, Cape Coast, Ghana
Henrique Pereira, German Centre for Integrative Biodiversity Research (iDiv), Martin Luther University Halle-Wittenberg, Leipzig, Sachsen, Germany
Alexander Prishchepov, Dept of Geosci & Natural Resource Mgmt, University of Copenhagen, Copenhagen, Denmark
Robert M. Scheller, North Carolina State University, Raleigh, NC, USA
Kalev Sepp, Inst. Agricultural & Environ. Sciences, Estonian University of Life Sciences, Tartu, Estonia
Anton Shkaruba, Inst. of Agricultural & Environ Sciences, Estonian University of Life Sciences, Tartu, Estonia
Ileana Stupariu, Dept. Regional Geography and Environment, University of Bucharest, Bucharest, Romania
Raymond Tutu, College of Humanities, Delaware State University, DOVER, DE, USA
Teiji Watanabe, Fac of Environmental Earth Sci, A-301, Hokkaido University, Sapporo, Hokkaido, Japan
Wei-Ning Xiang, University of North Carolina at Charlotte, Charlotte, NC, USA
Qing Zhao, Institute of Applied Ecology, Chinese Academy of Sciences, Shenyang, China

Springer's innovative Landscape Series is committed to publishing high quality manuscripts that approach the concept of landscape and land systems from a broad range of perspectives and disciplines. Encouraging contributions that are scientifically-grounded, solutions-oriented and introduce innovative concepts, the series attracts outstanding research from the natural and social sciences, and from the humanities and the arts. It also provides a leading forum for publications from interdisciplinary and transdisciplinary teams across the globe.

The Landscape Series particularly welcomes contributions around several globally significant areas for landscape research, which are anyhow non-exclusive:

- Climate and global change impacts on landscapes and ecosystems including mitigation and adaptation strategies
- Human Dimensions of Global Change

- Biodiversity and ecosystem processes linked to ecosystems, landscapes and regions
- Biogeography
- Ecosystem and landscape services including mapping, assessment and modelling
- Land System Science
- Regional ecology (including bioregional theory & application)
- Human-Environment Interactions and Social-Ecological Systems & Frameworks (SESF) - including theories, practice and modelling

Volumes in the series can be authored or edited works, cohesively connected around these and other related topics and tied to global or regional initiatives. Ultimately, the Series aims to facilitate the application of landscape research and land system science to practice in a changing world, and to advance the contributions of landscape theory and research and land system science to the broader scholarly community.

José Muñoz-Rojas • Roberto García-Ruiz
Editors

The Olive Landscapes of the Mediterranean

Key Challenges and Opportunities for their Sustainability in the Early XXIst Century

Springer

Editors
José Muñoz-Rojas
MED-Mediterranean Institute for Agriculture, Environment and Development
CHANGE-Associated Lab for Global Change & Sustainability Department of Landscape, Environment and Planning
Universidade de Évora
Valverde - Évora, Portugal

Roberto García-Ruiz
University Institute for Research on Olive Groves and Olive Oils
University of Jaén
Jaén, Spain

ISSN 1572-7742 ISSN 1875-1210 (electronic)
Landscape Series
ISBN 978-3-031-57955-4 ISBN 978-3-031-57956-1 (eBook)
https://doi.org/10.1007/978-3-031-57956-1

This Springer imprint is published by the registered company Springer Nature Switzerland AG
The registered company address is: Gewerbestrasse 11, 6330 Cham, Switzerland

Contents

Part I
Introduction: Why Are Olive Landscapes in the Mediterranean Worth a Book?

Chapter 1
Olive Groves (and Olive Oil) in the Mediterranean in 2023: A Heterogeneous and Dynamic Mosaic of Rural Landscapes and Ecosystems

José Muñoz-Rojas and Roberto García-Ruiz

Abstract Domesticated olive groves have been documented to be present in the Mediterranean basin for the past 8.00–6.000 years. Pollen and DNA records show how the wild olive or oleaster has persisted throughout the Holocene across the entire Mediterranean, expanding and retracting across Southern Europe, Northern Africa, and the Middle East in response to climate fluctuations and socio-political turns (Besnard et al., 2013; Breton et al., 2012). In this opening chapter, the reader will be introduced to the main stages and tipping points identified for the historical trajectories of olive groves and olive oil production across the Mediterranean, resulting in the complex olive landscape mosaic that we encounter today. Furthermore, the reader will also find hereby an introduction to the diverse social-ecological challenges they face in the era of sustainability, all of which are addressed along the book. Ultimately, we intend to provoque the reader to read the book, in whichever sequence and mode they may find best suited to their own interests and mood, and then question themselves; what do I know now about olive landscapes and their sustainabilty that I didn't know before?, and what else do I still not know nor understand that is relevant?. Whether you are an academic, practitioner, whether you are working in the public or private spheres, or just as some lay person with curiosity about the topics addressed in the book, we expect that you will follow us in the complex, even wicked, and yet nonetheless exciting quest for securing the sustainability of these ancient and yet highly innovative, valuable and yet largely threatened, most Mediterranean of all Mediterranean landscapes.

J. Muñoz-Rojas (✉)
MED-Mediterranean Institute for Agriculture, Environment and Development CHANGE-Associated Lab for Global Change & Sustainability Department of Landscape, Environment and Planning, Universidade de Évora, Valverde - Évora, Portugal
e-mail: jmrojas@uevora.pt

R. García-Ruiz
University Institute for Research on Olive Groves and Olive Oils, University of Jaén, Jaén, Spain
e-mail: rgarcia@ujaen.es

J. Muñoz-Rojas, R. García-Ruiz (eds.), *The Olive Landscapes of the Mediterranean*, Landscape Series 36,
https://doi.org/10.1007/978-3-031-57956-1_1

1.1 Olive Groves (and Olive Oil)—Their Relevance for Culture, Society, the Environment, and Economies in the Mediterranean; Past, Present, and (Presumable) Future

1.1.1 Olive Groves in the Mediterranean; Historical and Paleo-Geographic Evidence Stretching from the Holocene to Modern Times

Domesticated olive groves have been documented to be present in the Mediterranean basin for the past 8.00–6.000 years. Pollen and DNA records show how the wild olive or oleaster has persisted throughout the Holocene across the entire Mediterranean, expanding and retracting across Southern Europe, Northern Africa, and the Middle East in response to climate fluctuations and socio-political turns (Besnard et al., 2013; Breton et al., 2012).

It has been long acknowledged how it was the Phoenicians who originally brought olive oil and olives to Central and Western Mediterranean. Historians also agree that, since ancient times, the Berbers grafted oleaster trees long before the arrival of the Romans and the development of olive plantations by the Carthaginians. It is also possible that the Berber name 'Oleaster' may be at the origin of the Greek word ἐλαία [elaia] and the Latin olea used by the Romans (Loumou & Giourga, 2003).

Under Greek and Roman rule, olive cultivation intensified, throughout the Eastern and Western Mediterranean. Greeks and romans possessed knowledge about agronomic techniques, production, storage and marketing of oil, and thus olive oil trade flourished during such times, as in noted in the cultural records. Olives (Fig. 1.1) and their oil were the engine of growth for the maritime trade in the Mediterranean basin (Loumou & Giourga, 2003).

After the fall of the Roman Empire, the few sources available indicate that olive cultivation continued throughout Italy, south-eastern France, and the Mediterranean Iberian Peninsula. Planting of this crop and the production of its oil were mainly carried out for self-consumption with the main purpose was to provide sustenance (Loumou & Giourga, 2003).

During the Middle Ages, the growth of olive cultivation and the production of olive oil in the western and Christian regions of the Mediterranean appear to be rather small, even in decline. Some historians (Loumou & Giourga, 2003) put forward the hypothesis that this was due to a reduced presence of fat in the medieval diet, as the cookbooks of the time also attest.

In Spain, mention is made of the medicinal uses of olive oil in the medical works of Averroes, who recommended the addition of olive oil in culinary preparations. In the Iberian Peninsula, it is necessary to distinguish between the territories under Christian rule from those under Muslim rule, which concentrated most of the olive groves along the Almohad and Almoravid periods. The production and consumption

Fig. 1.1 The Olive tree of Vouves (Chania, Greece) is widely acknowledged to be the longest-living olive tree, dated at more than 3500 years bc. Now protected as a Natural Monument

of olive oil in the Iberian Peninsula were thus directly related to the Arab presence in the region. Due to animal fats being restricted within Arab religious culture, the consumption of fats of vegetable origin, such as olive oil, was favored (Guzmán Álvarez et al., 2020). During Muslim rule in Iberia, Sicily, and Eastern Europe, advances in agriculture were notable and many agricultural and olive cultivation treaties were written. The techniques applied in Al-Andalus are similar to those already used by the Romans, some of which are still practiced today in Andalusia, Extremadura, or North Africa. It was also in the Middle Ages that trade routes were created between the olive groves in North Africa and Marseille.

During the period of the great geographical discoveries and the maritime expansion of Spain and Portugal (especially in the fifteenth and sixteenth centuries), olive trees were introduced into these new territories in the Americas (Mexico, Peru, Argentina, etc.) (Guzmán Álvarez et al., 2020). However, there were periods of truly low demand during which time the exploitation of the olive tree and the consumption of its oil were put on hold. Consumption habits shifted rapidly, and the impact of weather phenomena, such as frost, for the growth of olive trees, cannot be ignored (Fig. 1.2).

In the nineteenth century, however, there was a renewed interest in olive oil. Its use and demand expanded, especially in the south of France, promoting economic growth across the entire region. It was not until the middle of the twentieth century that scientific and health-related studies on the properties of olives helped to boost, the consumption of oil for food purposes, particularly in Europe.

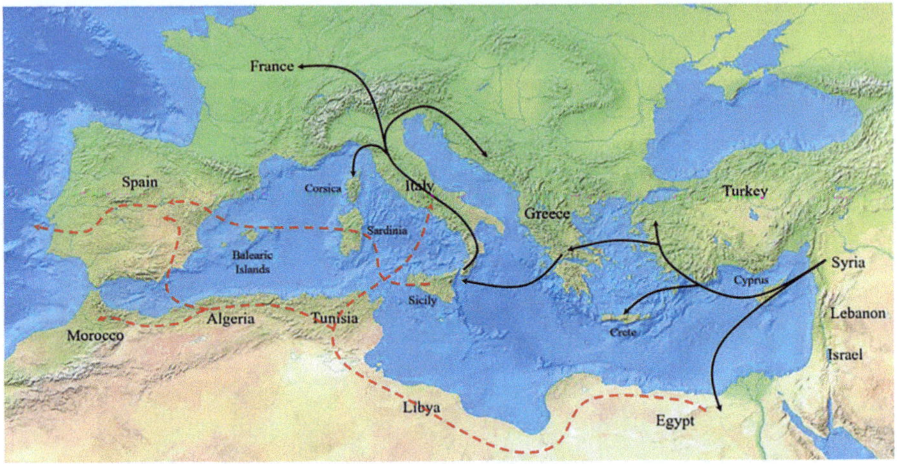

Fig. 1.2 Trajectories of the commerce and cultivation of olive groves, along history, according to Loumou and Giourga (2003)

Fig. 1.3 Tablet showing a Hand Clutching an Olive Branch. New Kingdom, Dynasty 18. Reign of Akhenaten Amarna Period. Middle Egypt, Hermopolis (Egypt) (dated 1353–1323 B.C). Currently in the collections of the British Museum

1.1.2 Olive Groves in the Mediterranean; Cultural Histories, and Symbolic Values

The Egyptians were among the first to use olive tree products in their daily lives but also in their rituals (for oil lamps in the temples or during funeral rites for example). This is well documented in classic Egyptian art, where hieroglyphs and tablets frequently show the importance of olive trees and olives as a sacred and ritual object of worship and devotion, symbolizing fertility and prosperity (Fig. 1.3).

Across all the Mediterranean polytheistic and monotheistic religions, we find this tree seen as a gift from the Gods. In general, the civilizations of the Mediterranean basin saw the olive tree as sacred. And since ancient times, olive oil has symbolized strength and light.

Attributed to Isis then Athena, the olive tree became the symbol of the city of Athens, having been created by Zeus' daughter. The goddess produced this tree, capable of feeding and healing for men (Loumou & Giourga, 2003).

In Christian legends and in the Bible, the olive tree is one of the trees that appears the most frequently: it is an olive branch in the beak of a dove that shows Noah that the Flood is over and that the first tree to have grown is an olive tree. In the *Book of Judges*, the olive tree refuses the title of king of trees. Mentioned seven times in the Quran, it is described as "a blessed tree (...) whose oil would almost glow forth (of itself), though no fire touched it."

This all has had an imprint on the cultures, traditions, peoples, and civilizations of the Mediterranean, which cannot be separated from the trajectories of olive grove cultivations and related landscapes.

1.1.3 Recent Trajectories of Change and Related Challenges for Societies, the Economy, and the Environment Across Spatial-Temporal Scales and Socio-cultural Contexts

As Infante-Amate (Infante-Amate, 2012) and others have clearly documented, the landscape dominated by olive groves that is now seen as characteristic of many areas of the Mediterranean, including in southern Spain (where it most clearly dominates current rural landscapes) is a relatively recent phenomenon. In the eighteenth, nineteenth, and much of the twentieth century, olives were not yet an industrial crop, as they currently are, grown on a large scale for the production of oil. Instead, olive trees were largely grown by small peasant farmers and used to produce timber and fodder as well as foodstuffs, forming one component of a diverse peasant economy.

Nowadays (Fig. 1.4), EU countries concentrated approximately 70% of the world's olive oil production in 2016. Spain (circa 43%), Italy (13%), Greece (10%), and Portugal (3%) altogether contribute to more than 95% of the total European production. Along with these EU countries, Tunisia (6%) and Morocco (4%) are also major olive oil producers within the Mediterranean basin. The total area occupied by cropped olive groves in these countries sums up to over 7 M ha and is considered a main component of their socio-economic and cultural life, shaping the rural natural landscapes of many of their regions.

According to the results of the SUSTAINOLIVE project (https://sustainolive.eu/), olive grove cultivation represents a key socio-economic asset for Mediterranean rural regions and helps preventing rural depopulation by generating more than 1000 million working days per year across the whole area.

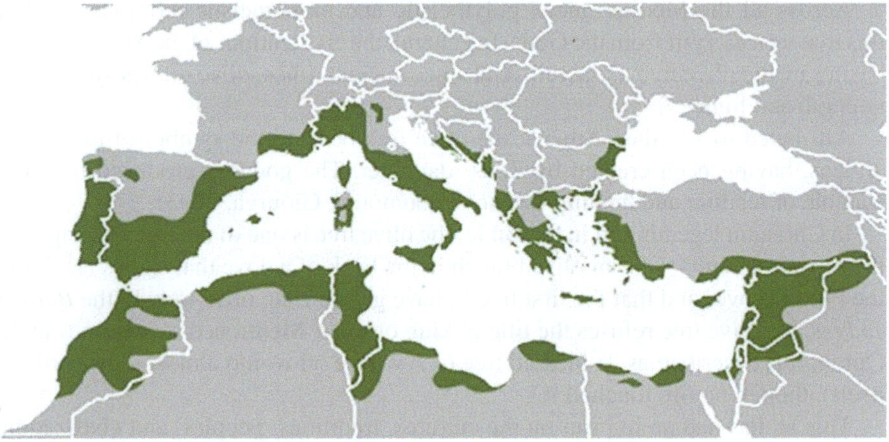

Fig. 1.4 Current biogeographic and bioclimatic distribution of olive groves across the Mediterranean basin

Indeed, the olive sector is a major economic player in terms of its capacity to generate employment and income for a large fraction of the Mediterranean population. As an example, the olive sector of Morocco and Tunisia already provides employment for 10–15% of the population and has greatly reduced unemployment for women in particular.

Beyond their productive value, they also constitute a rural tourist attraction with the presence of ancient olive trees or outstanding olive grove landscapes. Additionally, the olive oil sector represents an important source of healthy food and is highly appreciated worldwide as an integral element of the Mediterranean diet model, which since 2013 is designated as UNESCO Intangible cultural heritage, becoming an imprint for Southern European and Mediterranean countries.

Increased olive oil demand has acted as a catalyst for the intensification and expansion of olive groves, which bears inevitable territorial and social-ecological impacts (Silveira et al., 2018; Moreira et al., 2019).

The intensification of low-input traditional olive groves implies the systematic use of chemical fertilizers and pesticides with more impactful weed control and soil management practices. This has been underpinned by a clear trend to further intensify farming systems by means of irrigation, increased tree density, and mechanical harvesting. Poor management of both intensive and extensive live groves has frequently resulted in simplified olive landscapes with low nature value, driving greater negative environmental impacts, particularly in the form of soil erosion, run-offs to water bodies, increased rates of soil fertility loss, degradation of habitats and landscapes, and over-exploitation of scarce and vulnerable water resources (Fig. 1.5). The near absence of biological elements in too many (extensive and intensive) olive groves across the farm to landscape levels (other than olive trees), might lead to the inefficiency of ecological regulatory mechanisms, which potentially provide ecosystems with resistance and resilience to ever-changing environmental conditions (Moreira et al., 2019).

Fig. 1.5 Main impacts in the live groves resulting from unsustainable management techniques resulting in impacts and problems related to landscape homogenization and fragmentation (top left), groundwater overexploitation and ultimately exhaustion (top right), chemical contamination (bottom left), and soil erosion (bottom right)

1.1.4 A Highly Heterogeneous and Diverse Farming Landscape; Disentangling the Social-ecological Complexity of Scales, Functions, and Dynamics

As a result of the historical trajectories and driving forces above-described, the cultural landscapes of the Mediterranean are still partly marked by traditional olive groves and related production models linked to family farming and subsistence-oriented local economies. However, this is far from homogeneous across different regions and localities, with a diversity of farming and production systems co-existing that are synthesized in Table 1.1.

In contrast with persisting traditional farming systems, a rapid process of intensification and financialization has been identified across multiple Mediterranean regions, especially those with access to water resources and productive soils, that is in alignment with globalization; environmental, social, and cultural externalities (Silveira et al., 2018).

The resulting olive landscape mosaic is a complex palimpsest of social-ecological systems that largely varies across scales, functions, and dynamics of change, resulting in key challenges related to social-ecological resilience, multi-functionality, social fairness, landscape sustainability and integrity, and economic efficiency.

An increasing number of conceptual frameworks are in place to more effectively approach the increasing complexity, and externalities of farming systems are in place, including Ecosystem and Landscape Services and the Ecosystems Approach (Bastian et al., 2014), Landscape Approaches (Sayer et al., 2013), and Social-Ecological Resilience (Boyd & Folke, 2011), just to name the most popular ones.

Table 1.1 Traditionally accepted typology of olive grove cultivation models, used as a basis for current policies, funded and management strategies

	Traditional rainfed with low mechanization capacity	Traditional but irrigated and with some mechanization capacity	Intensified traditional groves	Intensive and super-intensive modern groves
Predominant landscape	Hilly and mountain areas. Also, in marginal lowlands and around villages. Highly natural olive groves.	Hilly and rolling plains. Monocultures.	Hilly and rolling plains. Highly simplified monocultures.	Rolling and flat plains. Highly simplified monocultures.
Tree density (n° ha-1)	20–100	100–300	200–400	400 up to 2000
Tree features and management	Old (> 80 years) or ancient with large tree volume canopies. Olive trees may be mixed with other fruit trees. Pruned biennially or infrequently.	Old and replanted young trees with annual/biannual tree pruning cycles. When irrigated tree density is increased by planting between existing rows.	Young trees annually pruned.	Short-stem "dwarf or bush" tree varieties. Young trees.
Soil and understorey management	Presence of spontaneous plant cover crops is frequent. Mowing and/or tillage may also be frequent or occasional. Grazing highly dependent on sheep/goat density in the surrounding areas.	Presence of spontaneous plant cover crops or bare soils by frequent rounds of tillage in combination with herbicides (e.g. simazine, glyphosate).	Bare soils driven by frequent rounds of tillage in combination with herbicides (e.g. simazine, glyphosate) or presence of spontaneous plant cover (typically 1–2 m stripe in the rows) crops controlled using tillage, herbicides or mowing.	Bare soils driven by frequent rounds of use of herbicides (e.g., simazine, glyphosate)
Fertilisation	None or only manure and/or chemical fertilizers (e.g., 1–2 kg combined fertilizer per tree).	Chemical fertilizers (e.g., 2–6 kg combined fertilizer per tree depending on plantation model, rainfall, irrigation, etc.). Nitrogen input 30–60 kg N ha^{-1}.	Chemical fertilizers (e.g., 2–6 kg combined fertilizer per tree depending on plantation, rainfall, irrigation, etc.). Nitrogen input 50–90 kg N ha^{-1}.	Chemical fertilizers usually applied through irrigation and/or leaf spraying. Annual nitrogen input 100–200 kg N ha^{-1}.

(continued)

Table 1.1 (continued)

	Traditional rainfed with low mechanization capacity	Traditional but irrigated and with some mechanization capacity	Intensified traditional groves	Intensive and super-intensive modern groves
Pesticide use	None or occasional. Sometimes only traditional products, such as Bordeaux mixture copper.	2–10 treatments per year depending on the area, pests, year, etc.	2–10 treatments per year depending on the area, pests, year, etc.	2–10 treatments per year depending on the area, pests, year, etc.
Irrigation	Not usual, although becoming common in certain specific areas, such as Crete.	Not usual but increasingly common (mainly drip systems)	Increasingly common (mostly drip irrigation)	Usual. Drip irrigation systems.
Harvest	Manual and also avoided years of little harvest.	Manual or mechanical.	Mechanical or manual	Mechanical

Nonetheless, a clear lag in the application of these novel approaches is clearly missing for the olive grove farming systems and olive oil production sector across the Mediterranean.

In response to such lags, this book aims to address olive groves across the Mediterranean as a complex mosaic of rural landscapes and ecosystems, and synthesize, from a critical and integrative perspective, the state-of-the-art knowledge about potential trajectories and actions for achieving more sustainable models of cultivation of Mediterranean olive groves, and production of olive oil. Expectedly, the book also fulfills some of the main knowledge gaps and limitations in the literature, whilst helping to move forward by disentangling heterogeneity and unraveling complexity through a social-ecological landscape lens.

1.2 What Can the Reader Learn Along This Book?; Contents, Structure, and Aspirations

1.2.1 Which Kinds of Readers Would We Like to Read this Book?

We must confess that the book was originally ideated, written, and edited from a largely academic perspective, as in clearly reflected in the fact that it is edited under a scientific collection of books in ecology by a scientific editorial (Springer).

Nonetheless, it is also largely influenced by its nature as subproduct of the research project SUSTAINOLIVE (https://sustainolive.eu/), funded by the PRIMA (Partnership for Research and Innovation in the Mediterranean Area—PRIMA) program of the European Commission between 2019 and 2023.

SUSTAINOLIVE aimed at enhancing the sustainability of the olive oil farming sector throughout the implementation and promotion of a set of innovative sustainable management solutions that are based on agro-ecological concepts, and on the exchange of knowledge and co-creation involving multiple actors and end-users.

As editors, we have attempted to imprint the current book with a clear language, highlighting key messages that can appeal to a larger audience, including; universities and research centers, sectoral associations and cooperatives, policy makers, farmers and land managers, agro-industrial agents, financial entities, natural resource managers, and lobbyists with interest in the future and sustainability of agriculture in the Mediterranean context.

Although it is clear that the main readership of the book will be conformed by those persons and institutions directly liaising with olive groves and olive oil, we expect that the relevance and pertinence of many of the issues addressed in the book will also attract others more generally dealing with the sustainability of farming systems in vulnerable regions, such as the Mediterranean, where trade-offs between climate change, food security, and biodiversity conservation are all at stake (Muñoz-Rojas et al., 2019). This is thus a context where implementing a landscape approach that adopts an integrated social-ecological research strategy may become crucial for the mid to longer term survival of current production systems.

1.2.2 What Will the Reader Find along the Book?, and How Is It Structured?

To achieve the aforementioned goals, the book is structured across the following series of sections, each of which contains multiple chapters:

Part I – Introduction: Why Are Olive Landscapes in the Mediterranean Worth a Book? Chap. 1

In this first section, the reader will find one only, introductory, chapter. In this chapter, the key arguments are presented that sustain the historical, socio-economic, cultural, and environmental relevance of olive groves and olive oil in the Mediterranean macro-region, and the challenges it faces in view of a rapidly changing social-ecological context where rural landscapes are facing novel, and increasingly demanding pressures.

Part II to VII – Regional/National Case Studies across the Northern and Southern Mediterranean. Olive Landscapes Across the Mediterranean: Regional Heterogeneity and Case Studies:

Along each section, the six Mediterranean countries condensing the greatest production of olive oil, and with the largest surface occupied with olive groves worldwide, are examined. This is performed from a critical perspective that is based on landscape and social-ecological perspectives, and that addresses issues related to the history and culture (Chap. 2 – Spain –, 11 – Italy –, 19 – Greece -, 27 – Portugal -, 35 – Tunisia –, 41 – Morocco), climate and soils (Chaps. 3 –Spain –, 12 – Italy –, 20 – Greece –, 28 – Portugal –, 36 – Tunisia, 42 – Morocco), cultivars and farming systems (Chap. 4 – Spain –, 13 – Italy –, 21 –Greece –, 29 – Portugal, 37 – Tunisia –, 43 – Morocco), soil management practices, prunning and harvesting (Chap. 5 – Spain –, 14 – Italy –, 22 – Greece –, 30 – Portugal, 38 – Tunisia –, 44 – Morocco –), pests and dieases (Chap. 7 – Spain –, 15 – Italy -, 23 – Greece –, 31 –Portugal, 45 – Morocco), water management (Chap. 8 – Spain -, 16 – Italy -, 24 – Greece –, 32 – Portugal, 40 – Tunisia –, 46 – Morocco –), ecosystem services (Chap. 9 –Spain –, 17 – Italy –, 23 – Greece –, 33 – Portugal, 47 – Morocco –), water management (Chap. 9 –Spain –, 17 – Italy –, 25 – Greece –, 33 – Portugal, 48 – Morocco), and market challenges (Chap. 10 – Spain –, 18 – Italy –, 26 – Greece –, 34 – Portugal, 49 – Morocco).

Part VIII – Olive Landscapes Across the Mediterranean: Regional Heterogeneity and Case Studies: Local Innovative Approaches to the Management of Olive Groves and Landscapes Across the Mediterranean.

In this section, seven institutions and experiences dealing with options for enhancing the sustainability of olive groves and olive oil are briefly described in representation of Spain (Chap. 50), Greece (Chaps. 51 and 52), Italy (Chaps. 53 and 54), and Portugal (Chap. 55). In these chapters different local approaches to olive grove sustainability can be read that are adapted to the national, regional, and local conditions across the N Mediterranean and that we envisage can provide with clues for those attempting to achieve equivalent results elsewhere. This is especially the case in the countries where our case studies are located, but also across other regions in the Mediterranean where changes in strategy and approaches to cultivation, management, transformation, and commercialization may be necessary.

Part IX – Olive Oil: Diversity, Heterogeneity and Quality Challenges

In this section, various key aspects related to olive oil production and its sustainability are examined, including:

Origin and Traceability of Olives and Olive Oils in the Mediterranean Area: Genomic, Metabolomic and Control Insights (Chap. 56).

Olive Oil Extraction and Table Olive Industries; Current State of Technological Applications and Novel Approaches to Value-Adding Methods for By-Products (Chap. 57).

Current State of Technological Applications and Novel Approaches to Value-Adding Methods for Byproducts from Oil Mills (Chap. 58).

We expect that this section will be of interest to those researchers, agro-industrial and food quality and regulatory agents operating in the field, both from a private or public point of view.

Part X – Olive Landscapes in The Mediterranean Transitioning Toward Enhanced Sustainability: Key Challenges and Opportunities.

This last section contains two last chapters that re-examine the questions originally posed in view of the information obtained mainly via two sources of information:

- The individual regional and thematic descriptions made in the chapters in sections II to IX of this book.
- The results and outputs obtained throughout the application of the multiple tasks and analyses conducted along the years 2019–2023 within the context of the SUSTAINLIVE project.

Resulting of this, we conclude the book by providing with an assessment of the Weaknesses, Threats, Strengths, and Opportunities of the Olive Oil Sector and Olive Groves and Landscapes Across Mediterranean Countries and Regions (Chap. 59), and a critical overview of future pathways and scenarios for improvement, which are provided in the form of key points to facilitate readership, clarity and underlying of the key lessons learned (Chap. 60).

1.2.3 What Would We Aspire for the Reader to Learn After Reading This Book?

Ultimately, what we expect the reader to know before reading this book can vary enormously. This will naturally depend largely upon their background, experiences, location, interests, approaches, and even their ideology, but hopefully, we expect that all of them will come out of their reading knowing more than when they started.

Due to the length and diversity of its contents, we cannot envisage that the book will be read throughout, and that it will be more likely used in the form of consultation for concrete aspects that are specific to the interest and geographic context of each reader.

It is inescapable that the book represents a relatively static view, in both space (six countries) and time (2019–2023). However, we as editors sincerely expect that the issues described will be have been addressed in such a critical way that its reading and interpretation will incite novel questions, thus also enlarging and enriching discussions. Whether this is effectively achieved is something that the reader might perhaps reflect upon by questioning whether our vision for the future (again, see Chap. 60) was ultimately realistic, achievable, and pertinent.

Acknowledgements and Funding This work was funded by the SUSTAINOLIVE project (Grant n° 1822; PRIMA EU Programme) for both co-authors, and by FCT-Portugal (UIDB/05183/2020 – funding for José Muñoz-Rojas, as member of MED-Universidade de Évora, and LA/P/0121/2020 as member of the CHANGE- Associated Lab-Universidade de Évora.

References

Bastian, O., Grunewald, K., Syrbe, R. U., et al. (2014). Landscape services: The concept and its practical relevance. *Landscape Ecology, 29*, 1463–1479. https://doi.org/10.1007/s10980-014-0064-5

Besnard, G., Khadari, B., Navascués, M., Fernández-Mazuecos, M., El Bakkali, A., Arrigo, N., Baali-Cherif, D., Brunini-Bronzini de Caraffa, V., Santoni, S., Vargas, P., & Savolainen, V. (2013). The complex history of the olive tree: From late quaternary diversification of Mediterranean lineages to primary domestication in the northern Levant. *Proceedings of the Royal Society B*, 2802012283320122833. https://doi.org/10.1098/rspb.2012.2833

Boyd, E., & Folke, C. (2011). *Adapting institutions: Governance, complexity and social-ecological resilience*. Cambridge University Press. isbn:978-052-189-750-1.

Breton, C. M., Warnock, P., & Berville, J. A. (2012). Origin and history of the olive. *INTECH*. https://doi.org/10.5772/51933

Guzmán Álvarez, J. R., Fernández Rodríguez, P., Gómez Calero, J. A., & Gnzález, A. L. (2020). *Olivares de España: Recorrido por la biografía del olivar español, su memoria y sus paisajes*. Almuzara; 1er edición (21 August 2020) ISBN: 978-841-808-934-3.

Infante-Amate, J. (2012). The ecology and history of the Mediterranean olive grove: The Spanish great expansion, 1750–2000. *Rural History, 23*(2), 161–184. https://doi.org/10.1017/S0956793312000052

Loumou, A., & Giourga, C. (2003). Olive groves: "The life and identity of the Mediterranean". *Agriculture and Human Values, 20*, 87–95. https://doi.org/10.1023/A:1022444005336

Moreira, F., Herrera, J. M., & Beja, P. (2019). Making olive oil sustainable. *Science, 365*, 873–873. https://doi.org/10.1126/science.aay7899

Muñoz-Rojas, J., Napoleone, C., & Pinto-Correia, M. T. (2019). Farm and land system dynamics in the Mediterranean basin: Integrating different spatial scales and management approaches. *Land Use Policy, 88*, 104082. https://doi.org/10.1016/j.landusepol.2019.104082

Sayer, J., Sunderland, T., Ghazoul, J., Pfund, J.-L., Sheil, D., Meijaard, E., Venter, M., Boedhihartono, A. K., Day, M., Garcia, C., Van Oosten, C., & Buck, L. E. (2013). Ten principles for a landscape approach to reconciling agriculture, conservation, and other competing land uses. *Proceedings of the National Academy of Sciences of the United States of America, 110*, 8356–8349. https://doi.org/10.1073/pnas.1210595110

Silveira, A., Muñoz-Rojas, J., Pinto-Correira, T., Guimarães, M. H., Ferrão, J., & Schmidt, L. (2018). The sustainability of agricultural intensification in the early 21st century: Insights from the olive oil production in Alentejo (Southern Portugal). In A. Delicado, N. Domingos, & L. de Sousa (Eds.), *Changing societies: Legacies and challenges* (The Diverse Worlds of Sustainability) (Vol. 3, pp. 247–275). Imprensa de Ciências Sociais. isbn:978-972-671-505-4.

Part II
Olive Landscapes Across the Mediterranean: Regional Heterogeneity and Case Studies: Olive Groves and Landscapes in Spain

Part II
Olive Landscapes Across the Mediterranean: Regional Perspectives and Case Studies: Olive Groves and Landscapes in Spain

Chapter 2
Notes for a Socio-ecological History of the Spanish Olive Grove

Juan Infante-Amate, Gloria Guzmán Casado, David Soto Fernández, and Manuel González de Molina

Abstract Spanish olive oil production has multiplied by about 7 during the last 120 years. During the first half of the twentieth century, the increase in production was mainly due to a marked increase in the olive grove area (doubled between 1850 and 1950) at expenses of the many natural areas such as the Mediterranean forest. Low planting olive tree density within a relatively diverse landscape matrix and with a secondary function but with a clear multifunctional role were characteristics of this olive grove. From the 1950s until now, area only has increased by 25%, and the rise in production is mainly due to an increase in productivity. Olive grove intensification based on breeding and changes in tree morphology, management practices, and an increase in tree density, all aimed at translocating as much photosynthesized energy as possible to the harvestable fruit, has resulted in homogenized monocultures with low natural and landscape values and a loss of olive grove multifunctional usage. These changes have required the use of large amount of external inputs manufactured and driven by fossil fuels and have been possible at the cost of a deterioration of the fund element.

In this section, the main drivers and strategies, with marked social-ecological, economic, and landscape consequences, during the recent history of the olive groves in Spain are assessed from a socio-ecological perspective.

J. Infante-Amate
Departmet of Economic Theory and History, Facultad de Ciencias Económicas y Empresariales, Universidad de Granada, Granada, Spain
e-mail: jinfama@ugr.es

G. G. Casado · M. G. de Molina (✉)
Agroecosystems History Lab – Departamento de Historia, Geografía y Filosofía, Pablo de Olavide University, Sevilla, Spain
e-mail: mgonnav@upo.es; giguzcas@upo.es

D. S. Fernández
Department of Economic History and Institutions, Facultade de Ciencias Económicas e Empresariais Avda, Universidad de Santiago de Compostela, A Coruña, Spain

J. Muñoz-Rojas, R. García-Ruiz (eds.), *The Olive Landscapes of the Mediterranean*, Landscape Series 36,
https://doi.org/10.1007/978-3-031-57956-1_2

2.1 Introduction

From a metabolic perspective, Spain has been the world's leading producer of olive oil since the end of the nineteenth century. This chapter briefly overviews the evolution of Spanish olive growing. The approach adopted is not a usual one in historical studies on this crop. Works generally focus on monetary aspects or production and consumption growth. Although these data are decisive to trace the crop's evolution, they are insufficient to explain how this growth has been possible or the changes and impacts entailed for Spain's agroecosystems. The latter are key to the crop's future sustainability, all in a context of climate change and environmental degradation in the Mediterranean basin. The metabolic approach allows us to appreciate the changes in the crop's management and functionality, as well as the impacts of these changes, both on the landscape and the environment.

2.2 Spain's Specialization in Olive Oil

With the exception of a few years in the mid-twentieth century, Spain has been the world's leading producer of olives and olive oil since the late nineteenth century, surpassing Italy. During this period, Spain took over between 25% and 50% of global olive oil production. Without this progress, we could hardly explain the planet's imposing expansion of olive oil production. Nor could we fully explain Spain's agricultural evolution, within which it has gained an increasingly significant role. As reported elsewhere (González de Molina et al., 2020), Spain's agricultural production efforts have concentrated since the end of the nineteenth century on expanding and intensifying cultivated surface areas and, within these, specializing in a small group of crops that includes the olive grove.

Indeed, Table 2.1 shows the evolution of biomass extracted from cultivated lands throughout the twentieth century. Farmers appropriated an increasing percentage of the overall net primary productivity (NPP) of agroecosystems.

If the total NPP grew as from 1900 by 38.2%, the extracted biomass increased more than twofold (González de Molina et al., 2020). Virtually all crops saw their production expand over the period, but the olive grove grew the most (Infante-Amate, 2014). It went from 6.6% of the extracted cultivation biomass to 9.7%, demonstrating its growing prominence within Spanish agriculture's specialization and intensification. In fact, the olive grove is the crop that has grown the most in volume, its production increasing threefold, largely more than cereals, Spain's major crop in terms of territory, and more than fruits and vegetables, despite having a much lower water content. Table 2.1 also shows that this growth mainly concerned the primary crop, i.e., olives, rather than residues. Olive extraction multiplied almost by six, while residues grew by only 61%. We will come back to the consequences of this later. We can thus explain that, measured in fresh matter and taking only olives into account, as is usually done, olive "production" grows from 1.09 million

Table 2.1 Evolution of domestic biomass extraction in Spain and olive grove shares in kt of ms

	1910	1933	1950	1970	1990	2008
Cropland	24.346	32.492	28.457	44.259	53.789	51.726
Primary crops	11.181	16.691	13.838	24.813	35.575	37.547
Olive	586	981	863	1.117	1.578	3.377
Residues	1.026	1.423	1.687	2.024	1.716	1.648
Olive groves	1.611	2.404	2.550	3.142	3.294	5.025
Interannual variation	**1900**	**1933**	**1950**	**1970**	**1990**	**2008**
Cropland	100	133	117	182	221	212
Primary crops	100	149	124	222	318	336
Olive	100	167	147	191	269	576
Residues	100	139	164	197	167	161
Olive groves	100	149	158	195	204	312

Source: Agricultural statistics

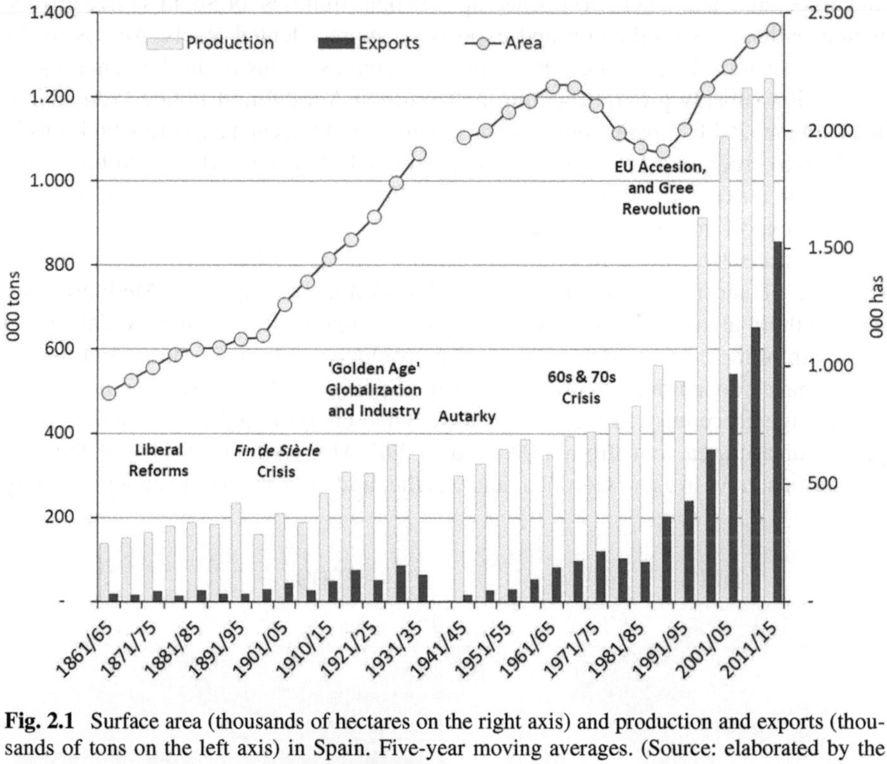

Fig. 2.1 Surface area (thousands of hectares on the right axis) and production and exports (thousands of tons on the left axis) in Spain. Five-year moving averages. (Source: elaborated by the authors based on agricultural statistics (MAPA)

tons to 7.35 tons in 2015, multiplying almost by a factor of seven. Figure 2.1 summarises the crop's evolution and its main magnitudes, expressed in a conventional way, that is, in fresh matter and deducting residues.

The expansion of the olive grove dates back to the early twentieth century, though it did indeed accelerate in the twentieth century (Fig. 2.1). During the first three decades, it experienced its so-called "Golden Age," with an unprecedented expansion of olive grove surface areas, production, and oil exports (Infante-Amate, 2014; Ramón-Muñoz, 2007, 2000; Zambrana, 1984, 1987). The Civil War (1936/39) and the autarchy of Franco's dictatorial regime resulted in a period of depression in the sector, characterized by a lack of inputs, black markets, and a general fall in exports (Christiansen, 2002, 2012). This crisis was followed by one that was deeper still, in the 1960s, when the Spanish economy was re-integrated into the world markets and the olive grove, which had been gradually abandoned under Francoism, was unable to compete with the rest of fats and oils (Tió, 1982:236–241). For the first time since records began, surface area dropped.

After this crisis, the sector underwent a transformation that paved the way to a new boom from the 1970s onwards, a second "Golden Age." Rising global demand and associated health benefits opened up new opportunities for Spain's olive groves, which built up its production and exports to unprecedented levels. Access to the European Union led, among other things, to Spain's inclusion in the generous oil production subsidy program within the Common Agricultural Policy (Scheidel & Krausmann, 2011). Spain went from producing 333 ktons and exporting 96 ktons in 1966/70 to producing 1243 ktons and exporting 643 ktons in 2011/15. Although the surface area expanded from 1.71 to 2.43 million hectares over the period, production and export growth was far greater.

The successive waves of crop expansion were highly localized geographically (Fig. 2.2). Historically, the olive grove had tended to develop in the Mediterranean and southern areas of the Spanish peninsula. Of note is the region of Andalusia, in the south of the country and, within it, the provinces of Jaén, Córdoba, and Seville which have concentrated almost 50% of the Spanish surface area since the end of the nineteenth century to the present day. Between 1890 and 2010, Spain's olive grove surface area went from 1.08 Mha to 2.47 Mha. A total of 44% of the new plantations grew in Jaén, Córdoba, and Seville, and 66% in Andalusia which today

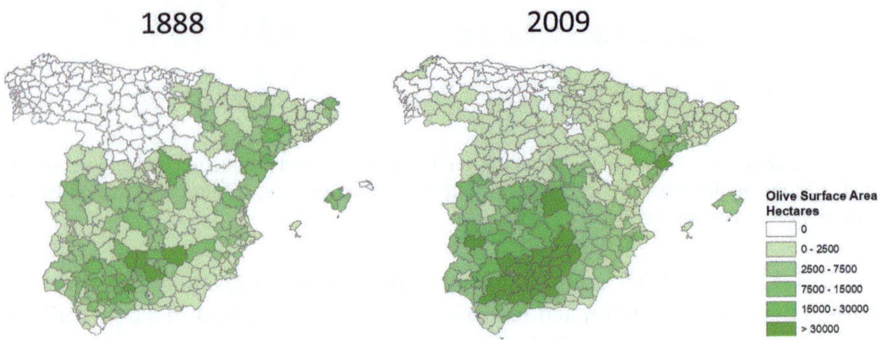

Fig. 2.2 Olive grove surface areas by region. (Source: elaborated by the authors based on agricultural statistics, MAPA)

accounts for 62% of Spain's olive grove surface area and 15% of the world's, representing the largest tree concentration in all of Europe. Nevertheless, the olive grove has also gained importance in areas in which it was traditionally absent. This latest trend has strengthened in recent years and has led to the emergence of super-intensive olive groves in the countryside and even in the inland fertile plains.

It is difficult to trace the varietal factor in the process described, but the fragmentary evidence at our disposal strongly leads us to believe that a varietal homogenization toward more productive forms with a higher fat yield has taken place. In 1888, a report on the JCA olive grove (1981) indicated 181 different types of varieties, although we do not know the spatial distribution of each. Today, we know that 58.2% of the olive groves in Andalusia, the region that concentrates most of Spain's olive groves, are occupied by a single variety, the Picual, characterized by its high productivity and fat yield. Together with Hojiblanca (18.1%), only two varieties occupy 76.3% of Andalusia's olive lands (de Andalucía, 2002: 26–27).

In recent decades, another decisive transformation has taken place with the application of irrigation, especially localized irrigation, to many traditional olive groves. At the end of the nineteenth century, only 70 thousand hectares of olive groves were irrigated. During the first half of the twentieth century, this figure remained stable, reaching around 100 thousand hectares. In the last years of the twentieth century, it grew exponentially, reaching 492 thousand hectares in 2010. All these transformations have been sustained by the expanding consumption of olive oil. Between 1961/63 and 2007/09, consumption grew in the major producing countries, from 707 Mtons to 1357 Mtons. In Spain, consumption per capita per year was 8.2 kg in 1961/3, reaching 11.5 kg in 2007/9. Since then, consumption has abated due to the economic crisis, reaching 7.5 kg in 2017 (MAPA, 2018a). This consumption is well above the European consumption average in magnitude, which has passed from 1,5 kg/inhabitant/year in 1961/3 to 2,4 kg/inhabitant/year in 2011/3 (FAO, 2020).

2.3 Management Changes

As we have seen, Spain's agricultural industrialization has led to production intensification and crop specialization. At a farm scale, these processes have entailed: a strong tendency to suppress crop and polyculture associations; the simplification of rotations and their later suppression; the quasi-elimination of fallows or their substantial reduction; the fostering of crop alternations governed by market demands; and significantly reduced genetic, structural and functional diversity. At the crop level, the aim of farmers has been to maximize the harvestable parts of plants, especially those with the highest commercial value. At the agroecosystem scale, cropland expanded and agrosilvopastoral integration collapsed. The subsequent segregation of land uses implied the loss of functional synergies between them. This phenomenon progressively reduced the capacity of agroecosystems for autonomous replacement fertility. Relative demand of fertilizers significantly increased, the spread of chemical fertilizers making this fundamental change possible.

These changes profoundly affected both the amount and the end use of biomass produced by the olive groves. They were due to ongoing efforts to increase the size of commercially oriented biomass production to meet growing demand for raw material, as well as human and animal feed. Olive production grew essentially in two different ways: on the one hand, the total amount of biomass per hectare increased and, on the other, a rising share of net primary productivity was appropriated. Improvements and changes to management practices consisted of translocating stem and leaf biomass to the harvestable fruit. Multifunctional usage has been superseded by a preference for the commercial use of fruits. From producing firewood, foliage, and olive kernel oil for livestock, table olives, domestic lighting, and edible oil, olive groves now almost exclusively produce oil, leading to changes in grove management and tree morphology (Infante-Amate, 2014; Infante-Amate & González de Molina, 2013). Another factor is the reduction of herbage that accompanies crops based on mechanical or chemical means. Accordingly, both production and technological efforts have been directed toward maximizing the share of biomass of the highest commercial value, entailing the reduction of crop multifunctionality.

2.4 Olive Grove Landscapes

The industrialization of olive grove cultivation has deeply transformed olive grove landscapes, although changes are not that visible at first glance. First, olive grove expansion has been a major driver of change of Spanish agroecosystem agricultural landscapes: arable crops have been widely replaced by olive groves in most of the country. But olive groves themselves have also been altered, reflecting the loss of crop multifunctionality. The olive grove has gone from an extensive farming crop, closer to models such as "dehesa" (pasturelands), to intensification under organic management throughout the nineteenth and early twentieth centuries, ultimately becoming an industrial monoculture.

Until well into the nineteenth century, the olive grove was a multifunctional crop, where wood extraction could be more important than the harvesting of fruit. Trees grew in a disorderly fashion on lands dedicated to other uses, animals grazing around them. Sometimes they were found in vineyards or vegetable fields. Somehow the olive grove was presented, as pointed out by the ecologist González Bernáldez (1981), as a landscape deriving from a Mediterranean forest shaped by man. Therefore, it was closer to systems such as pastureland. In fact, the olive grove often cohabitated with livestock, with a considerable height and a plantation density rarely reached above 100 trees/hectare (JCA, 1891).

The crop tended to become homogenized due to increasing intensification, although it continued to expand within the production limits of organic management that prevented high planting densities and big yields. What did occur was that by 1900 and to a much greater extent, in the middle of the twentieth century, dispersed, unproductive olive groves associated with livestock ceased to be widespread. In

1975, Agricultural Statistics counted around 1.7 million scattered trees, which, together, may have occupied a virtual area of 17 thousand hectares. At that time, the olive grove, as an associated crop, slightly exceeded 50 thousand hectares. However, the surface area of non-associated and non-dispersed olive groves exceeded 1.2 million hectares. In other words, types of agroforestry olive trees or trees integrated with other usages tended to disappear from the end of the nineteenth century, becoming residual by the mid-twentieth century. Monocultural olive groves accounted for most olive grove landscapes in the region. The massive import of animal feed, electrification and fossil fuel consumption all explain that olive trees lost certain traditional functions such as soap-production, lighting, fuel, or animal feed that characterized its traditional management and were key to peasant sustenance (Infante-Amate, 2013).

The massive addition of chemical fertilizers as well as expanding artificial irrigation allowed to increase planting density, driving what have perhaps been the most noticeable changes. In the end, traditional olive groves, which were rarely fertilized, could barely surpass production levels of 1000 kg/ha, largely because such fertility replenishment systems could not sustain more than 100 trees/ha (Naredo, 1981). A radical process of increase in tree density has taken place since the 1970s. The 80% of the olive groves had a density at the time that was below100 trees/ha, but by 2002, only 34% had a density below 100 trees/ha. The rest exceeded such a figure, i.e., almost 150 thousand hectares already had over 200 trees/ha. Furthermore, the increasing use of agricultural machinery and the chemical elimination of vegetation cover ended up turning the olive grove into a sort of woody desert.

2.5 The Degradation of the Olive Grove's Environmental Services

The industrialization process described has had negative impacts on agroecosystems, decreasing their ability to generate ecosystem services. A number of studies show high rates of erosion (Gómez et al., 2014) associated with intensive ploughing and the elimination of vegetation cover by herbicides, especially in hilly areas that concentrate most olive grove surface areas (Infante-Amate et al., 2016). Erosion rates accounting for over 80 tons/ha/year have been recorded. These figures and their association with intensive management have been confirmed by historical erosion works on Andalusian olive groves; in particular, on soil losses in Montefrío (Granada) between 1750 and 2000 (Vanwalleghem et al., 2011; Infante-Amate et al., 2013; Gómez et al., 2014) and in Baena during the twentieth century (Vanwalleghem et al., 2010).

Excess of fertilization has been shown to have one of the biggest environmental impacts of industrial olive grove management (Beaufoy, 2001; Tombesi et al., 1996). Only 40% of the applied nitrogen is extracted by crops, the remaining 60% being lost annually (Hermosín et al., 2009:131). Along with nitrogen, phosphorus is another nutrient dragged with sediments into watercourses (Gómez et al., 2009). Diffuse pollution by these nutrients when they reach the water causes eutrophication. The contamination of soil, water and, ultimately, food by pesticide residues also results from olive grove industrialization (Calderón et al., 2016; Tsatsakis et al., 2003).

As a consequence of intensive management, an ostensible loss of biodiversity has been detected in recent years (Duarte et al., 2009). Morgado et al. (2020) found that the intensification levels explained a large proportion of variations in bird communities (35%), while accounting for the significant contributions of orchard features (16%) and landscape (16%) in Portugal. Species richness declined from 29 species with traditional agriculture, to 25 with intensive production and to 20 with super-intensive groves. Additionally, the probability of occurrence of many species also declined significantly during the transition from traditional to intensive production. Nevertheless, the growth of certified organic olive groves reached 209,288.32 ha in 2019, accounting for 8% of the total surface area (MAPA, 2018b), marking the beginning of a trend reversal. The de-intensification of certified organic olive groves is leading to a certain recovery of biodiversity, both of arthropods that operate as natural enemies of olive tree pests (Álvarez et al., 2019), and of pollinators (Martínez-Núñez et al., 2020). The abandoning of pesticides and the recovery of green cover and natural vegetation areas, similar to traditional organic management, facilitate this improvement. There are, however, additional factors, such as the shape of the tree trunks. The presence of cavities in the trunks of less intensive olive groves, as opposed to their disappearance in more intensive olive groves, positively affects the presence of cavity-nester insectivorous birds (Morgado et al., 2020).

To finish, olive grove intensification has also led to a drop in soil organic carbon (SOC) stocks. According to Aguilera et al. (2018), the SOC stock (at 30 cm depth) per hectare has been falling in Spanish olive groves since the 1930s, first as a result of the fall in biomass productivity due to the Civil War, and second, from the 1960s onwards, because of lower amounts of unharvested olive grove biomass (the use of herbicides being key) and of rising SOC decomposition rates, especially in rainfed crops, due to climate change. In a similar way to what occurs with biodiversity, the process of deintensification proper to certified organic olive groves is altering the trend, leading to a notable increase in unharvested organic olive grove biomass (Guzmán & Astier, 2017) and net CO_2eq sequestration (Aguilera et al., 2015). However, organic olive groves in Spain accounted for only 7.6% of the olive grove surface area in Spain in 2017 and even less in Andalusia (4.9%) (MAPA, 2018b). Therefore, it fails to compensate for the serious loss of ecosystem services provided by Spanish olive groves as a whole.

References

Aguilera, E., Guzmán, G. I., & Alonso, A. M. (2015). Greenhouse gas emissions from conventional and organic cropping systems in Spain. II. Fruit tree orchards. *Agronomy for Sustainable Development, 35*, 725–737.

Aguilera, E., Guzmán, G. I., Alvaro-Fuentes, J., Infante-Amate, J., García-Ruiz, R., Carranza-Gallego, G., Soto, D., & González de Molina, M. (2018). A historical perspective on soil carbon in Mediterranean cropland (Spain, 1900–2008). *Science of the Total Environment, 621*, 634–648.

Álvarez, H. A., Morente, M., Oia, F. S., Rodríguez, E., Campos, M., & Ruano, F. (2019). Semi-natural habitat complexity affects abundance and movement of natural enemies in organic olive orchards. *Agriculture, Ecosystems and Environment, 285*, 106618.

Beaufoy, G. (2001). *EU policies for olive farming. Unsustainable on all counts.* BirdLife Internacional-WWF.

Calderón, M. J., De Luna, E., Gómez, J. A., & Hermosín, M. C. (2016). Herbicide monitoring in soil, runoff waters and sediments in an olive orchard. *Science of the Total Environment, 569–570*, 416–422.

Christiansen, T. (2002). Intervención del Estado y mercado negro en el sector oleícola durante el primer franquismo. *Historia Agraria, 27*, 221–246.

Christiansen, T. (2012). *The reason why. The post civil-war agrarian crisis in Spain.* Prensas Universitarias de Zaragoza.

Junta de de Andalucía (2002). *El olivar andaluz*. Junta de Andalucía.

Duarte, J., Campos, M., Guzmán, J. R., Beaufoy, G., Farfan, M. A., Cotes, B., Benítez, E., Vargas, J. M., & Muñoz-Cobo, J. (2009). Olivar y biodiversidad. In J. de Andalucía (Ed.), *Sostenibilidad de la Producción de Olivar en Andalucía*. Sevilla.

FAO (2020). *Faostat. Production. Crops.* http://www.fao.org/faostat/en/#data/QC

Gómez, J. A., Guzmán, M. G., Giráldez, J. V., & Fereres, E. (2009). The influence of cover crops and tillage on water and sediment yield, and on nutrient, and organic matter losses in an olive orchard on a sandy lo- soil. *Soil and Tillage Research, 106*, 137–144.

Gómez, J. A., Infante-Amate, J., de Molina, M. G., Vanwalleghem, T., Taguas, E. V., & Lorite, I. (2014). Olive cultivation, its impact on soil erosion and its progression into yield impacts in southern Spain in the past as a key to a future of increasing climate uncertainty. *Agriculture, 4*(2), 170–198.

González Bernáldez, F. (1981). Ecología y paisaje. Madrid: Blume

González de Molina, M., Soto Fernández, D., Guzmán Casado, G., Infante-Amate, J., Aguilera Fernández, E., Vila Traver, J., & García-Ruiz, R. (2020). The social metabolism of Spanish agriculture, 1900–2008. In *The Mediterranean way towards Industrialization*. Springer.

Guzmán, G. I., & Astier, M. (2017). Organic farming: Between the relocation of energy flows and input replacement. In G. I. Guzmán & M. González de Molina (Eds.), *Energy in agroecosystems: A tool for assessing sustainability*. CRC Press.

Hermosín, M. C., Rodríguez-Linaza, L. C., Cornejo, J., & Ordóñez-Fernández, R. (2009). Efecto de uso de agroquímicos en el olivar sobre la calidad de las aguas. In J. A. Gómez (Ed.), *Sostenibilidad de la producción de olivar en Andalucía* (pp. 127–160). Consejería de Agricultura y Pesca.

Infante-Amate, J. (2013). ¿Quién levantó los olivos? La expansión olivarera decimonónica como estrategia campesina (1750–1930). *Historia Social, 76*(II), 25–44.

Infante-Amate, J. (2014). *¿Quién levantó los olivos? Historia de la especialización olivarera en el sur de España (ss. XVIII–XX)*. Ministerio de Agricultura, Alimentación y Medio Ambiente.

Infante-Amate, J., & González de Molina, M. (2013). The socio-ecological transition on a crop scale. The case of olive orchards in S. Spain (1750–2000). *Human Ecology, 41*(6), 961–969.

Infante-Amate, J., Soto, D., Cid, A., Guzmán, G., & González de Molina, M. (2013). *Nuevas interpretaciones sobre el papel del olivar en la evolución agraria española. La gran transformación del sector (1880–2010)*. XIV Congreso Internacional de Historia Agraria, Badajoz, 7–9 de noviembre de 2013.

Infante-Amate, J., Villa, I., Aguilera, E., Torremocha, E., Guzmán, G., Cid, A., & de Molina, M. G. (2016). The making of olive landscapes in the south of Spain. A history of continuous expansion and intensification. In *Biocultural diversity in Europe* (pp. 157–179). Springer.

JCA (Junta Consultiva Agronómica) (1891). *Avance estadístico sobre el cultivo y producción del olivo, Junta Consultiva Agronómica*. Dirección General de Agricultura, Industria y Comercio.

MAPA (Ministry of Agriculture, Fisheries and Food). (2018a). *Food consumption report in Spain 2017*. MAPA.

MAPA (Ministry of Agriculture, Fisheries and Food). (2018b). *Organic Agriculture. Statistics 2017*. MAPA.

Martínez-Núñez, C., Manzaneda, A. J., & Rey, P. J. (2020). Plant-solitary bee networks have stable cores but variable peripheries under differing agricultural management: Bioindicator nodes unveiled. *Ecological Indicators, 115*, 106422.

Morgado, R., Santana, J., Porto, M., Sanchez-Oliver, J. S., Reino, L., Herrera, J. M., Rego, F., Beja, P., & Moreira, F. (2020). A Mediterranean silent spring? The effects of olive farming intensification on breeding bird communities. *Agriculture, Ecosystems and Environment, 288*, 106694.

Naredo, J. M. (1981). La incidencia del «estraperlo» en la economía de las grandes fincas del Sur. *Agricultura y Sociedad, 19*, 81–129.

Ramón-Muñoz, R. (2000). Specialization in the international market for olive oil before World War II. In S. Pamuk & J. G. Williamson (Eds.), *The Mediterranean response to globalization before 1950* (pp. 159–198).

Ramón-Muñoz, R. (2007). La producción y el comercio de aceite de oliva en los países del Mediterráneo (1850–1938): Competencia y especialización. *Mediterráneo Económico, 7*, 329–354.

Scheidel, A., & Krausmann, F. (2011). Diet, trade and land use: A socio-ecological analysis of the transformation of the olive oil system. *Land Use Policy, 28*, 47–56.

Tió, C. (1982). *La política de aceites comestibles en la España del siglo XX*. Ministerio de Agricultura, Pesca y Alimentación.

Tombesi, A., Michelakis, N., & Pastor, M. (1996). Recommendations of the working group on olive farming production techniques and productivity. In *Olivae, n. 63*. International Olive Oil Council.

Tsatsakis, A. M., Tsakiris, I. N., Tzatzarakis, M. N., Agourakis, Z. B., Tutudaki, M., & Alegakis, A. K. (2003). Three-year study of fenthion and dimethoate pesticides in olive oil from organic and conventional cultivation. *Food Additives & Contaminants, 20*(6), 553–559.

Vanwalleghem, T., Laguna, A., Giráldez, J., & Jiménez-Hornero, F. J. (2010). Applying a simple methodology to assess historical soil erosion in olive orchards. *Geomorphology, 114*, 294–302. https://doi.org/10.1016/j.geomorph.2009.07.010

Vanwalleghem, T., Infante-Amate, J., González de Molina, M., Soto, D., & Gómez, J.A. (2011). Quantifying the effect of historical soil management on soil erosion rates in olive orchards over the last 250 years. *Agriculture, Ecosystems and Environment, 142*, 341–351.

Zambrana, J. F. (1984). El aceite de oliva y su dependencia del mercado internacional de grasas vegetales. Un análisis histórico 1861-1935. *Agricultura y Sociedad, 33*, 159–196.

Zambrana, J. F. (1987). *Crisis y modernización del olivar*. Ministerio de Agricultura, Pesca y Alimentación.

Chapter 3
Pedoclimatic and Landscape Conditions of the Spanish Olive Groves

Julio Calero, Tomás Fernández, and Francisco Moya

Abstract Due to its eco-physiological characteristics, the olive tree is well adapted to Mediterranean-type climates, mainly by its ability to compensate the summer water deficit through effective regulation of its transpiration. Thus, avoiding only the Atlantic climate zones with rainfall >800–900 mm, the olive tree is widely distributed by the Mediterranean Spain. Despite this, olive groves define fairly heterogeneous landscape units, with very different features in terms of relief and soils. Many olive groves are sited in plateaus (e.g. La Mancha) or river valleys (e.g. Guadalquivir basin) with gentle to moderate slopes (2–15%), while up to a third of Spanish olive groves are settled in mountainous areas with average slopes >15%. The altitude is also highly variable; it is possible to find olive orchards from the sea level to altitudes above 1500 m (e.g. Sierra Nevada); however, the average range could be established around 500 m. Up to 84% of the Spanish olive orchard soil is included in map units with predominance of five taxonomic groups: Calcic Cambisols, Chromic Luvisols, Chromic Vertisols, Eutric Regosols, and Dystric Regosols. Broadly, soils tend to be fine grained and relatively stony, being remarkable for the low content of organic carbon of the topsoil (< 2%). All the representative profiles of olive grove soils analyzed were marginally suitable (S3 class) in the FAO Land Evaluation systems. The main pedological constraints that have been

J. Calero (✉)
Department of Geology, University of Jaén, Jaén, Spain

Centre for Advanced Studies in Earth Sciences, University of Jaén, Jaén, Spain
e-mail: jcalero@ujaen.es

T. Fernández
Centre for Advanced Studies in Earth Sciences, University of Jaén, Jaén, Spain

Department of Cartographic, Geodetic and Photogrammetric Engineering, University of Jaén, Jaén, Spain
e-mail: tfernan@ujaen.es

F. Moya
Department of Cartographic, Geodetic and Photogrammetric Engineering, University of Jaén, Jaén, Spain

J. Muñoz-Rojas, R. García-Ruiz (eds.), *The Olive Landscapes of the Mediterranean*, Landscape Series 36,
https://doi.org/10.1007/978-3-031-57956-1_3

detected derive from steepness, heavy texture, and organic carbon depletion and, in some soils, the acid pH and low base saturation (Dystric Regosols) or, on the contrary, high pH and $CaCO_3$ content (Calcic Cambisols). Moreover, some climatic constraints can also be found: Some olive groves in mountain areas and high plateaus showed mean annual temperatures close to the lower limit of suitability (13 °C). In general, the Spanish olive grove is well adapted to the biophysical possibilities of the land, although it has intrinsic limitations for sustainable use that reduce profitability and increase the required inputs and degradation risks.

3.1 Introduction

Olive groves occupy in Spain an area of 2,733,620 ha, which makes it the second crop with the largest area in the country, a very short distance from an important herbaceous crop such as barley (2,837,810 ha). This accounts for 5.4% of the country's total area and 16% of all Spanish agricultural lands (ESYRCE, 2019).

The concentration of the crop is especially important in Andalusia, where it occupies 19% of the total surface of the region, mainly in the provinces of Jaén (592,848 ha, 44% of the provincial surface), Córdoba (374,081 ha, 14% of the provincial area), and Seville and Granada (231,868 and 209,332 ha, respectively, 8% of both). After Andalusia, only two Spanish administrative regions ("Autonomous Communities") exceed 10% of the national olive groves: Castilla La Mancha (16%) and Extremadura (11%), the others assuming less than 5% of the area (Basque Country, Galicia, Asturias and Cantabria, in the Atlantic northern quarter of the country, reported negligible olive grove surfaces in 2019, ESYRCE, 2019). These data highlight the huge economic, social, and environmental prominence that olive groves have in Spain, as well as the significant environmental pressures it can generate on some territories.

One of the basic issues of sustainable agriculture is the optimal use of the land's capacities to support crops (Guzman & González de Molina, 2017). Adapting to the maximum potential of a territory (soil, climate, water, and genetic resources) to a specific crop, the need for inputs is significantly reduced (irrigation, mechanization, and chemicals), which translates into less pressure on the ecosystem functions and less risk of degradation (Keestra et al., 2018).

The FAO system (FAO, 1976) for land evaluation focuses on assessing the potential of a land for specific crops. These requirements can be grouped into three categories: climatic, topographic, and pedological. Land evaluation allows knowing the degree of suitability or, on the contrary, the marginality of the crop, as well as the limiting factor or factors that cause suboptimal productivity. The extension of crops to suboptimal or marginal areas is usually compensated with unsustainable intensification by means of mechanization, irrigation, and chemical fertilization, increasing the risks of degradation which, in the case of olive groves, mainly results in water erosion.

The aim of this work is the pedoclimatic characterization of the Spanish olive groves from 20 topographic, climatic, and pedological attributes. Since only three of the 17 Spain regions, Andalusia, Castilla La Mancha, and Extremadura, represent almost 90% of the olive groves, a detailed analysis of these regions will be carried out, comparing their olive groves with each other and with the Spanish olive grove as a whole. Finally, a quantitative assessment of the suitability of Spanish olive groves will be performed, revealing the limiting factors of cultivation and the intrinsic capacity of the lands to support a sustainable crop from the economic and environmental point of view.

3.2 Material and Methods

3.2.1 Material

A spatial analysis of the Spanish olive groves and its three regions with the most olive grove area (Andalusia, Castilla-La Mancha, and Extremadura) was carried out by means of geographic information systems (GIS). This analysis tried to characterize the spatial distribution of the olive grove in Spain based on three types of parameters: (1) topographic, (2) climate, and (3) pedological. The Geospatial Data Sources used are described below:

(1) Topography. A total of 1106 digital elevation models (DEM) of 25 m pixel size were downloaded in raster format (.asc) from the National Centre of Geographical Information (CNIG) (http://centrodedescargas.cnig.es/CentroDescargas/index.jsp), which were treated in ArcGIS software (version 10.5) to attain three basic terrain attributes: Slope (%), Altitude (m) and Orientation class.
(2) Climate. One layer in vector format (.shp) was downloaded from the CNIG (http://centrodedescargas.cnig.es/CentroDescargas/buscar.do?filtro.codFamilia=RTANE) for each climate attribute: Mean annual temperature (T, °C), Mean annual rainfall (P, mm) and Average absolute minimum temperature of the coldest month (ATC, °C).
(3) Pedology. Digitized cartographic information of soils, on the highest resolution available, was downloaded in vector format (.shp) for the whole of Spain, as well as for the regions of Castilla La Mancha and Extremadura, from the European Soil Data Centre ESDAC (ESDB v2.0, 2004; Panagos, 2006). These data are mainly based on the EU Soil map at scale 1:1,000,000 (CEC, 1985). ESDAC data included: (i) the Soil Geographical Database of Eurasia (SGDE), a wide database, generated by the digitalization of the EU soil map, enclosing qualitative attributes about the nature and properties of the soils as texture, moisture regime, stoniness, etc. The SGDE is composed of soil map units (SMU), each of which is characterized by one dominant Soil Typological Unit (STU). From the SGDE, we obtained the map of the main soil types classified

according to the FAO (1975) system; (ii) the Pedotransfer Rules Database (PTRD), reached from meta-analysis of various sources of information and related to the SGDE through SMU and STU. From the PTRD, which also contains qualitative information, we extracted the attributes texture (TX), organic carbon (OC), stoniness (VS), and rooting depth (RD); and (iii) the Soil Profile Analytical Database of Europa (SPADE), connected to the EU Soil map and with more than 1000 representative soil profiles for the 28 EU countries (49 from Spain). The SPADE profiles were linked through the Soil Typological Units (STU) with the soil map of Spain. Therefore, a typical, geographically located profile was available, with all the horizons and morphological, physical, and chemical properties necessary for a quantitative analysis.

For Andalusia, a higher resolution 1: 400,000 scale soil map (CSIC-IARA, 1989) was employed. The map was freely downloaded in vector format (.shp) from the Andalusian Environmental Network (REDIAM) (http://descargasrediam.cica.es/repo/s/RUR). Each SMU of the 1: 400,000 map is associated with one or more representative profiles, which can be accessed in the Andalusian Soil Profile Catalogue (De La Rosa, 1984), summarizing relevant morphological and analytical soil properties.

(4) Olive groves distribution. Olive groves polygons have been extracted from the Information System on Land Occupation of Spain (SIOSE). SIOSE is a land use map (1: 25,000 scale) for the whole of Spain that can be downloaded for free from the website of the National Centre of Geographical Information (CNIG) (https://www.siose.es/web/guest/productos). In this case, SIOSE files have been downloaded for each administrative region in Spain.

Finally, the average soil loss of the olive grove was estimated based on the Map of Erosive States of Spain, scale 1: 1,000,000, downloadable in vector format (.shp) in the web of the "Ministerio para la transición ecológica y el desafío demográfico," MITECO (2017):

(https://www.miteco.gob.es/es/biodiversidad/servicios/banco-datos-naturaleza/informacion-disponible/mapas_estados_erosivos.aspx).

3.2.2 Methods

3.2.2.1 Raster Layers

DEMs with pixel size of 25 m have been uploaded in the ArcGIS 10.5 software for each Spanish region. Then basic statistics has been calculated, and they have been reclassified in height intervals of 400 m. The areas of these intervals have been also determined. Moreover, basic statistics and areas of height areas have been calculated in olive groves using GIS tools and spreadsheets (after exporting tables from GIS), for each Administrative Region and the whole country. From DEMs, slope in % and aspect maps have been obtained by means of surface analysis. As in DEMs, basic statistics and areas of the intervals in which these maps have been classified are also calculated. Slope maps have been classified in 5 intervals (0-2-4-8-15-25- %) and aspect maps have been done in the 8 main orientations. Again, basic statistics and areas of height areas have been calculated in olive groves areas for the different administrative regions and whole Spain.

3.2.2.2 Vector Layers

Different vector layers, climatic, pedological, and erosional have been uploaded in the GIS software. These maps have been analyzed by means of GIS tools to calculate areas of the different classes distinguished in them for the different Spanish regions and the whole country. Moreover, extracting the areas of these layers corresponding to olive groves, the former analyses have been made also in these areas. The calculations have been completed in spreadsheets after export tables from the GIS.

3.2.2.3 Soil Evaluation

The FAO land evaluation method for olive groves (Sys et al., 1991) was applied to nine representative profiles of the main STU of this crop. Sixteen parameters of landscape and soil characteristics: climate (P, T, ATC), Topography (S, Drainage class), physical properties (D, Sand, Clay, VS), and chemical properties (pH, OC, Cation Exchange Capacity, Base saturation, Sum of exchange basic cations, Electric Conductivity, Exchange Sodium Percentage) were assessed according to four limitation degrees to classify profiles into five suitability classes: S1 (Highly Suitable Land), on which sustained use of the soil is expected to yield benefits without unacceptable risk of degradation; S2 (Moderately Suitable), land having moderately severe limitations for sustained use that reduce productivity and increase required inputs and degradation risks; S3 (Marginally Suitable) land having severe limitations for sustained use that reduce productivity or increase the required inputs and degradation risks so that the cultivation will be only marginally justified; N1 (Currently Not Suitable), land having limitations which can be temporally corrected but at expenses of non-assumable costs and degradation risks; and N2 (Permanently Not Suitable) land having limitations so severe that prevents the cultivation of olive groves. For each suitability class, the more limiting parameter (the one that yields a greater limitation degree) is pointed out. This would allow us to interpret which is the most sensitive productivity factor and/or the most active process of environmental degradation.

3.3 Results and Discussion

3.3.1 Climate Constraints

Olive trees need minimum temperatures of between 0 and 7 °C during the winter dormancy for flowering bud differentiation in spring (Therios, 2002). Minimum temperatures above 7 °C cause a failure of flowering. On the other hand, the olive tree does not tolerate absolute temperatures in winter below −7 °C, because intense frosts, especially sudden ones, cause irreversible tissue damage (Connor & Fereres, 2005). As can be seen in Table 3.1, 95% of Spanish olive groves are located in areas with minimum winter temperatures which are optimal for their development (from −2 to 5 °C), while only 5% are in ranges close to their tolerance limit (> 7 °). In addition to minima, average temperatures are also an important factor in the physiology of olives groves (Table 3.2, Fig. 3.1). If temperatures do not rise above 7 °C or fall below 16 °C at any time of the year, there is also a high risk of failure in flowering (Sibbett & Fergusson, 2005). For this reason, annual mean temperatures higher than 22–24 °C can be considered as problematic from the profitability point of view (Sys et al., 1991), even more so in climate change future scenarios with increases in temperatures (Ponti et al., 2014), a plausibility which has already been studied for Morocco (Tabet, 2020), where tipping points for desertification and impacts on olive cultivation have already set the baseline for what may soon become the climate reality for Spanish olive groves.

In total, 31% of the Spanish olive grove is under irrigation (ESYRCE, 2019), so the average annual precipitation (Table 3.3) seems a less limiting factor than the temperature. Almost 65% of the Spanish olive grove is located in areas with optimal rainfall for its development (between 500 and 800 mm). However, 35% of the Spanish olive grove (> 70% in the region of Castilla La Mancha) receives suboptimal rainfall, between 300 and 500 mm per year, which can mean a loss of yield in rainfed orchards and an increased pressure regarding water resources. Moreover, the fall in rainfalls predicted by some climate change scenarios (Cramer et al., 2018) can be very damaging for the Spanish olive grove in the future, even more so combined with the effects on temperature.

3.3.2 Topography

The Spanish olive groves present an average slope of 11% and an average altitude of 520 masl (Fig. 3.2), the highest olive grove being located at 1603 m in Trevélez (Sierra Nevada, Andalusia). While altitude is related with temperatures (discussed above), mean slope (Table 3.4) is a factor that strongly conditions orchard suitability. On the one hand, both the soil workability and the possibilities of intensification are seriously undermined on slopes greater than 25%. On the other hand, the harnessing of rainfall is up to tenfold higher in flat areas. In addition, the slope is one

Table 3.1 Average absolute minimum temperature classes of the coldest month of Spanish olive groves

	Andalucia olive groves		Castilla La Mancha olive groves		Extremadura olive groves		Spain olive groves		Total Spain	
Classes (°C)	Ha	%	ha	%	Ha	%	ha	%	Ha	%
< −7.5	0	0.00	0	0.00	0	0.00	0	0.00	111,843	0.22
−7.5 to −5	0	0.00	0	0.00	0	0.00	0	0.00	348,477	0.69
−5 to −2.5	0	0.00	0	0.00	0	0.00	55	0.00	2,990,927	5.91
−2.5 to 0	4511	0.27	144,081	33.06	173	0.06	121,846	4.46	15,952,884	31.53
0–2.5	489,623	29.63	286,836	65.82	61,125	21.28	926,414	33.89	16,393,380	32.40
2.5–5	1,082,216	65.49	4898	1.12	225,115	78.38	1,560,690	57.09	10,754,001	21.25
5–7.5	75,556	4.57	0	0.00	794	0.28	123,940	4.53	3,665,716	7.25
> 7.5	584	0.04	0	0.00	0	0.00	675	0.02	378,473	0.75
Total	1,652,489	100.00	435,816	100.00	287,207	100.00	2,733,620	100.00	50,595,700	100.00

Data from the National Centre of Geographical Information, CNIG

Table 3.2 Mean annual temperature classes of Spanish olive groves

Classes (°C)	Andalucia olive groves		Castilla La Mancha olive groves		Extremadura olive groves		Spain olive groves		Total Spain	
	ha	%	ha	%	Ha	%	ha	%	ha	%
< 12.5	143	0.01	2970	0.68	117	0.04	3295	0.12	16,751,046	33.11
12.5–15	192,312	11.64	292,845	67.19	12,584	4.38	540,526	19.77	14,730,385	29.11
15–17.5	958,638	58.01	140,001	32.12	274,280	95.50	1,606,383	58.76	14,815,315	29.28
17.5–20	501,396	30.34	0	0.00	225	0.08	583,416	21.34	4,298,954	8.50
> 20	0	0.00	0	0.00	0	0.00	0	0.00	0	0.00
Total	1,652,489	100.00	435,816	100.00	287,207	100.00	2,733,620	100.00	50,595,700	100.00

Data from the National Centre of Geographical Information, CNIG

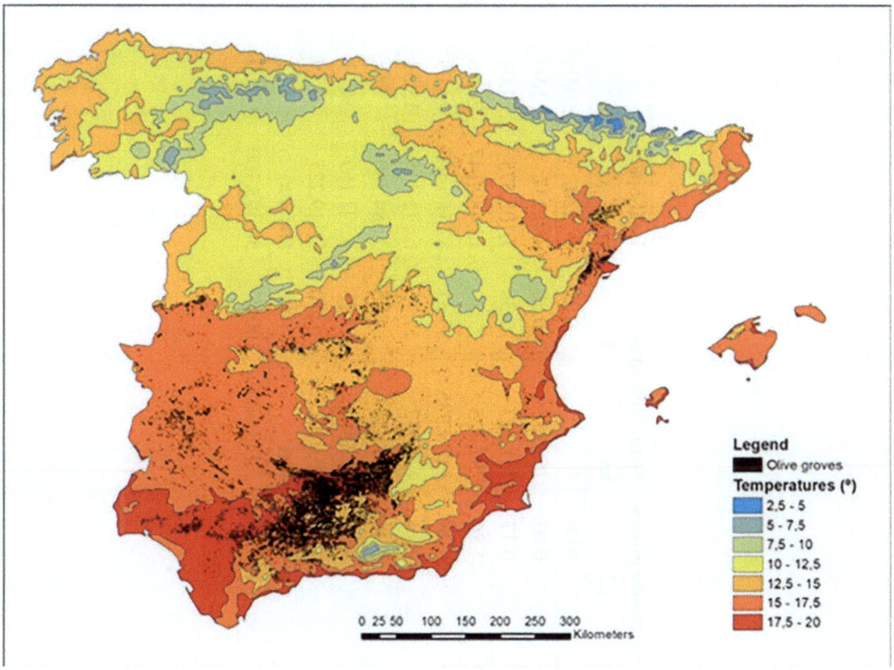

Fig. 3.1 Average absolute minimum temperature classes of the coldest month of Spanish olive groves. (Data from the National Centre of Geographical Information, CNIG)

of the primary factors that condition the soil water erosion, which is one of the major degradation problem that affects Spanish olive groves.

Orchards are relatively well distributed in terms of slope classes, with Andalusians having the highest steepness (8–15% prevailing). Despite this, almost a third of Spanish olive groves are located on marginal (15–25%) or unsuitable (> 25%) slopes for cultivation.

3.3.3 Soil Groups

Table 3.5 and Fig. 3.3 show the distribution of the Spanish soils and the olive groves, according to the soil map of Europe 1: 1,000,000. Eight types of soils, out of more than 40 present in Spain, accounted 75% of the total surface. In general, the most abundant soils in Spain, such as Calcic Cambisols or Eutric Regosols, are also predominant in olive groves, although olive cultivation is lacking in some important groups, such as Humic Cambisols, which is typical of Atlantic climates.

The main soil group in Andalusia and Castilla la Mancha is the Calcic Cambisols (Bk), accounting for 73% of all soils. In Andalusia, they occupied large extensions of hilly reliefs of the Guadalquivir Basin over Miocene marls and sandstones

Table 3.3 Mean annual rainfall classes of Spanish olive groves

	Andalucia olive groves		Castilla La Mancha olive groves		Extremadura olive groves		Spain olive groves		Total Spain	
Classes (mm)	ha	%	ha	%	Ha	%	ha	%	ha	%
< 200	6	0.00	0	0.00	0	0.00	7	0.00	17,220	0.03
200–300	3441	0.21	635	0.15	0	0.00	7586	0.28	960,137	1.90
300–400	26,962	1.63	137,927	31.65	479	0.17	201,217	7.36	7,105,575	14.04
400–500	390,449	23.63	198,204	45.48	120,219	41.86	757,325	27.70	13,489,147	26.66
500–600	771,094	46.66	79,983	18.35	75,483	26.28	1,079,935	39.51	12,186,506	24.09
600–700	338,654	20.49	15,421	3.54	34,861	12.14	476,932	17.45	6,771,445	13.38
700–800	90,245	5.46	2840	0.65	17,851	6.22	129,822	4.75	3,884,780	7.68
800–1000	28,671	1.74	806	0.18	24,606	8.57	61,585	2.25	3,397,169	6.71
1000–1400	2857	0.17	0	0.00	13,707	4.77	19,076	0.70	2,119,225	4.19
> 1400	110	0.01	0	0.00	0	0.00	136	0.00	664,496	1.31
Total	1,652,489	100.00	435,816	100.00	287,207	100.00	2,733,620	100.00	50,595,700	100.00

Data from the National Centre of Geographical Information, CNIG

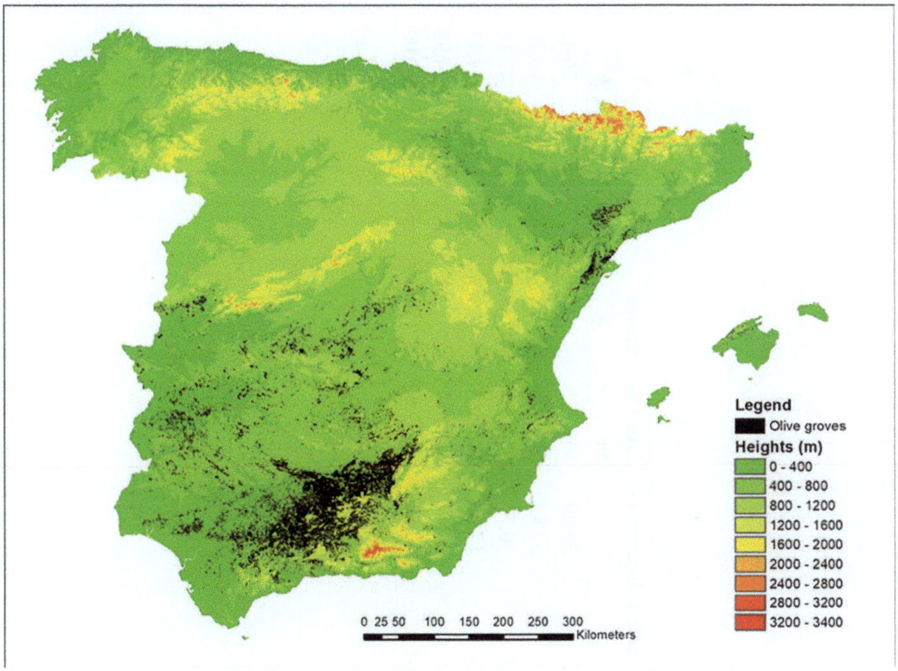

Fig. 3.2 Slope classes of Spanish olive groves. (Data from the National Centre of Geographical Information, CNIG)

(*albarizas*), as well as marls, marly limestones, and limestones of the Jurassic and Cretaceous in piedmont zones of the Betic mountain ranges, normally in landscape association with Calcareous Regosols, Lithosols or Calcareous Fluvisols. Calcic Cambisols, in Miocene and Mesozoic clays, and in Triassic (Keuper) clays and marls, gave rise to Chromic Vertisols (Vc).

In Castilla La Mancha, Calcic Cambisols were mainly associated with Miocene marls, marly limestones, and clays, in areas of flat reliefs (e.g., "La Mancha plateau" which occupies much of the provinces of Ciudad Real and Albacete). Calcic Cambisols over carbonated materials alternate with Plio-Pleistocene clays and gravels, where Chromic Luvisols (Lc) and Eutric Regosols (Re) were developed in the so-called *rañas* formations (eg. "Campos de Calatrava" region in the Ciudad Real province). Eutric Regosols were more abundant toward the northeast of Castilla La Mancha (province of Toledo), due to the outcrop of Paleozoic slates and the increase in mean slope (zones of the "Toledo plateau" and Montes de Toledo mountain range), although in these landscapes the olive grove tends to be less and less abundant.

Unlike in Andalusia and Castilla La Mancha, in Extremadura there is a prevalence of olive groves over siliceous materials, which are concentrated in the eastern part of the Montes de Toledo and the Paleozoic slates of Sierra Morena mountains (Badajoz province), being dominated by Eutric Regosols (17%) and Dystric

Table 3.4 Slope classes of Spanish olive groves

	Andalucia olive groves		Castilla La Mancha olive groves		Extremadura olive groves		Spain olive groves		Total Spain	
Classes (%)	ha	%	ha	%	ha	%	ha	%	Ha	%
0—4[a]	268,687	16.26	194,662	44.67	88,462	30.80	592,955	21.69	13,366,524	26.42
4—8[b]	315,228	19.08	115,199	26.43	79,924	27.83	581,338	21.27	8,451,483	16.70
8—15[c]	445,088	26.93	76,620	17.58	59,473	20.71	684,907	25.05	8,684,331	17.16
15—25[d]	354,477	21.45	34,825	7.99	34,625	12.06	505,585	18.50	7,453,816	14.73
> 25[e]	269,009	16.28	14,510	3.33	24,723	8.61	368,835	13.49	12,639,546	24.98
Total	1,652,489	100.00	435,816	100.00	287,207	100.00	2,733,620	100.00	50,595,700	100.00

Data from the National Centre of Geographical Information, CNIG
FAO slope classes
[a]Flat
[b]Level
[c]Gently sloping
[d]Steep
[e]Sloping

Table 3.5 Main soil types distribution in Spain olive groves

Soil group[a]		Andalucía olive groves		Castilla La Mancha olive groves		Extremadura olive groves		Spain olive groves		Total Spain	
		ha	%	ha	%	ha	%	ha	%	ha	%
Calcic Cambisol[b]	Bk	1,203,717	72.84	205,995	47.27	61,334	21.36	1,750,138	64.00	16,691,577	32.99
Humic Cambisol	Bh	1363	0.08	98	0.02	13,261	4.62	155,03	0.57	5,036,658	9.95
Calcaric Lithosol	Ic	1714	0.10	5466	1.25	0	0.00	49,257	1.80	3,472,902	6.86
Calcaric Fluvisol	Jc	73,654	4.46	9583	2.20	6321	2.20	114,979	4.21	28,80,166	5.69
Chromic Luvisol[c]	Lc	16,503	1.00	86,779	20.31	21,708	7.55	97,620	3.58	1,313,422	2.60
Dystric Regosol[d]	Rd	16,929	1.02	13,666	3.14	76,531	26.65	106,390	3.89	4,301,199	8.50
Eutric Regosol[e]	Re	91,656	5.55	79,769	18.30	49,431	17.21	209,975	7.68	3,639,347	7.19
Chromic Vertisol[f]	Vc	99,424	6.02	0	0.00	0	0.00	115,033	4.21	488,549	0.97
Other soil groups		147,529	8.93	34,460	7.51	58,621	20.41	27,4725	10.06	12,770,747	25.25
Total		1,652,489	100.00	435,816	100.00	287,207	100.00	2,733,620	100.00	50,594,567	100

Data from the European Soil Data Centre, ESDAC, Panagos (2006)
[a]In Soil Taxonomy classification (Soil Survey Staff, 2010)
[b]Calcixerepts and Calcic Haploxerepts
[c]Rhodoxeralf and Palexeralf
[d]Typic Xerorthent
[e]Dystric Xerorthent
[f]Chromic Calcixerets and Chromic Haploxerets

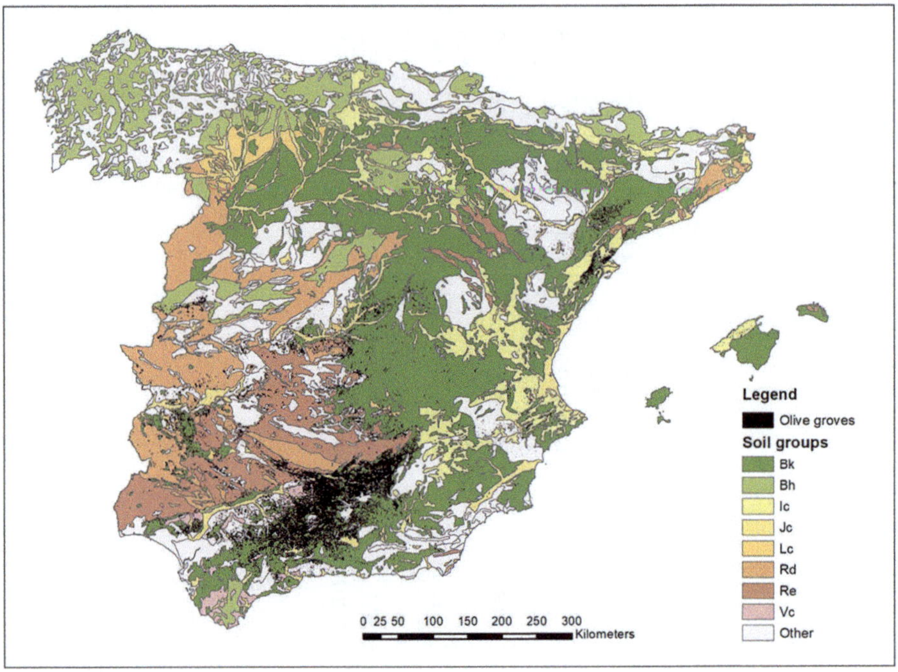

Fig. 3.3 Main soil types distribution in Spain olive groves. (Data from the European Soil Data Centre, ESDAC, Panagos, 2006)

Regosols (Rd, 27%). Dystric Regosols were frequently associated to Paleozoic quartzites and intrusive igneous rocks such as granodiorite. In contrast, Calcic Cambisols are concentrated in Neogene and Quaternary materials of the Guadiana and the Tajo Basins, as well as the dispersed but common Paleozoic recifal limestones.

3.3.4 Pedological Settings

Tables 3.6, 3.7, 3.8 and 3.9 show four key soil properties that determine the economic profitability and the susceptibility of soils to degradation. Given the abundance of marly materials, the main texture in the Andalusian olive groves was fine (54%), while in Castilla La Mancha and Extremadura was medium (around 65%). Stoniness is also an important factor because it controls the volume of soil that roots can explore to uptake water and nutrients. Furthermore, a high stoniness (> 20%) hinders the workability of the soil. Around a third of the olive groves have stoniness >20%, which means severe limitations for cropping.

Most of Spanish olive grove soils (63%) do not present significant limitations in terms of rooting depth (> 80 cm). This is correlated, as stoniness, with the

Table 3.6 Texture classes (% weight) of Spanish olive groves

	Andalucía olive groves		Castilla La Mancha olive groves		Extremadura olive groves		Spain olive groves		Total Spain	
Classes (%)	ha	%	ha	%	ha	%	ha	%	ha	%
No data	2066	0.13	342	0.08	440	0.15	4139	0.15	284,441	0.56
Coarse[a]	75,393	4.56	39,848	9.14	79,147	27.56	195,417	7.15	5,167,489	10.21
Medium[b]	475,449	28.77	282,415	64.80	183,917	64.04	1,079,038	39.47	31,825,860	62.90
Medium-Fine[c]	210,267	12.72	27,417	6.29	0	0.00	279,269	10.22	4,412,223	8.72
Fine[d]	889,314	53.82	85,794	19.69	23,703	8.25	1,175,757	43.01	8,905,686	17.60
Very Fine[e]	0.00	0.00	0.00	0.00	0.00	0.00	0.00	0.00	34,177	0.07
Total	1,652,489	100.00	435,816	100.00	287,207	100.00	2,733,620	100.00	50,595,700	100.00

Data from the European Soil Data Centre, ESDAC, Panagos (2006)

[a]18% < clay and > 65% sand

[b]18% < clay <35% and > 15% sand, or 18% < clay and 15% < sand <65%

[c]< 35% clay and < 15% sand

[d]35% < clay <60%

[e]clay >60%

Table 3.7 Stoniness of the topsoil (% vol.) of Spanish olive groves

	Andalucía olive groves		Castilla La Mancha olive groves		Extremadura olive groves		Spain olive groves		Total Spain	
Classes (%)	ha	%	ha	%	ha	%	ha	%	ha	%
No data	0	0.00	0	0.00	0	0.00	7	0.00	1133	0.00
0–10%	933,481	56.49	185,251	42.51	155,954	54.30	1,408,161	51.51	24,757,527	48.93
10–15%	611	0.04	5559	1.28	4083	1.42	52,249	1.91	3,959,712	7.83
15–20%	147,080	8.90	107,621	24.69	70,434	24.52	330,261	12.08	11,140,608	22.02
>20%	571,317	34.57	137,385	31.52	56,736	19.75	942,942	34.49	10,736,720	21.22
Total	1,652,489	100.00	435,816	100.00	287,207	100.00	2,733,620	100.00	50,595,700	100.00

Data from the European Soil Data Centre, ESDAC, Panagos (2006)

Table 3.8 Rooting depth (cm) classes of Spanish olive groves

	Andalucía olive groves		Castilla La Mancha olive groves		Extremadura olive groves		Spain olive groves		Total Spain	
Classes (%)	ha	%	ha	%	ha	%	ha	%	ha	%
No data	3558	0.22	1123	0.26	21,442	7.47	28,606	1.05	1,400,732	2.77
> 80 cm	1,203,100	72.81	231,760	53.18	74,331	25.88	1,732,783	63.39	18,917,282	37.39
60–80 cm	47,627	2.88	21,564	4.95	828	0.29	94,085	3.44	4,178,129	8.26
40–60 cm	331,591	20.07	122,060	28.01	176,759	61.54	649,439	23.76	12,941,216	25.58
20–40 cm	66,613	4.03	59,309	13.61	13,847	4.82	228,707	8.37	13,158,341	26.01
Total	1,652,489	100.00	435,816	100.00	287,207	100.00	2,733,620	100.00	50,595,700	100.00

Data from the European Soil Data Centre, ESDAC, Panagos (2006)

Table 3.9 Organic carbon (OC) of the topsoil (% weight) of Spanish olive groves

Classes[a] (%)	Andalucía olive groves		Castilla La Mancha olive groves		Extremadura olive groves		Spain olive groves		Total Spain	
	ha	%	ha	%	ha	%	ha	%	ha	%
No data	0	0,00	0	0,00	0	0,00	7	0,00	1133	0,00
High	0	0,00	0	0,00	0	0,00	0	0,00	46,336	0,09
Medium	50,786	3,07	1502	0,34	13,261	4,62	74,209	2,71	19,409,964	38,36
Low	1060,561	64,18	175,732	40,32	133,003	46,31	1,563,526	57,20	9,990,718	19,75
Very low	541,141	32,75	258,582	59,33	140,943	49,07	1,095,878	40,09	21,147,550	41,80
Total	1,652,489	100,00	435,816	100,00	287,207	100,00	2,733,620	100,00	50,595,700	100,00

Data from the European Soil Data Centre, ESDAC, Panagos (2006)

[a]OC classes: *High* > 6.0%; *Medium* 2,1 – 6.0%; *Low* 1.1–2.0%; *Very Low* < 1.0%

abundance of unconsolidated parent materials like marls or clays. However, the rooting depth distribution in olive groves is different from that of the total Spanish soils, preferring deeper or less depth-limited soils (53% of Spanish soils have a limitation from 60 cm, compared to only 32% in olive groves). The shallower soils were those of Extremadura (67% with obstacles inside 60 cm), which is explained by the abundance of competent rocks such as slates, quartzites or recifal limestones in this region.

Topsoil organic carbon is one of the soil properties that has the most impact on physical, biological, and chemical soil fertility, because it relates to an adequate structure and porosity, and to a sufficient content of nutrients. However, the relatively low organic carbon content of olive groves in Spain stands out. Thus, 98% of Spanish olive groves, with few differences between regions, have organic carbon contents of less than 2% (low and very low classes), compared to the 62% of soils in the whole of Spain.

3.3.5 *Suitability Assessment*

Tables 3.10 and 3.11 show the characteristics, suitability classes, and limiting factors of the representative profiles of the Spanish olive grove: Bk1 to Bk4 and Vc were, respectively, a Calcic Cambisols and a Chromic Vertisol from Andalucía, Bk5 and Lc were a Calcic Cambisol and a Chromic Luvisol from Castilla La Mancha, and Re and Rd. were a Eutric Regosol and a Dystric Regosol from Extremadura.

As can be seen, all representative profiles are classified into the marginally suitable class (S3), which indicates some severe limitations that affect the economic and/or environmental sustainability of the crop. The main limiting factor affecting all groups except the Chromic Luvisol was the slope > 15% (sloping or steep classes). These high slopes indicate a significant struggle in the mechanization, which is translated into high management costs, mainly harvesting, and a decrease in benefits that often jeopardizes the profitability of the olive grove (Fleskens, 2011). Clay amount (> 40%) is the second most limiting factor, affecting Cambisols (Bk1, Bk4) and the Vertisol. Heavy textures, in combination with low organic carbon content, can cause poor structure development, mechanical damage, and tree drainage and aeration problems (Ortega et al., 2016). Regarding organic carbon, although it showed values below what is recommended in all profiles (low or very low classes), it was explicitly limiting in Regosols; in the case of the Dystric Regosol, in addition to an acidic pH and a low nutrient content (base saturation < 50%), which reveals a severe chemical fertility concern in this soil.

Acidic pH also is considered a limiting factor in Chromic Luvisols. In the case of the Cambisol Bk3, typical *Albariza* soil, the high carbonate content (average equivalent $CaCO_3$ in the first 80 cm of 50%) determines a pH of 8.3, which can affect the absorption of micronutrients. Finally, the two representative profiles of Castilla La Mancha (Bk5 and Lc) showed a climatic limitation, with annual average temperatures close to 13 °C, the minimum acceptable for profitable cultivation (Sys

Table 3.10 Profile characteristics and suitability classes (SC) for representative Spanish soil profiles

	Profile characteristics							Soil Suitability for olive groves	
			Climate			Landscape			Limiting
Profile	Soil type	Parent material	P	T	ATC	S	Dr	SC	factor
Bk1	Calcic Cambisol	Miocene marl	450	17,0	3,5	17	Good	S3	Slope
Bk 2	Calcic Cambisol	Cretaceous Marly limestone	650	16,0	3,0	9	Moderate	S3	Texture
Bk 3	Calcic Cambisol	Cretaceous marl	550	16,0	3,0	16	Good	S3	Slope; pH
Bk 4	Calcic Cambisol	Jurasic limestone	650	16	3,0	15	Moderate	S3	Texture
Bk 5	Calcic Cambisol	Miocene marl	450	13,5	1,5	8	Good	S3	T
Vc	Chromic vertisol	Miocene marl	450	16,0	3,0	18	Moderate	S3	Texture; slope
Lc	Chromic Luvisol	Pliocuaternary conglomerate	450	13,5	2,0	0	Good	S3	T; pH
Re	Eutric Regosol	Ordovician shale	500	16	2	17	Good	S3	Slope; OC
Rd	Dystric Regosol	Permian shale	650	16	3,5	20	Good	S3	Slope; pH; OC; BS

Profiles characteristics were obtained from the European Soil Data Centre, ESDAC and the National Centre of Geographical Information, CNIG

P annual precipitation (mm); *T* mean annual temperatura (° C); *ATC* Average absolute minimum temperature of the coldest month (°C); *S* slope (%); *Dr* drainage; *SC* Suitability class (*S1* Highly Suitable; *S2* Moderately Suitable; *S3* Marginally Suitable; *N1* Currently Not Suitable; *N2* Permanently Not Suitable)

Table 3.11 Physical and chemical properties for representative Spanish soil profiles

	Physical and chemical fertility[a]										
Profile	D	Sand	Clay	VS	CEC	BS	V	pH	OC	EC_{25}	ESP
Bk1	80	15	32	0	16.35	100	16.35	7.8	0.69	1.0	1.2
Bk 2	80	25	41	7	11.28	99	11.28	7.6	0.94	0.7	3.2
Bk 3	100	29	35	8	11.21	100	11.21	8.3	0.91	0.1	1.3
Bk 4	80	6	66	20	17.10	95	17.10	7.9	1.70	0.6	2.6
Bk 5	180	29	27	1	16.40	100	16.40	8.1	1.10	1.0	3.7
Vc	100	17	50	0	14.04	100	19.00	7.4	0.87	0.8	7.2
Lc	150	41	34	2	24.50	60	16.90	6.1	0.99	1.0	2.1
Re	75	24	24	2	13.67	100	13.67	7.8	0.34	1.0	5.4
Rd	75	24	25	3	14.60	45	5.2	6.1	0.35	1.0	0.7

Soil properties were obtained from the Soil Profile Analytical Database of Europa, SPADE, and the 1: 400,000 scale soil map, CSIC-IARA, (1989)

D thickness (m); VS volume of stones (> 2 mm fraction) (% vol,); *CEC* cation exchange capacity (cmol+ kg^{-1}); *BS* base saturation (%); *V* sum of basic cations (cmol+ kg^{-1}); *OC* organic carbon (%); EC_{25} electric conductivity (dS m^{-1}); *ESP* Exchange sodium percentage (%)

[a]Weighted average at 80 cm depth (except OC, averaged at 20 cm)

Table 3.12 Potential risk of erosion in Spanish olive groves

	Andalucía olive groves		Castilla La Mancha olive groves		Extremadura olive groves		Spain olive groves		Total Spain	
Erosión t ha^{-1} $yr.^{-1}$	ha	%	ha	%	Ha	%	Ha	%	Ha	%
0–5	228,808	13.85	69,415	15.93	25,878	9.01	400,860	14.66	15,423,189	30.48
5–12	58,401	3.53	133,225	30.57	124,502	43.35	292,113	10.69	11,415,861	22.56
12–25	76,362	4.62	97,690	22.42	58,269	20.29	254,439	9.31	10,950,964	21.64
25–50	361,058	21.85	55,273	12.68	32,225	11.22	546,848	20.00	5,776,173	11.42
50–100	127,028	7.69	62,896	14.43	34,789	12.11	235,219	8.60	3,540,098	7.00
100–200	634,164	38.38	10,888	2.50	8026	2.79	796,592	29.14	2,300,745	4.55
<200	159,231	9.64	5774	1.32	2784	0.97	1976,36	7.23	716,601	1.42
Water	7335	0.44	649	0.15	734	0.26	9686	0.35	360,378	0.71
Urban	103	0.01	6	0.00	0	0.00	228	0.01	111,691	0.22
Total	1,652,489	100.00	435,816	100.00	287,207	100.00	2,733,620	100.00	50,595,700	100.00

Data from map of Erosive States of Spain, scale 1: 1000,000, MITECO (2017)

et al., 1991). Less suitable olive grove soils for a sustainable management were, therefore, Dystric Regosols, because they presented severe limitations both at the level of the terrain (slope) and at the level of chemical properties. However, Dystric Regosols accounted for a very small percentage of the total olive grove (3.8%).

Finally, the combination of steep slopes, fine-textured soils, and low OC content, characteristic of many of olive grove soils, conditions a huge environmental (and economic) concern in Spain: Water erosion. Thus, low infiltrations in steeply sloping soils can generate high runoff, which may explain the noteworthy risk of erosion of more than 25 t ha^{-1} year^{-1} of losses affecting 65% of Spanish olive groves (Table 3.12).

3.4 Conclusions

Although the olive tree is fairly adapted to the climate, relief, and soils of Spain, many of the national olive groves showed significant limitations that threaten their economic profitability and environmental sustainability. This is probably due to the expansion of the crop to marginal areas (high slopes, low temperatures, and acidic soils) with a vocational use for forestry or livestock rather than agriculture, as well as degradation processes accelerated by conventional management. The latter could be stated in the case of organic carbon losses and water erosion in Spanish olive groves. Although the high inputs applied in conventional cultivation offset, at least in part, the marginal tendency of many Spanish olive groves, a truly sustainable planning of the territory would do well to consider the biophysical potential of the territory.

References

CEC. (1985). *Soil map of the European Communities at 1: 1,000,000* (p. 124p). CEC-DGVI.

Connor, D. J., & Fereres, E. (2005). The physiology of adaptation and yield expression in olive. *Horticultural Reviews, 31*, 155–229.

Cramer, W., Guiot, J., Fader, M., Garrabou, J., Gatusso, J. P., Iglesias, A., Lange, M., et al. (2018). Climate change and interconnected risk to sustainable development in the Mediterranean. *Nature Climate Change, 8*, 972–980.

CSIC-IARA. (1989). *Memoria del mapa de suelos de Andalucía, escala 1: 400,000*. Consejo Superior de Investigaciones Científicas e Instituto de Recursos Naturales y Agrobiología.

De la Rosa, D. (1984). *Catálogo de Suelos de Andalucía. AMA*. Junta de Andalucía. http://www.juntadeandalucia.es/medioambiente/site/portalweb/menuitem.7e1cf46ddf59bb227a9ebe205510e1ca/?vgnextoid=0799bf6d57857010VgnVCM1000000624e50aRCRD&vgnthirdoid=fdb4760edc7a7010VgnVCM1000000624e50aRCRD. Accessed 7 May2020.

Encuesta sobre Superficie y Rendimiento, ESYRCE. (2019). *MAPAMA, Gobierno de España*. (In Spanish). https://www.mapa.gob.es/es/estadistica/temas/estadisticas-agrarias/boletin2019_tcm30-536911.pdf. Accessed 7 Apr 2020.

ESDB v2.0. (2004). The European soil database distribution version 2.0. In *European Commission and the European Soil Bureau Network, CD-ROM, EUR 19945 EN*. https://esdac.jrc.ec.europa.eu/resource-type/european-soil-database-soil-properties. Accessed 7 May 2020.

FAO. (1975). *Soil map of the world at 1: 5,000,000* (Vol. I, p. 62). Europe. UNESCO.

FAO. (1976). *A framework for land evaluation*. FAO Soils Bull., n° 32.

Fleskens, L. (2011). A typology of sloping and mountainous olive plantation systems to address natural resources management. *Annals of Applied Biology, 153*, 283–297.

Guzmán, G. I., & González de Molina, M. (2017). Energy in agroecosystems: A tool for assessing sustainability. In *Advances in Agroecology*. CRC Press.

Keestra, S., Nunes, J., Novara, A., Finger, D., Avelar, D., Kalantari, Z., & Cerdà, A. (2018). The superior effect of nature based solutions in land management for enhancing ecosystem services. *Science of the Total Environment, 610–611*, 997–1009.

MITECO. (2017). Ministerio para la Transición Ecológica y Reto Demográfico. In *Map of erosive states of Spain, scale 1: 1,000,000*. Gobierno de España.

Ortega, E., Lozano, F. J., Martínez, F. J., Bienes, R., Gallardo, J. F., & Asensio, C. (2016). Soils of the Mediterranean areas. In J. F. Gallardo (Ed.), *The soils of Spain*. Springer.

Panagos, P. (2006). The European soil database. *GEO Connexion, 5*(7), 32–33.

Ponti, L., Gutierrez, A. P., Ruti, P. M., & Dell'Aquila, A. (2014). Fine-scale ecological and economic assessment of climate change on olive in the Mediterranean Basin reveal winners and losers. *PNAS, 111*(15), 5598–5603.

Sibbett, G. S., & Fergusson, L. (2005). *Olive production manual* (2nd ed., p. 180). University of California Agriculture and Natural Resources.

Soil Survey Staff. (2010). *Keys to soil taxonomy* (11th ed.). USDA – NCRS.

Sys, C., Van Rast, E., Debaveye, J., 1991. Land evaluation (Parts I, II and III). Agricultural Publications, N° 7. University of Ghent.

Tabet, O. (2020). *Description of the state of olive groves in Morocco. Masc Thesis. Universidad de Jaén (Spain).* Msc Analysis, conservation and restoration of physical and biological components of habitats. Available: https://crea.ujaen.es/bitstream/10953.1/13172/1/OUIAME-TABET-TFM.pdf. Feb 2024.

Therios, I. (2002). *Olives. Crop production science in horticulture n° 18* (p. 409). CAB International.

Chapter 4
Cultivars and Olive Growing Systems in Spain

Raul de la Rosa, Lorenzo León, Angjelina Belaj, and Ignacio Lorite

Abstract Traditional olive orchards in Spain have been characterized by low planting densities, long centennial life span, and high number of cultivars and are typically located in marginal areas. This traditional olive growing is characterized by low productivity, low inputs, and a very high biennial bearing but with a high cultivar biodiversity. However, from the mid-twentieth century, a new trend of intensification is taking place in olive orchards. In the last 20 years, olive growing intensification led to the development of the hedgerow growing system, characterized by a very high planting density where trees formed hedgerows to be harvested by straddle machines. This intensification was accompanied by a reduction of cultivar's diversity, which has boosted the activity of some breeding programs aimed at producing modern cultivars specifically adapted to this new growing system. This is the case of "Sikitita," the first breeding cultivar produced in Spain, with a weeping habit. In the future, both olive growing systems and cultivars should evolve together in order to fit the changing demands in the olive oil market as well as potential unknown challenges emerging from new scenarios such as climate change.

R. de la Rosa (✉)
Plant Improvement and Biotechnology Unit, IFAPA-Alameda del Obispo, Córdoba, Spain
Instituto de Agricultura Sostenible, Consejo Superior de Investigaciones Científicas, Córdoba, Spain
e-mail: raul.rosa@ias.csic.es

L. León · A. Belaj · I. Lorite
Plant Improvement and Biotechnology Unit, IFAPA-Alameda del Obispo, Córdoba, Spain
e-mail: lorenzo.leon@juntadeandalucia.es; angjelina.belaj@juntadeandalucia.es; ignacioj.lorite@juntadeandalucia.es

J. Muñoz-Rojas, R. García-Ruiz (eds.), *The Olive Landscapes of the Mediterranean*, Landscape Series 36,
https://doi.org/10.1007/978-3-031-57956-1_4

4.1 Past

Olive is one of the oldest crops in the Mediterranean basin (Zohary & Spiegel-Roy, 1975). Cultivation started around 4.000–6.000 years ago by doing an empirical selection of the most outstanding wild genotypes. This initial selection was based on increasing fruit size and oil content (Belaj et al., 2011) and started in the Fertile Crescent in Near East Asia (Besnard et al., 2017). Then olive cultivation was extended toward the West of the Mediterranean basin up to the Iberian Peninsula (Jiménez-Ruiz et al., 2020). During centuries of cultivation, growers have selected a wide number of traditional olive cultivars that were confined to their area of origin, thus producing a "mosaic" of olive cultivars from Iran to Portugal and from the North of Italy to South Algeria. This mosaic of olive cultivars has been slowly evolving from ancient times up to the mid-twentieth century. The result of this is a very wide genetic biodiversity that encompasses an enormous variability for most of the agronomic traits of interest on olive (Rallo et al., 2005). The olive genetic patrimony is estimated to include more than 1200 cultivated in 54 countries. Over two-thirds of the olive cultivars are present in the southern European countries such as Italy (538), Spain (272), France (88), and Greece (52) (Bartolini, 2008). In addition, a high number of olive cultivars is still preserved on farms in traditional olive-growing countries and outside Europe, such as in Syria, Turkey, Tunisia, Morocco, and Iran (Belaj et al., 2003, 2016). The use of seedlings as rootstocks, or their spreading and further selection both in the traditional area of cultivation and in countries where olive cultivation has been recently introduced, like Argentina, Australia, Brazil, California, China, Colombia, Mexico, and Uruguay, might have contributed to a "new-emerging" genetic variability (Koehmstedt et al., 2011; Beghé et al., 2015). In both traditional and new olive growing areas, the survival of very ancient olive trees has been reported throughout the entire Mediterranean area and beyond (Díez et al., 2011; Barazani et al., 2014; Ninot et al., 2018). The ancient olives are the living proof of the very long tradition of olive cultivation and, in most cases, they include different and unknown local cultivars constituting thus an unexploited reservoir of olive diversity.

Olive growing systems have evolved very little (Fernández-Escobar et al., 2013). In traditional olive orchards, trees were planted at low densities, up to 100 trees/ha, and even as low as 16 trees/ha in some desert areas as in South Tunisia. They are typically located in marginal areas and always rainfed. The main drawback of this cultivation is the slow and high labor demanding at harvesting. The extended period for harvesting negatively influenced oil quality and promoted the biennial bearing, i.e., the alternation of years with high and low yields. This biennial bearing was also caused by environmental stresses, mainly water stress, very typical of the Mediterranean climate.

4.2 Present

Traditional olive orchards still represent more than half of the olive cultivation surface in Spain (Penco, 2020), including Andalucía. Therefore, they still have a paramount economic and social importance of some rural areas of the country (Barranco et al., 2017). Those traditional orchards are conserving the wide range of olive biodiversity above mentioned (Fig. 4.1).

However, since the 1970s, olive growing systems have evolved rapidly. The first innovations were the development of irrigation, use of synthetic pesticides and fertilizers, and the inclusion of machinery on the harvesting (Fernández-Escobar et al., 2013; Pastor & Humanes, 1990). This led to more intensive olive growing, with plantations in new areas with higher water availability and better edaphoclimatic conditions (Barranco et al., 2017). Those new intensive olive orchards are also characterized by an increase of density, up to 400 trees/ha, with final spacing determined by water availability, edaphoclimatic conditions, harvest system, and cultivar (Pastor et al., 2007). Few cultivars are used in these new plantations, leading thus to the genetic erosion of the wide olive germplasm. For example, in Andalucía, most of the orchards established in the last decades of the XX Century were planted with cultivar "Picual." Therefore, to promote the conservation and the use of this endangered biodiversity, the Worldwide Olive Germplasm Bank of Cordoba

Fig. 4.1 Examples of different growing systems: traditional in marginal area (**a**), traditional able to be reconverted (**b**), intensive (**c**), and hedgerow (**d**). Reproduced with permission from Fernández-Escobar et al. (2013)

Fig. 4.2 Cultivar collection of the World Olive Germplasm Bank of Cordoba, IFAPA

(Fig. 4.2) was established, 50 years ago, at the Center of the Agricultural, Fishery, Food and Organic Farming Research & Training Institute (IFAPA), "Alameda del Obispo" (WOGB-IFAPA). Its establishment represented the first international attempt of conservation and management of the olive germplasm through a FAO-INIA project and with the IOC support (IOC, 2022; Belaj et al., 2016). In addition, the WOGC-IFAPA collection belongs to the INIA network of olive germplasm collections representing the national olive germplasm collection of reference (ESP-046; see Rallo, 2019). From its initials, this collection has been nourished by cultivars collected from Spanish and international prospecting surveys (Barranco & Rallo, 2000; Belaj et al., 2003; Ninot et al., 2018) as well as accessions provided by the IOC Network of Olive Germplasm Collections (Caballero et al., 2006; IOC, 2022). Nowadays, this collection reaccounts more than 1000 accessions from 29 countries in the field, being 35.5% of them of national (Spanish) origin. Besides, the backup collection established at IFAPA Centre "Venta del Llano" accounts around 800 accessions in the field. Besides, recently the WOGB-IFAPA is being enlarged through prospecting surveys in Andalusia (Gómez-Gálvez et al., 2024).

With the beginning of the new century, this olive intensification process had a new step forward with the hedgerow growing system. It was initially based on densities around 2000 trees/ha and spacing of around 3.75 × 1.35 m, always with drip irrigation (www.olint.com). Plants were initially trained in a monocone shape from the beginning of plantation. Three or four years after plantation, continuous hedgerows are formed on the tree lines (Gomez-del-Campo et al., 2009). The main advantage of this system is the fully mechanized harvesting, done by straddle machines developed for vineyards with some modifications (Fig. 4.3). Other advantages of this system are the early entrance into commercial production, from three years

Fig. 4.3 Straddle harvesting machine in action in a hedgerow olive orchard

after planting, and the reduced biennial bearing due to early harvest typically done in this system and to the low tendency of the cultivars used to alternate production (de la Rosa et al., 2007; Centeno et al., 2019). Additionally, the faster and earlier harvesting of this growing system, respect to the intensive one, could have a positive influence on the final olive oil quality. However, when large areas are planted with this system, enough milling capacity should be available to guarantee a good quality of the olive oil. The main disadvantages of this system are the high cost of setting up and the reduced number of cultivars with canopy vigor low enough to be suitable for this system (de la Rosa et al., 2007; Rallo et al., 2008; Centeno et al., 2019).

In the case of table olive, low level of mechanization has been achieved up to now and production systems are still very traditional. The main problem for the mechanization of this system is the browning produced on the fruits (de la Rosa et al., 2007; Morales-Sillero et al., 2014).

4.3 Future

Future prospects for traditional olive growing have different possible ways. One is their transformation in a more intensive and efficient growing system, when possible. In the case where the field slope and other environmental conditions make it difficult, growers may look for indirect income alternatives such as landscape or social uses. Organic farming is proposed as another option to increase the economic profitability of traditional olive growers. Protected Denomination of Origin (DOP) should also focus on the promotion of the peculiarities of this type of olive growing, especially the wide range of olive oils produced by the wide variability of traditional cultivars.

The main near-future advantage of the intensive olive growing is the wide range of cultivars that can be planted including those specifically determined in the DOP existing in Spain. Modern trunk shakers have greatly reduced the cost of harvesting in this type of orchards, although are still highly labor consuming (Barranco et al., 2017). Besides, there are some current attempts to fully mechanize harvesting in vase-formed olives, but, so far, there are not any solution commercially available. The main challenge of intensive orchards is to achieve a production cost as low as hedgerow olive orchard.

The growing system having the greater potential for the forthcoming olive growing is, without any doubt, the hedgerow system. The fully mechanized harvesting, early bearing, and low biennial bearing are the main reasons for the widespread and success of this growing system. However, the large investment required by this planting system has promoted a reduction in tree density to around 1200 trees/ha, with 4 × 2 m spacing. The two main challenges of this system in the near future are the pruning and to widening the cast of suitable cultivars (de la Rosa et al., 2007; Rallo et al., 2008; Centeno et al., 2019). Pruning is the most expensive management practice of hedgerow and the most critical one. If it is not well performed, it could rapidly ruin the viability of the orchard. More on this issue could be found in the pruning chapter of this book.

On the other side, few cultivars are currently adapted to this system, namely, the traditional “Arbequina” and “Arbosana” and the new breed “Sikitita” (Rallo et al., 2008) and “Lecciana” (www.olint.es) and stability on their oils use to be low to medium. Due to the fact that most of the traditional cultivars are too vigorous for the high densities of the hedgerow systems, new olive cultivars, obtained by systematic breeding, are needed to widen the offer of cultivars with improved oil composition, as those under selection on the olive breeding program of IFAPA at Córdoba (Pérez et al., 2018).

Another option to widen the number of cultivars suitable for hedgerow orchards is the obtention of dwarfing rootstocks. However, the vigor of a cultivar is largely influenced by the environment, so specific trials should be developed to test the suitability of a given cultivar in a specific environment. This is especially important in those areas outside the Mediterranean Basin, where climatic conditions could dramatically affect this trait. Besides, some experiments of hedgerow orchards on dry farming at wider spacing are currently under way with promising results. However, this new strategy should be tested in several environments, and with a wide number of cultivars before it can be considered efficient.

In the case of table olive, the low market prices during the last years in some countries have promoted research on high tree-density plantations (Morales-Sillero et al., 2014). The success of this could lead on a change in the density and training of those orchards, boosting their profitability. New cultivars will also be needed for this intensification (de Castro et al., 2019).

In summary, both growing systems and cultivars should evolve together in order to cope with the enormous challenges that Spanish olive growing is currently facing. Probably there is not a single strategy that will be successful. Highly efficient growing systems producing olive oil at low cost might coexist with more traditional strategies that could obtain an added value due to a peculiar quality or environment conservation.

Finally, climate change will have a critical impact in the Mediterranean olive orchards, promoting the cultivar selection with early flowering and low chilling requirements (Belaj et al., 2020). However, the limited genetic variability for olive cultivar's flowering date (Belaj et al., 2020) limits the breeding programs for new cultivars with these characteristics. Other objectives as the obtention of new cultivars more resilient to water and heat stress events are planned at medium-large term.

References

Barazani, O., Westberg, E., Hanin, N., et al. (2014). A comparative analysis of genetic variation in rootstocks and scions of old olive trees – a window into the history of olive cultivation practices and past genetic variation. *BMC Plant Biology, 14*, 146. https://doi.org/10.1186/1471-2229-14-146

Barranco, D., & Rallo, L. (2000). Olive cultivars in Spain. *Hort Technology, 10*, 4.

Barranco, D., Fernández-Escobar, R., & Rallo, L. (2017). *El cultivo del olivo, 7o*. Mundiprensa.

Bartolini, G. (2008). *Olive germplasm (Olea europaea L.): Cultivars, synonyms, cultivation area, collections, descriptors*. http://www.oleadb.it/

Beghé, D., Molano, J. F. G., Fabbri, A., & Ganino, T. (2015). Olive biodiversity in Colombia. A molecular study of local germplasm. *Scientia Horticulturae, 189*, 122131. https://doi.org/10.1016/j.scienta.2015.04.003

Belaj, A., Caballero, J. M., Barranco, D., et al. (2003). Genetic characterization and identification of new accessions from Syria in an olive germplasm bank by means of RAPD markers. *Euphytica, 134*, 261–268. https://doi.org/10.1023/B:EUPH.0000004954.93250.f5

Belaj, A., León, L., Satovic, Z., et al. (2011). Variability of wild olives (Olea europaea subsp europaea var. sylvestris) analyzed by agro-morphological traits and SSR markers. *Scientia Horticulturae, 129*, 561–569. https://doi.org/10.1016/j.scienta.2011.04.025

Belaj, A., Gurbuz, V., Sikaoui, H., et al. (2016). Olive genetic resources. In *The olive tree Genome* (pp. 27–54).

Belaj, A., de la Rosa, R., León, L., et al. (2020). Phenological diversity in a world olive germplasm Bank: Potential use for breeding programs and climate change studies. *Spanish Journal of Agricultural Research, 18*, e0701. https://doi.org/10.5424/sjar/2020181-15017

Besnard, G., Terral, J.-F., & Cornille, A. (2017). On the origins and domestication of the olive: A review and perspectives. *Annals of Botany, 385–403*. https://doi.org/10.1093/aob/mcx145

Caballero, J. M., Del Río, C., Barranco, D., & Trujillo, I. (2006). The olive world germplasm bank of Cordoba, Spain. *Olea, 25*, 14–19.

Centeno, A., Hueso, A., & Gómez-del-Campo, M. (2019). Long-term evaluation of growth and production of olive cultivars in super high-density orchard under cold-weather conditions. *Scientia Horticulturae, 257*, 108657. https://doi.org/10.1016/j.scienta.2019.108657

de Castro, A. I., Rallo, P., Suárez, M. P., et al. (2019). High-throughput system for the early quantification of major architectural traits in olive breeding trials using UAV images and OBIA techniques. *Frontiers in Plant Science, 10*, 1472. https://doi.org/10.3389/fpls.2019.01472

de la Rosa, R., Leon, L., Guerrero, N., et al. (2007). Preliminary results of an olive cultivar trial at high density. *Australian Journal of Agricultural Research, 58*, 392–395. https://doi.org/10.1071/ar06265

Díez, C. M., Trujillo, I., Barrio, E., et al. (2011). Centennial olive trees as a reservoir of genetic diversity. *Annals of Botany, 108*, 797–807. https://doi.org/10.1093/aob/mcr194

Fernández-Escobar, R., de la Rosa, R., Leon, L., et al. (2013). Evolution and sustainability of the olive production systems. In N. Arcas, F. N. Arroyo López, J. Caballero, R. D'Andria, M. Fernández, R. Fernández-Escobar, A. Garrido, J. López-Miranda, M. Msallem, M. Parras, L. Rallo, & R. Zanoli (Eds.), *Present and future of the Mediterranean olive sectorPresent and future of the Mediterranean olive sector* (Options méditerranées. Series A. Mditerranean Seminars. Num. 106) (p. 11.42). CIHEAM. isbn:2-85352-512-0.

Gomez-del-Campo, M., Centeno, A., & Connor, D. J. (2009). Yield determination in olive hedgerow orchards. I. Yield and profiles of yield components in north-south and east-west oriented hedgerows. *Crop & Pasture Science, 60*, 434–442.

Gómez-Gálvez, F. J., Ninot, A., Rodríguez, J. C., Compañ, S. P., Andreva, J. U., Rubio, J. A. G., Aragón, I. P., Viñuales-Andreu, J., Casanova-Gascón, J., Šatović, Z., Lorite, I. J., De la Rosa-Navarro, R., & Belaj, A. (2024). New insights in the Spanish gene pool of olive (*Olea europaea* L.). preserved ex situ and in situ based on highthroughput molecular markers. *Frontiers in Plant Science, 14*. https://doi.org/10.3389/fpls.2023.1267601

IOC. (2022). El proyecto resgen: 16 años después. 2011. Olivae 116 (Editorial IOC).

Jiménez-Ruiz, J., Ramírez-Tejero, J. A., Fernández-Pozo, N., et al. (2020). Transposon activation is a major driver in the genome evolution of cultivated olive trees (*Olea europaea* L.). *The Plant Genome, e20010*, 1–19. https://doi.org/10.1002/tpg2.20010

Koehmstedt, A. M., Aradhya, M. K., Soleri, D., et al. (2011). Molecular characterization of genetic diversity, structure, and differentiation in the olive (Olea europaea L.) germplasm collection of the United States department of agriculture. *Genetic Resources and Crop Evolution, 58*, 519–531. https://doi.org/10.1007/s10722-010-9595-z

Morales-Sillero, A., Rallo, P., Jimenez, M. R., et al. (2014). Suitability of two table olive cultivars ("Manzanilla de Sevilla" and "Manzanilla Cacereña") for mechanical harvesting in superhigh-density hedgerows. *HortScience, 49*, 1028–1033.

Ninot, A., Howard, W., José, M., et al. (2018). Survey of over 4,500 monumental olive trees preserved on-farm in the northeast Iberian Peninsula, their genotyping and characterization. *Scientia Horticulturae, 231*, 253–264. https://doi.org/10.1016/j.scienta.2017.11.025

Pastor, M., & Humanes, J. (1990). Increased densities in traditional rainfed adult olive groves in Andalusia. *Acta Horticulturae, 286*, 291–294.

Pastor, M., Garcia, M., Soriano, M. A., et al. (2007). Productivity of olive orchards in response to tree density. *Journal of Horticultural Science Biotechnology, 82*, 555–562.

Penco, J. M. (2020). Aproximación a los costes de cultivo del olivo. In *Asociación Española de Municipios del Olivo*. https://www.aemo.es/slides/slide/estudio-de-costes-aemo-2020-241/download. Accessed 2 Feb 2020.

Pérez, A. G., León, L., Sanz, C., & de la Rosa, R. (2018). Fruit phenolic profiling: A new selection criterion in olive breeding programs. *Frontiers in Plant Science, 9*, 1–14. https://doi.org/10.3389/fpls.2018.00241

Rallo, L (2019). *Book of the IOC network of Germoplasm banks*. International Seminar. 21–24 October 2019. Córdoba. IOC & Universidad de Córdoba. Spain. Available at: https://www.internationaloliveoil.org/wp-content/uploads/2021/08/ANNEXE-1-A-BOOK-OF-THE-IOC-NETWORK-OF-GERMPLASM-BANKS-Completa.pdf. 02/2024.

Rallo, L., Barranco, D., Caballero, J. M., et al. (2005). *Variedades de Olivo de España*. Mundiprensa, Junta de Andalucia.

Rallo, L., Barranco, D., de la Rosa, R., et al. (2008). "Chiquitita" olive. *HortScience, 43*, 529–531.

Zohary, D., & Spiegel-Roy, P. (1975). Beginnings of fruit growing in the old world. *Science, 187*, 319–327.

Chapter 5
Olive Grove Fertilization in Spain

Juan Carlos Hidalgo, Javier Hidalgo, Victorino Vega, Ignacio Lorite, Raul de la Rosa, and Lorenzo León

Abstract Fertilization is widely used in order to increase or maintain the quantity and quality of crop yield and to correct nutritional deficiencies. Spanish olive orchards are not an exception, where fertilization is usually implemented both under rainfed and irrigated conditions. However, the amount of fertilizer, composition, and application schedules are not usually subjected to technical criteria. This fact can lead to excessive application of some elements, or, in other cases, do not meet crop requirements. Works carried out in the SE of Andalusia revealed the great variability in the fertilizer contributions that are applied in olive groves with similar characteristics.

In recent decades, fertilization advances have led to an improvement of the application efficiency, thus increasing the productivity of Spanish olive orchards. That improvement has contributed to increase sustainability of farming practices and to reduce the negative environmental impact derived from the abusive use of fertilizers. A rational fertilization must consider key elements such as the potential yield on the plantation, water availability, soil fertility, and nutritional status of the olive trees. Thus, fertilization is highly conditioned by rainfall and irrigation, allowing the dissolution of fertilizers applied to the soil and achieving satisfactory efficiency in their use.

In Spanish rainfed olive groves (approx. 65% of the total olive groves surface) planted on limestone and/or clay soils, conventional soil fertilization usually is not very efficient at short term, although valuable results can be obtained in olive groves with middle-high productivity in the long term. On the other hand, we can identify a greater vegetative development and productive capacity in irrigated olive groves. In addition, the amount of the soil volume available for the olive tree, with high root activity and density, is significantly reduced, being mainly located in the wet bulbs, especially in dry years. Then, fertilization must be adjusted to these characteristics.

J. C. Hidalgo (✉) · J. Hidalgo · V. Vega · I. Lorite · R. de la Rosa · L. León
Plant improvement and Biotechnology Unit, IFAPA-Alameda del Obispo, Córdoba, Spain
e-mail: juancarlos.hidalgo@juntadeandalucia.es; javier.hidalgo@juntadeandalucia.es; victorino.vega@juntadeandalucia.es; ignacioj.lorite@juntadeandalucia.es; lorenzo.leon@juntadeandalucia.es

J. Muñoz-Rojas, R. García-Ruiz (eds.), *The Olive Landscapes of the Mediterranean*, Landscape Series 36,
https://doi.org/10.1007/978-3-031-57956-1_5

Fertigation becomes an essential practice in order to maintain high production in the olive grove and to reduce the alternate bearing. However, the doses of the main macro-elements (N, P, K) must be adjusted according to the nutritional requirements of the crop to achieve a maximum efficiency. Excessive doses of some elements do not improve the production and can reduce the quality of the olive oil by reducing total polyphenols content and oxidative stability.

The application of manure is not widespread in Spanish olive groves, due to the large surface area of these groves, which exceeds 2.5 million hectares, making the availability of manure scarce. Manure is used mainly in organic orchards. Proper management of composted oil mill pomace is an interesting alternative for the coming years, solving the growing problem of olive grove by-products and allowing to obtain a product with a high capacity to improve the physical-chemical properties of the soil.

Nowadays, digital agriculture is imposing new practices in olive grove fertilization, such as nutrients control through direct measurements of sensors installed in irrigation heads or the use of drones equipped with multispectral cameras to identify tree deficiencies. These improvements are contributing to the agricultural development, increasing the fertilization efficiency and thus improving the profitability and environmental sustainability of the Spanish olive orchards.

5.1 Introduction

Fertilization is a very important agronomical practice in olive groves, being an important factor in olive oil production and its quality, in the incidence of frosts and diseases, and in the contamination of water resources. The lack of technical criteria in the fertilization planning of olive groves leads, under specific conditions, to excessive application of some elements, such as nitrogen, and in other cases, a deficit fertilization in some other elements such as potassium, generating crop deficiencies. A rational fertilization must optimize the application of fertilizers, avoiding the adverse effects caused by excessive fertilization. To achieve this rational use, the consideration of important aspects such as the **potential productive capacity** of the orchard, the **water availability,** and the possible nutrients dissolved in it is required. In addition, the **soil fertility** and the **nutritional state** of the olive trees are also determining factors to carry out a rational fertilization.

The productive capacity of the olive grove is a decisive factor to achieve a correct productive response to fertilizer applications. In olive groves with low productivity, the nutritional requirements could be contributed by the organic matter mineralization in the soil, by the nutrients dissolved in rainfall and irrigation water, and by the nutritional elements present in the soil exchange complex. However, in olive groves with high productive capacity, frequently under irrigation, the nutritional requirements can be higher than the nutrients supply from these sources, and therefore, sustainability of these productive levels could be at risk in short and medium term.

Fertigation improves the productivity in irrigated orchards by increasing efficiency in the use of water (Pastor et al., 2005).

From the point of view of fertilization scheduling, **soil and foliar analysis** are common in the Spanish olive groves. **Soil analysis** allows us to know the availability of elements that can be assimilated by the tree, as well as the physico-chemical characteristics of the soil that can affect the behavior of fertilizers. However, in irrigated olive groves, soil fertility has lower relevance due to the application of water, and fertilizers are located into very specific areas (wet bulbs) under the drip irrigation emitters. **Foliar analysis** is the most appropriate method to diagnose the nutritional status of an olive orchard, evaluating the actual absorption of nutrients by the tree, the existence of nutritional deficiencies or excesses, as well as the occurrence of possible antagonism between elements. In addition, foliar analysis allows assessing the available reserves of those elements that are mobile in the tree. In general, in olive groves, it is recommended to carry out an annual foliar analysis, taking samples during the month of July. Adequate levels of foliar N, P, and K have been reported to be between 1.5 and 2.0%, above 0.1%, and above 0.8%, respectively (Freeman et al., 1994; Pastor et al., 2005). According to these authors, deficient levels are below 1.4%, 0.06%, and 0.4% for N, P, and K, respectively.

5.2 Fertilization in Rainfed and Irrigated Olive Groves

In **rainfed olive groves**, the practice of fertilization is highly conditioned by the rainfall (Ortega-Nieto, 1964). Rainwater is necessary to dissolve the fertilizers applied to the soil and to achieve an acceptable use efficiency. Therefore, the soil fertilization must be carried out when rainfall is foreseen in the following days to the application. However, rainfall incidents with high intensity could generate runoff and losses of fertilizers, especially in the case of nitrogen. On the other hand, to achieve a high fertilization efficiency, root activity is required in order to promote a good absorption of the fertilizers by the tree. With low temperatures, roots significantly slow down their growth and absorption capacity, and then, fertilization will be less efficient. In addition, depending on the fertilizer nature and chemical composition, the transformation into assimilable forms for the olive tree will be different. In short, mild temperatures, moderate rainfall forecasts, and the selection of a suitable form of fertilizer are the main qualities for a high fertilization efficiency under rainfed conditions.

In addition, and especially during dry years and very limestone soils, foliar fertilization can be a very efficient procedure for supplying nutrients to the olive tree (Pastor et al., 2005). Thus, N and K are very well absorbed using foliar fertilization, and P is absorbed by the leaf in an acceptable way. However, not all the elements have this behavior; elements such as Mg, Zn, Mn, B, Cu, and Mo have a medium or low absorption, and Ca and Fe are very poorly absorbed in foliar form.

Irrigated olive groves tend to increase their production significantly compared to rainfed olive groves (Iniesta et al., 2009; Pastor et al., 2005). In this type of olive

Table 5.1 Different strategies for fertilization in organic orchards in Spain

Option	Management	Fertilization
1	Handled with recent spontaneous cover crop and without chopping of pruning waste	Sheep manure: 9.000 kg/ha
2	Handled with recent spontaneous cover crop and without chopping of pruning waste	Sheep manure: 4.500 kg/ha Potassium sulfate: 100 kg/ha
3	Handled with recent spontaneous cover crop and with chopping of pruning waste	Composted olive mill pomace: 2.500 kg/ha
4	Handled with established spontaneous (or legume) cover crop and with chopping of pruning waste	Composted olive mill pomace: 2.500 kg/ha (once each 3 years) Potassium sulfate: 130 kg/ha

Considering:
Sheep manure: 0,83%N, 0,23%P_2O_5 y 0,67% K_2O (% on wet matter)
Composted olive mill pomace: 1,08%N, 0,33% P_2O_5 y 1,90% K_2O
Potassium sulfate: 28–30% K_2O

groves, fertilization is essential to maintain high yields over time and decrease alternate bearing. Irrigation allows to apply the fertilizer simultaneously with the water, using the fertigation technique. It is convenient to split the fertilizer doses as much as possible, trying to distribute them throughout the irrigation season and during practically the whole duration of each irrigation event. In this way, the efficiency in the use of fertilizers, especially in the case of nitrates, is improved compared with the traditional fertilization procedures (Baena et al., 2005).

In organic olive groves that actually represent 8% of the olive grove in Spain (209.288 ha), the sustainability of the production is conditioned by the application of mineral and organic products, due to the prohibition of using chemical products. For organic olive groves with an average production of 3.000 kg/ha, Guzmán et al. (2014) propose several fertilization alternatives with different cover crops management and pruning remains (Table 5.1). At present, the management of composted olive mill pomace is the alternative that raises the best prospects for the near future, due to its greater availability with respect to manure and its nutritional characteristics.

5.3 Nitrogen Fertilization

Nitrogen (N) is an essential element in the fertilization of the olive groves that induces an acceleration in the vegetative activity and in the tree development. This element is very mobile within the tree, being quickly translocated to stems, leaves, flowers, and fruits.

A correct availability of N in the soil is necessary to obtain high fruit yield and olive oil quality. An excess of N could cause a decrease in yield and quality (Morales et al., 2007; Fernández-Escobar et al., 2006), a greater sensitivity to diseases, and delays in maturation. On the opposite, a deficiency of N could cause a reduction in the number of fruits per tree, a small size of the olives, an excessively early ripening, and a premature fall of fruits.

In agricultural soils, a fraction of the available N comes from the mineralization of organic matter and the rainwater. However, for highly productive olive groves these sources could not meet the olive N requirements. Although under specific conditions the nitrogen deposited in the soil by the rain could have an important role, usually this source has a small relevance in the fertilization of olive groves with high yields. Previous studies carried out in the Guadalquivir Valley (Ordóñez et al., 1992; Polo et al., 1997; Fernández-Escobar et al., 2012) have evaluated this annual contribution ranging between 8 and 10 kg N/ha, but with a marked interannual variability. This fact generates a great uncertainty in the quantification of this contribution, and therefore in the development of the fertilization scheduling for this type of olive groves. The amount of mineralized N from the soil organic matter is directly and positively related to the total amount of organic matter, the textural and structural characteristics of the soil, and by the climatic conditions. A wide range of observed values has been found due to the different temperatures, soil moisture, and previous nitrogen fertilization practices. For example, for contents of 1% of soil organic matter, average values for the edaphoclimatic conditions identified in Andalusia varied from 10 to 12 kg N/ha for fine-textured soils (clay) to 22–25 kg N / ha for those with sandy texture (Espada, 2004). These values could be altered by fertilization practices and even could be reduced by denitrification (and then, negative balance) in overfertilized fields (Fernandez-Escobar et al., 2011).

In previous studies carried out in rainfed olive groves in Andalusia, Navarro (2003) and Fernández Escobar et al. (2009) found no response to the contribution of N when the leaf levels of this element in July were above the correct threshold (> 1.5%) (Freeman et al., 1994; Pastor et al., 2005). The studies concluded that in such cases the application of N fertilization was not necessary. On the other hand, Ortega Nieto (1964) and Ferreira et al. (1986) applied moderate doses of N in different Andalusian olive-growing regions and obtained positive yield responses to N in orchards with a medium/high productive level, especially in the years of adequate rainfall. For traditional rainfed olive groves, these authors recommended nitrogen doses of between 0.6 and 1.0 kg N per tree, depending on the yield; the lowest values for yields under 25 kg of fruit per tree and the highest ones for yields greater than 35 kg of fruit per tree.

Frequently, foliar application of N in rainfed olive groves has provided more satisfactory results than conventional soil fertilization (Pastor et al. 2005). Foliar applications of urea at 2–3% have shown interesting results when the soil has not been fertilized with N. In addition, a synergistic effect on the foliar absorption of other nutrients such as potassium has been identified. Thus, low urea concentrations (around 0.5–1.0%) are enough to achieve a good response to potassium fertilization. Other studies carried out by Hartmann et al. (1986) in irrigated olive groves in California (USA) did not find an evident response to the contribution of nitrogen when the previous nutritional condition of the tree was adequate. These authors recommended to cancel or limit nitrogen fertilization until the leaf content drops below the indicated threshold level (Freeman et al., 1994; Pastor et al., 2005). However, for low fertile soils, omitting the maintenance nitrogen fertilization cannot be recommended. Other studies in high-density orchards (with more than 200

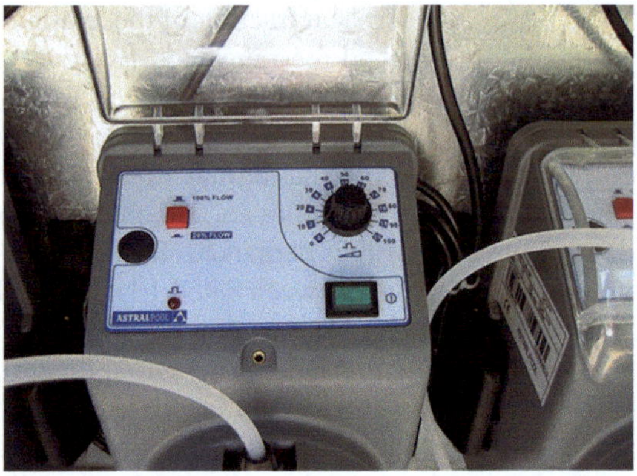

Fig. 5.1 Injection pump that allows to fractionate the total amount of fertilizer applied by fertigation

trees/ha) using drip irrigation, found good results using 580 g of N in fertigation per tree, fractionating the dose throughout the entire irrigation season (Troncoso et al., 1997).

Since 2003, IFAPA has developed a long-term field experiment in an intensive highly productive irrigated adult olive grove with "Picual" variety, following the methodology proposed by Pastor et al. (2005). This experiment has established the optimal doses and periods of nitrogen application in fertigation. The results obtained have shown that N application close to 5–6 g N per kg of fruit olive (i.e. an average application of 67–81 kg N per ha) achieved increases in yield of 18%, although the leaf nitrogen levels in July were above the appropriate thresholds in all treatments (Freeman et al., 1994; Pastor et al., 2005). However, higher doses of nitrogen have not been able to improve yield and, on the other hand, the quality of the olive oil obtained decreased mainly due to a reduction of the content of total polyphenols and oxidative stability. This fact was confirmed by Fernández-Escobar et al. (2006) and Morales et al. (2007).

Among the different fertilization methods, fertigation in drip irrigated olive groves allows nitrogen applications depending on the tree requirements with the irrigation water (Fig. 5.1). This fact avoids the dependence to weather conditions (such as erratic rainfalls), achieving higher N use efficiencies Baena et al. (2005), Haberman et al. (2019). However, it is highly recommended to distribute the fertilization application throughout the crop season, setting the amount of fertilizers according to the crop requirements for each crop stage. Following these rules, the fertilization application efficiency will increase significantly, avoiding N losses and water contamination due to nitrate leaching (Baena et al., 2005).

5.4 Phosphoric Fertilization

Phosphorus (P) is an essential element for olive tree, involved in the processes of cell division, tissue development, and transport of the energy captured in the photosynthesis. Although phosphorus deficiency is rare in olive groves, sometimes in

Andalusia low leaf P contents have been observed in acidic or in limestone soils (Soria, 2002). These deficiencies could be caused by low levels of assimilable phosphorus in the soil. Equally, very productive trees could show low levels of this element in leaf when foliar analysis is performed. These deficiencies could generate intense tree defoliation, more frequent in the oldest branches of the tree and in poor soils (with high limestone content and low depth). This symptomatology is especially boosted during dry years. Simultaneous low levels of P in leaf and relatively low levels of N are observed.

Traditionally, it has been suggested that P fertilization should be carried out when the minimum level of sufficiency in leaf P content on dry matter, measured in July, is below 0.1% (Freeman et al., 1994; Fernández Escobar et al., 2009). However, other authors found a positive response of olive trees to P fertilization even when its leaf content was above 0.1% (Erel et al., 2013). Based on these results under controlled conditions, the same authors suggested the need of revision of the above minimum values.

In limestone soils, conventional applications of P to the soil are usually not effective, achieving valuable results uniquely at long-term (Ferreira et al., 1986). In these situations, correction of P deficiency could be carried out by foliar applications of monoammonium phosphate or monopotassium phosphate (2.0%) at the short-term.

In irrigated olive groves of the "La Loma" region (Jaen), where poor soils in P and high productions are frequent, fertigation with phosphoric acid (500–700 g/ tree and year) has made possible to quickly correct the P deficiency, maintaining an adequate nutritional level and high yields (Pastor et al., 2005).

5.5 Potassium Fertilization

At the physiological level, potassium plays an important role in the accumulation of carbohydrates and fats in olive fruits, as well as in the processes of transpiration and water movement in the tree, regulating the opening and closure stomata mechanism. The highest K demand occurs during the fruit development. Thus, olive fruits accumulate large amounts of K during the ripening period, which can cause temporary deficiencies, even in soils relatively well supplied with this nutrient (Fig. 5.2). Soil type has a special relevance in the identification of K deficiencies. Thus, in Andalusia in a large percentage of olive groves cultivated on limestone and clay soils, nutritional anomalies have been identified, being more frequent after successive years with high yields.

In addition to the physico-chemical characteristics of the soil, the response of the olive grove to the application of potassium fertilizers will depend on the rainfall, the olive grove typology, and the procedure of fertilizer application. Thus, in rainfed olive groves cultivated in limestone or clay soils, the conventional K fertilization is not highly efficient at the short term (Ferreira et al., 1986), but at long term, interesting results could be achieved in olive groves with intermediate productive level. Thus, after long-term field experiments carried out in various locations in Andalusia,

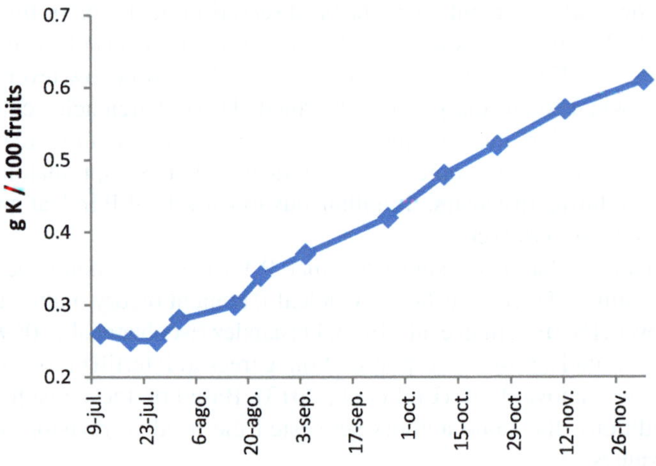

Fig. 5.2 Evolution of the potassium content in olive fruits in olive trees with drip irrigation. Irrigated olive orchard "Picual" variety. Finca Alameda del Obispo – Córdoba (Spain). Plantation density: 278 trees/ha

higher yields were obtained when using N-K fertilizer compared with those fields fertilized only with N (Ferreira et al., 1986).

Foliar applications can correct potassium deficiencies in limestone and clay soils, providing an important part of the tree requirements. Foliar applications are usually carried out with potassium nitrate, potassium chloride, or potassium sulfate at concentrations of 1.5 to 2.5%. Usually, the applications are carried out in spring, summer, and autumn, simultaneously to the phytosanitary treatments. A positive response to foliar applications of K was observed in a field experiment carried out for 6 years in an intensive olive grove (6 × 6 m) located at Córdoba (Southern Spain), irrigated with 1500 m^3/ha and without soil fertilization (Fig. 5.3).

Although potassium is less mobile in the soil than N, when applied in fertigation is able to move deeply and to be absorbed by the roots once the exchange complex is saturated. This movement is relatively fast considering that a large amount of fertilizer is supplied in a relatively small volume of soil (the wet bulb) throughout the irrigation season. Troncoso (1994) recommended an annual K fertigation of between 250–350 gr of K_2O equivalent per olive tree.

Since 2006, IFAPA Alameda del Obispo, following the methodology proposed by Pastor et al. (2005), carries out a long-term field experiment under irrigation in adult olive orchard (208 olive trees per ha) with "Manzanilla de Sevilla" table olive variety. In this experiment, optimal doses and scheduling for the application of potassium with fertigation are being evaluated. The results show a positive response of the yield with different K fertilizer doses compared with the non-fertilized control (Fig. 5.4). Thus, the optimal dose (around 12.1 g K_2O per kg of fruit olives) achieved an annual yield increase of 24% or 1.248 kg olives/ha. The application of half of this optimal dose also generated a positive response compared to the non-fertilized control, but higher doses of potassium did not increase yield.

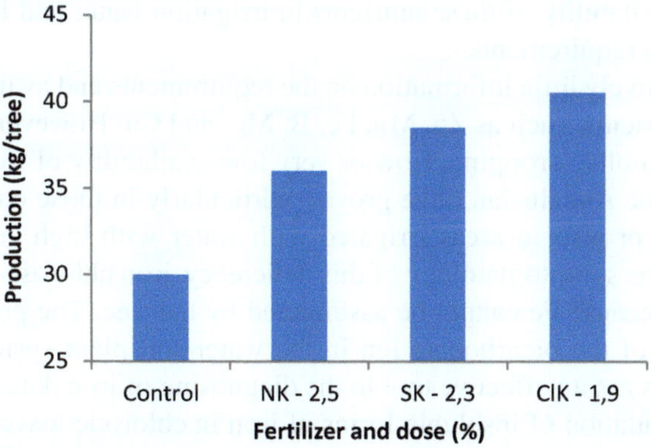

Fig. 5.3 Response of the irrigation olive grove to foliar contributions with different potassium fertilizers. The fertilizers considered were potassium nitrate (NK), potassium sulfate (SK), and potassium chloride (ClK), with 3 applications per year during spring, summer, and autumn. Doses represent % of fertilizer in the treatment tank

Fig. 5.4 Medium-term (2006–2012) responses of olives production to increases in the dose of potassium applications by fertigation. Trial located in La Rambla—Córdoba (Spain). Plantation density: 208 trees/ha

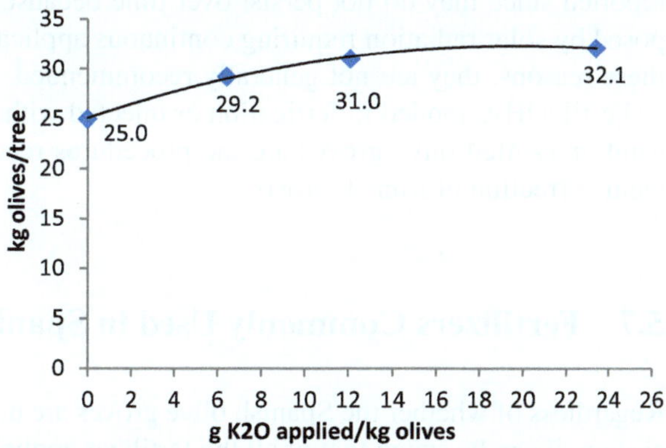

The previous results were obtained for table olive groves, in which the olive fruits do not complete the maturation and oil formation processes, and then, the consumption of potassium by the fruit is not as high as for olive oil groves. Due to this circumstance, the optimal dose of potassium for olive oil groves ranges between 15 and 18 gr of K_2O per kg of fruit olives.

5.6 Other Elements

For other macronutrients, such as Ca, Mg and S, the results of the foliar analyses could be considered as a preliminary assessment. Generally, the decision to apply these macronutrients should be done only when the level of these nutrients in leaf are below the recommended thresholds (Pastor et al., 2005), as in the vast majority of situations availability of these nutrients in irrigation water and in soil is enough to meet the crop requirements.

There is relatively little information on the requirements and availability for most of the micronutrients, such as Zn, Mn, Fe, B, Mo, and Cu. However, iron has a special relevance in olive cropping. Low or very low availability of iron is frequent in some areas of the Andalusian olive groves, particularly in those cultivated on very limestone soils or also in areas irrigated with water with high concentrations of bicarbonates. The symptomatology of this deficiency, iron chlorosis, is very specific and is caused because Fe cannot be assimilated by the tree. The problem increases with an excess of the bicarbonate ion in the water-soil-plant system. In addition, foliar analysis is not an effective tool in the diagnosing of iron deficiency due to the frequent accumulation of insoluble forms of iron in chlorotic leaves due to the low mobility of this element in the trees. Therefore, the diagnosis of this deficiency should be made by experienced personnel based on visual symptoms, by measurements of the chlorophyll content in leaves, or by Soil Plant Analysis Development (SPAD) diagnostic tools.

Unsatisfactory results with foliar applications with iron compounds have been reported since they do not persist over time because these compounds are decomposed by solar radiation requiring continuous applications (Pastor et al., 2005). For these reasons, they are not generally recommended. Then, the use of iron chelates or Fe-EDDHA applied in fertigation or injected with water into the ground near the trunk in rainfed olive groves are the procedures recommended for a fast and efficient correction of iron chlorosis.

5.7 Fertilizers Commonly Used in Spanish Olive Grove

Regardless of whether the Spanish olive groves are under drip- or rain-fed irrigation system, it can be stated that of all the fertilizer applied to the soil, 60% corresponds to solid products and 30% to liquid products for fertigation. Within the solids, the most used are Ammonium sulfate (aprox.50,000 Tn/year), and NPK olive-specific formulations. The NPK olive special complexes are normally developed with a specific formula by each company and its use for fertilization is around 45,000 Tn/year. In minor quantities (15,000 Tn/year), it is being used in the traditional NPK 15–15-15 complex. While the use of NPK olive special blendings counts around 10,000 Tn/year.

For liquids, there are also olive-specific commercial NPK formulations (50% of the total of fertilizers) and the other half corresponds to personalized formulations according to the customers' needs as well as nitrogen liquids such as N-32.

On the other hand, together with other treatments against pests and diseases, foliar applications of fertilizers are widely used. In this sense, the commonly applied products are urea (0.5–1%), monoammonium phosphate or monopotassium phosphate (1–2%) and potassium nitrate, sulfate or chloride (2.2–2.5%). Two to three foliar treatments are usually applied in a year.

Finally, in case of any microelement deficiency, a micronutrient cocktail is usually added either by fertigation or foliar spraying.

References

Baena, G., Ordóñez, R., Pastor, M., & González, P. (2005). *Comparación de niveles de nutrientes en la solución de suelo y la planta entre fertirriego continuo y fraccionado*. Actas XXII Congreso Nacional de riegos (en prensa).

Erel, R., Yermiyahu, U., Van Opstal, J., Ben-Gal, A., Schwartz, A., & Dag, A. (2013). The importance of olive (Olea europaea L.) tree nutritional status on its productivity. *Scientia Horticulturae, 159*, 8–18.

Espada, J. L. (2004). La fertilización razonada del olivar. *Inf. Técnica*, N° 140/2004.

Fernández-Escobar, R., Beltrán, G., Sánchez-Zamora, M. A., García-Novelo, J., Aguilera, M. P., & Uceda, M. (2006). Olive oil quality decreases with nitrogen over-fertilization. *HortScience, 41*(1), 215.

Fernández-Escobar, R., Parra, M. A., Navarro, C., & Arquero, O. (2009). Foliar diagnosis as a guide to olive fertilization. *Spanish Journal of Agricultural Research, 7*(1), 212–223.

Fernández-Escobar, R., et al. (2011). Use and abuse of nitrogen in olive fertilization. *Acta Horticulturae, 888*(888), 249–258.

Fernández-Escobar, R., García-Novelo, J. M., Molina-Soria, C., & Parra, M. A. (2012). An approach to nitrogen balance in olive orchards. *Scientia Horticulturae, 135*, 219–226.

Ferreira, J., García-Ortiz, A., Frías, L., & Fernández, A. (1986). Los nutrientes N, P, K en la fertilización del olivar. *Olea, 17*, 141–152.

Freeman, M., Uriu, K., & Hartmann, H. T. (1994). Diagnosing and correcting nutrient problems. En: Olive production manual. In L. Ferguson, G. S. Sibbett, & G. C. Martin (Eds.), *Division of agriculture and natural resources* (Vol. 3353, pp. 77–86). University of California. Publ.

Guzmán G. I., Foraster, L., & Sánchez, J. L. (2014). *Cultivo ecológico del olivar*. Consejería Agricultura, Pesca y Desarrollo Rural. Junta de Andalucía

Haberman, A., Dag, A., Shtern, N., Zipori, I., Erel, R., Ben-Gal, A., & Yermiyahu, U. (2019). Significance of proper nitrogen fertilization for olive productivity in intensive cultivation. *Scientia Horticulturae, 246*, 710–717.

Hartmann, H. T., Opitz, K. W., & Bentel, J. A. (1986). La producción oleícola en California. *Olivae, 11*, 24–49.

Iniesta, F., Testi, L., Orgaz, F., & Villalobos, F. (2009). The effects of regulated and continuous deficit irrigation on the water use, growth and yield of olive tres. *European Journal of Agronomy, 30*(4), 258–265.

Morales, A., Jiménez, R., Fernández, J. E., Troncoso, A., & Beltrán, G. (2007). Influence of fertigation in 'Manzanilla de Sevilla' olive oil quality. *HortScience, 42*(5), 1157–1162.

Navarro, C. (2003). Estrategias de fertilización en el olivar compatibles con el medio ambiente. In *Actas XI Simposium Internacional del Aceite de Oliva. Foro del Olivar y el Medio Ambiente*.

Ordóñez, R., Giráldez, J. V., & González, P. (1992). *Evolución temporal y estacional del nitrógeno depositado por la lluvia en el Valle del Guadalquivir. Actas de las X Jornadas Técnicas sobre Riegos* (pp. 11–16). Asociación Española de Riegos y Drenajes.

Ortega-Nieto, M. (1964). *Valor fertilizante del abono mineral y orgánico y sus relaciones con el agua en el olivo*. INIA Ministerio de Agricultura.

Pastor, M., Vega, V., Hidalgo, J. C., & Nieto, J. (2005). *Cap. Fertilización en el olivar de riego*. El cultivo del olivo con riego localizado.

Polo, M. J., Ordóñez, R., & Giráldez, J. V. (1997). Uso agrícola de los Lodos de Depuradora. In J. de Andalucía (Ed.), *Consejería de Agricultura y Pesca*. Comunicación I+D 23/97.

Soria, L. (2002). *Fertilización y riego en el olivar de la provincia de Jaén: Comarcas de la Lola y Sierra Morena*. Tesis doctoral. Universidad de Jaén.

Troncoso, A. (1994). Fertirrigación del olivo. Fruticultura Profesional. *Suplemento del, 2*, 65–68.

Troncoso, A., Liñán, J., Cantos, M., Zárate, R., & Lavee, S. (1997). Influencia de la fertirrigación con urea sobre la disponibilidad de N-NO3 y el desarrollo del olivo. *Fruticultura Profesional, 88*, 83–87.

Chapter 6
Pruning and Harvesting of Olive Groves in Spain

Daniel Pérez Mohedano

Abstract Three types of olive growing systems coexist in Spain, each of them with specific pruning and harvesting requirements. The traditional system is based on wide plantation frames (usually around 100 trees/ha) and usually with several trunks per olive tree. The intensive system is based on an open center or vase shape with one trunk per tree and higher tree density, up to 400 trees/ha. Finally, the hedgerow growing system has planting densities of up to 2000 trees/ha, where trees are trained to form continuous hedgerows. In both traditional and intensive systems, initial pruning aims to obtain a permanent tree structure with several main branches, producing canopies with a large surface. After several years, renovation of the main branches is carried out when trees reach a very low leaf/wood ratio.

In the hedgerow growing system, harvesting is made by straddle machines that work over the hedgerow. Therefore, a tree size limited by the dimensions of the machinery is required. This imposes a very different pruning strategy. A particularity of the pruning of this system is that there are different approaches to carry out the formation and maintenance of the hedgerow.

6.1 Pruning in the Main Olive Growing Systems in Spain

Nowadays, the olive groves in Spain are grouped into three olive growing systems; (a) the traditional one with several trunks per olive tree and wide plantation frames, (b) the intensive system based on an open center or vase shape with one trunk per olive tree and higher tree density, and (c) the hedgerow growing systems.

Currently in Spain, the traditional system represents around 71%, the intensive system 26%, and the hedge system 3% (Penco 2020).

D. Pérez Mohedano (✉)
Plant improvement and Biotechnology Unit, IFAPA-Alameda del Obispo, Córdoba, Spain
e-mail: daniel.perez.mohedano@juntadeandalucia.es

J. Muñoz-Rojas, R. García-Ruiz (eds.), *The Olive Landscapes of the Mediterranean*, Landscape Series 36,
https://doi.org/10.1007/978-3-031-57956-1_6

6.1.1 The Traditional System

This is characterized by a wide plantation frame (around 100 trees per hectare) being almost exclusively found in adult olive orchards. In fact, in the last decades, new plantations have used the intensive growing system. Traditional systems produced relatively low yield (around 35% less production) (Pérez-Mohedano et al., 2012), and harvest is barely mechanizable. In most cases, this system is the one with the lowest economic profitability (Pérez-Mohedano et al., 2012). These limitations have promoted the conversion of many orchards with traditional plantation frames to more intensive systems.

The pruning of the traditional olive systems has many similarities with the pruning for intensive systems since both aim to obtain permanent tree structures able to intercept most of the solar radiation. Moreover, in both systems, similar renovation pruning of the main branches is carried out when their leaf/wood ratio is very low.

Harvesting procedure determines the pruning recommendations. In both olive growing systems, tree trunk shakers are the most convenient and profitable harvesting procedure (Fig. 6.1), although a greater harvesting efficiency is usually achieved in the intensive systems. Alternatives to trunk shakers are found in many Spanish regions and typically include branch shakers, mechanical combs, or canes. These alternatives are forced by the small field size of the farms and inadequate orography of the traditional growing systems, which limit the acquisition of machinery.

Fig. 6.1 Trunk vibration machinery

6.1.2 Intensive Growing System

This is based on an open center or vase shape with one single trunk per olive tree. The initial pruning is focused on achieving, as quickly as possible, a robust structure of the olive tree in relation with the trunk and main branches. For that, tree canopy can grow freely during the first three years after planting, allowing a maximum development of the canopy, roots, and trunk (Barranco et al., 2017). After achieving a correct development of the canopy during this initial stage, annual maintenance pruning operations are done. This maintenance pruning is focused on those branches with incorrect locations that would avoid a good development to the future main branches. This pruning will promote the growth of the future main branches that would be those forming an angle of around 60° with the horizontal (Fig. 6.2).

6.1.2.1 Pruning for Tree Formation

One common mistake during the first years of tree formation is to cut the main axis of the tree at a certain height to promote the development of a round canopy. However, this is a wrong practice since it causes a concentration of branches in the same point of the trunk, at a few centimeters from where it was pruned. Moreover, associated to a wrong pruning, new branches tend to be very horizontal in their first sections (Fig. 6.3), which promotes the emergence of secondary vertical branches to counterbalance the horizontality of the first ones. Another common mistake in

Fig. 6.2 Open center or vase shape formation scheme

Fig. 6.3 Pruned olive tree with excessive horizontal main branches

pruning practice is to form the canopy at very low height in the main trunk thus hindering mechanical harvesting by trunk shaking of the trees.

6.1.2.2 Pruning for Tree Maintenance

Once the main tree branching structure is defined, the pruning progresses from a formation to a production-maintenance strategy.

The production or maintenance pruning of the adult olive grove begins once the tree structure is correctly defined. Thus, when the olive orchard reaches a canopy volume balanced with the environment in which it is located, the objective of the pruning must be to maintain an equilibrated productive level with the available resources (mainly related with water and soil), as long as possible (Pastor & Humanes, 2010). The canopy volume adapted to the available resources and compatible with maximizing the intercepted solar radiation is known as the optimal canopy volume, and must be maintained by pruning (Fig. 6.4). The optimal volume in irrigated land should range between 12,000 and 15,000 m^3/ha and in rainfed between 8000 and 10,000 m^3/ha.

Therefore, the maintenance pruning procedure carried out in the Spanish olive systems by trained farmers is as follows: Inside the tree canopy, those vigorous and vertical branches, and any other branch with great vigor or that crosses the canopy from one side to the other, are pruned. The rest of the branches are kept inside of the canopy in order to: (1) preserve the productivity inside of the tree canopy, (2)

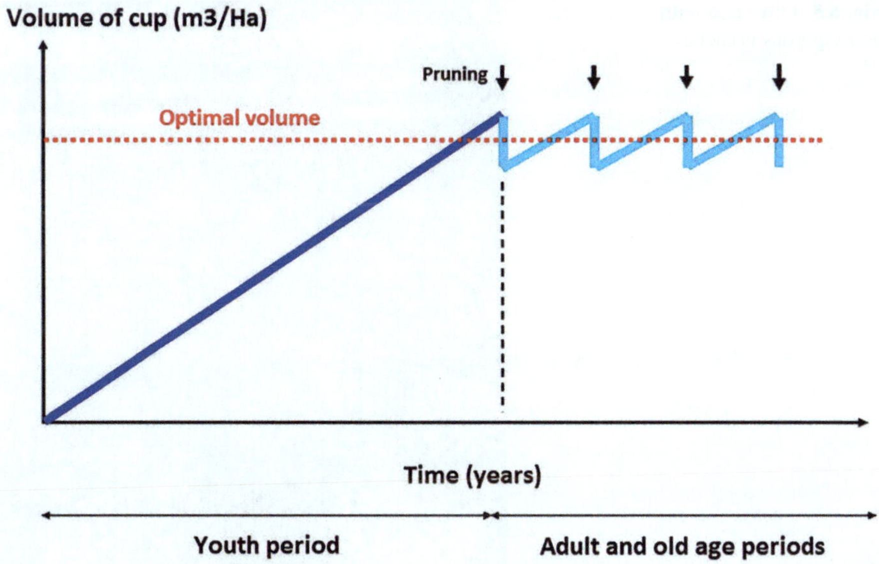

Fig. 6.4 Optimal canopy volume

protect the wood of the main branches from an excess of solar radiation in the central hours of summer that could cause burns, and (3) avoid an excessive emergence of vigorous vertical branches, with the energy cost that this implies. On the opposite side, a very common wrong practice that still persists in Spain is to maintain the entire interior of the olive tree free of branches and leaves (see Fig. 6.5), thus generating severe tree damage by summer sun burn.

Other good-pruning practices observed in the Spanish olive orchards is that the most vertical and vigorous branches that exceed the main branches in height, at the top of the olive tree, are pruned. If this is not done, the olive tree will have an excessive growth in height, which would mean a deficient tree structure (with an excess of unproductive wood) and a problem of shading of the adjacent trees. A correct pruning in this upper section should result in a canopy with a balanced height and size.

On the other hand, most of the productive branches are located in external areas of the canopy. In these areas, branches of medium size could be very close to each other, obstructing their development and reducing the intercepted solar radiation of the rest of the branches. Pruning in these external areas consists of cutting the branches that are occupying the same space with others, maintaining those with the best development. This practice will reduce the roundness of the tree canopy, causing recesses and protrusions that will increase the surface exposure to solar radiation (Fig. 6.6). The main difficulty of this practice is assessing the intensity of the pruning, caused by the subjectivity in the interpretation by the farmer of the spatial distribution of the branches. If the pruning is excessive, a loss of productivity capacity will be promoted by the vegetative response of the olive tree to compensate the

Fig. 6.5 Olive tree with non-vigorous branches

Fig. 6.6 Pruned olive tree with correct exposed surface

excess of pruning. Besides, an excess of solar radiation within the canopy will be generated. In the case of insufficient pruning, the absence of solar radiation within the canopy would cause defoliation, reducing the external leaf area in the vase. Currently, in many Spanish regions, olive pruning is excessive and then, the first group of consequences is more frequently observed.

At the lowest section of the canopy, all the branches below 0.5 meters must be removed to facilitate the application of the herbicide, to allow access of the machinery for trunk vibration, and to prevent some branches reach the ground, thus hindering the harvest.

6.1.2.3 Pruning for Tree Renovation

Eventually, main branches of the adult olive tree will begin to show less vegetative growth, caused by a poor branch conservation, or/and by a low leaf-wood ratio. This situation will fundamentally occur due to the absence of solar radiation and space during the branch development, which will be directly related to the orchard tree density. To avoid this, the renovation of the main branches by pruning is a widespread practice in Spain.

The renovation pruning is carried out sequentially, starting by the branch with the worse situation. The second branch will not be renewed until the first one has been replaced by another that occupies a significant part of the available space (Ortega Nieto, 1969). When the branch to be renewed is totally pruned, sap is accumulated under the cut. This, together with sunlight, promotes the dormancy break of the buds next to the cutting area and produces the emission of shoots (see Fig. 6.7). These shoots are left to grow freely for two years, and then, they are gradually eliminated starting from those with worse location and development until only one remains. Therefore, the selection of the shoot that will finally occupy the space of the old branch will be done in the following years among those that have the best development and orientation. The branch renovation may be done at undetermined number of times. However, repeated renovations on the same trunk could cause an accumulation of dead wood that could hamper the evacuation of rainwater, thus resulting in rots that could jeopardize the integrity of the trunk. In cases where the trunk shows significant problems with sap circulation and low consistency for trunk vibration, renovation pruning must be discarded, and then, the complete renovation of the orchard should be considered.

6.1.3 Hedgerow Growing System

Hedgerow olive orchards have a very different pruning procedure than the previously described for traditional and intensive systems. These differences are caused by the specific harvesting machinery used for orchards of hedgerow growing systems. The harvesting machine has internal shakers that cover the whole hedgerow

Fig. 6.7 Pruning of the main branch with shoots

(Fig. 6.8). The requirements of this machinery are related to the tree size and flexibility of the upper and outer branches. The control of vigor must be addressed through olive cultivars with reduced vigor, or through controlled deficit irrigation and fertilization strategies, although many of these alternatives are currently under evaluation. In any case, these strategies will facilitate pruning needed to maintain the hedgerow size within the requirements of the harvesting machinery.

Due to lack of long-term studies and the limited experience of technicians and farmers, different criteria for the initial formation and the later maintenance of the hedge are currently coexisting.

6.1.3.1 Pruning for Hedge Formation

The objective is to form an open branch structure in the same orientation of the hedgerow. This promotes the maintenance of the tree canopy within the size requirements of the harvesting machinery. Two strategies have been used to achieve this. In the first one, a central axis is vertically oriented and, in the second one, tree can grow freely from the beginning of the plantation. For both systems, the harvesting machinery size requirements must be met and then, the lower part of the canopy

Fig. 6.8 Harvesting machinery for olive systems in hedgerow

must be formed at 50 and 60 cm height. The central axis is guided to a height of approximately two meters by a tutor of these dimensions, allowing free growth for the rest of the branches. The second system allows the olive tree a free growth from 50 to 60 cm height, requiring a tutor of around 1 meter tall. With the aim of reducing the growth in height mechanical pruning in the top of the canopy by disc pruners is carried out several times from planting. With the free growth system, the pruning in height causes an initial reaction of hardening of the branches located in the lower section of the canopy, and a subsequent reaction of vertical vegetative growth with a vigor proportional to the canopy reduction carried out by pruning. In some cases, lateral pruning is done with discs to concentrate branching in that area of the canopy.

6.1.3.2 Pruning for Hedge Maintenance

Regarding the productive maintenance pruning in hedgerow growing systems, the objective is to keep the canopy flexible, within the limits defined by the harvesting machinery, and with a high leaf–wood ratio, at the lowest cost. Currently, different pruning strategies coexist, some based on the use of chainsaws and others based on alternated cuts with disc pruners at the top and on the sides of the canopy. The most mechanized option is the use of a disc pruner to reduce the width and height (topping) up to 2.5 meters (see Fig. 6.9). The cuts promote vigorous growth that varies with the thickness of the branches pruned. To avoid this vigorous growth, it is necessary perform a manual pruning with a chainsaw to fully remove half-cut thick branches. The most common manual strategy is the annual pruning with chainsaw. This pruning is focused on all the branches of a certain thickness located in the

Fig. 6.9 Topping pruning in olive groves in hedgerows

upper and outer part of the hedge, that are fully remove. Only flexible branches are left in these areas, that permit straddle machinery to pass. Again, the lowest section of the hedge (around 50 cm from ground) requires to be maintained free of vegetation using a disc pruner or chainsaw.

Independently of the strategy of pruning considered, it is very important to ensure that thick branches do not accumulate inside of the hedge. To achieve this, a constant and gradual renovation pruning is required in both the outer and the inner part of the hedgerow.

6.2 Conclusions

In summary, in the Spanish olive groves, the tendency is to adapt pruning to the available harvesting machinery, with the final objective of maintaining maximum productivity considering the fundamental principles of pruning based on balance leaf/wood and leaf/root ratios, with the lowest possible costs. For all this, we must avoid pruning in excess and try to adapt the intensity of the pruning to the productive potential of the land. It is necessary to have balanced olive trees without excess height, without vigorous branches in the inner part of the canopy and with a number of branches that allow light to enter the interior.

References

Barranco, D., Fernández-Escobar, R., & Rallo, L. (2017). *El Cultivo del Olivo*. Editorial Mundiprensa.

Ortega Nieto, J. M. (1969). *La poda del olivo. Ministerio de Agricultura*. Dirección General de Agricultura.

Pastor, M., & Humanes, J. (2010). In S. A. Agrícola Española (Ed.), *Poda del olivo: moderna olivicultura*.

Penco, J. M. (2020). *Aproximación a los costes del cultivo del olivo*. AEMO.

Pérez-Mohedano, D., Hidalgo, J. C., Vega, V., Hidalgo, J., & Arriaza, M. (2012). Mejoras en productividad y costes en plantaciones intensivas respecto a plantaciones tradicionales. *Revista Vida Rural, 351*, 44–47.

Chapter 7
Pests and Infectious Diseases of Olive Trees in Spain, and Main Control Strategies

Ramón González-Ruiz, José Alfonso Gómez-Guzmán, and María Pilar Cordovilla

Abstract Olive cultivation is one of the main agricultural assets in Spain, especially in Andalusia, where the largest extension of olive groves is found. This chapter focuses on the main phytophagous arthropods and infectious diseases that affect olive trees in Spain. These are grouped into categories according to their economic importance damages. In the case of pests, the different species are listed according to the different resources affected: olive development, production, fruit, and olive oil quality. Likewise, the different types of management in current olive growing are considered, highlighting the role of herbaceous cover in maintaining the community of beneficial insects, and the factors that influence their entomophagous activities. In the case of infectious olive diseases, the causal agents and the characteristic symptoms they cause in olive trees are described, resulting in three categories, depending on their economic relevance. For the most frequent ones,—peacock spot, verticillium wilt, olive knot, anthracnose, and cercosporiosis—the detection methods are described, along with the factors that favor the severity of their attacks (climatology, cultivation practices, etc.) as well as the susceptibility of the most common cultivars in Spain.

Keywords Olive pest and diseases · Olea europaea · Integrated pest management (IPM) · Beneficial insects

R. González-Ruiz · J. A. Gómez-Guzmán (✉) · M. P. Cordovilla
Departamento de Biología Animal, Biología Vegetal y Ecología, Universidad de Jaén, Jaén, Spain
e-mail: ramonglz@ujaen.es; jgguzman@ujaen.es; mpilar@ujaen.es

J. Muñoz-Rojas, R. García-Ruiz (eds.), *The Olive Landscapes of the Mediterranean*, Landscape Series 36,
https://doi.org/10.1007/978-3-031-57956-1_7

7.1 Introduction

A third of the surface of Spain is used for agriculture (Martin Cerdeño, 2014), of which 2.58106 h holds most of this surface (60.38%), which is equivalent to 13.8% of the olive groves (MAPA, 2017).

The control of phytoparasites is largely based on use of pesticides, with direct and indirect environmental impacts, especially over animal communities (Vives de Quadras, 1988), the arising of other opportunistic pests, and of lineages resistant to insecticides (Carrera, 1996). Moreover, traditional intensive tillage practices have led to a negative impact on the diversity of soil flora and fauna (Hodkinson & Hughes, 1993), affecting soil fertility. The suppression of key functional plant communities has led to a sharp reduction of entomophagous activity, and a greater dependence on synthetic pesticides, leading to a vicius cycle in the use and dependence of agrochemicals. In recent years, the awareness policy by public agricultural authorities hasfostered and promoted a change of mindset for farmers, based on valuing the importance of natural enemies in maintaining agro-ecological balances (Arambourg, 1986: Luckmann & Metcalf, 1990), a target that is relevant across scales from the plot to the landscape. Universities are enhancing training efforts on the buffering effects of biodiversity against insect pests. A key focus being on devising agronomic practices that may prove decisive to guarantee natural enemies, and thus also their functionalities.

7.2 Olive Arthropod Pests

A wide range of arthropods finds refuge in olives (Arambourg, 1986), (Viggiani, 1977; De Andrés Cantero, 2001), including multiple species associated to different trophic functions. Among the phytophagous, 40–60 species (Table 7.1) have been recorded across the Mediterranean (Arambourg, 1986, De Andrés Cantero, 2001). Olive fly (*Bactrocera oleae*) and moth (*Prays oleae*) stand out as the main pests, economically, although this is strongly influenced by factors including climate conditions (e.g., extreme winter/summer temperatures).

A second group (B) includes species of secondary economic impact, although they may gain greater relevance, depending on certain environmental conditions or agronomic practices. This often confers them a relative endemic character. These include the Jasmine moth (*Margaronia unionalis*), the olive violet scale (*Parlatoria oleae*), the olive bark beetle (*Phloeotribus scarabaeoides*), and the olive branch borer (*Euzophera pinguis*).

A third group (C) includes species of secondary relevance and local outreach and depends on factors such as soil types, olive varieties, spring rainfall regimes, use of insecticides, patches of natural vegetation, pruning irrigation, and fertilization practices, amongst others. Among these, we highlight the olive black scale (*Saissetia oleae*), a species whose economic impact has decreased considerably lately. This is

Table 7.1 Phytophagous arthropod species associated with olive trees in Spain. An assessment of their frequency index is presented (column 1), along with the resources affected (columns 2–5), and classification according to their decreasing order of economic relevance (last column)

			freque ncy index *	oliv e gro wth	oil produ ction	frui t qua lity	oil qua lity	Gl ob al Sc or e
Bactrocera oleae	(Gmelin, 1788)	olive fruit fly	5			0	0	A
Prays oleae	(Bernard, 1788)	olive moth	5		0			A
Margaronia unionalis	(Hübner, 1796)	jasmine moth	4	0	0	0		B
Parlatoria oleae	(Colvée, 1880)	olive violet scale	4			0	0	B
Hylesinus taranio	(Danthione, 1788)	olive black bark beetle	4	0	0			B
Euzophera pinguis	(Haworth, 1811)	olive branch borer	5	0	0			B
Phloeotribus scarabaeoides	(Bernard, 1788)	olive bark beetle	5	0	0			B
Oxycenus maxwelli	(Keifer, 1939)	olive bud mite	1	0				C
Vesperus xatarti	(Dufour, 1839)	olive longhorn beetle	1	0	0			C
Closterotomus trivialis	(Costa, 1853)	olive flower bug	1		0			C
Kalotermes flavicollis	(Fabricius, 1793)	yellownecked dry-wood termite	1	0				C
Polyphylla fullo	(Linnaeus, 1758)	white root worm	1	0	0			C
Ceramida cobosi	(Báguena, 1955)	white root worm	2	0	0			C
Melolontha papposa	(Illiger, 1803)	olive cockchafer	2	0	0			C
Anoxia villosa	(Fabricius, 1781)	dung beetle	2	0	0			C
Aspidiotus nerii	(Bouche, 1833)	oleander scale	2		0	0	0	C
Cossus cossus	(Linnaeus, 1758)	goat moth	2	0	0			C
Lepidosaphes ulmi	(Linnaeus, 1758)	apple mussel scale	2	0	0	0	0	C
Zeuzera pyrina	(Linnaeus, 1761)	leopard moth	2	0	0			C
Otiorhynchus cribricollis	(Gyllenhaal, 1834)	olive weevil	3	0	0	0		C
Saissetia oleae	(Olivier, 1791)	olive black scale	3	0	0			C
Tetranychus urticae	(Koch, 1836)	red spider mite	3	0	0			C
Resseliella oleisuga	(Targ.-Tozz., 1986)	olive gall midge	3		0			C
Lichtensia viburni	(Signoret, 1873)	cottony scale	3	0	0			C
Pollinia pollini	(Costa, 1857)	globose scale	3	0	0			C
Polydrusus xanthopus	(Gyllenhal, 1834)	leaf green weevil	3	0	0			C
Cicada barbara lusitanica	(Boulard, 1982)	olive cicada	4		0			C

(continued)

Table 7.1 (continued)

Aceria oleae	(Nalepa, 1900)	olive rust mite	4	O			C
Dasineura oleae	(Angelini, 1831)	olive gall midge	4		O		C
Agalmatium flavescens	(Olivier, 1791)	olive planthopper	4		O		C
Euphyllura olivina	(Costa, 1839)	olive psyllid	5		O		C
Liothrips oleae	(Costa, 1857)	olive thrips	5		O	O	C
Acherontia atropos	(Linnaeus, 1758)	death's-head hawkmoth	1				D
Anoplotettix fuscovenosus	(Ferrari 1882)	leafhopper	1				D
Cionus fraxini	(De Gueer, 1775)	ash weevil	1				D
Erannis defoliaria	(Clerck, 1759)	mottled umber	1				D
Hyles livornica	(Esper, 1780)	striped hawk-moth	1				D
Lagorina sericea	(Waltl, 1835)	olive blister beetle	1				D
Lytta vesicatoria	(Linnaeus, 1758)	olive blister beetle	1				D
Sphinx ligustri	(Linnaeus, 1758)	privet hawk moth	1				D
Ceresa bubalus	(Fabricius, 1794)	buffalo treehopper	2				D
Aleurolobus olivinus	(Silvestri, 1911)	olive whitefly	3				D
Metriochroa latifoliella	(Millière, 1886)	olive leaf miner	3				D

Adapted from Gómez (2022)
(*) General frequency index (5 as maximum value)
A Main pest, of economic relevance
B Secondary pest, of medium economic relevance
C Secondary pest, of medium economic relevance and localized character
D Phytophagus of little or negligible economic relevance

likely due to the improvements in pruning techniques, and to the decrease in environmental impacts by the use of synthetic insecticides.

Finally, a large group of phytophagous (D) species, of which only a few examples are listed (Table 7.1). This includes a few within the Oleaceae family, such as the olive white fly (*Aleurolobus olivinus*) and the ash bark beetle (*Leperisinus varius*). The presence of polyphagous species is usually sporadic and its damage is of little or no relevance. Among the factors that favor them are the absence of insecticides, distance from irrigated crops, and proximity to areas of native vegetation, thus dominating in mountain olive groves.

7.2.1 The Role of Plant Diversity

Tackling soil erosion has been the key driver behind the adoption of alternative management practices, such as the introduction of an herbaceous cover. However, the substitution of conventional management practices will also bear positive impacts on biodiversity (Hernández et al., 2005; Martínez et al., 2006). Collateral effects of soil and cover conservation practices include the conservation of animal biodiversity, as well as the natural control of pests/pathogens. This is a complex set of processes with multiple ecosystem effects, including:

Plants as Shelter Herbaceous and shrub vegetation provide an apt environment and habitat for microclimatic conditions that are favorable for numerous animal species. *Chrysoperla carnea* overwinter at their adult stage between the cracks, under detached bark, and mainly in evergreen shrubs (Gepp, 1967; Ickert, 1968), enhancing their chances for winter survival (Sengonca & Frings, 1987). Four-related species have been reported in shrub areas adjacent to olive groves, especially *Chrysoperla agilis* (Bozsik & González-Ruíz, 2006; González-Ruiz et al., 2008).

Plants as Source of Prey/Alternative Hosts Herbaceous cover allows the settlement of phytophagous primary consumers, which are key for the maintenance of beneficial insects (Hodkinson & Hughes, 1993). The carpophagous larvae of the olive moth are lodged inside fruits during summertime, becoming inaccessible to predators and parasitoids. In such situation, beneficial insects still can still feed on aphids (*Aphis fabae*, *Aphis nerii*, *Myzus persicae*) (Pappas et al., 2007), or several microlepidopterae (*Ephestia kuehniella*, *Cacoecia pronubana*) from herbaceous species, securing their summer survival (Viggiani, 1977).

Plants as a Source of Essential Trophic Resources Vegetation in olive grove provides a food source to a wide range of invertebrates and vertebrates. Beneficial insects require a diet rich in proteins and sugars, essential to ensure longevity and fecundity (Campos & González, 1990). The importance of flower nectar and pollen (*Sinapis* spp., *Daucus carota*, *Stachys rectum*) in the sustenance of parasitoids (Manojlovic et al. 2001) and predators, such as common green lacewings, is well demonstrated (Bozsik, 2000; Canard et al., 2001). Volatile aromatic compounds such as phenylacetaldehyde, released from flowering plants (*Tagetes erecta*, *Lathyrus odoratus*, etc), have proven attractive to green lacewings (Tóth et al., 2006). The conservation of natural habitats o is essential for the sustainability of olive groves. The International Organization for Biological Control (IOBC) (Baur et al., 2011) establishes the urgency for ecological compensation areas, set aside from any chemical control practice, with a surface of 5% of each olive grove, excluding adjacent forest areas, indicated as a minimum threshold.

7.2.2 *Main Approaches to Pest Control*

From a chronological stand point, one should consider conventional management practices as the starting point. This is generally characterized by the absence of associated vegetation in olive groves, which is accompanied by the systematic use of synthetic pesticides (herbicides, fungicides, insecticides). Most olive groves

(70%; > one million Ha. in Andalusia) are still managed conventionally (Leyva et al., 2017). Pest control is thus almost exclusively enacted with insecticides.

Alternatively, integrated pest management (IPM) (ca. 26% of olive groves in Andalusia; 400,000 Ha.) is based on a precise knowledge of the biology and ecology of pests, flora, and beneficial insects (Ehi-Eromosele et al., 2013). IPM encourages the development of herbaceous cover (Calabrese et al., 2012; Gómez et al., 2018), and the introduction of pest population thresholds to minimize the need for pesticide application. The combined use of capture devices baited with semiochemicals is also a frequent practice.

A minor area of olive groves in Andalusia (ca. 4% of olive groves, ca. 75,000 Ha.) is certified as organic (Leyva et al., 2017). This is characterized by the decline to apply synthetic pesticides and by the presence of a continuous herbaceous stratus. Pest control depends almost exclusively on the entomophagous activity of the beneficial insects, combined with the use of products of organic origin. Table 7.2 shows

Table 7.2 Characterization of the different models of olive grove management according to the chemical treatments applied against the most frequent pests/pathogens

Pest/pathogen	Pesticide (Active Principles)	Conventional volume-weight (avg/Ha and year) date of application	IPM volume-weight (avg/Ha and year) date of application	Organic volume-weight (avg/Ha and year) date of application
Herb vegetation	Glyphosate (20%) + Oxifluorfen (3%)	0.3 L/Ha 4 L/Ha March/October	–	–
Spilocaea oleagina	Copper oxychloride (20%) + Propineb (15%)	0.8 kg/Ha April	0.8 kg/Ha April	0.8 kg/Ha April
Colletotrichum gloesporides	Copper oxychloride (20%) + Propineb (15%)	0.8 kg/Ha October	0.8 kg/Ha October	0.8 kg/Ha October
Pseudomonas savastanoi	Copper oxychloride (20%) + Propineb (15%)	0.8 kg/Ha October	0.8 kg/Ha October	0.8 kg/Ha October
Margaronia unionalis	Dimethoate 40%	0.9 L/Ha April & September	0.3 L/Ha April	–
Prays oleae	Dimethoate 40%	0.9 L/Ha April & June	0.3 L/Ha May	–
Euzophera pinguis	Chlorpyrifos 48%	1.5 L/Ha April & May	0.75 L/ha April	–
Phloeotribus scarabaeoides	Dimethoate 40%	0.9 L/Ha June & July	0.45 L/Ha May	–
Bactrocera oleae	Dimethoate 40%	1.8 L/Ha September & October	0.9 L/Ha September	–

Period 2007–2017; Ministry of Agriculture of Spain and Spanish Committee of Organic Agriculture
Adapted from Gómez (2022)

the average volume of pesticides to control some of the main phytoparasites in olive groves across the various management regimes abovedescribed.

When compared with conventional management, IPM favors a reduction of 50–66% of the average volume of agrochemicals applied, resulting in a 50% reduction in their frequency of application.

Figure 7.1 shows the relative abundance of the main species of beneficial insects captured in groves under different management regimes. The minimum values are recorded for conventional grove, and the maximum values in organic ones, where the abundance of most of the beneficial insects is two or three times higher. The effects of vegetation (herbaceous) cover can be evaluated by comparing the abundance values across IPM and conventional olive groves. The lack of plant flora negatively affects, e.g., *Pnigalio mediterraneus*, *Agenispis fuscicollis,* and *Tetrastichus cesirae*. The effects of an insecticide application are verified by comparing organic olive groves with IPM ones, becoming the main limiting factor, with only a few exceptions (e.g., *Pnigalio mediterraneus*). Among beneficial olive insects that are especially sensitive to insecticides, predatory species such as *Anthocoris nemoralis* and *Harraphidia laufferi* stand out. It seems relatively surprising the

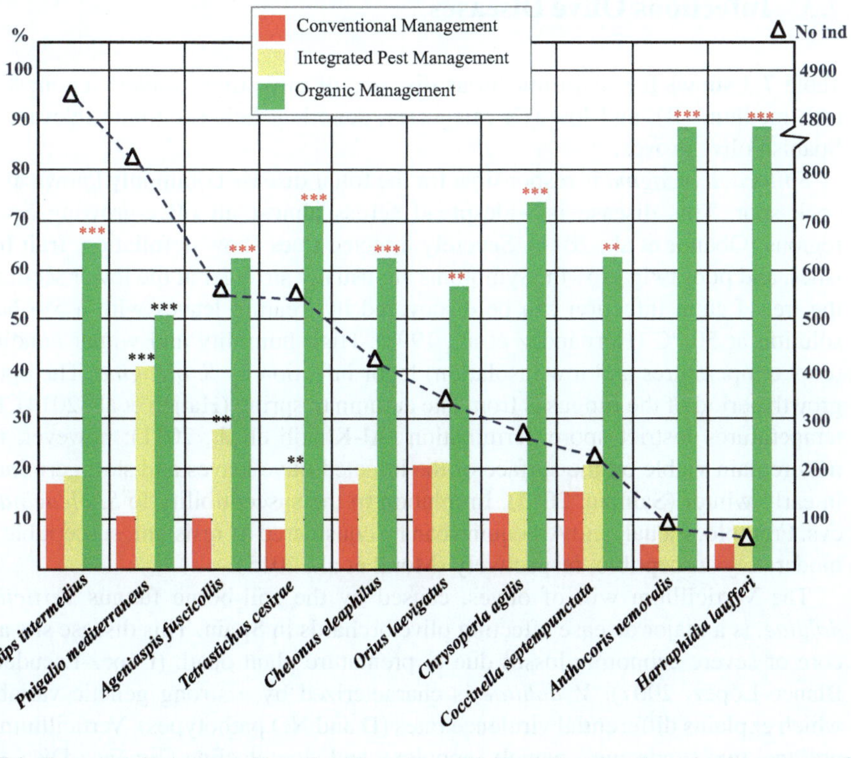

Fig. 7.1 Frequency histogram of beneficial insects, and their relative abundance in olive groves of the three types of management (conventional, IPM, and organic). (Adapted from González-Ruiz & Gómez-Guzmán, 2019)

similarity detected between IPM and conventional oliv groves, since IPM ones tend to be presented as an alternative to reduce the environmental impacts of conventional management, and are mainly oriented to prioritize beneficial insects. However, it could be argued that the active substances most commonly applied in IPM still slightly vary from those applied in conventional management. This suggests the need to review the IPM control protocols, fostering the implementation of less environmentally aggressive tactics. Among the alternatives at stake, it seems to us self-evident the relevance to promote biological insecticides, which are much more tightly adjusted to IPM practices and standards (Nikolova et al., 2015).

In addition to the diversity and abundance of natural enemies of pests, the evolution of olive growing practices for pest control in Andalusia along recent decades can be assessed through the entomophagous activity and impact on natural pest control. From this angle common green lacewings (Chrysopidae) can be considered as indicators. During the oviposition period of the olive moth in the fruit *Prays oleae*, predation values range between 50% and 70% in conventional, between 70% and 85% in IPM, and exceed 90% in organic olive groves.

7.3 Infectious Olive Diseases

Table 7.3 shows the main infectious diseases of olive trees, ranked amongst high (A), medium (B), and low (C) categories, denoting their economic relevance for Spanish olive groves.

Spilocaea oleagina is responsible for the foliar disease commonly known as peacock spot. This disease is widespread across almost all olive-growing Spanish regions (Obanor et al., 2011). Severely infected trees show defoliation, fruit blemishes, and poor twig growth. Symptoms are usually stronger at the lower sections of the tree. Latent infection can be discovered by treating leaves with a 5% NaOH solution at 50 °C (Hernández et al., 1998). High humidity and winter conditions (cool temperatures and low insolation) favor infection by *S. oleagina*. The optimal growth period of the fungus is from late autumn to spring (Hajjeh et al., 2014). High temperatures restrict spore germination (Al-Khatib et al., 2011); however, these may remain viable on the surface of the infected olive leaves and start germinating in early winter (Salman, 2017). In relation to the susceptibility to *S. oleagina*, the cvs. Frantoio, Picual, and Arbequina can be considered as resistant, susceptible, and moderately susceptible, respectively (Moral et al., 2015).

The Verticillium wilt of olives, caused by the soil-borne fungus *Verticillium dahliae*, is a major disease affecting olive orchards in Spain. This disease sits at the core of severe economic losses due to premature plant death (López-Escudero & Blanco-López, 2007). *V. dahliae* is characterized by a strong genetic variability which explains differential virulence rates (D and ND pathotypes). Verticillium wilt presents two syndromes, namely, apoplexy and slow decline (Jiménez-Díaz et al., 1998) (Table 7.3). This disease is favored by temperatures in the range of 20–25 °C during spring days followed by summer days with peaks not higher than 30–35 °C

Table 7.3 Principal infectious olive tree diseases in Spain

Disease	Causal agent	Type of pathogen (Phylum)	Principal symptoms	(*)
Peacock spot	*Spilocea oleagina*	Fungus (Ascomycota)	Dark brown circular spots (2-15 mm) surrounded by concentric yellowish or pale brown haloes on the upper side surface of the leaf.	A
Verticillium wilt	*Verticillium dahliae*	Fungus (Ascomycota)	a) Apoplexy: rapid dieback of twigs/branches. Death of young trees. b) Slow decline: heavy defoliation, foliar chlorosis mummification/death of inflorescences.	A
Olive knot	*Pseudomonas savastanoi*	Bacterium (Proteobacteria)	Overgrowth formation (Tumors or warts) in the aerial part of the trees, mainly on stems and branches.	A B
Anthracnose	*Colletotrichum acutatum s.l. C. gloeosporioides s.l.*	Fungus (Ascomycota)	Fruit circular rot; mummification dieback of branches	B
Leaf mould	*Pseudercospora cladosporoides*	Fungus (Ascomycota)	Yellowish and necrotic spots on the upper-side surface of the leaf. Leaden-grey areas on the under-side surface.	B
Phytophthora root rot	*Phytophthora megasperma Phytophthora inundata*	Fungal-like organisms (Oomycota)	Foliar wilting, dieback and death of young trees. Root necrosis.	B C
Armillaria root rot	*Armillaria mellea*	Fungus (Basidiomycota)	General decline and premature defoliation.	C
Olive fruit rot	*Camarosporium dalmaticum*	Fungus (Ascomycota)	Foliar wilting, dieback and death of young trees. Root necrosis	C
Olive leprosy	*Phlyctema vagabunda*	Fungus (Ascomycota)	Rounded necrotic lesions (2-3 mm) depressed, dark brown, limited by a darker edge and surrounded by chlorotic halo in fruit.	C
Sooty mould	*Capnodium elaeophilum*	Fungus (Ascomycota)	Black layer on leaves and branches (structures of the fungus).	C
Crown gall	*Agrobacterium tumefaciens*	Bacterium (Proteobacteria)	Swellings and galls on stems and roots near ground level.	C
Olive quick decline syndrome	*Xylella fastidiosa*	Bacterium (Proteobacteria)	Leaf scorch.	C

Adapted from Gómez (2022)

(*) economic relevance

A Main pest, of economic relevance

B Secondary pest, of medium economic relevance

C Secondary pest, of medium economic relevance and localized character

(López-Escudero & Mercado-Blanco, 2011). Over the last decades, this disease has increased markedly due to the introduction of irrigation in olive orchards (Pérez-Rodríguez et al., 2015). Planting drought-resistant cultivars is the most effective tool for reducing the spread of the disease (López-Escudero et al., 2004). However, the most common Spanish cultivars are susceptible or extremely susceptible to defoliating and nondefoliating *V. dahliae* isolates: Picual, Arbequina, Manzanilla sevillana, and Hojiblanca are extremely susceptible; whilst Koroneiki is moderately susceptible; and finally, Frantoio is strongly resistant (López-Escudero & Blanco-López, 2007).

The olive knot is caused by *Pseudomonas savastanoi* (Table 7.3). This disease is favored by the mild temperatures and rains of autumn and spring. *P. savastanoi* is able to survive as an epiphyte on the olive phyllosphere, where its population is high in spring and autumn, and declines in summer and winter (Quesada et al., 2010). Accordingly, olive knot is considered a chronic disease as its symptoms may persist and recur for many years. *P. savastanoi* requires plant-released signals from the wound itself to activate tumor formation (Surico, 1993). According to Valverde et al. (2020) cv. Frantoio is highly susceptibile, whereas cv. Arbequina shows intermediate susceptibility. In contrast, Picual is moderately susceptible and Lechín de Granada shows low sensitivity (Moral & Moral, 2018).

Anthracnose in Spain is particularly severe in those areas where the susceptible cvs Hojiblanca and Picudo are prevalent (Moral & Trapero, 2009). The cvs described as more highly resistant to anthracnose are Frantoio and Picual (Moral et al., 2015). In favorable conditions such as rain periods with mild temperatures, high inoculum pressures, and severe attacks by *Bactrocera oleae* during fruit ripening, may result in up to 100% loss of fruit production for susceptible cultivars (Moral et al., 2008). Anthracnose can also negatively affect olive oil quality.

Pseudocercospora cladosporioides, the agent causal of cercosporiosis (leaf mold), has been cited in the most important olive-growing areas worldwide, including naturally Spain and Andalusia (Farr & Rossman, 2019). *P. cladosporioides*, as *Spilocea oleagina*, spread mainly through rainfall periods and temperatures between 5 and 25 °C. Incubation and conidial production periods for these are usually long (up to 11 months) (Romero et al., 2020). The fungus mainly survives on the infected leaves that remain on the tree canopy. Injuries are then enlarged during autumn, winter, and spring, whilst rainfall initiates conidial spread and fosters new infections, particularly in young, susceptible leaves. Diseased leaves mainly fall in summer (Triki & Rhouma, 2008). The cvs Picual and Arbequina are described as moderately susceptible, whereas the cv. Frantoio is considered as susceptible (Moral et al., 2015).

References

Al-Khatib, M., Alhussaen, K., El-Banna, N., et al. (2011). Biological control of olive leaf spot (peacock spot disease) caused by *Cycloconium oleaginum* (*Spilocea oleaginea*). *Journal of Microbiology and Antimicrobials, 3*(9), 64–67. https://doi.org/10.5897/JMA.9000018

Arambourg, Y. (1986). *Traité d'entomologie oléicole.* Conseil oléicole international.

Baur, R., Baur, R., Wijnands, F., et al. (2011). *Integrated production: Objectives, principles and technical guidelines*. INRA, Unité de pathologie végétale.

Bozsik, A. (2000). Nahrungsanalytische untersuchungen an einigen mitteleuropäischen chrysopiden-imagines (Neuroptera: Chrysopidae). *Beiträge zur Entomologie Contributions to Entomology, 50*(1), 237–246. https://doi.org/10.21248/contrib.entomol.50.1.237-246

Bozsik, A., & González-Ruíz, R. (2006). *First data on the sibling species of the common green lacewings in Spain (Neuroptera: Chrysopidae):(The taxonomic status of the most important cryptic species of Chrysoperla carnea complex in Spain)*. In Anonymous 4th International Plant Protection Symposium at Debrecen University and 11th Trans-Tisza Plant Protection Forum, 18–19 October, 2006, Debrecen, Hungary Debreceni Egyetem, Agrártudományi Centrum, Mezögazdaságtudományi Kar (pp. 3–11).

Calabrese, G., Tartaglini, N., & Ladisa, G. (2012). *Study on biodiversity in century-old olive groves*. CIHEAM-Mediterranean Agronomic Institute.

Campos, M., & González, R. (1990). Influence of breeding conditions on longevity and fecundity of *Raphitelus maculatus* [Hym.: Pteromalidae] reared under standard laboratory conditions. *Entomophaga, 35*(3), 411–420. https://doi.org/10.1007/BF02375265

Canard, M., McEwen, P., New, T., et al. (2001). Natural food and feeding habits of lacewings. *Lacewings in the Crop Environment*, 116–129.

Carrera, J. M. (1996). *Lucha integrada contra las plagas agrícolas y forestales*.

De Andrés Cantero, F. (2001). *Enfermedades y Plagas del olivo* (4ª Edición, 646 pp.). Riquelme y Vargas Ediciones.

Ehi-Eromosele, C. O., Nwinyi, O., & Ajani, O. O. (2013). Integrated pest management. https://doi.org/10.5772/54476.

Farr, D. F., & Rossman, A. Y. (2019). No title. Fungal Databases, US National Fungus Collections, ARS, USDA. http://nt.ars-grin.gov/fungaldatabases/. Accessed 1 Jul 2020.

Gepp, J. (1967). Uberwinterung von *Chrysopa carnea* Stephens (Planipennia: Chrysopidae). *Entomol. Z. (Frankf.a.M.), 77*, 113–114.

Gómez, J. A. (2022). *Una metodología práctica para el diagnóstico de olivares con manejo Orgánico, basada en la determinación de efectos subletales sobre la entomofauna beneficiosa*. Phd tesis (biology). Universidad de Jaén (Spain). https://ruja.ujaen.es/bitstream/10953/1240/1/TesisJos%c3%a9AlfonsoG%c3%b3mezGuzm%c3%a1n_Ruja.pdf. 02/2024.

Gómez, J. A., Campos, M., Guzmán, G., et al. (2018). Soil erosion control, plant diversity, and arthropod communities under heterogeneous cover crops in an olive orchard. *Environmental Science and Pollution Research, 25*(2), 977–989. https://doi.org/10.1007/s11356-016-8339-9

González-Ruiz, R., & Gómez-Guzmán, J. A. (2019). Agricultural management greatly affects the beneficial entomofauna of the olive groves. *American Journal of Biomedicinal Science and Research*, 141–151. https://doi.org/10.34297/AJBSR.2019.01.000530

González-Ruiz, R., Al-Asaad, S., & Bozsik, A. (2008). Influencia de las masas forestales en la diversidad y abundancia de los crisópidos (Neur.:" Chrysopidae") del olivar. *Cuadernos de la Sociedad Española de Ciencias Forestales, 26*, 33–38.

Hajjeh, H., Salman, M., Abuamsha, R., et al. (2014). Latent infection of olive leaf spot disease on Palestinian olives. *Annual Research & Review in Biology*, 2517–2524. https://doi.org/10.9734/ARRB/2014/9845

Hernández, M. S., Davila, A. R., De Algaba, A. P., et al. (1998). Occurrence and etiology of death of young olive trees in southern Spain. *Europoen Journal of Plant Pathology, 104*(4), 347–357. https://doi.org/10.1023/A:1008624929989

Hernández, A. J., Lacasta, C., & Pastor, J. (2005). Effects of different management practices on soil conservation and soil water in a rainfed olive orchard. *Agricultural Water Management, 77*(1–3), 232–248. https://doi.org/10.1016/j.agwat.2004.09.030

Hodkinson, I. D., & Hughes, M. K. (1993). *La fitofagia en los insectos*. Oikos-Tau.

Ickert, G. (1968). *Beiträge zur biologie einheimischer Chrysopiden (Planipennia, Chrysopidae)*.

Jiménez-Díaz, R. M., Tjamos, E. C., & Cirulli, M. (1998). *Verticillium wilt of major tree hosts: Olive. A compendium of Verticillium wilts in tree species* (pp. 13–16).

Leyva, A., Hidalgo, J., Vega, V., et al. (2017). *Introducción al Olivar Ecológico en Andalucía. Consejería de Agricultura, Pesca y Desarrollo Rural* (pp. 1–23). Instituto de Investigación y Formación Agraria y Pesquera.

López-Escudero, F. J., & Blanco-López, M. A. (2007). Relationship between the inoculum density of Verticillium dahliae and the progress of Verticillium wilt of olive. *Plant Disease, 91*(11), 1372–1378. https://doi.org/10.1094/PDIS-91-11-1372

López-Escudero, F. J., & Mercado-Blanco, J. (2011). Verticillium wilt of olive: A case study to implement an integrated strategy to control a soil-borne pathogen. *Plant and Soil, 344*(1–2), 1–50. https://doi.org/10.1007/s11104-010-0629-2

López-Escudero, F. J., Del Río, C., Caballero, J. M., et al. (2004). Evaluation of olive cultivars for resistance to Verticillium dahliae. *European Journal of Plant Pathology, 110*(1), 79–85. https://doi.org/10.1023/B:EJPP.0000010150.08098.2d

Luckmann, W. H., & Metcalf, R. L. (1990). *Introducción al manejo de plagas de insectos*. Limusa.

Manojlovic, B., Zabel, A., Stankovic, S., et al. (2001). Additional diet of the parasitoids (Hymenoptera: Braconidae) and the parasitizing of the Elm Bark Beetle (Coleoptera: Scolytidae). *Anzeiger für Schädlingskunde Journal of Pest Science, 74*(3), 66–71. https://doi.org/10.1046/j.1439-0280.2001.01014.x

Martin Cerdeño, V. J. (2014). *Veinticinco años de alimentación en España*. De dónde venimos y hacia dónde vamos en los hábitos de consumo de alimentos y bebidas. MAGRAMA (2014): Alimentos de España 25.

Martínez, J. R. F., Zuazo, V. H. D., & Raya, A. M. (2006). Environmental impact from mountainous olive orchards under different soil-management systems (SE Spain). *Science of the Total Environment, 358*(1–3), 46–60. https://doi.org/10.1016/j.scitotenv.2005.05.036

Ministerio De Agricultura, Pesca Y Alimentación (MAPA). (2017). *Encuesta sobre plantaciones de árboles frutales, olivar y uva de mesa.* https://www.mapa.gob.es/es/estadistica/temas/estadisticas-agrarias/agricultura/encuestas-plantaciones-arboles-frutales/. Accessed 20 Jul 2020.

Moral, J., & Moral, J. (2018). Sanidad del cultivo del olivo. *Phytoma, Valencia,* 330.

Moral, J., & Trapero, A. (2009). Assessing the susceptibility of olive cultivars to anthracnose caused by *Colletotrichum acutatum. Plant Disease, 93*(10), 1028–1036. https://doi.org/10.1094/PDIS-93-10-1028

Moral, J., Bouhmidi, K., & Trapero, A. (2008). Influence of fruit maturity, cultivar susceptibility, and inoculation method on infection of olive fruit by *Colletotrichum acutatum. Plant Disease, 92*(10), 1421–1426. https://doi.org/10.1094/PDIS-92-10-1421

Moral, J., Alsalimiya, M., Roca, L. F., et al. (2015). Relative susceptibility of new olive cultivars to Spilocaea oleagina, Colletotrichum acutatum, and *Pseudocercospora cladosporioides. Plant Disease, 99*(1), 58–64. https://doi.org/10.1094/PDIS-04-14-0355-RE

Nikolova, I., Georgieva, N., & Tahsin, N. (2015). Toxicity of neem and pyrethrum products applied alone and in combination with different organic products to some predators and their population density. *Romanian Agricultural Research, 32*(32), 291–301.

Obanor, F. O., Walter, M., Jones, E. E., et al. (2011). Effects of temperature, inoculum concentration, leaf age, and continuous and interrupted wetness on infection of olive plants by *Spilocaea oleagina. Plant Pathology, 60*(2), 190–199. https://doi.org/10.1111/j.1365-3059.2010.02370.x

Pappas, M. L., Broufas, G. D., & Koveos, D. S. (2007). Effects of various prey species on development, survival and reproduction of the predatory lacewing *Dichochrysa prasina* (Neuroptera: Chrysopidae). *Biological Control, 43*(2), 163–170. https://doi.org/10.1016/j.biocontrol.2007.07.006

Pérez-Rodríguez, M., Alcántara, E., Amaro, M., et al. (2015). The influence of irrigation frequency on the onset and development of Verticillium wilt of olive. *Plant Disease, 99*(4), 488–495. https://doi.org/10.1094/PDIS-06-14-0599-RE

Quesada, J. M., Penyalver, R., Pérez-Panadés, J., et al. (2010). Dissemination of *Pseudomonas savastanoi pv. Savastanoi* populations and subsequent appearance of olive knot disease. *Plant Pathology, 59*(2), 262–269. https://doi.org/10.1111/j.1365-3059.2009.02200.x

Romero, J., Ávila, A., Agustí-Brisach, C., et al. (2020). Evaluation of fungicides and management strategies against Cercospora leaf spot of olive caused by Pseudocercospora cladosporioides. *Agronomy, 10*(2), 271. https://doi.org/10.3390/agronomy10020271

Salman, M. (2017). Biological control of *Spilocaea oleagina*, the causal agent of olive leaf spot disease, using antagonistic bacteria. *Journal of Plant Pathology, 99*(3), 741–744. https://doi.org/10.4454/jpp.v99i3.3958

Sengonca, C., & Frings, B. (1987). *Artificial overwintering habitat for the predative Chrysoperia carnea* [German]. DLG Mitteilungen.

Surico, G. (1993). Scanning electron microscopy of olive and oleander leaves colonized by *Pseudomonas syringae* subsp. *savastanoi*. *Journal of Phytopathology, 138*(1), 31–40. https://doi.org/10.1111/j.1439-0434.1993.tb01358.x

Tóth, M., Bozsik, A., Szentkirályi, F., et al. (2006). Phenylacetaldehyde: A chemical attractant for common green lacewings (*Chrysoperla carnea* sl, Neuroptera: Chrysopidae). *European Journal of Entomology, 103*(1), 267. https://doi.org/10.14411/eje.2006.033

Triki, M. A., & Rhouma, A. (2008). First report of *Pseudocercospora cladosporioides*, the causal agent of Cercospora leaf spot of olive trees, in Tunisia. *Phytopathologia Mediterranea, 47*(3), 262–265.

Valverde, P., Zucchini, M., Polverigiani, S., et al. (2020). Olive knot damages in ten olive cultivars after late-winter frost in Central Italy. *Scientia Horticulturae, 266*, 109274. https://doi.org/10.1016/j.scienta.2020.109274

Viggiani, G. (1977). *Lotta biologica ed integrata*. Liguori.

Vives de Quadras, J. M. (1988). Control de plagas de insectos. Problemas y alternativas. In *Insecticidas Biorracionales* (pp. 3–13).

Chapter 8
Challenges for Sustainable Water Resource Management in the Spanish Olive Orchards

Ignacio Jesús Lorite, Raúl de la Rosa, Javier Hidalgo, José Manuel Cabezas, Juan Carlos Hidalgo, and Lorenzo León

Abstract Olive in Spain has traditionally been a rainfed crop. However, this has changed over the last 30 years, and there are now 840,000 ha of irrigated olive orchards, about 30% of the cropping area. The province of Jaén in south-eastern Spain is an example of this shift. In this region, the implementation of deficit irrigation strategies involving the application of less than 1500 m^3/ha (much less than the optimum crop requirements), fixed rotation supply, and drip irrigation has shown excellent results. Despite the advances made in water management in Spanish olive orchards in recent years, these systems are threatened by low profitability (associated with high energy costs) and growing competition for water resources with other sectors, accelerated by the effects of climate change, which could jeopardize the economic and social sustainability of olive cultivation. Therefore, some strategies have been proposed to increase the resilience and sustainability of Spanish olive orchards, which include developing integrated water resources management at

I. J. Lorite (✉) · J. M. Cabezas
Natural Resources Unit, IFAPA - Centro Alameda del Obispo, Córdoba, Spain
e-mail: ignacioj.lorite@juntadeandalucia.es; josem.cabezas@juntadeandalucia.es

R. de la Rosa
Plant Improvement and Biotechnology Unit, IFAPA-Alameda del Obispo, Córdoba, Spain

Instituto de Agricultura Sostenible, Consejo Superior de Investigaciones Científicas,, Córdoba, Spain
e-mail: raul.rosa@ias.csic.es

J. Hidalgo · J. C. Hidalgo
Agri-Food Engineering and Technology Unit, IFAPA - Centro Alameda del Obispo, Córdoba, Spain
e-mail: javier.hidalgo@juntadeandalucia.es; juanc.hidalgo@juntadeandalucia.es

L. León
Plant improvement and Biotechnology Unit, IFAPA-Alameda del Obispo, Córdoba, Spain
e-mail: lorenzo.leon@juntadeandalucia.es

J. Muñoz-Rojas, R. García-Ruiz (eds.), *The Olive Landscapes of the Mediterranean*, Landscape Series 36,
https://doi.org/10.1007/978-3-031-57956-1_8

basin scale; increasing the irrigated area using reclaimed, desalinated, or groundwater resources; or developing site-specific assessment promoting deficit irrigation scheduling to achieve significant water savings with a limited impact on yield.

8.1 Description of the Olive Systems in Spain in Terms of Water Management

Irrigation is of huge economic and social relevance to Spanish agriculture. Although the irrigated area in Spain constitutes around 14% of its total agricultural area, it contributes more than 50% of the final crop production and 2.4% of Gross Domestic Product (GDP), while accounting for 4% of total employment (MAPA, 2020). Spain devotes around 67% of its total water resources to irrigated agriculture, mainly olive (0.84×10^6 ha), vineyards (0.40×10^6 ha), non-citrus orchards (0.38×10^6 ha), maize (0.34×10^6 ha), citrus orchards (0.29×10^6 ha), and alfalfa (0.16×10^6 ha) (MAPA, 2019a).

According to the most recent available data, 2.73×10^6 ha are under olive cultivation in Spain. Of this area, around 31% is under irrigation (MAPA, 2019a). Andalusia, the leading producer of olive oil in the world, has around 1.65×10^6 ha of olive, 38.5% of which is under irrigation (0.64×10^6 ha), representing 76% of the area cultivated with olive in Spain. Olive yield in irrigated orchards is around 106% higher than that of rainfed olive groves (MAPA, 2019a). However, these differences are not exclusively due to irrigation, as irrigation is usually implemented in olive groves that have better agronomic characteristics and potential response, whereas low-yield mountainous olive groves are rainfed. In any case, the introduction of olive irrigation is considered the driver of the economic and social improvement in large areas of Spain (Berbel et al., 2019) (Fig. 8.1).

Optimum irrigation requirements for olive (amounts that prevent crop water stress throughout the whole crop cycle) depend on weather conditions and orchard characteristics and could exceed 6000 m³/ha in southern Spain (Moriana et al.,

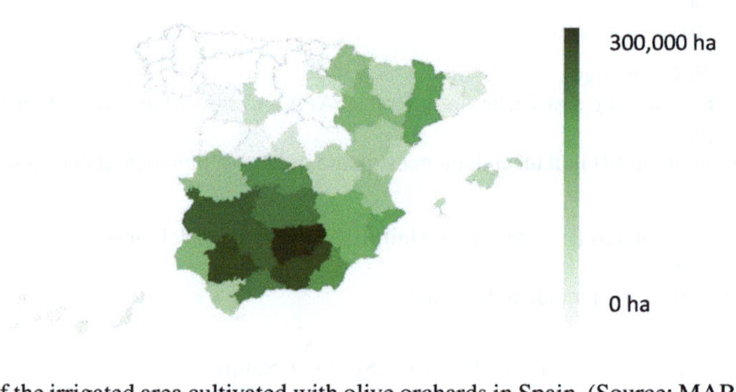

Fig. 8.1 Map of the irrigated area cultivated with olive orchards in Spain. (Source: MAPA, 2019a)

2003). However, this amount is not currently supplied by the farmers, even when water availability is high (Lorite et al., 2004, 2012). Thus, irrigation in Spanish olive orchards are mainly managed using deficit irrigation strategies (DIS) (Lorite et al., 2004, 2012), with an increasing percentage using regulated deficit irrigation, where the reduction of irrigation depends on the phenological stage, avoiding severe water stress during flowering and maturity stages (Fereres & Soriano, 2007).

The advantages of applying DIS in Spanish olive orchards are related to the significant increase in yield compared with rainfed systems, with reductions in alternate bearing and interannual yield variability. Moreover, compared with full irrigation strategies, DIS increase water use efficiency (Moriana et al., 2003) and irrigation water productivity (Lorite et al., 2004); allow the irrigated area to be extended (as was done in La Loma de Úbeda area, in Jaén province) due to the lower allocation per hectare and the resulting water savings; and improve the olive oil quality, and increase the oil content (Moriana et al., 2003) and the polyphenol concentration (Fernandes-Silva et al., 2013; Caruso et al., 2014).

The most common irrigation method in Spain is drip irrigation (accounting for 53.1% of the total irrigated area), followed by surface irrigation (with 23.5%) and sprinklers (23.4%). However, those percentages change in olive systems, where 94.7% are under drip irrigation, 5.1% under surface irrigation, and 0.3% under sprinklers (MAPA, 2019b). This is clear evidence of the efforts made by olive farmers in Spain to improve irrigation efficiency. The selection of drip irrigation systems for the newly irrigated olive areas is based on several advantages, such as the higher irrigation efficiency (reaching values higher than 80% under proper maintenance and operation), low possibility of non-point source contamination, possibility of automatization, and lower negative impacts of irrigation with saline water resources. However, some disadvantages have also been identified, including higher energy costs (as pumping is required) and higher infrastructure and maintenance costs to prevent obturation and leakages.

About 75% of the water consumed for irrigation in Spain comes from surface water sources, 23.1% from groundwater, and only about 1.9% from desalinated and reclaimed water (INE, 2018). The use of groundwater resources for irrigation has increased the productivity of some olive areas in southern Spain, ensuring the survival of trees during drought periods (González-Ramón et al., 2013). However, in recent years, groundwater resources for olive irrigation have been used without proper control, resulting in significant water withdrawals in some areas. For instance, as much as 35 hm^3 was extracted in a hydrological year for the irrigation of about 24,000 ha of olive orchards in La Loma de Úbeda in Jaén (González-Ramón et al., 2013). The third source, reclaimed water, has not been a commonly used water source in Spanish olive orchards. However, this source is becoming increasingly relevant as farmers have identified it as the only option for introducing irrigation in their fields. Thus, new irrigation districts have started to promote the use of reclaimed water resources. For example, in Jaén province, around 3000 ha cultivated of olive orchards will be irrigated with this water resource in the coming years.

The water costs associated with the olive irrigation in Spain vary according to the water source. In modern irrigation districts with pressurized systems, on-demand delivery, and water sourced from surface water, the water costs are around 0.10–0.12 €/m^3 (Lorite et al., 2012). For areas using dams and pumping systems, and with low allocations (such as those located in Jaén province), the costs rise to 0.5 €/m^3. Finally, in areas using groundwater or reclaimed sources, water costs could exceed 1 €/m^3. Irrigation water productivity (IWP) is a critical indicator to identify the sustainability of the irrigation. IWP is defined as the profit obtained by each cubic meter of irrigation water supplied and is usually measured in €/m^3. Some values of IWP for olive in irrigated areas in southern Spain were calculated to be around 0.8 €/m^3 (Lorite et al., 2012), ranging between 0.5 and 1.1 €/m^3 for wet years and dry years, respectively (Lorite et al., 2004). By comparing the irrigation costs and IWP, it is possible to determine the sustainability of the irrigation, with its viability being threatened when water costs are higher than IWP.

8.2 Threats and Weaknesses of Olive Orchards in Spain Related to Water Management

The threats facing Spanish olive orchards related to water management include external factors that can negatively affect these olive systems. Some of the threats identified are:

- In pressurized systems, high energy costs lead to high costs associated with irrigation, leading to severe reductions in the farmer's profit.
- Non-differentiated added-value olive oil, promoting competition with other olive-producing Mediterranean areas, leading to reductions in the yield prices and thus in the thresholds of profitability of irrigation.
- Effects of climate change that could reduce the availability of water resources due to the increasing frequency of drought events.
- Competition with other sectors (such as environmental, urban, or tourism) for water resources, resulting in lower water availability for agriculture.
- Increasing social pressure to improve irrigation efficiency in agriculture, requiring irrigation to generate environmental and social benefits.
- Increasing legal restrictions on the use of reclaimed water related to the preservation of ecological flows.

The weaknesses of irrigated systems in relation to water management in Spain stem from the inherent limitations of these systems in terms of infrastructure and agricultural practices. Some of the weaknesses are:

- Soil erosion and salinization are caused by inadequate management of soil and irrigation.
- Low technological level at field scale is reflected in the lack of water meters or plant-soil sensors for irrigation scheduling.

- Limited knowledge of relevant agronomic information (such as water requirements, fertilization, and irrigation scheduling) by farmers.
- Limited use of cover crops caused by a lack of knowledge of how to properly manage them to prevent competition for water resources with olive trees.
- Rigid irrigation infrastructures that limit the adaptability of these systems to changes in olive orchards or water availability.
- Inadequate structure of irrigation districts and farmers' associations, solely focused on administrative functions without providing technical advisory services.
- Generation of negative environmental impacts related to groundwater pollution or river ecology.

8.3 Strategies to Improve the Sustainability of Olive Systems in Spain Related to Water Management

As a result of the changes made in Spanish olive orchards in recent decades, the water management has improved significantly. However, there are still limitations that could impact the economic, social, and environmental sustainability of these systems. To address these concerns, the following strategies to improve the sustainability of Spanish olive orchards are proposed:

1. Promotion of controlled deficit irrigation strategies in all the irrigated areas (Fereres & Soriano, 2007), even when water availability is high.
2. Development of irrigation scheduling based on accurate knowledge of crop water requirements and leaching fraction for the maintenance of salt balance and prevention of salinization (Fereres & Villalobos, 2016).
3. Improvement of irrigation infrastructure, building reservoirs for storing irrigation water, and promoting drip irrigation, water meters in each field (Lorite et al., 2013), and semi-demand delivery options (where the irrigator requests a specific depth of water and flow rate to be supplied on one or several days; Mateos et al., 2002) in an attempt to improve energy efficiency.
4. Development of new irrigated areas using reclaimed water or groundwater resources with very low allocation (irrigation support or deficit irrigation strategies) and high irrigation efficiency. For olive, this small increase in water supply would lead to high increases in yield (Moriana et al., 2003) and IWP (Lorite et al., 2004), boosting the economic development of rural areas.
5. Site-specific assessment to increase irrigation efficiency provided by local and regional irrigation advisory services, promoting the use of sensors at field scale with user-friendly communication and maintenance (such as sensors of soil moisture and/or crop water status with technology based on FIWARE), and the development of automatic weather station networks (such as the SIAR network, available at http://eportal.mapa.gob.es/websiar/SeleccionParametrosMap.aspx?dst=1).

6. Development and promotion of decision support systems for irrigation management (http://www.servifapa.es, http://santaella.crgenilcabra.es or http://eportal.mapa.gob.es/websiar/NecesidadesHidricas.aspx) developed by public institutions, cooperatives or farmers' associations.
7. Reallocation of water resources at basin scale aimed at balancing agronomic, economic, social, and environmental components. Previous studies have confirmed the satisfactory performance when water resources are distributed based on rules prioritizing terms such as IWP (Lorite et al., 2007).
8. Raising farmers' awareness of the need for proper irrigation management taking into account the recommendations provided by the local advisory services (Lorite et al., 2012).
9. Implementation of sustainable intensification practices, increasing the use of inputs such as irrigation, but also increasing environmental protection by avoiding non-point source contamination, groundwater overexploitation, erosion or water waste.
10. Promotion of effective cover crops in olive orchards, ensuring appropriate management to avoid negative interactions with the olive trees.
11. Preserving the soil fertility, biodiversity, and landscape of the irrigated ecosystems by promoting strategies such as organic farming, conservation agriculture and encouraging biodiversity through the use of borders between fields.
12. Fostering collaboration between public research institutions, private companies, and farmers' associations/cooperatives in order to provide user-friendly advisory tools adapted to farmers' requirements.

8.4 Conclusions

Although there have been significant improvements in water management in Spanish olive orchards in recent decades, new concerns related to the impact of climate change (reduced water availability for crops due to more frequent droughts and higher temperatures), heightened awareness of the need to protect the environment, and lower profits from agriculture, are all driving the need for substantial changes in Spanish olive orchards. Strategies such as the development and promotion of irrigation advisory services, implementation of deficit irrigation strategies, improvements to water management at the field, irrigation district, and basin scales to boost irrigation efficiency, and the mitigation of adverse impacts on the environment will ensure the future sustainability of Spanish olive orchards.

Spanish olive orchards will therefore require a major transformation, in which all the stakeholders will play a relevant role. Coordination between farmers, cooperatives, private companies, public research institutions, and authorities will be essential in order to pull together and ensure the sustainable development of Spanish olive systems, while enabling the preservation of the environment.

References

Berbel, J., Expósito, A., Gutierrez-Martín, C., & Mateos, L. (2019). Effects of the irrigation modernization in Spain 2002-2015. *Water Resources Management, 33*, 1835–1849.
Caruso, G., Gucci, R., Urbani, S., Esposto, S., Taticchi, A., Di Maoio, I., Selvaggini, R., & Servili, M. (2014). Effect of different irrigation volumes during fruit development on quality of virgin olive of cv. Frantoio. *Agricultural Water Management, 134*, 94–103.
Fereres, E., & Soriano, M. A. (2007). Deficit irrigation for reducing agricultural water use. *Journal of Experimental Botany, 58*(2), 147–159.
Fereres, E., & Villalobos, F. J. (2016). Agronomy and the sustainability of crop production. In F. Villalobos & E. Fereres (Eds.), *Principles of agronomy for sustainable agriculture* (pp. 527–542). Springer.
Fernandes-Silva, A. A., Gouveia, J. B., Vasconcelos, P., Terreira, T. C., & Villalobos, F. J. (2013). Effects of different irrigation regimes on the quality attributes of monovarietal virgin olive from cv. "Cobrançosa". *Grasas y Aceites, 64*(1), 41–49.
González-Ramón, A., Rodríguez-Arévalo, J., Martos-Rosillo, S., & Gollonet, J. (2013). Hydrological research on intensively exploited deep aquifers in the 'Loma de Úbeda' area (Jaén, southern Spain). *Hydrology Journal, 21*, 887–903.
INE. (2018). *Encuesta sobre el uso del agua en el sector agrario (EUASA) Año 2016.* Instituto Nacional de Estadística.
Lorite, I. J., Mateos, L., & Fereres, E. (2004). Evaluating irrigation performance in a Mediterranean environment. II. Variability among crops and farmers. *Irrigation Science, 23*, 85–92.
Lorite, I. J., Mateos, L., Orgaz, F., & Fereres, E. (2007). Assessing deficit irrigation strategies at the level of an irrigation district. *Agricultural Water Management*, 51–60.
Lorite, I. J., García-Vila, M., Carmona, M. A., Santos, C., & Soriano, M. A. (2012). Assessment of the irrigation advisory services' recommendations and farmers' irrigation management: A case study in southern Spain. *Water Resources Management, 26*, 2397–2419.
Lorite, I. J., Santos, C., García-Vila, M., Carmona, M. A., & Fereres, E. (2013). Assessing irrigation scheme water use and farmers' performance using wireless telemetry systems. *Computers and Electronics in Agriculture, 98*, 193–204.
MAPA. (2019a). *Encuesta sobre superficies y rendimientos de cultivos 2019.* Resultados 2019. Ministerio de Agricultura, Pesca y Alimentación. https://www.mapa.gob.es/es/estadistica/temas/estadisticas-agrarias/agricultura/esyrce/
MAPA. (2019b). *Encuesta sobre superficies y rendimientos de cultivos 2019. Informe sobre Regadíos en España.* Ministerio de Agricultura, Pesca y Alimentación. https://www.mapa.gob.es/es/estadistica/temas/estadisticas-agrarias/agricultura/esyrce/informes-sectoriales/
MAPA. (2020). *Gestión Sostenible de regadíos.* Ministerio de Agricultura, Pesca y Alimentación. https://www.mapa.gob.es/es/desarrollo-rural/temas/gestion-sostenible-regadios/
Mateos, L., López-Cortijo, I., & Sagardoy, J. A. (2002). SIMIS: The FAO irrigation decision support system for irrigation scheme management. *Agricultural Water Management, 56*, 193–206.
Moriana, A., Orgaz, F., Pastor, M., & Fereres, E. (2003). Yield responses of a mature olive orchard to water deficits. *Journal of the American Society for Horticultural Science, 128*(3), 425–431.

References

Chapter 9
Management Practices Boosting Ecosystem and Landscape Functions and Services in Spain

Juan Antonio Torres-Cordero, Roberto García-Ruiz, Gema Siles Colmenero, and Antonio García-Fuentes

Abstract Olive oil cultivation shapes the landscape, culture, and way of life of many rural areas of Spain, especially southern. Increased olive oil demand has acted as a catalyst for the intensification of olive groves. The intensification of low-input traditional olive groves and the implementation of new intensive and super-intensive plantations with irrigation and increased tree density have resulted in simplified landscapes with olive groves with low-nature-value, driving greater negative environmental impacts (e.g., soil erosion, run-offs to water bodies, increased rates of soil fertility loss, degradation of habitats and landscapes, and over-exploitation of scarce water resources). The near absence of biological elements other than olive trees in many olive groves across the farm to landscape levels has led to the inefficiency of ecological regulatory mechanisms, which potentially provide ecosystems with resistance and resilience to ever-changing environmental conditions. The reversal of the negative impacts of modern olive farming on the environment is a key concern at the European level and European policies are now in place that promote eco and wildlife-friendly management models. Increasingly, research indicates that the level of internal regulation of ecological functions in agroecosystems is largely dependent on their level of diversity and complexity. Biodiversity delivers a variety of ecosystem services that span beyond the production of food, and that include the recycling and retention of nutrients, regulation of micro-climates and local hydrological processes, and control or mitigation of pests and diseases. This leads to improving benefits such as crop protection and soil fertility even when external inputs are limited, such as is frequently the case in the Mediterranean context.

J. A. Torres-Cordero (✉) · G. S. Colmenero · A. García-Fuentes
Department of Animal and Plant Biology, and Ecology, Universidad de Jaén, Jaén, Spain
e-mail: jatorres@ujaen.es; gsiles@ujaen.es; agarcia@ujaen.es

R. García-Ruiz
University Institute for Research on Olive Groves and Olive Oils, University of Jaén, Jaén, Spain
e-mail: rgarcia@ujaen.es

J. Muñoz-Rojas, R. García-Ruiz (eds.), *The Olive Landscapes of the Mediterranean*, Landscape Series 36,
https://doi.org/10.1007/978-3-031-57956-1_9

This chapter analyses a relatively wide range of technically and economically feasible synergistic management practices, at both farm and landscape levels, that promote and enhance the biotic interactions, functional traits, creation of biophysical structures and ecological processes, and includes, among others: (i) temporarily wild or seeded cover crops, (ii) tree pruning management, (iii) organic fertilization, and (iv) livestock integration.

9.1 Current Olive Oil Farming Contributes to Ecosystem Disservices

Olive grove cultivation in Spain has experienced a strong intensification during the last decade, and this has been especially the case of the Andalusian region (southern Spain), the highest producer worldwide. The increased demand for olive oil combined with the need to lower the production cost have driven changes in the production systems and management resulting in the transformation of complex agrarian landscapes with patches of semi or natural Mediterranean forest to monocultures of high tree densities and very few olive tree varieties (Ramos & Santos, 2010). At the scale of olive farms, soils are usually kept bare by combining herbicides and tillage, resulting in a high impact on runoff and soil erosion (Duran Zuazo et al., 2009), a low biodiversity and soil fertility. In addition, many of the components which provide diversity and complexity at a landscape scale, such as patches of natural or semi-natural Mediterranean forests, hedge, and adjacent crops, such as cereals in rotation, have typically been replaced by olive groves plantations.

Clearly, the intensification of the olive grove in Spain has been aimed, above all, at promoting one element of the ecosystems service: the yield of olive fruit to produce olive oil, and firewood which results from olive tree pruning. And this has been so in detrimental to other ecosystem services (ESs) such as those of supporting, regulating and cultural-aesthetic. Soldati et al. (2023) indicate to the convenience to address ecosystem services in these agro-ecosystems via the implementation of life cycle tools. Ultimately, it is becoming increasingly clear that olive oil production is based on an increasingly unsustainable model of production which contributes to disservices or loss of services, despite having acknowledged benefits for our health of the olive oil. These disservices come at a significant cost to society, but there is often a mismatch between the benefits, which accrue to the olive oil sector, and the costs, which are typically borne by society at various scales, from local communities impacted by pesticides in drinking water or rural infrastructure damages to the global commons affected by global warming. Olive groves were unaware of these tradeoffs or simply considered them less important in favor of a collective mindset to maximize olive yield.

This goes against the current EU and Spanish politics and strategies aimed to boost the sustainability of the agricultural sectors, such as the recent Science for

Environment Policy or the European Green Deal. Indeed, there is increasing emphasis on securing wider ecosystem service benefits from agricultural land, beyond provisioning services of food and fiber (Power, 2010).

The agroecology provides a long-standing and well-recognized conceptual framework from which a specific set of management practices aimed to enhance positive synergies among multiple ESs and adapted to olive cropping can be delineated.

Among, but not exclusively, those management practices that are technically and economically feasible and boost the ESs stand out: temporary spontaneous or seeded cover crops, shedder of the tree pruning, diversity, and complexity of landscape elements, integration of livestock and organic fertilization (Fig. 9.1 and Table 9.1).

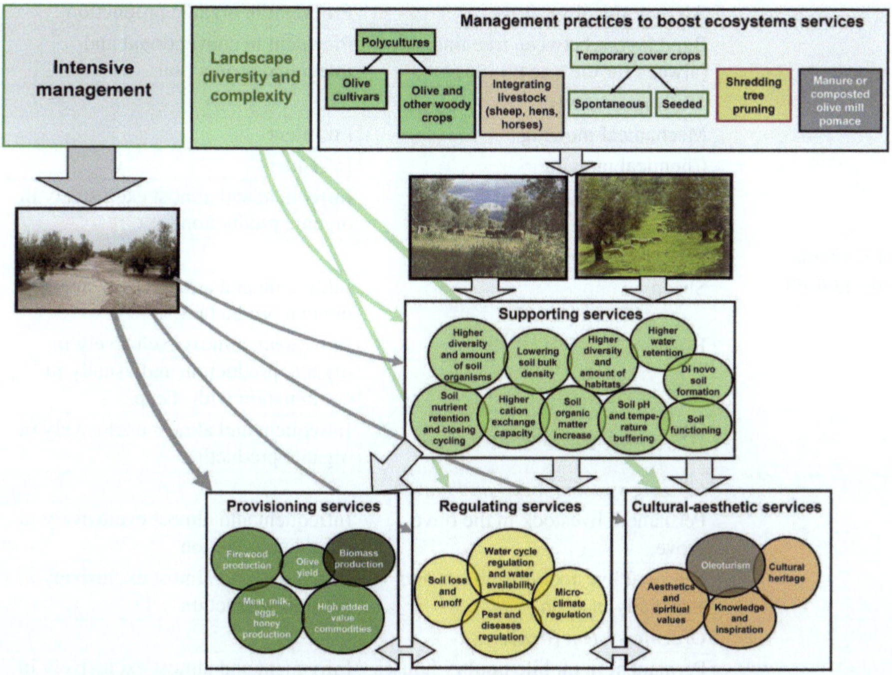

Fig. 9.1 Single and combined management practices in olive farming that enhance ecosystem services related to sustainability. Red arrow stands for ecosystem disservices of intensive olive farming. Items with a grey background in the provisioning and cultural-aesthetic ecosystem services refer to those of economic importance for the olive farmer

Table 9.1 Summary of the variability and applicability of some of the management practices that drive ecosystem and landscape functions and services in olive groves

Management practice	Variability	Applicability
Cover crop and tree pruning	*Implementation type*	
	Spontaneous: Seed bank	Prevalent
	Spontaneous: Grasses selection by using a selection of herbicides.	Prevalent
	Seeded grasses	Prevalent
	Seeding a mix of grasses and legumes	Prevalent
	Seeding with cruciferous	Prevalent only in areas affected by *Verticillium dahliae*
	Seeding with a mixture of high diversity seeds	Infrequent
	Tree pruning	Prevalent and usually combined with spontaneous cover crops
	Spatial distribution on the farm	
	All over the olive grove	Prevalent in organic production
	Band layout between tree lines (always for the tree pruning)	Prevalent in conventional and integrated production
	Mowing methods	
	Mechanical mowing	Prevalent
	Chemical mowing	Infrequent
	Livestock mowing	Infrequent and almost exclusively in organic production
Livestock integration	*Type of livestock*	
	Sheep	Infrequent and almost exclusively in organic production
	Horse	Infrequent, almost exclusively in organic production and usually in combination with sheep.
	Hen	Infrequent and almost exclusively in organic production
	Grazing system (sheep and horse)	
	Permanent livestock in the olive grove	Infrequent and almost exclusively in organic production
	Seasonal livestock in autumn and especially in spring	Infrequent and almost exclusively in organic production
	Grazing system (hen)	
	Permanent in mobile poultry houses	Infrequent and almost exclusively in organic production
	Permanent in mobile enclosures	Infrequent and almost exclusively in organic production

(continued)

Table 9.1 (continued)

Management practice	Variability	Applicability
Organic fertilization	*Fertilizers*	
	Manure (sheep, hen and horse)	Infrequent and almost exclusively in organic production
	Composted olive mil pomace	
	Green manure rich in legumes	
	Organic liquid and solid fertilizers	Infrequent and almost exclusively in organic production
Boosting landscape diversity	Diverse olive tree varieties and ages, and polycultures	Infrequent
	Diversification of the natural landscape: Recovery of diverse natural vegetation structures	Infrequent with exceptions in mountainous olive groves

9.2 Temporary Spontaneous or Seeded Cover Crops and Crushing Olive Tree Pruning

Both managements play an essential role in delivering the conditions for providing many of the ESs that promote sustainability.

Temporary spontaneous or seeded cover crops are based on seeding, or allowing growing, of herbaceous vegetation in the band between tree lanes during the rainfall season (autumn/fall and winter) controlling chemically or mechanically the cover crop in early spring (from the end of March to the end of April, depending on rainfall predictions) to prevent losses of soil water by transpiration, and maintaining its residues over the surface until next fall when, ideally, it will regrow from seeds produced during the previous year. This management practice is increasingly widespread in the Spanish olive groves.

On the other hand, the disposal of the olive tree pruning remains after crushing on top of the soil as an alternative to its burning, has also become an increasingly common practice and it is complementary to cover crops. Usually, olive tree pruning is performed every 2 years rendering the equivalent of between 1.0 and 2.0 Mg ha^{-1} year^{-1}. In the vast majority of olive groves, the tree pruning is disposed between the tree lines and they are crushed using mechanical grinders coupled to mechanical traction (Fig. 9.2).

The presence of a temporary cover crop (TCP) in combination with the crushing of the tree pruning remains (TPR) generates and reinforces many key elements of provisioning, supporting, and regulating ecosystem services, which boost the long-term sustainability (Fig. 9.1). Productivity of the whole agroecosystem, and not only of the commercial olive fruit and firewood, is increased with TCP. Vicente-Vicente et al. (2017) found that annual aboveground net primary production of spontaneous TCP of ten olive oil groves averaged 1.48 Mg dry weight ha^{-1}, which a slightly lower of that found for seeded TCP (Repullo-Ruibérriz et al., 2012). Taking into account the belowground biomass, this figure is close to the Spanish

Fig. 9.2 (**a–c**) Examples of the operations and the final result of the shedder of the olive tree pruning. (**a**) Arrangement of the tree pruning between the tree lines on an olive grove for their subsequent crushing. (**b**) Tree pruning crushed using a crusher coupled to a tractor. (**c**) Detail of how the crushed tree pruning remains on top of the soil

annual olive yield, which averages about 2.1 Mg dry weigh ha^{-1}. This extra biomass increases the provisioning ES if the biomass is controlled by livestock, which then is converted into meat and milk or if added values are produced from seeded TCP. TCP and TPR remain increasing at short-to-medium terms for many of the components related to supporting services. Indeed, the soil content of organic matter and other indicators of soil fertility, such as cation exchange capacity, soil aggregate stability, soil functional fertility, and bulk density, among others, have been shown to increase after several years (Garcia-Ruiz et al., 2008; Ordóñez-Fernández et al., 2015). In addition, TCP and TPR are good strategies to retain nutrients not taken by olive trees, contributing to the closure of biogeochemical cycles of nutrients.

TPR and TCP play a vital role in some of the ES related to regulating. Both, but especially TCP, have a clear significant effect on reducing soil losses, and thus on losses of soil fertility, in olive orchard and other woody crops, with an average reduction close to 60% (Gómez, 2017). Both TCP (Vicente-Vicente et al., 2017) and TPR (Repullo et al., 2012; Gómez-Muñoz et al., 2016) also contribute to soil carbon sequestration and the more negative olive farming C footprint, and thus in regulating global climate. TPR and TCP provide micro and macro-habitats, which added to the organic carbon they provided, allow an optimal development of biodiversity and microflora, micro, meso and macrofauna, which, when interacting with each other, play a key role in the regulation of pest and diseases in olive groves (González-Ruiz & Gómez-Guzmán, 2019). Finally, the TCP, especially when integrated with livestock, along with other landscape elements, also reinforces some of the ESs related to non-material cultural-aesthetic values. During the last decade, these have started to be exploited economically and now there is a growing market of role tourism where the know-how and a sustainable concept is commercialized adding an added value to the olive oil.

Spontaneous TCP based on in situ soil seed bank is the most frequent strategy mainly due to the lower cost, the scarcity of commercial seed, and the capacity of renewal of the seed bank (Fig. 9.3c). The greatest success has been obtained in low-slope olive grove of valleys with surface tillage and abundant seed banks. However, in olive groves with a recent history of bare soils with herbicides and/or tillage, where soil has suffered some degradation and seed bank is poor (Carpio et al., 2020), the success has been uneven. In these most disturbed soils , strategies aimed at fast remediation of soil, which might include sources of refractory organic carbon locally available (such as composted olive mil pomace and biochar), which improve at the short-term the physical fertility of the soils are desirable.

The seeding of target herbaceous annual species on the soil surface or at minimum soil depth to reinforcement specific ESs is a current tendency. However, success in the use of cereal species and mixture of forage species to convert biomass into meat and milk have been uneven due to the relatively poor soil, variability in the Mediterranean climate, and the necessity of continuous replanting (Hernández González et al., 2015). In olive groves with fairly eroded soils, seeded TCP composed of monospecific native grass is recommended because of the soil loss reduction by their residues which decompose slowly and the low needs for soil preparation. However, at the short-to-medium term, this monospecific native grass tends to show little competitiveness and it is recommended its combination with other creeping legumes with greater germination and covering capacities (Guzman & Mielgo, 2001; Siles et al., 2016), lowering the risk for plant exotic invasions and soil compaction (Fig. 9.3d). Some studies have shown that the persistence of the legumes in this type of mixture is obtained exclusively by grazing or mechanical cutting (Rodríguez et al., 2013). During the last decade some *Brassicaceae* has been seeded due to its growing rate and high net primary production (Alcántara et al., 2009). Species like *Sinapis alba* subsp. *mairei* y *Eruca vesicaria* have shown high adaptability potential to low organic matter and high calcium carbonate soils, typically found in many areas of Andalusian. The majority of the olive farmers use these

Fig. 9.3 (**a-d**) Olive groves with different strategies of temporary spontaneous cover crops. (**a**) Arrangement of spontaneous cover crops between olive tree lines perpendicular to the maximum slope of the terrain. (**b**) Detail of a monospecific seeded cover crop of Brassicaceae. (**c**) Detail of a mature, high-diverse spontaneous cover crops. (**d**) detail of a spontaneous cover crops with predominance of grasses (*Bromus* sp.) and legumes (*Medicago* sp.)

brasicáceas for their role as biological control of some diseases, such as *Verticilosis* (Fig. 9.3b). Under these, tillage is the main control strategy; otherwise, they might become permanent.

Finally, the seeding of a mixture of heterogeneous wild species has become a long-term proposal (Gomez et al., 2014; Jiménez-Alfaro et al., 2020), as not only enhance the reduction of soil loss and improve soil nutrient retention, C sequestration and biological control of pest and diseases (Alcalá et al., 2019) but also improve other services such as biodiversity (Carpio et al., 2018; Alignier et al., 2020) and pollination. The selection of mixture of annual herbaceous wild species of short life cycles and with capabilities for adaption to a wide range of soil properties and environmental conditions and the characterization of their growing pattern and their capacity to provide nectar and pollen are becoming lines of research to be developed in the coming years.

TCP is usually located in bands between the tree lines, and it is recommended against the maximum slope (Fig. 9.3a). However, in organic production, the TCP are usually spread all over the land. The most common control method is by mechanical weeding, and to a lesser extent, by chemical control, or where appropriate, by sheep (Fig. 9.4a, c, d).

Fig. 9.4 (**a–d**) Examples of integrating livestock into olive farming. (**a**) Comparing an olive grove with bare soil with one with spontaneous cover crops controlled by sheep. (**b**) Using tree pruning branches for sheep. (**c**) Mobile fences with spontaneous herbaceous controlled by chickens. (**d**) Permanent use of spontaneous cover crop by asturcón horse

9.3 Integrating Livestock to Olive Oil Farming

The management of the olive grove as an agrosilvopastoral system where the tree coexists with a stable TCP and livestock (especially sheep, but also hen and horse) is an increasing model of olive farming (Fig. 9.4). Under this management, not only the provisioning ESs is enhanced, converting plant biomass into meat, milk or eggs providing a diversified production, but also the supporting (e.g., a further enhance of soil organic matter and a locally closed nutrient balances), regulating (e.g., biogeochemical cycles and soil erosion), and cultural (e.g., oleotouristic and aesthetic) ESs. In soils of olive farming that have integrated sheep, total and available nutrient deficiencies have not been found in comparison with comparable olive farms without livestock (García Fuentes et al., 2008, 2015; Cebadero et al., 2020). In addition, no significant differences in olive fruit yield have been recorded by integrating sheep. Other observed benefits include: (i) economic savings in the removal of twigs that grow annually in the lower end of the olive trunk, (ii) control of the TCP, and (iii) savings in the fertilizer bill. Sheep stocking density in mature spontaneous TCP is about 2.5 head ha^{-1} (Torres et al., 2013; Rodríguez et al., 2014), although variability can be ample according to pedoclimatic conditions and tree density. Currently, the integration of sheep is more frequent in organic traditional rainfed mountain olive farming of areas where livestock is a tradition. The relatively lower yield potential of these agrosilvopastoral systems is partially compensated with the meat. In addition, relatively high landscape complexity and diversity is also a feature of these models of olive farming, allowing to the regular movement of the sheep between the olive grove and complex matrix of surrounding natural landscape, mainly for an extra fed, but also for resting. A good management includes a minimum of infrastructures such as farm fencing and watering holes (Fig. 9.4).

The integration of the livestock in extensive areas of simplified low-slope olive groves is rather limited due to the long distance and scarcity of diverse grazing areas, especially in periods of low available food, such as summer. In these olive areas, sheep are typically integrated only organic olive oil farming.

9.4 Organic Fertilization

Organic fertilization in olive farming is key to promote the increase in abundance and diversity of soil organisms by enhancing habitat conditions and resources (i.e., organic carbon and macro and micronutrients). In their role as regulators of soil ecosystem processes, soil organisms perform a number of vital functions which include; (i) organic matter processing to render available nutrients, (ii) promotion of P acquisition by mycorrhizas, (iii) improve stabilization of soil aggregates and water infiltration, (iv) soil chemical weathering, v) atmospheric N fixation, (vi) promotion of humification, and (vii) pests and diseases control, among others. In addition, the balance of nutrients often found in organically fertilized soils promotes optimum conditions for plant growth, prevents pulses of available nutrients in plant tissues, and reduces palatability or preference by pests (Altieri & Nicholls, 2003).

Despite the role of organic fertilization in enhancing the supporting and regulating ESs throughout the promotion of soil microbial activity and closing nutrient cycling at landscape scale, the majority of the olive farms of Andalusia are not organically fertilized. This is mainly due to the high degree of dispersion of livestock farms relative to the high olive grove extension resulting in low manure availability, the low degree of mechanization that makes it difficult to spreading the manure, the relatively high transport cost of the manure, and the lack of clear protocols. Organic fertilization is almost exclusively spread in organic olive farming, which accounts for about 8% of the Spanish olive area (MAPA, 2019), especially in mountainous areas with livestock tradition (Fig. 9.5). There is high variability in the type of organic fertilizer, period, doses and frequency and there are no statistics at these respects. Typically, sheep, hen, and to a minor extent horse, manures are applied below tree canopies when mechanization is not possible, or it is distributed more or less homogeneously in the band between tree lanes when mechanization is possible. Organic fertilization is typically done each 2 years with an equivalent annual dose of between 2 and 5 Mg $year^{-1}$ ha^{-1}.

Olive mill pomace co-composted with other locally available raw materials, such as olive leave, manure, and straw, is currently being use not only for recycle,

Fig. 9.5 Composted olive mil pomace applied to an olive groves and piles of manure before their spreading

recirculate, and reuse nutrients of byproducts of the olive oil industry but also as organic fertilizers in olive groves near by the composting plants. About 150,000 Mg of composted olive mill pomace is being currently produced in Andalusia annually, which accounts for less than 10% of the olive mill pomace produced in this region. The application of composted olive mill pomace (Fig. 9.5) has been demonstrated to enhance many of the supporting and regulating ESs in olive groves (García-Ruiz et al., 2012).

9.5 Promoting Landscape Diversity

As demonstrated by Landis (2017), landscape features that permit olive production whilst simultaneously enhancing soil function and associated ecosystem services are rare, and thus highly sought after. Currently, the dominant olive landscapes in Andalusia are characterized by simplified, low-diversity, and low-complexity landscapes with olive groves with semi and intensive tree densities. Most of the current olive cultivation landscapes have emerged as the result of policy and market forces that drive farmer decisions about what to grow and how to grow it. The expansion and intensification of the monoculture models of olive cultivation have taken place at expense of others crops, mainly cereals, driven from the EU subsidiary agrarian politics, resulting in the regional specialization of olive cultivation and landscape simplification. While individual olive farms may be highly efficient gronomically speaking, the simplified landscapes that emerge are losing functionality and contributing to disservices.

On the other hand, many areas of the north and east of Andalusia remain with complex landscapes which include terraced traditional olive groves within a matrix of woody hedgerows and patches of natural or seminatural Mediterranean of forest scrublands and prairies found in mountainous or semi-mountainous terrain. The latter are considered marginal olive groves on steeper slopes and with relatively low yield.

Given the need for productive and sustainable forms of olive oil cultivation and the evidence that intensified systems are failing to conserve key ecological functions, future olive oil landscapes will likely need to be explicitly designed to more effectively support biodiversity and ESs. To stem the loss of functions will require actions that alter landscape structure at scales larger than individual olive farms and suggest that mechanisms for planning and coordination will be required. For example, even if many farmers were willing to make individual changes, it is unlikely that the negative aspects of landscape-level intensification can be mitigated merely through uncoordinated olive farmer decisions. In contrast, analyzing specific landscapes and implementing a coordinated landscape design presents the opportunity to alleviate structural deficits in an efficient manner.

The goals and methods for design of particular olive landscapes will vary according to their degree of intensification and the mix of desired ESs (Rey et al., 2019). While highly intensified and simplified landscapes rank high in provisioning services (yield and firewood), they frequently provide only low levels of supporting, regulating and cultural services (Landis, 2017). For example, in areas of the NE of the city of Jaén, yields can be exceptionally high but associated rates of soil loss and nutrient export may become unsustainable, whilst pollination and pest suppression are reduced, and recreational opportunities limited. In such landscapes, a design goal may be to restore their integrity so that production rates can remain high while mitigating disservices. It seems plausible to suggest that moderately intensified landscapes may provide a more balanced set of services. While yields are typically more modest, other services such as soil retention, water infiltration, and recreational opportunities can become relatively high (Yuan et al., 2009; Van Vooren et al., 2017). The goal for such landscapes may be to increase their overall multifunctionality. Finally, in less intensified landscapes the agronomic yields may be comparatively low but balanced by increased supporting, regulating, and cultural services. The goal for these landscapes may be to increase provisioning services while maintaining current levels of other services; however, some tradeoffs are likely.

References

Alcalá, R., Ruano, F., Gálvez, C., Frischie, S., & Campos, M. (2019). Attraction of green lacewings (Neuroptera: Chrysopidae) to native plants used as ground cover in woody Mediterranean agro-ecosystems. *Biological Control, 139*, 1–8.

Alcántara, C., Sánchez, C., Pujadas, S. A., & Saavedra, M. (2009). *Brassica* species as winter cover crops in sustainable agricultural Systems in Southern Spain. *Journal of Sustainable Agriculture, 33*(6), 619–635.

Alignier, A., Solé-Senan, X. O., Robleño, I., et al. (2020). Configurational crop heterogeneity increases within-field plant diversity. *Journal of Applied Ecology, 57*(4), 654–663.

Altieri, M. A., & Nicholls, C. I. (2003). Soil fertility management and insect pests: Harmonizing soil and plant health in agroecosystems. *Soil and Tillage Research, 72*, 203–211.

Carpio, A. J., Castro, J., & Tortosa, F. S. (2018). Arthropod biodiversity in olive groves under two soil management systems: Presence versus absence of herbaceous cover crop. *Agricultural and Forest Entomology, 21*(1), 58–68.

Carpio, A., Lora, A., Martín-Consuegra, E., et al. (2020). The influence of the soil management systems on aboveground and seed bank weed communities in olive orchards. *Weed Biology and Management, 20*(1), 12–23.

Cebadero, M., Torres-Cordero, J. A., Siles, G., & Fernández-Ondoño, E. (2020). Cambios en el suelo por la utilización de ganado ovino en olivar ecológico. *Spanish Journal of Soil Science, 10*(1), 16–28.

Duran Zuazo, V. H., Rodríguez Pleguezuelo, C.R., Arroyo Panadero, L., Martínez Raya, A., Francia Martínez, J. R., & Cárceles Rodríguez, B. (2009). Soil conservation measures in rainfed olive orchards in South-Eastern Spain: Impacts of plant strips on soil water dynamics. *Pedosphere 19*(4), 453–464. https://doi.org/10.1016/S1002-0160(09)60138-7

García-Fuentes, A., Torres, J. A., & Valenzuela, L. (2008). Aprovechamiento con ganado de la cubierta vegetal del olivar ecológico: estudios de diversidad florística y parámetros edáficos. In: Fernández et al (eds) *Pastos, clave en la gestión de los territorios: Integrando disciplinas, Córdoba* (pp. 137–142).

García-Fuentes, A., Torres, J. A., Siles, G., & Valenzuela, L. (2015). Uso del ganado para el control de cubiertas herbáceas en olivar ecológico. *Revista Pastos, 45*(2), 33–46.

García-Ruiz, R., Ochoa, V., Hinojosa, B., & Carreira, J. A. (2008). Suitability of enzyme activities for the monitoring of soil quality improvement in organic agricultural systems. *Soil Biology and Biochemistry, 40*, 2137–2145.

García-Ruiz, R., Ochoa, V., Hinojosa, B., & Gómez-Muñoz, B. (2012). Improved soil quality after 16 years of olive mil pomace application in olive oil groves. *Agronomy for Sustainable Development, 32*, 803–810.

Gómez, J. A. (2017). Sustainability using cover crops in Mediterranean tree crops, olive and vines – Challenges and current knowledge. *Hungarian Geographical Bulletin, 66*, 13–28.

Gómez, J. A., Campos, M., Guzman, G., et al. (2014). Use of heterogeneous cover crops in olive orchards to soil erosion control and enhancement of biodiversity. In *The earth living skin: soil, life and climate changes*. https://digital.csic.es/handle/10261/159778

Gómez-Muñoz, J., Valero-Valenzuela, D., Hinojosa, M. B., & García-Ruiz, R. (2016). Management of tree pruning residues to improve soil organic carbon in olive groves. *European Journal of Soil Biology, 74*, 104–113.

González-Ruiz, R., & Gómez-Guzmán, J. A. (2019). Agricultural management greatly affects the beneficial entomofauna of the olive groves. *American Journal of Biomedical Science & Research, 1*(3), 144–151.

Guzmán, G. I., & Mielgo, A. M. (2001). *El uso de abonos verdes en agricultura ecológica*. Boletín n° 4.7/01. Ed. CAAE. Sevilla.

Hernández González, M., Jiménez-Alfaro, B., & Galvez, C. (2015). *Are the taxa used as 506 ground covers in Mediterranean olive groves suitable for this conservation agriculture 507 practice?*. In SER-6th world conference on ecological restoration, Manchester, pp. 23–27.

Jiménez-Alfaro, B., Frischie, S., Stolz, J., et al. (2020). Native plants for greening Mediterranean agroecosystems. *Nature Plants, 6*, 209–214.

Landis, D. A. (2017). Designing agricultural landscapes for biodiversity-based ecosystem services. *Basic and Applied Ecology, 18*(2017), 1–12.

MAPA. (2019). *Anuario de Estadística: estadísticas agrarias y de alimentación*. Secretaría Técnica.

Ordoñez-Fernández, R., Repullo, M., Román-Vázquez, J., & González-Fernández, P. (2015). Macronutrients released during the decomposition of pruning residues used as plant cover and their effect on soil fertility. *The Journal of Agricultural Science, 153*(4), 615–630.

Ramos, A. F., & Santos, F. L. (2010). Yield and olive oil characteristics of a low-density orchard (cv. Cordovil) subjected to different irrigation regimes. *Agricultural Water Management 97*(2), 363–373. https://doi.org/10.1016/j.agwat.2009.10.008

Power, A. G. (2010). Ecosystem services and agriculture: Tradeoffs and synergies. *Philosophical Transactions of the Royal Society B, 365*, 2959–2971.

Repullo, M. A., Carbonell, R., Hidalgo, J., Rodríguez-Lizana, A., & Ordóñez-Fernández, R. (2012). Using olive pruning residues to cover soil and improve fertility. *Soil and Tillage Research, 124*, 36–46.

Repullo-Ruibérriz, M. A., Ordóñez-Fernández, R., Moreno-García, M., et al. (2019). Soil nitrate and organic carbon improvement by use of leguminous plants as cover crops in an organic olive orchard. *AGROFOR International Journal, 4*, 41–49.

Rodríguez, V., Díaz, C., Sánchez, M., et al (2014). Integración agroecológica de ovino y olivar de sierra. In: SEAE (Eds.), *Actas XI Congreso de SEAE: Agricultura ecológica Familiar, Soc. Española de Agricultura Ecológica* (pp. 48–51).

Rodrigues, M. A., Pires, J. M., Claro, M., et al. (2013). Instalação de leguminosas anuais de ressementeira natural como cobertos vegetais em olivais de sequeiro. *Pastagens e Forragens, 32*, 85–97.

Rey, P. J., Manzaneda, A. J., Valera, F., et al. (2019). Landscape-moderated biodiversity effects of ground herb cover in olive groves: Implications for regional biodiversity conservation. *Agriculture, Ecosystems and Environment, 277*, 61–73.

Siles, G., Torres, J. A., Ruiz-Valenzuela, L., & García-Fuentes, A. (2016). Germination trials of annual autochthonous leguminous species of interest for planting as herbaceous cover in olive groves. *Agriculture, Ecosystems and Environment, 217*, 119–127.

Soldati, C., De Luca, A. I., Iofrida, N., et al. (2023). Ecosystem services and biodiversity appraisals by means of life cycle tools: State-of-art in Agri-food and forestry field. *Agriculture & Food Security, 12*, 33.

Torres, J. A., García-Fuentes, A., Ruiz, L., Siles, G., et al. (2013). Ganado ovino como herramienta para el control de la cubierta vegetal en olivar ecológico: diversificación de la riqueza. *Ganadería, Revista Técnica Ganadera, 88*, 60–63.

Van Vooren, L., Reubens, B., Broekx, S., et al. (2017). Ecosystem service delivery of Agri-environment measures: A synthesis for hedgerows and grass strips on arable land. *Agriculture, Ecosystems and Environment, 244*, 32–51.

Vicente-Vicente, J. L., Gómez-Muñoz, B., Hinojosa, B., Smith, P., & García-Ruiz, R. (2017). Carbon saturation and assessment of soil organic carbon fractions in Mediterranean rainfed olive orchards under plant cover management. *Agriculture, Ecosystems and Environment, 245*, 135–146.

Yuan, Y., Bingner, R. L., & Locke, M. A. (2009). A review of effectiveness of vegetative buffers on sediment trapping in agricultural areas. *Ecohydrology, 2*, 321–336.

Chapter 10
Market Challenges in Spanish Olive Groves

Manuel Parras Rosa and Francisco José Torres Ruiz

10.1 Introduction

Two particular features have characterized the evolution of the global olive sector in recent years. On the one hand, world demand for olive oils stagnated during the last few seasons, after a notable expansion in consumption from the beginning of the 1990s until 2013–2014. This stagnation is mainly due to a decrease in the consumption of olive oils in the main producing countries, Spain, Italy, and Greece. On the other hand, we have witnessed the expansion of modern, intensive/super-intensive olive groves, which are much more competitive than traditional olive groves (the majority of which are in Spain).

In this context, the Spanish olive oil sector faces significant and varied challenges. Its main challenge is to adequately exercise its global leadership so that it contributes, effectively and sustainably over time, to increasing the wealth and well-being of the society that supports it (Parras & Gómez-Limón, 2017).

This chapter analyses the national and international context of the olive oil market and proposes a set of strategies that the Spanish olive oil sector should implement to address the challenges it faces.

M. P. Rosa (✉) · F. J. T. Ruiz
Department of Business Organization, Marketing, and Sociology,
University of Jaén, Jaén, Spain
e-mail: mparras@ujaen.es; fjtorres@ujaen.es

J. Muñoz-Rojas, R. García-Ruiz (eds.), *The Olive Landscapes of the Mediterranean*, Landscape Series 36,
https://doi.org/10.1007/978-3-031-57956-1_10

10.2 National and International Context

10.2.1 Low Prices

In Spain, olive oils have experienced a 46% drop in prices at origin in the last 2 years, falling from 3.71 euros/kg in January 2018 to below 2 euros/kg at the end of 2019. These prices are below the production costs of most Spanish olive groves.

Several factors explain the current price level. The following stand out: (a) the proliferation of substitute products that "imitate" the healthy characteristics of olive oils, such as high-oleic sunflower, colza, and canola oils, (b) an unbalanced value chain, made up of many sellers and few buyers, (c) the absence in the sector of an efficient self-regulation mechanism to balance supply and demand, (d) consumer ignorance and confusion about the differences between the various olive oils, which does not allow them to fairly assess those of the highest quality, (e) the poor market orientation of the producers, which consigns them to be mere providers of raw materials (bulk oil) that have high price volatility, and (f) stagnation of the world demand and a fall in domestic demand in the main producing countries. There has been oversupply, as the European Commission has pointed out: "due to oversupply after the 2018/19 harvest, prices in recent months in the Spanish, Greek and Portuguese markets have been particularly low." This oversupply has pushed prices down. In our opinion, this imbalance between supply and demand is the key explanatory factor for the current price level.

Supply strategies, which will be discussed later, can and must be implemented, but there is one essential: policies should be designed to increase demand, both in emerging foreign markets and in the important mature markets. We highlight among the latter the main producing countries of the European Union, for example, Spain.

10.2.2 Demand Situation and Development Strategies

It is obvious that the development of demand in the main producing countries must be a strategic priority. In the context of worldwide growth in the consumption of oils and fats, stagnant olive oil consumption means the sector's market share decreases year on year (Table 10.1).

Table 10.1 Comparison of the global annual consumption of olive oils and oils and fats

Global consumption (1000s of tonnes)	2013/14	2014/15	2015/16	2016/17	2017/18	2018/19
Olive oils	3.237	3.069	3.164	3.100	3.187	3.262
Oils and fats	199.120	203.470	210.570	217.170	225.810	234.520
Olive oils/ Oils and fats	1.63	1.51	1.50	1.43	1.41	1.39

Mielke (2019). Own design

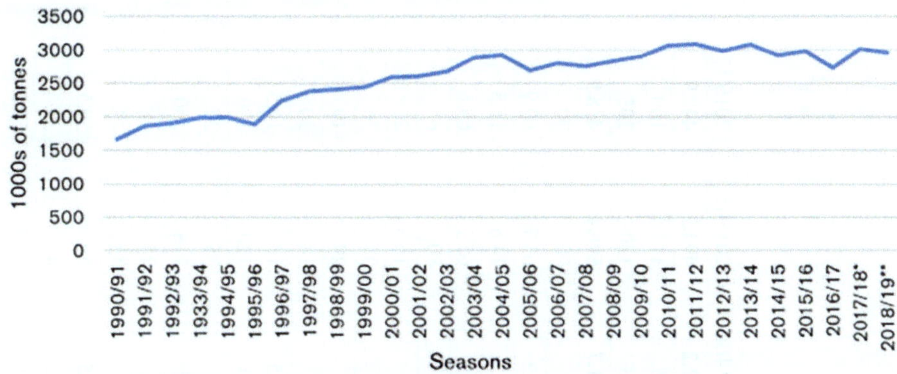

Fig. 10.1 Global evolution of olive oil consumption. (COI, 2019). Own design

This tendency toward stagnation ended the significant increase in demand experienced between the beginning of the 1990s and 2013/2014 (Fig. 10.1).

An analysis of the demand for olive oils in individual countries from season 2008/09 to season 2016/17 (last verified data) confirms that the stagnation in the market is due to a decrease in consumption in the main producing countries, while in emerging countries demand continues to grow steadily (Table 10.2). Indeed, consumption in the EU declined by 460,000 tonnes, mainly due to the fall in Spain (90,700 tonnes), Greece (124,000 tonnes), and Italy (271,100 tonnes). France and Portugal joined the decline, albeit with lower-digit falls.

The general decrease in the consumption of oils, particularly olive oils, in Spanish households is shown also in data published by the Ministry of Agriculture, Fisheries and Food (Table 10.3).

It will be critical for the sector to rapidly reverse this trend and recover the nearly 500,000 tonnes drop in EU consumption. Many hypotheses can be posted on the main causes of this decline: the population has aged and now consumes fewer fats; more and more people eat outside the home and the HORECA channel uses less olive oil than normal households; young people consume many processed products that do not use olive oil, etc.

However, regardless of these factors, something else seems obvious. To increase the demand, the sector will first have to invest in understanding the relevant issues, and then in communication. Making olive oils more attractive to young people, and even incorporating them into the production of processed foods, might be solutions, as would further penetrate the HORECA channel. It should be noted that Spain annually consumes more than 300,000 tons of sunflower oil.

The responsibility for designing demand development strategies lies more with the EU than with Spain alone. In Spain, the Spanish Olive Oil Interprofessional organization, along with public bodies and the private sector, can play a fundamental role. But we believe that the International Olive Council has a huge opportunity to regain its former place as a worldwide promoter of olive oil consumption, a role that it played some years ago with magnificent results.

Table 10.2 Global olive oil consumption in 1000 s of tonnes

Country/Season	2008/09	2009/10	2010/11	2011/12	2012/13	2013/14	2014/15	2015/16	2016/17
Spain	**533.6**	**539.4**	**554.2**	**574**	**486.9**	**524.8**	**492.2**	**494.5**	**442.9**
France	113.5	114.8	112.8	112	113.1	110.6	106	113.4	109
Greece	**229**	**228.5**	**227.5**	**200**	**180**	**140**	**130**	**140**	**105**
Italy	**710**	**675.7**	**660**	**610**	**550**	**641.1**	**571.7**	**598.1**	**438.9**
Portugal	87.5	87.8	82	78	74	75	70	70	70
Germany	47.7	50.1	58.8	61	60.5	66	64.2	62.7	60.9
Holland	11.7	27.2	16.3	14.6	14.4	13.9	13.3	14.6	15.5
United Kingdom	56.4	55.3	69.5	59.2	62	61.3	62.9	65.1	69.6
Algeria	55	33.5	59	42.5	60.5	48.5	65	80	67
Egypt	5	7	5	7.5	12	18.5	20	16.5	22
Israel	31	24.5	33	29	33	35	37	37	36
Jordan	23.5	20	20	17	20	25	22	29	19
Lebanon	20	9.5	20	20	20	18	18	18	20
Libya	15	15	15	15	15	15	15.5	18	16.5
Morocco	70	90	100	122	129	120	120	120	120
Tunisia	21	30	30	35	40	37	30	35	21
Turkey	108	110	131	150	150	105	125	116	150
Saudi Arabia	6.5	6.5	15	16	21	22	25	25	28
Australia	37	44	44	40	37	37	37	42	45
Brazil	42	50.5	61.5	68	73	72.5	66.5	50	59.5
Canada	30	37	40	39.5	37	40.5	37.5	41	39.5
China	15	22	34.5	45	45	40.5	41.5	45.5	50.5

USA	256	258	275	300	287	301.5	295	321	315
Japan	30	40.5	35.5	43	51	54	59	53.5	54.5
Mexico	9.5	9	10	11.5	14	14	15.5	14.5	14
Russia	15	22	21	24	27	30	19	19.5	19.5
Syria	110	120.5	130.5	135.5	160.5	170.5	126	104	98
Total EU-28	**1.863**	**1.852.5**	**1.873.5**	**1.796.3**	**1.627.4**	**1.730.9**	**1.604.7**	**1.660.4**	**1.402.2**
Total	2.831.5	2.902	3.061	3.085.8	2.989.4	3.075.4	2.916.2	2.979.9	2.726.2

COI (2019). Own design

Table 10.3 Consumption of oils in Spanish households. 1000s of kgs

Products	Years					
	2013	2014	2015	2016	2017	2018
Total oils	606,967.61	594,232.90	558,774.00	555,818.15	534,798.86	547,083.41
Total olive oils	422,041.09	412,741.45	372,787.70	373,524.93	341,846.03	355,225.61
Virgin olive oil	n.a.	n.a.	51,626.07	46,357.58	52,539.45	57,371.33
Extra virgin olive oil	112,030.94	109,577.13	101,530.19	106,829.42	114,188.91	122,358.41
Olive oil	237,591.62	223,250.79	211,514.54	220,337.97	175,117.66	175,495.85
Girasol oil	155,746.38	142,360.87	138,978.23	140,775.21	170,500.15	172,920.64
Semilla oil	23,331.39	35,195.80	39,152.53	33,494.89	15,292.76	13,552.83
Orujo oil	4897.17	3043.09	7261.40	7445.17	6461.84	4695.69

MAPA (2019)

10.3 The Challenges Faced by Spanish Oleiculture and Strategies to Boost the Offer

The olive sector has faced significant challenges for years, even when prices were significantly above 3 euros/kg. These challenges require the adoption of supply side strategies and stronger regulatory enforcement by public administrations in relation to the value chain and fraud avoidance.

In this section, some of these challenges are identified and some supply side strategies are proposed. These are based on Gómez-Limón and Parras (2017).

Main challenges include:

1. Increase the competitiveness of the producers' olive groves, oil mills and packaging/bottling plants.
2. Achieve generational replacement (an issue linked to competitiveness).
3. Rebalance the control and influences in the value chain, through inter-company co-operation: mergers, co-operative integration, strategic alliances, etc.
4. Achieve more sustainable olive growing, which will contribute to the challenges of climate change and the energy model. Olive farmers as creators of carbon sinks and renewable resources.
5. Achieve greater market orientation and internationalization. To get closer to consumers, having a greater presence in end markets.
6. The digitalization at the field scale and in the value chain.
7. Professionalization of the management.
8. Diversification of the olive production process, through concentric diversification strategies: oleotourism, oleiculture, agro-energy, etc.

To meet these challenges, we propose the following viable strategies:

(a) *Olive grove intensification*. The conversion of traditional, extensive olive groves into intensive where the terrain is favorable, even in rainfed conditions (intensive rainfed olive farming), or where it is limited (deficient irrigation intensive olive farming). Conversion of traditional, extensive olive groves to super-intensive where the terrain is favorable and there is water available. These measures are not easy to adopt because they collide with emotional issues, they need financial investment and, more importantly, they need available irrigation water. Furthermore, they are difficult to adopt due to the high fragmentation level of the olive farms (an issue we discuss later).

(b) *Assisted cultivation.* Management of olive farms through service provision by specialized companies, with the aim of forming larger agricultural units through which to manage production resources more efficiently. This is key in territories with small, highly geographically fragmented olive farms. Added to this is the increasing age of olive growers and the lack of generational replacement. The creation of "section or farm management units" within the current olive-producing cooperatives is proposed. Through this method, that is, cultivating continuous farms of many hectares, economies of scale could be obtained, thereby reducing costs. In addition, in medium or large-size olive farms, intensification strategies, organic cultivation, etc., should be considered.

(c) *Gaining size.* Company mergers that enable both horizontal and vertical integration (between companies operating at different levels of the value chain) and the establishment of inter-company agreements and strategic alliances are a strategy to gain size. These measures are essential to reduce costs and, above all, to be able to undertake larger commercial transactions. In recent years, we have witnessed various kinds of integration processes: mergers between cooperatives, inter-company joint sale and purchase agreements, and also continued support for, and empowerment of, second-degree cooperatives.

(d) *Differentiate the offer to increase revenue.* This include: (i) the increase of the implementation of quality management and certification systems (public and private), which will facilitate the promotion of differentiation strategies and access to markets that require accreditations: ISO 14001, BRC Food, GLOBALG.A.P, IFS (International Food Standard), etc., (ii) the production of high quality extra virgin olive oils differentiated by their place of origin: Protected Designations of Origin (PDO) and Protected Geographical Indications (PGI), (iii) create higher quality extra virgin olive oils differentiated by their production processes: production based on integrated, organic and biodynamic olive cultivation methods, (iv) production of extra virgin olive oils that link territorial olive cultivation to social externalities, such as territorial vitality, the maintenance of the landscape and heritage, boosting biodiversity and other environmental services, etc., and (v) to increase the general quality of the olive oils produced: bulk qualification.

In this sense, special attention should be taken to increase the importance of the values that companies project to their consumers. We highlight the so-called "hidden attributes" of olive oils. The contribution of the product to the revitalization of rural areas and the role of territorial population "attachment," etc. have gained importance as differentiators; consequently, consumer groups, based on these intangible assets, are willing to pay the producers for oils.

(e) *Related and unrelated diversification.* Many activities can be developed along these lines. The most outstanding includes: (1) Extend "agro-energy olieculture": energy production and the valorisation of biomass, pips, pellets, orujillo and other olive by-products, and elayotechnics; expand the use of biomass derived from the olive groves as compost for the restitution of agricultural soils (increased organic matter content), (2) Extract high added-value compounds from the olive leaves and alperujo (squalene, hydroxytyrosol, maslinic acid, etc.) for use as raw materials in the cosmetic, canning and meat (cold, cured) industries, (3) Develop oleotourism linked to the cultural (historical-artistic, ethnographic, gastronomic, etc.) and the natural heritage (landscape, hiking, ornithology, etc.) of the olive groves and associated industries, (4) Make olive growing compatible with other traditional land uses: livestock (sheep and goats, mainly), other woody crops (fruit), herbs (including more profitable alternatives, such as aromatic and medicinal plants) and hunting activities, (5) Implement, in a general way (in all olive grove types), olieculture based on friendly environmental practices which contribute to reduce erosion (minimisation of runoff), climate change (soil carbon fixing) and pollution (minimization of agrochemical-based contaminants), and the maintenance of cultivation-based soil fertility and biodiversity (flora and fauna), and (6) Increase the acreage of organic olive groves and other environmentally sustainable management models (integrated production, biodynamic and other environmental certifications).

(f) *Digitalization.* Promote olive oil-based electronic commerce by developing more virtual stores and internet-based customer service strategies (CRM, *Customer Relationship Management*); manage social networks to inform and promote olive oil consumption among their users.

(g) *Professionalization.* Although this issue is placed last in the list it should be highlighted that without professionals it will be impossible to implement any of the above strategies. This is as simple as incorporating knowledge into the production process, from specialists in agronomy to marketing experts. It is indicative to refer to what Antonio Gascón and Miramón wrote about this matter in 1929 in their splendid book "Spanish Olive Oil Cooperative Organisation" (Gascón & Miramón, 1998): "The personnel must be well chosen and well paid. Stinginess in this matter is extremely harmful, because it causes a decrease in quality of service. Hitherto, there have been abuses of free services in Spanish cooperatives and, in some instances, overreliance on too many poorly-paid, low-quality employees who, in the event, turn out to be expensive."

10.4 Conclusion

This work presents the challenges facing the Spanish olive oil sector. In addition to promoting strategies for developing domestic demand, olive growers need to reduce costs, and increase income through the differentiation of supply, greater market orientation and internationalization, concentric diversification, digitalization, sustainability and, above all, professionalization.

References

Consejo Oleícola Internacional –COI. (2019). Series Estadísticas del Aceite de Oliva Mundial. https://www.internationaloliveoil.org/. Accessed 24 Sept 2020.

Gascón, & Miramón, A. (1998). *Organización cooperativa de la venta de los aceites españoles*. Sobrinos de Sucesora de M.

Gómez-Limón, J. A., & Parras-Rosa, M. (Eds.). (2017). *Economía y comercialización de los aceites de oliva*. Factores y perspectivas para el liderazgo español del mercado global.

Mielke, T. (2019). *Oil World Annual 2019*. ISTA Mielke GmbH.

Ministerio de Agricultura, Pesca y Alimentación -MAPA- Panel de Consumo Alimentario. (2019). *Bases de datos de consumo en hogares*. https://www.mapa.gob.es/app/consumo-en-hogares/consulta11.asp. Accessed 24 jul 2020.

Parras-Rosa, M., & Gómez-Limón, J. A. (2017). Los retos del sector de los aceites de oliva en España: estrategias para el liderazgo global. In J. A. Gómez-Limón & M. Parras Rosa (Eds.), *Economía y comercialización de los aceites de oliva* (pp. 15–33). Factores y perspectivas para el liderazgo español del mercado global, CAJAMAR.

Part III
Olive Landscapes Across the Mediterranean: Regional Heterogeneity and Case Studies: Olive Groves and Landscapes in Italy

Chapter 11
A Historical and Social Insight of Olive Groves and Landscapes in Italy

Anna Irene De Luca, Teodora Stillitano, Giovanni Gulisano, and Tullia Gallina Toschi

Abstract The first evidences about the extraction of oil from olives date back to the sixth century BC in central regions of Italy. Over the centuries, olive oil became a rare and precious good, and its consumption was mainly destined for liturgical or sacral uses, during masses and sacramental celebrations, while at the end of the medieval period, different uses of olive oil began to assert. Only from the twelfth century, the olive groves assumed dignity and importance as an "income crop." Around the nineteenth century, Italian olive growing took on the appearance of monoculture, and while many territories suffered the poverty caused by wars and the dominance of *latifundium*, olive oil production remained almost the only economic resource for these rural areas. In 1966, the European policies for the agricultural sector and the economic aids partially solved the crisis of the sector and pushed olive farmers to pay more attention to quality, singularity, and biodiversity issues, together with the need to modernize olive groves and make olive oil production more profitable and efficient will be the challenge of the coming years. In this context, olive growing represents one of the most important sources of income and employment for rural economy, although in many territories it is not expressing itself to full potential, generating; therefore, a mix of urgencies and issues to solve in order to make the most of this natural asset, also as expression of significant potentiality in terms of tourism and landscape. In the following paragraph, a brief history about the diffusion of the olive in Italy and its social implications for the national economy and territorial development is provided.

A. I. De Luca (✉) · T. Stillitano · G. Gulisano
Department of Agriculture, Mediterranean University of Reggio Calabria, Reggio Calabria, Italy
e-mail: anna.deluca@unirc.it; teodora.stillitano@unirc.it; giovanni.gulisano@unirc.it

T. G. Toschi
Department of Agricultural and Food Sciences, Alma Mater Studiorum – University of Bologna, Cesena, Italy
e-mail: tullia.gallinatoschi@unibo.it

J. Muñoz-Rojas, R. García-Ruiz (eds.), *The Olive Landscapes of the Mediterranean*, Landscape Series 36,
https://doi.org/10.1007/978-3-031-57956-1_11

11.1 Chronicle of Olive Grove Diffusion in Italy

Olive tree was introduced in Italy in different periods and different areas, with genetically heterogeneous plants (Morettini, 1950). Archaeobotanical findings from Emilia-Romagna identified olive fossil leaves in the Pliocene sands (from 5300 to 2500 million of years ago), pollen and wood charcoal of olive tree in Sicily and Liguria in the Mesolithic period (7000–6000 BC) (Cavara, 1886), while Olea pollen evidences from Puglia and Toscana dated the olive presence in these regions to the 5000 BC (Caracuta, 2019). In the Bronze Age (2000 BC), the olive begins to spread more and more throughout the Italian peninsula, gradually moving from wild to domestic type olive. The discovery of traces of olive oil founded in small crockery containers in Sicily, Calabria, and Puglia confirms the importance of the use of olive also later (1200–1100 BC) when, thanks to the relations with the Aegean populations, the extraction of olive oil spreads in several Italian territories.

However, scholars dated to the sixth century BC the first findings of manufacturers for the extraction of olives located in areas of Etruscan origin, namely, Lazio, Tuscany, and northern Campania (Caracuta, 2019). The Romans extended the olive crop to all the territories of the empire also thanks to the introduction of technological improvements (Bartolini & Petruccelli, 2002; Brun, 2004).

This is testified also by several Latin literary works, written during the third century BC and after, from Marco Porcio Catone, Lucio Giunio Moderato Columella, and Marco Terenzio Varrone. The first two authors paid great attention especially to operational activities such as nursery, fertilization, and pruning—with particular attention to its periods—as well as to the extraction systems such as the mola olearia and trapetum. Varrone considered the olive crop not so expensive, not so challenging, but very profitable and provided suggestions to the owners of olive growers of the time about the manual pruning in order to preserve olive and oil quality, and to protect the health of the trees (Ciuffoletti et al., 2004; Nisard, 1856; Vannucci, 2009).

In the same period, the great progress of olive growing techniques leads to the codification of different species (the current "cultivars"), also thanks to Virgilio, Plinio, Macrobio, and Palladio. At the same time, also different categories of olive oils were defined on the basis of their quality and intended use. Five typologies were distinguished: "Oleum ex albis ulivis," derived from still unripe olives; "Oleum viride" from freshly turned olives; "Oleum maturum," from ripe olives; "Oleum caducum," from olives harvested from the ground; "Oleum cibarium," from olives attacked by parasites, principally for non-food uses (Vannucci, 2009).

After the Roman Empire's decay, the olive growing and the oil trade suffered a serious dip and the production remained totally in the hands of small producers, who, over time, began to opt for other crops. Many olive groves were abandoned, disappeared or became wild, and in other cases, they were destroyed by fire, water and, war devastations. However, a minor contraction was in those territories where soil and climate conditions—e.g., in coastal areas—allowed olive tree to survive spontaneously, even without human care (Pinto, 2002).

Between the fourth and seventh centuries, the political and economic crisis of the European West continued to worsen the Italian olive-growing panorama, leading to a sparse presence of olive trees and a very limited oil production. Olive oil became a rare and precious good, and its consumption was mainly destined for liturgical or sacral uses, during masses and sacramental celebrations. This is the reason for the persistence and care of the olive groves mostly in the land holdings of ecclesiastical entities. The survival of olive cultivation around the year 1000 was due certainly to the commitment of some religious orders who operated to the remediation of marshland and the planting of new trees, giving a new impulse to the olive growing and contributing to shape the Italian agricultural landscape (e.g., with terracing by Benedictines and specialized olive grove by Cistercians) (Ciuffoletti et al., 2004).

According to Pinto (2002) toward the end of the medieval period, different uses of olive oil, from the liturgical one, begin to assert. These included therapeutic and cosmetic use, for lighting house, and successively also for food purposes—until then not widespread due to the high consumption of animal fats—related to new eating habits favoring the consumption of vegetables. Furthermore, olive oil was used in several industries, such as soap production and leather or wool processing (Cortonesi, 2005). Starting from the twelfth century, the olive assumes dignity and importance of "income crop." A strong development occurred especially in southern Italy, due to the high demand by north-central cities for oil employed in the manufacture of textile and soap. In central Italy, particularly in Tuscany, public authorities supported the increase of olive crop extensions with specific laws, but the boost also came from the upper citizen class who saw the land ownership as a right completion of business activities and the olive oil production useful for self-consumption and market reason (Nanni, 1992). In the wake of this great recovery of the olive trees all over Italy (from Val d'Aosta to Sicily), the Renaissance era until the eighteenth century represented the golden age for the national olive growing (Pini, 1980) characterized by technical innovations, studies, treaties, and incentives. A strong specialization related to the different areas of production led to differentiate the most valuable olive oil from Tuscany, Veneto, and Emilia Romagna more specifically for food use, and the so-called "lampante" olive oil—for industrial uses—mainly from the areas of south Italy with higher quantitative production (Vannucci, 2009).

Around the nineteenth century, Italian olive growing took on the appearance of monoculture, and while many territories suffered the poverty caused by wars and the dominance of *latifundium*, olive oil production remained the only economic resource for its cultivation areas (Bigongiali, 2000). In the mid-nineteenth and the beginning of the twentieth century, there was a general decline in Italian olive-growing mainly due to unfavorable weather conditions, which led to the abandonment of many olive groves. For example, along the Reno river and the Bologna hills the olive oil production dropped from about 254 kg in 1811 to 165.5 kg in 1841, when the local population (70,000 inhabitants of the city) consumed about 680.4 tons of olive and seeds oil, thus massively importing it from Puglia, Marche and, above all, Tuscany (Baldini, 2003). At the same time, the processing companies increased, and the production of olive oil underwent a real "industrialization" process aimed, above all, at exporting the product. Shortly afterward, the refining technique also extended to all other vegetable oils, substitutes for olive oil, very competitive in price for the industry, also due to the lower production of olives caused by the frequent frosts of those years (Vannucci, 2009).

In 1966, with the advent of European policies for the agricultural sector and the first regulation for the olive sector, the economic aids partially solved the crisis of the sector, on the one hand. On the other hand, it pushed olive farms to pay more attention to quantity and the compromise of maintaining high quality, singularity, and biodiversity (number of monovarietal, PDO, and PGI oils), together with the need to modernize olive groves and make olive oil production more profitable and efficient, will be the challenge of the coming years.

11.2 Socio-economic Aspects of Italian Olive Growing Farms

As already mentioned, the numerous ancient and modern documentary sources, as well as all the scholars related to olive cultivation show that the presence of olive tree and its cultivation in Italy is steeped in history and tradition and, undoubtedly, its role in the development of heritage, culture, landscape, diet, and health cannot be underestimated. Italy is among the world's strongest producers, with special features in the high number of varieties and the highest number of PDOs and PGIs. Today Italian olive production covers approximately 1,700,000 ha, 80% of which are located in the southern of Italy, where Apulia is responsible for around 37, 000 ha, followed by Calabria and Sicily. These three regions account for more than 60% of Italian olive oil production, even if 17 out of 20 Italian Regions have a relevant production, while in terms of Protected Designation of Origin (PDO)/ Protected Geographical Indication (PGI), the highest amount produced comes from the Tuscany PGI.

The small-scale Italian olive farms represent the main characteristic of the sector, as well as the traditional olive orchards (with less than 200 trees/ha) still covering a high quota of olive area, even in the most suitable olive-growing areas. Of the 907,197 olive farms (for table and for oil), 60.2% have an average farm size of less than two ha of Utilized Agricultural Area [https://www.istat.it/it/agricoltura?dati]. Due to the limited enterprise dimensions, the majority of Italian olive farms are often family owned oriented almost exclusively toward local or national markets (Tudisca et al., 2014).

In this context, olive growing represents one of the most important sources of income and employment for rural economy, although in many territories it is not expressing itself to full potential, generating therefore, a mix of urgencies and issues to solve in order to make the most of this natural asset. This traditional or extensive olive system, which according to Eurostat represents about 53% of the total olive surface cultivated in Italy, is characterized by scarce inputs of labor and resources, structurally limiting factors (slopes, old trees, unfertile soils, and no irrigation), alternation of production, low productivity and then poor profitability (Russo et al.,

2016). Semi-intensive and super-intensive systems account for about 42% and 5% of the Italian olive area, respectively. The Italian olive groves are distributed in two distinct area, marginal olive-growing and suitable olive-growing areas, both differentiated by different territorial contexts, and strictly linked to the evolution of olive tree production through the centuries.

However, in an attempt to describe the Italian olive groves reality from its protagonists' point of view —namely, the olive growing farms —cannot ignore the complexity enclosed in the plethora of different business realities. In this sense, it can be useful to consider multiple factors that allow explaining the diversity between "competitive" olive farms and "marginal" olive farms, the first market-oriented, and the second that meet strategies addressed toward social discourses and of safeguarding the territory. To this aim, an analysis of census data for 902,075 olive farms allowed to draw a picture of the main types of companies in the Italian olive sector (ISMEA, 2014). The methodology considered several indicators, among which: physical size and land use; economic dimension measured by standard output; geographic location; management system; revenue breakdown; related activities; recourse to passive subcontracting; accounting; work of the tenant and family members; gender and age of the manager; off-farm work activities; and soil and agricultural landscape conservation.

Results of ISMEA's study led to identify six clusters of olive farms subsequently classified into three main categories: "marginal" farms or those with low competitive potential (63% of the total); potentially "multifunctional small" business enterprises (26% of the total); and "structured" companies (11% of the total) with high or improved competitive potential. In particular, the first category covers about 52% of the total national olive area and involves the most specialized olive growing farms characterized by extreme marginality, with little probability of medium-term survival, strongly dependent on EU direct payments (and principally located in Umbria, Marche, Abruzzo, Molise, Campania, Basilicata e Calabria, and Sardinia). The second group includes small and multifunctional farms (most localized in Apulia and Sicily), covering about 27% of the national olive-growing area. The last group is related to "structured" farms, market-oriented, with low self-consumption and sales as the main source of revenues, high rate of work dedicates for activities connected to agricultural production, and subcontracting used for agricultural activities on-farm. These companies cover 21% of the national olive area are localized prevalently in center and southern Italy and are defined by ISMEA with high or improved competitive potential (ISMEA, 2014). Regarding more specifically social-economic aspects, according to the last data available from the sixth Agricultural General Census, olive-growing specialized farms are characterized by an advancing aging of their entrepreneurs and a low generational turnover. In particular, 41% of olive companies have a farmer over 65 years old, and only 3% have a tenant lesser to 34 years. Added to this is the low level of education of farmers, which is equal to or lower the middle school (for around 72% of tenant) (ISMEA, 2014). This situation has not changed much over time being still the professionalism of farmers is not adequate and who are not considered real entrepreneurs, and which involves a significant delay in the adoption of technology innovations and missed or few investments (Andrioli & Castellani, 2019; ISMEA, 2019). All this is reflected in a strong difficulty for the Italian olive sector in directing olive growth toward more productive and profitable groves. Only 37% of the 825,000 small and medium olive oil companies present in Italy are considered capable to support the competitiveness of the market (ISMEA, 2019).

11.3 Tourism and Landscape: Key Aspects for Italian Olive Growing Resilience

As argued by Pulido-Fernández et al. (2019), "olive-oil tourism is a booming tourism typology" in all Mediterranean basin countries. Through a deepening review, authors underscore the importance of this specific tourism correlating it with gastronomy, technological and industrial aspect, rural environment, nature, culture, territorial development, and local economy. During the last 10 years, this phenomenon has aroused great interest in Italy, so much to be officially regulated by the state budget law (Repubblica Italiana, 2019), by extending the same provisions already existing, from 2017, for the better known "enotourism activities."

According to the Italian Budget Law in 2020, oil tourism – or oleotourism – enters the Italian legislative framework and is thus equated with wine tourism regulated by Law no. 205 of 27 December 2017. The term "oleotourism" encompasses a range of activities, including the acquisition of knowledge about olive oil production, visits to locations offering insights into the cultural and industrial aspects of olive oil, the observation of tools utilized in olive cultivation, the tasting and marketing of olive oil, and the participation in educational and recreational initiatives situated within the context of olive cultivation and production. From an empirical and recent survey. Garibaldi (2019) has emerged a strong interest of tourists toward different experience typologies linked to the olive oil production process. In particular, it is recorded that 61% of tourists would be very motivated to take part in the harvesting activities in the field, but also during processing to mill oil for self-consumption. In addition, more than half of the respondents to the survey showed interest in the possibility of carrying out artistic activities in the olive groves; as well as, the interest toward the knowledge of the companies, their owners and territories are very relevant. The culinary and gastronomic value turns out to be an attractive element for tourists who want to taste different types of quality olive oils, and also appreciate it with local dishes and specialties (79% of the sample interviewed), thanks to specific suggestions and combinations offered by restaurateurs (Garibaldi, 2019). Other useful insights coming out of the survey are related to the communications activities related to olive oil-themed experiences showing a lack of effective information on the offer, probably also not very attractive and therefore negatively affecting travelers' motivations (Garibaldi, 2019).

According to Pulido-Fernández et al. (2019), olive oil tourism can respond to multiple needs of the modern world. On the one hand, it could provide an income integration for farmers as a useful business diversification strategy; on the other hand, it could answer to an increasing interest about olive oil culture linked to the sustainable use of territorial resources and typical products PDO/PGI of olive-oil regions.

In addition, the survey above mentioned showed a strong appreciation expressed by tourists toward the historic mills (76%) and the landscape of ancient olive trees (70%). This reveals how the landscape value of olive oil areas can be favorably appreciated, thus becoming an economic asset to be enhanced. To confirm this, it is noteworthy that the great variety of Italian rural landscapes involves also beautiful territories linked, from millennia to traditional olive growing practices. According to the official Register of "rural landscapes traditions or historical interest, related traditional practices and knowledge" supported by the Italian Minister of Agricultural, Food and Forestry Policies (MIPAAF), the historical persistence of traditional olive growing, as well as the cultural value are recognized until now for nine Italian sites: Oliveti a bosco di Lucinasco (Liguria); Mosaico paesistico del Montalbano (Toscana); Gole del Farfa and Oliveti terrazzati di Vallecorsa (Lazio); Oliveti della Coroncina (Marche); Oliveti di Loreto Aprutino (Abruzzo); Oliveti di Venafro (Molise); Oliveti delle serre salentine (Puglia); and Oliveti monumentali di Gioia Tauro (Calabria) (Rete Rurale Nazionale, 2019).

11.4 Conclusions

Olive trees were introduced in Italy in different periods, in different areas, with genetically heterogeneous plants. Archaeobotanical findings have occurred over time testifying to a very ancient presence of olive trees in Sicily, Liguria, Calabria, Puglia, Emilia Romagna, Toscana, and Lazio. The presence of olive tree and its cultivation in Italy is steeped in history and tradition and, undoubtedly, its role in the development of heritage, culture, landscape, diet, and health cannot be underestimated. Italy is among the world strongest producers, with special features in the high number of varieties, PDOs, and PGIs. Today Italian olive production covers approximately 1,700,000 ha, 80% of which are located in the southern zone of Italy, where in particular Puglia is responsible for around 370,000 ha followed by Calabria and Sicily. These three regions account for more than 60% of Italian olive oil production. The small-scale Italian olive farms (with an average farm size of less than 2 ha of Utilized Agricultural Area) represent the main characteristic of the sector, as well as the traditional olive orchards (with less than 200 trees/ha) still covering a high quota of olive area, even in the most suitable olive-growing areas. Due to the limited enterprise dimensions, the majority of Italian olive farms are often family owned oriented almost exclusively toward local markets. In this context, olive growing represents one of the most important sources of income and employment for rural economy, although in many territories it is not expressing itself to full potential, generating therefore, a mix of urgencies and issues to solve in order to make the most of this natural asset. This traditional or extensive olive system, which according to Eurostat represents about 53% of the total olive surface cultivated in Italy, is characterized by scarce inputs of labor and resources, structurally limiting factors (slopes, old trees, unfertile soils, no irrigation), alternation of production, low productivity and then poor profitability (Russo et al., 2016). In total, 42% and 5% are the Italian olive surface quotas, respectively, represented by semi-intensive system and super-intensive systems. In other terms, and according to FAO we can refer to the Italian olive groves as two different typologies, marginal olive-growing areas, and suitable olive-growing areas differentiated by territorial contexts and strictly linked to the evolution of olive-tree production through the centuries. Starting from the consideration that Italian olive growing has a multifaceted role as pivot of a complex network of ecological, socio-cultural and economic meanings (Colella et al., 2019), this paragraph tried to deepen and synthetically discuss its chronological evolution in the national territory as well as some its main modern features.

References

Andrioli, A., & Castellani, N. (2019). *L'olivicoltura italiana va rifondata, altrimenti non ha futuro.* L'Informatore Agrario n. 28–29/2019.

Baldini, E. (2003). *Notizie sull'Olivicoltura Bolognese*. Accademia Nazionale di agricoltura.

Bartolini, G., & Petruccelli, R. (2002). *Classification, origin, diffusion and history of the olive.* Food and Agriculture Organization of the United Nations.

Bigongiali, G. (2000). *Il libro dell'olio e dell'olivo*. Calderini Edagricole.

Brun, J. P. (2004). *Archéologie du vin et de l'huile dans l'Empire romain*. Editions Errance.

Caracuta, V. (2019). Olive growing in Puglia (southeastern Italy): A review of the evidence from the mesolithic to the middle ages. *Vegetation History and Archaeobotany*. https://doi.org/10.1007/s00334-019-00765-y

Cavara, F. (1886). *Sulla flora fossile di Mongardino. Studi stratigraficie paleontologici*. Memorie della Reale Accademia delle Scienze di Bologna, Tip. Gamberini e Parmeggiani.

Ciuffoletti, Z., De Palma, F., & Cerere, G. (2004). *Olivo tesoro del Mediterraneo* (p. 208). Firenze, Alinari.

Colella, C., Carradore, R., & Cerroni, A. (2019). Problem setting and problem solving in the case of olive quick dedline syndorme in Apulia, Italy: A sociological approach. *Phytopathology, 109*, 187–199. https://doi.org/10.1094/PHYTO-07-18-0247-FI

Cortonesi, A. (2005). L'olivo nell'Italia medievale. *Reti Medievali Riv, 6*(2), 1–30. https://doi.org/10.6092/1593-2214/190

Garibaldi, R. (2019). *Olio e turismo: un legame da potenziare e valorizzare per diffondere la cultura della qualità.* https://www.robertagaribaldi.it/olio-e-turismo-un-legame-da-potenziare-e-valorizzare-per-diffondere-la-cultura-della-qualita/

ISMEA. (2013). *Le aziende olivicole nel 6° Censimento Generale dell'Agricoltura.*

ISMEA. (2014). *Le aziende olivicole nel 6° Censimento Generale dell'Agricoltura: un'analisi delle tipologie di aziende*. Piano di Settore Olivicolo-Oleario.

ISMEA. (2019). *Schede di Settore. Olio di Oliva.*

ISTAT. (2010). *6th General census of agriculture.*

Morettini, A. (1950). *Trattato di Olivicoltura.* REDA.

Nanni, P. (1992). L'olivo e l'olio nelle proprietà dei Medici (sec. XV). *Riv Stor Agr, XXXII*(2), 143–156.

Nisard, M. (1856). *Les agronomes latins: Caton, Varron, Columelle, Palladius* (p. 676). Firmin Didot frères.

Pini, A. I. (1980). Due colture specialistiche del Medioevo: la vite e l'olivo nell'Italia padana. In V. Fumagalli & G. Rossetti (Eds.), *Medioevo rurale. Sulle tracce della civiltà contadina* (pp. 119–138). Il Mulino.

Pinto, G. (2002). Olivo e Olio. In G. Pinto, C. Poni, & U. Tucci (Eds.), *Storia dell'agricoltura Italiana. Volume II. Il medioevo e l'età moderna* (pp. 489–498). Firenze, Edizioni Polistampa.

Pulido-Fernández, J. I., Casado-Montilla, J., & Carrillo-Hidalgo, I. (2019). Introducing olive-oil tourism as a special interest tourism. *Heliyon, 5*(12), e02975. https://doi.org/10.1016/j.heliyon.2019.e02975

Repubblica Italiana. (2019). *Legge 27 dicembre 2019, n. 160. Bilancio di previsione dello Stato per l'anno finanziario 2020 e bilancio pluriennale per il triennio 2020–2022*. GU Serie Generale n.304 del 30-12-2019—Suppl. Ordinario n. 45.

Rete Rurale Nazionale. (2019). La valorizzazione dei paesaggi rurali del catalogo nazionale dei paesaggi rurali storici: schede sintetiche dei paesaggi rurali per la creazione delle mappe attraverso l'applicazione google my maps.

Russo, C., Cappelletti, G. M., Nicoletti, G. M., Di Noia, A. E., & Michalopoulos, G. (2016). Comparison of European olive production systems. *Sustainability, 8*(8), 825. https://doi.org/10.3390/su8080825

Tudisca, S., Di Trapani, A. M., Donia, E., Sgroi, F., & Testa, R. (2014). The market reorientation of farms: The case of olive growing in the Nebrodi area. *Journal of Food Products Marketing, 21*(2), 179–192. https://doi.org/10.1080/10454446.2013.843484

Vannucci, S. (2009). Storia dell'olio. In M. Pisante, P. Inglese, & G. Lercker (Eds.), *L'ulivo e l'olio* (p. 768). Bayer CropScience.

Chapter 12
Pedoclimatic and Landscape Conditions of Italian Olive Groves

Annunziato Scaramozzino, Giorgio Pace, Enrico Valli, and Patricia García Salas

Abstract The Italian territory is characterized by a large variability in climate, landscape, contour, and soil. The context is highly heterogeneous, and olive cultivation, particularly in the South-Center but almost everywhere, since thousands of years, is highly diversified. It is therefore difficult to define a single "olive cultivation model" for Italy.

12.1 Extension of Olive Cultivation in Italy

The olive tree is an element that has characterized the natural and anthropic Italian landscape for thousands of years. Italian olive production covers approximately 2,320,000 ha, 80% of which are located in southern Italy (http://dati.istat.it/ accessed March 2021). Besides this, Italy ranks second in the world (after Spain) for production, with an average oil quantity (over the last 4 years) of 550,000 tons, mainly represented by extra-virgin and virgin olive oils. Puglia represents the most important region, with about 250,000–300,000 tons of olive oil per year (Calabrese et al., 2016).

The cultivation of the olive tree in Italy, characterized by a pluri-millennial history, represents an essential element of the landscape as well as of the economy and rural societies of the Center-South and the islands, especially in marginal areas. Although it is very widespread in almost all regions, there are significant differences

A. Scaramozzino (✉) · G. Pace
UNAPROL Consorzio Olivicolo Italiano, Rome, Italy
e-mail: a.scaramozzino@unaprol.it

E. Valli · P. G. Salas
Department of Agricultural and Food Sciences, Alma Mater Studiorum – University of Bologna, Cesena, Italy
e-mail: enrico.valli4@unibo.it; patricia.garciasalas@unibo.it

J. Muñoz-Rojas, R. García-Ruiz (eds.), *The Olive Landscapes of the Mediterranean*, Landscape Series 36,
https://doi.org/10.1007/978-3-031-57956-1_12

in its local distribution due to the wide latitudinal, altitude, and pedoclimatic conditions that characterize the Peninsula.

Thanks to a long selection action carried out by olive growers and its ability to adapt even to very variable ecological conditions, the olive tree is currently present in cultivation in 19 Italian regions (with the almost total exception of Valle d'Aosta), giving rise in each of these cultivation systems specifically adapted landscapes and very diversified in varietal, structural and territorial distribution terms.

The Italian UAA (utilized agricultural area) devoted to olive oil production represents around 8.7% of the total. However, this average figure varies considerably if the macro-regions are examined separately. In the North, olive cultivation only represents the 1.5% of the UAA, whereas values are 8.4% in the Center, 15.7% in the South, and 7.7% in the islands. From the point of view of surface altimetric distribution, 61% of the olive tree UAA is found in the hills, 28% in the plains, and 11% in the mountains (Dal Ferro & Borin, 2017).

Italy has an olive growing area of over two million ha, of which 21% is irrigated while the rest is rain-fed. The most important olive growing areas are in the Puglia region, which accounts for 45% of the total olive growing area: Calabria (19%), Sicilia (10%), Campania (7%), Lazio (5%), Toscana and Sardegna (3%, respectively), and Basilicata and Umbria (2%, respectively); with the rest of the surface area found in Molise, Liguria, Marche, and Veneto.

The most important olive cultivation areas in Central-South Italy belong to the province of Bari, which stands out from the rest with almost 200,000 ha of olive groves. Bari is followed by other provinces in the Southeast of the country, such as Lecce, Brindisi, and Foggia, all of them belonging to the Puglia region. Another important region is Calabria in the Southern limits of the Italian Peninsula, where the province of Cosenza accounts for more than 100,000 ha of olive groves (http://dati.istat.it).

The classified organic area covers approximately 176,000 hectares, 16% of the national total (in 2017). The region with the largest surface area is Puglia with 55,000 hectares, 32% of the national total and 15% of the total regional surface area covered by olive trees. The second region with the greatest surface area converted to organic farming is Calabria with 51,000 hectares (ISMEA, 2019).

The olive farms practicing the organic system represent 2.8% of the total olive farms and cover about 12% of the national olive UAA. They represent 57.8% of the total of Italian organic farms and cover an area equal to 17.2% of the national organic UAA. The area that olive farms on average dedicate to organic farming is 5.4 ha, much higher than the average area of olive farms in general, but much lower than the national average organic area (18 ha).

12.2 Heterogeneity for Climates and Soils

The Italian territory is characterized by climatic variability, with conditions ranging from temperate to Mediterranean, due to the interaction of continental climate (northern and central-northern area) with that of the Mediterranean basin (central, southern, and islands). Olive groves extend mainly in the humid subtropical regions (e.g., Lombardy, Veneto, Emilia-Romagna, Apulia, Umbria), hot-summer Mediterranean regions (e.g., Liguria, Tuscany, Lazio, Apulia, Calabria, Sicily), and Oceanic climate areas (e.g., Abruzzo, Molise) of Italy.

The production of North of Italy is concentrated in areas characterized by particular climatic conditions, such as Liguria and the areas surrounding the lakes of Garda and Iseo, while in the Center and—mostly—in the South, the crop is almost ubiquitous from sea level to the medium hilly belt. Over half of the Italian oil production comes from Puglia (52%), followed by Calabria (13%) and Sicily (11%), characterized by a hot summer Mediterranean climate (Orlandi et al., 2020).

The territory has multiple altitudinal bands (plains, hills, and mountains), starting from sea level up to the highest mountains in the north. This contributes to the territorial variability in which olive trees are cultivated (Righini et al., 2001). The soils include Cambisols and Andosols with Regosols in Lazio and Campania, Leptosols with Podzols and Luvisols in Sardinia, Cambisols and Luvisols with Leptosols in Calabria, and Cambisol, Vertisol, Luvisol with Fluvisols in Puglia (Costantini et al., 2012). Olive growing is mainly developed on soils with low organic matter content, silty-loamy soils, and clay-loamy with good water retention (see Chap. 14) (Costantini, 2015). It is possible to distinguish between those areas where the olive groves take advantage of the best conditions for climate and soil (average texture, good water retention, clay below 40%), and the areas where the environment for olive growing is marginal, with low fertility or other unfavorable conditions. Cultivation of olive is possible also on soils that are superficial and rich in skeleton (coarse), calcareous, with steep slopes. On the other hand, it is in such areas that olive growing shows its multifunctionality, defining the landscape and preserving the rural environment. The tolerance of olive trees toward pedological features allows the extreme diversity of Italian landscapes, thus potentially contributing to food security (Kapur & Erşahin, 2014). According to average data, about 60% of olive-growing areas occur in the hills, 30% in plains, and 10% in mountains (Loreti, 2012).

Lowland olive-growing areas of Southern Italy, for example, are today characterized by an olive monoculture derived from an original polyculture system that saw the promiscuous cultivation of the olive prevail with other fruit trees or annual herbaceous plants alternating with rest and with pasture. There, the presence of old traditional olive groves is often characterized by very wide planting layouts and predominance of monumental plants up to over 15 m high, alongside specialized high productivity cultivation systems (intensive and super-intensive plants).

By overlaying the climate and orographic conditions with the soil characteristics of Italy, it has been identified 43 unique soil and climatic units (Dal Ferro & Borin,

2017). In this context, the northern regions, with an average of 7 soil and climate units, show the lowest territorial diversification, about half of that of the Center (an average of 14.3), south (11), and islands (12). The degree of diversification is not related to the territorial extensions of the regions.

This study indicates that irrespective of the management, olive groves in the Umbria region of Italy are characterized by a high level of soil C storage, if compared to those growing in other areas and to forest ecosystems, indicating that the practices adopted in the area are not negatively affecting SOC (soil organic carbon) stock. Therefore, maintaining olive cultivation would be recommended in the region in order to maintain the high soil C stock in the area, avoiding its return to atmosphere as a consequence of land use conversion to arable land or the adoption of different soil management practices such as soil tillage (Bateni et al., 2019).

12.3 Heterogeneity of Italian Olive Agroecosystems

Currently, this historical landscape is widely jeopardized by the need to increase profitability through intensive cultivation system. Furthermore, the spread of *Xylella fastidiosa* in some key production areas such as Salento, in the Puglia region, is questioning the survival of the olive sector itself. At the same time, an alarming trend toward crops forsaking is occurring in the marginal hilly and foothill areas of the whole country. Those same areas have been qualified for centuries by promiscuous cultivation and by the simultaneous presence of capillary hydraulic agricultural arrangements. Such trend began in the second half of the 1900s and continues today, along with the progressive urbanization of the population: the consequence is the intensification of fires and the spread of phenomena of hydrogeological degradation, desertification, or re-naturalization with consequent loss of production area. In the hill areas where instead, especially in Central Italy, the agronomic operating conditions (slope, soil, climate, etc.) and the socio-economic context result more favorable to production rationalization processes, traditional low-input olive growing is progressively, albeit very slowly, moving toward semi-intensive systems aimed to reduce costs and increase productivity.

This process, whose impact on the traditional landscape becomes more and more evident, involves the intensification of the plants, the adoption of new soil management, fertilization, and defense techniques, sometimes the introduction of irrigation, the reduction of the volume and the height of the crown through appropriate pruning operations, in order to rationalize and mechanize as far as possible all agronomic practices. Therefore, it is still possible today to find in the suitable hilly areas with a few dozen plants per hectare where the olive trees are cultivated in free form and still resist traces of cultivation promiscuity and at the same time to plants in specialized cultivation of olive trees of maximum 4–5 m in height bred in obligatory form like the polyconic pot, in densities reaching 500 plants/ha.

As can be seen from this brief overview, in Italy there is not a single olive tree agroecosystem (Fig. 12.1), but rather cultural models coexist along the Peninsula as

Fig. 12.1 Heterogeneity of olive groves in Central Italy

pieces of a heterogeneous and complex puzzle that declines along latitudinal, altitude, and also socio-economic gradients and has diversification as the only common denominator (Barbera et al., 2004).

12.4 The Italian Olive Landscape Between Challenges and Opportunities

Today, the olive tree landscape in Italy is facing a challenge that involves the entire olive sector: on the one hand, there is the growing pressure of the market that pushes toward the intensification of cultivation, the introduction of nonnative varieties, simplification and massive use of chemical and mechanical inputs; on the other hand, there is the need to safeguard the multifunctionality of traditional olive growing, the biodiversity of the territories, the environmental benefits of low-intensity agricultural practices, the typical and organoleptic richness of the Italian varietal heritage, the cultural value that the Italian olive landscape, in its multifaceted heterogeneity, still represents today.

The Italian olive sector is facing the complex challenge of improving the performance of production while simultaneously preserving the features of this landscape and the integrity of its ecosystem and cultural services it provides.

This challenge can be overcome, on the one hand, by means of the development of techniques, models, and systems that make it possible to rationalize traditional olive growing without dissolving its identity, on the other hand through the valorization of the diversity, territoriality, and heterogeneity that are the identifying traits of the Italian olive landscape. In fact, in the context of an increasing attention to sustainability and to the certainty of high-quality food-origin products, Italy plays a key role in Europe and in the world because of its historical vocation for the production of foods closely related to territories of origin. It is not a coincidence that Italy is the country that boasts the highest number of designations of origin (PDO and PGI) products, of which 48 are extra virgin olive oils and 4 table olives.

Therefore, it increases the number of European consumers who consider important purchase drivers not only the quality of the product but also its association with a close territorial link and—possibly—with virtuous agricultural practices that contribute to the maintenance of biodiversity, the uniqueness of the landscapes, environment, and cultures of origin (Dal Ferro & Borin, 2017).

It is also well known that the sensory attributes of food products can be influenced by a number of variables, such as the technological process, cultivars, and agronomic techniques, but also the physical environment of production plays an important role. The enhancement of the cultivation vocation of a territory can be one of the most effective tools for the protection of the quality and "typicality" of food, the landscape tout-court, as well as the soil resource from the risks of degradation. Therefore, the awareness that the quality of a food product is closely linked to a particular environment can and must constitute the basis for a general concern about the conservation of the characteristics and quality of this environment, both by consumers and by farmers who intend to add a further value to the image of the product sold on the market. For this, the potential contribution of products with designation of origin to the protection of the Italian olive landscape can be decisive. In fact, these products can improve the economic performance of olive farms by allowing access to markets where the specific quality characteristics and reputation of the oil are recognized, appreciated, and finally rightly remunerated, making even more traditional production systems economically sustainable, which otherwise would not be able to withstand international competition.

Therefore, the premium price paid by consumers for products with designation of origin, together with the specific prizes provided by the Community Agricultural Policy, can provide an incentive for farmers not to forsake traditional olive growing practices completely and to adopt more environmentally respectful practices, effectively acting as compensation for the ecosystem services provided.

Finally, although the system of designations of origin itself cannot be considered as an instrument of environmental or landscape protection, nevertheless it could potentially play a positive role in Italy in this sense, acting as a barrier to the growing intensification of the olive oil sector and preserving traditional agricultural systems with their corollary of positive externalities (Belletti et al., 2015).

References

Barbera, G., Inglese, P., & La Mantia, T. (2004). *La tutela e la valorizzazione del paesaggio colturale dei sistemi tradizionali dell'olivo in Italia.* In Atti del Convegno Europeo: "Il futuro dei sistemi olivicoli in aree marginali: aspetti socio-economici, conservazione delle risorse naturali e produzioni di qualità", Matera 12–13 October 2004, pp. 59–76.

Bateni, C., Ventura, M., Tonon, G., & Pisanelli, A. (2019). Soil carbon stock in olive groves agroforestry systems under different management and soil characteristics. *Agroforestry Systems.* https://doi.org/10.1007/s10457-019-00367-7

Belletti, G., Marescotti, A., Sanz-Cañada, J., & Vakoufaris, H. (2015). Linking protection of geographical indications to the environment: Evidence from the European Union olive-oil sector. *Land Use Policy, 48*, 94–106. https://doi.org/10.1016/j.landusepol.2015.05.003

Calabrese, G., Abdellatif, O., & Perrino, E. (2016). Correlations between organic and conventional management, on-field biodiversity and landscape diversity, in olive groves in Apulia (Italy). *Advances in Plants & Agriculture Research, 5*, 568–583.

Costantini, E. A. C. (2015). La componente pedologica del terroir. *Italus Hortus, 22*, 15–30.

Costantini, E. A. C., L'Abate, G., Barbetti, R., Fantappiè, M., Lorenzetti, R., & Magini, S. (2012). *Carta dei suoli d'Italia 1:1,000,000.* Firenze 2012. www.soilmaps.it. Accessed Mar 2021.

Dal Ferro, N., & Borin, M. (2017). Environment, agro-system and quality of food production in Italy. *Italian Journal of Agronomy, 12*(793), 133–143.

ISMEA. (2019). *Scheda di settore 'olio di oliva'.* http://www.ismeamercati.it. Accessed July 2020.

Kapur, S., & Erşahin, S. (Eds.). (2014). *Soil security for ecosystems management.* Springer. https://doi.org/10.1007/978-3-319-00699-4

Loreti, F. (2012). *La gestione del suolo.* Volume VIII Collana divulgativa dell'Accademia. Accademia Nazionale dell'Olivo e dell'Olio, Spoleto. ISSN 2281–4930.

Orlandi, F., Rojo, J., Picornell, A., Oteros, J., Perez-Badia, R., & Fornaciari, M. (2020). Impact of climate change on olive crop production in Italy. *Atmosphere, 11*, 595. https://doi.org/10.3390/atmos11060595

Righini, G., Costantini, E. A. C., & Sulli, L. (2001). La banca dati delle regioni pedologiche italiane. *Bollettino della Società Italiana Scienza del Suolo, 50*, 261–271.

Chapter 13
Varieties and Olive Grove Production Systems in Italy

Annunziato Scaramozzino, Giorgio Pace, Nelson Marmiroli, Caterina Agrimonti, and Elena Maestri

Abstract One of the strengths in Italian olive cultivation is the unique varietal heritage. Currently, it is estimated a number of 540 olive cultivars is 42% of the world's total. Considering the cultivation systems, there is a sharp prevalence of traditional or extensive and semi-intensive olive groves with respect to super-intensive models.

One of the main strengths of Italian olive growing is the unique varietal heritage in the world, the result of a multi-millennial work of selection, propagation, and diffusion of cultivars based on their productivity, agronomic characteristics, resistance to pests and diseases and adaptation to environmental stresses (Fig. 13.1).

The extreme variability of the Italian territory in terms of orography, latitude, soil, and climate characteristics has also contributed to a widespread diversification of the genotypes making the variety composition in the regions highly differentiated. This complexity creates problems in the exact determination of the number of Italian cultivars. It is estimated the existence of more than 1300 synonyms and the number of varieties stands at around 540 (42% of the world total), of which about 50 contribute to over 80% of national production (Caruso et al., 2009).

This differentiation means that the Italian olive biodiversity is by far the largest on the planet, far exceeding that of the other world production areas. It allows the country to boast 48 oils with PDO (42) or PGI (6) certification as well as countless multi- and mono-variety oils expressing an inimitable range of hints, flavors, sensory nuances, and degrees of intensity.

A. Scaramozzino (✉) · G. Pace
UNAPROL Consorzio Olivicolo Italiano, Roma, Italy
e-mail: a.scaramozzino@unaprol.it; g.pace@unaprol.it

N. Marmiroli · C. Agrimonti · E. Maestri
University of Parma, Interdepartmental Center for Energy and Environment (CIDEA), Parma, Italy
e-mail: nelson.marmiroli@unipr.it; caterina.agrimonti@unipr.it; elena.maestri@unipr.it

J. Muñoz-Rojas, R. García-Ruiz (eds.), *The Olive Landscapes of the Mediterranean*, Landscape Series 36,
https://doi.org/10.1007/978-3-031-57956-1_13

Fig. 13.1 The number of Italian olive cultivars is very high

Although some varieties originate from limited areas, they have a national diffusion today. However, most of the cultivars remain closely linked to the territory of origin or else they have been introduced in areas that show a substantial homogeneity of climatic and pedological conditions with those of origin.

About 42% of the total Italian olive-growing area, equal to about 1.1 million ha, is occupied by the ten most widespread varieties in the country: Coratina, Ogliarola Salentina, Cellina di Nardò, Carolea, Frantoio, Leccino, Ogliarola Barese, Moraiolo, Bosana, and Cima di Mola. Coratina, in particular, with 80 thousand hectares in Puglia and around 10 thousand in the other regions, makes up about 9% of the national olive-growing area; the areas occupied by Ogliarola Salentina (70 thousand ha) and Cellina di Nardò (60 thousand ha) are almost equally large, but they are widespread exclusively in Puglia. Among the other Italian cultivars with a diffusion > 10,000 ha, we mention Dolce di Rossano, Ogliarola Messinese, Ottobratica, Sinopolese, Nocellara del Belice, Canino, Carboncella, Itrana, Biancolilla Moresca, Rotondella, Taggiasca, Tondina, Grossa by Gerace, and Nocellara Etnea (Alfei et al., 2013; Pannelli & Perri, 2012; Perri et al., 2011).

Besides those listed above, many other varieties that are almost exclusively local, and often not yet fully studied, contribute with their genetic background to constitute a valuable potential reservoir of useful characteristics for productivity, adaptability to new cultivation techniques, and—above all—resistance to abiotic and biotic stresses.

This reservoir is invaluable, especially in today's context, in which the trend toward an increase in average temperatures, as well as changes in rainfall regimes, can actually represent a significant threat to Mediterranean olive-growing, together with the risk of contamination by pathogenic organisms which did not co-evolve—made more plausible than ever by the intensification of global trade (see the case of *Xylella fastidiosa* in Puglia).

13.1 Main Features of the Most Renowned Italian Olive Cultivars

Some of the most representative cultivars (cv) of the various olive-growing areas of the country are briefly described in Table 13.1. For the organoleptic profile of each cv, the reference is to the website www.olimonovarietali.it [Accessed July 2020].

13.2 Production Systems and Current Trends

The heterogeneity of the Italian varietal heritage and the adaptation over thousands of years of cultivation techniques to environmental and soil and climatic conditions and to the economic and social dynamics have determined in Italy such a level of diversification and variability of the olive system that it is not possible to define a homogeneous Italian olive growing model (Fig. 13.2 and 13.3) (Russo et al., 2016; Stillitano et al., 2017).

The diffusion in the Country of different production model systems could be summarized in Table 13.2.

The growing need to increase the national production of oil and at the same time to guarantee the economic sustainability of olive-growing, raising the quality standards of the product to meet the needs of the market, have led in recent years to an increase in exhortations to renew the groves and to replicate where possible the super-intensive model.

At the moment, however, the application of this model, in addition to the concerns it raises from the environmental impact point of view, to the limited number of cultivars that can be used and to its long-term economic sustainability, has serious structural limitations almost everywhere in Italy: it requires the availability of adequate surfaces, of mills equipped to transform large quantities of olives in a short time, of a structured subcontracting system to perform some operations such as harvesting by straddle harvesters, of a strong organization of the territorial production chain, aimed at the concentration and enhancement of the oil offer (Fiorino, 2003).

At the same time in the hilly areas especially in central Italy, where conditions are favorable for the rationalization of production (thickening, change of training systems to encourage greater mechanization of operations, grassing, establishment of new groves, recovery of abandoned olive groves), an effort has been made with greater success to promote intensive olive growing that can combine cost reduction and plant productivity increase with the protection of biodiversity, the conservation of the organoleptic features of the product and the preservation of the traditional landscape.

Table 13.1 Main Italian cultivars of olive with information on geographic distribution, oil organoleptic profile, agronomy, and biology

Cultivar (PDO or PGI oil)[a]	Main Range	Regional distribution	Organoleptic profile of the oil	Fertility	Productivity	Alternate bearing	Tree vigor	Tree canopy habit
Casaliva (PDO Laghi Lombardi, PDO Garda)	North	Lombardia	Medium/high fruity, bitter, and peppery, prevalence of fresh almond and hints of grass and artichoke	Self-sterile	High	Low	Medium	Spreading
Taggiasca (PDO Riviera Ligure)	North	Liguria	Medium fruity, medium bitter, and peppery, with nuances of fresh almond, grass, artichoke	Self-sterile	High	Low	Medium	Weeping
Frantoio (PDO Umbria, PDO Lucca)	Center	Toscana-spread throughout Italy	Medium-high fruity, bitter, and peppery, prevalence of fresh almond and hints of grass and artichoke	Self-fertile	High	Low	High	Spreading
Leccino (PGI Toscana)	Center	Toscana-spread throughout Italy	Medium fruity, medium-low bitter and peppery, prevalence of fresh almond and hints of grass and artichoke	Self-sterile	High	Low	High	Spreading
Moraiolo (PDO Umbria)	Center	Toscana-Umbria-Lazio	Medium fruity, bitter, and peppery, prevalence of grass and hints of fresh almond, artichoke, and tomato	Self-sterile	Medium	High	Low	Upright
Canino (PDO Canino)	Center	Lazio	Medium fruity, medium-low bitter and peppery, prevalence of fresh almond and hints of grass and artichoke	Self-sterile	High	Medium	High	Upright
Itrana (PDO Colline Pontine)	Center	Lazio	Medium-high fruity, medium-low bitter and peppery, prevalence of grass and tomato, and hints of fresh almond and artichoke	Self-sterile	High	Medium	High	Upright
Dritta PDO Aprutino Pescarese)	Center	Abruzzo	Medium-high fruity, bitter, and peppery, prevalence of fresh almond and hints of grass and artichoke	Self-sterile	High	Low	Medium	Spreading

Ortice	Center	Campania	Medium-high fruity, medium/low bitter and peppery, prevalence of grass and tomato and hints of artichoke and fresh almond	Self-sterile	High	High	Medium	Upright
Cellina di Nardò	South	Puglia	Medium fruity, bitter, and peppery and a particular sensation of forest fruits	Self-sterile	Medium	Low	Medium	Upright
Coratina (PDO Terra di bari)	South	Puglia	Medium/high fruity, bitter, and peppery, prevalence of fresh almond and hints of grass and artichoke	Self-sterile	High	Medium	Medium	Spreading
Ogliarola Salentina	South	Puglia	Medium fruity, bitter, and peppery with a particular sensation of forest fruits	Self-sterile	High	High	High	Spreading
Carolea (PDO Lametia)	South	Calabria	Medium fruity, bitter, and peppery, prevalence of grass and hints of fresh almond, artichoke, and tomato	Self-sterile	High	High	Medium	Upright
Dolce di Rossano	South	Calabria	Medium fruity, medium-low bitter and peppery, prevalence of fresh almond and hints of grass and artichoke	Self-sterile	High	High	High	Upright
Ottobratica (PGI Olio di Calabria)	South	Calabria	Medium fruity, bitter, and peppery, with prevalence of grass and hints of fresh almond, artichoke, and tomato	Self-sterile	High	High	Medium	Spreading
Nocellara del Belice (PGI Sicilia)	Major Isles	Sicilia	Medium-high fruity, medium/low bitter, and peppery, prevalence of grass and tomato and hints of artichoke and fresh almond	Self-sterile	High	Low	Medium	Spreading
Bosana (PDO Sardegna)	Major Isles	Sardegna	Medium fruity, bitter, and peppery, prevalence of grass and hints of fresh almond, artichoke, and tomato	Self-fertile	High	Medium	Medium	Spreading

Where relevant, the name of the olive oil with protected designation has been reported
[a]*PDO* Protected Denomination of Origin, *PGI* Protected Geographical Indication

Fig. 13.2 Traditional olive grove with old trees

Fig. 13.3 Intensive olive grove in Central Italy

Table 13.2 Production model systems for olive growing in Italy

Production model	Total ha	% of total area	Trees/ha	Position	Irrigation	Other typical features
Traditional or extensive olive groves	521,095	48.66%	less than 140	Hills, steep slopes	None	Low labor and resources inputs. Common presence of large and old olive trees arranged according to irregular layouts, poorly fertile soils, fragmentation of properties. Productivity and profitability generally low, alternate bearing high. Olive groves often located in marginal areas and subject to abandonment processes. High diversity of cultivars.
Intensive olive groves	507,967	47.44%	140 to 399	Hills, medium slopes	Possible	Use of chemical fertilizers and pesticides to control pests; control of weeds by soil procedures or chemical weeding; large use of machines for cultivation and harvesting. Regular planting schemes. Good productivity and profitability. High diversity of cultivars.
Super-intensive olive groves	41,771	3.90%	up to 2500	Foot hills and flat areas	Required (around 3000 m^3/ha)	Accurate horticultural management with a high supply of fertilizers and pesticides; harvesting and pruning fully mechanized, very high productivity, and high planting costs. Use of a limited range of non-vigorous olive cultivars required

Surface data: Eurostat statistics, accessed on March (2020)

13.3 Main Challenges

Olive growing represents in Italy an enormous reservoir of biodiversity, with more than 500 olive varieties spread all over the Country.

This biodiversity should be protected and enhanced to us coping with climate change issues and improving the resilience of the future olive agroecosystems. It could be a key point for the sustainable development of the sector.

In recent years, various factors have endangered this heritage; the most critical are climate change, evolution of cultivation techniques, and market demands. All these three elements, combined with the need for an increased production and a cost reduction, may induce to propagate and spread only a few particularly productive, resistant, and appreciated varieties, neglecting the others.

The increased intensity and frequency of extreme climatic phenomena can weaken or damage olive trees, provoke shifts in the cultivation area, and lead to the spreading of new pathogens and pests or changes in existing ones.

New planting models, such as those with high plantation density and/or with significant demand for mechanization, could gradually direct the choice of olive grower toward specific cultivars.

Finally, market demands driven by the evolving tastes of consumers could induce olive growers to choose only most popular and requested cultivars, to the detriment of traditional and autochthone ones.

At the same time, it is precisely a well-preserved diversity that can offer us the opportunity to cope with all the previous challenges and to satisfy the need for a sustainable olive growing.

By way of example, scientific research against the spread of *Xylella fastidiosa* has identified only two resistant varieties so far, but still few cultivars have been tested among all those available in the Country, so it is legitimate to expect others to be found: olive biodiversity thus could help fighting emerging diseases.

As for the intensification of production, there are many Italian low-vigor cultivars that could adapt to the new super-intensive models. Productivity levels could thus be increased, without losing the peculiar qualitative aspects and the biodiversity that characterize Italian olive oil.

Finally, a well-planned and well-planned and executed information and educational campaigns can help consumers to discover and appreciate the distinctive characteristics of the many varieties of Italian olives. This, together with the enhancement of PDO and PGI oils and of the touristic appeal of historical and landscape character of some olive growing areas, can direct the market toward greater attention to the productions obtained from local and native cultivars.

References

Alfei, B., Pannelli, G., & Ricci, A. (2013). *Olivicoltura. Coltivazione, olio, territorio*. Edagricole.

Caruso, T., Rotundo, A., & Sebastiani, L. (2009). Germoplasma e varietà. In M. Pisante, P. Inglese, & G. Lercker (Eds.), *L'Ulivo e l'Olio. Collana Coltura e Cultura* (pp. 555–583). Bayer Crop Science.

Eurostat. (2017). *Agricultural production—Orchards*. https://ec.europa.eu/eurostat/statistics-explained/index.php/Agricultural_production_-_orchards. Accessed July 2020

Fiorino, P. (2003). *OLEA Trattato di olivicoltura*. Edagricole.

Pannelli, G., & Perri, E. (2012). *Scelte varietali in olivicoltura*. Accademia nazionale dell'olivo e dell'olio, Volume XVIII, Spoleto. ISSN 2281-4930.

Perri, E., Benincasa, C., & Muzzalupo, I. (2011). Following olive footprints in Italy, In M. El-Kholy (Ed.), *Following Olive Footprints (Olea europaea L.). Scripta Horticulturae n. 13, International Society For Horticultural Science Edition* (pp. 176–191). ISBN 978 90 6605 734 0

Russo, C., Cappelletti, G. M., Nicoletti, G. M., Di Noia, A. E., & Michalopoulos, G. (2016). Comparison of European olive production systems. *Sustainability, 8*, 825. https://doi.org/10.3390/su8080825

Stillitano, T., De Luca, A., Iofrida, N., Falcone, G., Spada, E., & Gulisano, G. (2017). Economic analysis of olive oil production systems in Southern Italy. *Quality - Access to Success, 18*, 107–112.

Chapter 14
Soil Management Practices, Fertilization, Pruning, and Harvesting of Italian Olive Groves

Annunziato Scaramozzino, Giorgio Pace, Alessandro Triantafyllidis, and Sara Petrucci

Abstract In Italy, olive cultivation is carried out in a complex and fragile environmental context, where landscape variability, anthropic pressure, and climate dynamics are continually evolving. The adoption of agronomic practices with low environmental impact, assisted by correct exploitation of modern technologies, are here more than elsewhere necessary to bring together productivity and provision of invaluable ecosystem services.

The adoption of efficient and environmentally friendly agronomic practices is a crucial aspect in the development of a truly sustainable olive-growing model. This statement is particularly true in countries where orographic vulnerability, anthropogenic pressure, and climatic dynamics interact, outlining a highly complex and extremely fragile context, as in the case of Italy.

In particular, in Italy, the soil resource represents, more than elsewhere, the element on which to focus attention.

According to studies about soil loss by water erosion in Europe, the highest mean annual soil loss rate at country level is found in Italy (8.46 t/ha) due to a combination of high rainfall erosivity and steep topography (Panagos et al., 2015). Furthermore, according to the Italian Ministry of the Environment, about 27% of the national territory is threatened by soil desertification processes (Caliandro & Stelluti, 2005).

A. Scaramozzino (✉) · G. Pace
UNAPROL Consorzio Olivicolo Italiano, Rome, Italy
e-mail: a.scaramozzino@unaprol.it; g.pace@unaprol.it

A. Triantafyllidis · S. Petrucci
AIAB Associazione Italiana Agricoltura Biologica, Genova, Italy
e-mail: a.trianta@tin.it; info@sarapetrucci.it

J. Muñoz-Rojas, R. García-Ruiz (eds.), *The Olive Landscapes of the Mediterranean*, Landscape Series 36,
https://doi.org/10.1007/978-3-031-57956-1_14

Therefore, the need for Italian olive sector to qualitatively and quantitatively improve its production and compete on markets should be pursued by enhancing and not reducing its multifunctional vocation. Olive groves rationally managed, with the adoption of low-impact practices and a correct exploitation of modern technologies, do not in fact only perform merely productive functions, but, in addition to their undisputed landscape function, play useful roles for the whole community. They offer indeed invaluable ecosystem services such as the containment of erosive, instability, and desertification phenomena and the mitigation of climate change by storing carbon in plants and soil (see also Chap. 17).

14.1 Crop Management and Soil Protection

In the complex and heterogeneous Italian olive grove landscape, mosaic of numerous different cultivation models, the interventions of soil management and fertilization, together with the other interventions on plant and soil such as pruning, sanitary defense, and irrigation, must be adapted and calibrated from time to time to contexts also very different from each other. However, the common denominator of all cultivation techniques must always remain the trend toward a balance with the environment, aimed at maintaining the productivity of the trees over time, guaranteeing the quality of the final product, limiting environmental impacts, and minimizing resources' depletion.

In olive growing, there are two main possible strategies for soil management: one is based on the constant removal of spontaneous flora throughout spring and summer period through mechanical or chemical weeding, the other instead consists in keeping the soil permanently or temporary, completely or partially covered by vegetation. Combinations of the two aforementioned strategies are also commonly applied (Fig. 14.1).

Both approaches have advantages and disadvantages, but the numerous experimental tests carried out suggest that, in the case of the removal of spontaneous flora, the disadvantages are by far prevalent, so much so that the grass cover of the tree groves, together with the other measures aimed at preserving the soil organic matter, is also incentivized by the Common Agricultural Policy at EU level and is part of the practices included among the agro-environmental schemes envisaged under measure 10 of the RDP 2014–20 "Agro-climatic-environmental payments."

The use of plant coverings is currently recommended to protect the soil in the olive grove management, as is also the case in other types of orchards. The presence of a live cover, in fact, not only has beneficial effects on the properties of the soil but also determines a better biochemical fertility and favors the accumulation of greater biomass and microbial diversity in respect to the soils more intensively cultivated. Permanent cover decreases erosion and compaction and increases water infiltration and the accumulation of organic matter in the soil profile.

Currently in Italy, the most common practice of soil management in the olive grove is based on the use of periodical cultivation, typically harrowing or milling,

Fig. 14.1 Olive-growing landscape in Central Italy with tilled olive groves next to olive groves covered by vegetation

mainly aimed at burying fertilizers and reducing evapotranspiration by eliminating weeds and interrupting soil capillarity that leads to direct water losses from the soil deepest layers. However, this "traditional" approach has many undesirable effects on the physical, chemical, and biological properties of the soil and can worsen its structure and fertility (Alfei et al., 2013).

Conventional tillage (Fig. 14.2) increases mineralization of the organic matter and soil compaction, reduces porosity, and increases surface runoff, erosion, degradation of the soil structure, and soil losses. Compact layers decrease the infiltration of water which, in turn, increases runoff in sloping land and stagnation in flat land.

The running of time and the conditions of the soil at the moment of cultivation, as well as the machines used, affect the intensity of the processes described, but these inevitably determine the onset of stress conditions for the trees, the depletion of soil fertility and the increasing dependence on chemical inputs for defense and fertilization, with unavoidable consequences also on the environmental impact of the crop and on the quality of the product (Caliandro & Stelluti, 2005).

Furthermore, the extent of the negative repercussions of cultivation on the soil system has been accentuated in recent years by the modification of the pluviometry regime connected to the current climate change, which has led to the increase in high-intensity precipitation events that aggravate erosive and degradative phenomena (Costantini & Lorenzetti, 2013).

For these reasons, in Italy, the practice of growing plant covers is increasing, albeit rather slowly spreading. Its application derives from the evidence that

Fig. 14.2 Olive grove where conventional tillage is applied

spontaneous flora—if properly managed to reduce its competitive power can represent a resource capable of increasing soil fertility and biodiversity (Gucci et al., 2011). Temporary or permanent cover is in fact an alternative system of intervention on the soil and can significantly contribute to balancing the complex physical-chemical and biological phenomena that characterize the tree-soil system.

The maintenance of the presence of spontaneous herbaceous flora, especially if combined with the return to the soil of organic residues produced during the year, such as pruning residues (Fig. 14.3) and mowings as well as composted soil amendments obtained from pomace and other by-products of the olive oil supply chain, contribute in a decisive way to increase the soil's organic matter with positive consequences on the soil physical, chemical and biological fertility (Celano et al., 2011; Gucci & Cantini, 2012). The improved efficiency of biological processes in the soil, in particular, also allows a significant reduction in the external supply of nitrogen fertilizers, contributing to the containment of costs and environmental impacts (Calabrese et al., 2015).

Last but not least, grassy olive groves (Fig. 14.4) can become agroecosystems able of absorbing more CO_2 than is emitted in the overall olive oil production process, contributing on the one hand to mitigate climate change and on the other hand to maintain the competitiveness of economically and agronomically marginal areas through the generation of carbon credits for the voluntary market (Simeone et al., 2013).

Olive growing and olive oil production can play an important role in atmospheric CO_2 emission and fixation, especially by affecting the soil carbon (C) storage.

Fig. 14.3 The return of pruning residues to the soil could positively affect the soil carbon (C) storage

Fig. 14.4 Old-growth grassy olive grove

Particularly, olive groves are agricultural systems that under certain conditions (local origin and traditional practices) are able to remove all the emissions associated to the olive oil production chain (Proietti et al., 2017).

In summary, the advantages of the plant cover in relation to the agro-ecological characteristics of the olive grove, which vary according to the context and the form of application (temporary/permanent, total/partial) can be enumerated as follows (Vignozzi et al., 2013):

- decreasing of the risks of erosion and of the loss of fertile soil;
- improvement of the soil water holding capacity, reduced water lodging and run-off, and increase of water reserves in the soil;
- facilitating the development of the root system of trees and increasing the possibility to explore more fertile layers of the soil;
- increase in the bearing capacity of the ground and reduction of the compaction caused by machines;
- increase in functional biodiversity, in particular, there is a general increase in the useful entomofauna and a reduction of the harmful one;
- stimulation of a better balance between vegetative-productive phase of the tree, which reduces the alternation of production and decreases the susceptibility to diseases;
- increase of physical and chemical fertility connected to increase in the supply of organic matter thanks to the decomposition of the material resulting from periodic mowing and the continuous renewal of the roots of the grassy turf which also leads to an increase in the soil biodiversity and presence of species such as earthworms, which improve the structure of the soil and promote humification;
- increase of the carbon stocking capacity of the soil.

In the face of the undoubted benefits of covers, there are nevertheless some critical aspects to consider, especially in the case of recently planted olive groves; we refer in particular to the competition between trees and the herbaceous layer for water and nutrients when these are available in limited quantities, and to possible interferences due to the secretion of allelopathic substances in the soil by some plant species (Proietti & Regni, 2019).

To mitigate the competition and the allelopathy above mentioned, a programmed cover can also be considered, to be achieved with the sowing of mixtures of species deemed particularly beneficial to the achievement of the positive objectives (Campiglia et al., 2005). With the sown cover, species can be chosen that last over time and are resistant to the passage of mechanical vehicles, such as white clover (*Trifolium repens*) among leguminous species and *Poa pratensis*, *Festuca rubra* and *Festuca ovina* among grasses. The latter settles slowly but then prevails and resist, showing little water competition with the tree crop. On poor soils, the choice can also fall on a mixture of self-reseeding leguminous species, such as different varieties of underground clovers capable of adapting to the most diverse soils and climatic conditions, or annual species such as the ubiquitous *Medicago polymorpha*.

These and other species develop in the autumn-winter period, bear fruit, and die before the onset of summer and this determines a marked ecological complementarity that minimizes the phenomena of competition but at the same time allows an efficient use of environmental resources.

In the case of sown covers, of course, companies must be able to bear the cost for the purchase of about 50 kg/ha of seeds and the cost for the sowing and any subsequent rolling, which is why the choice is to be carefully evaluated.

14.2 Improvement of Efficiency Through Smart Farming

European farmers can now have technologies and tools that until a few years ago were not even imaginable. Remote sensing, advanced sensor networks, Internet of Things, Precision Farming, advanced forecasting models, vast and updated databases, and decision support systems (DSS) are the main tools of the so-called Smart Farming, a novel approach to agriculture that uses modern technology to improve produce and the environmental performance of crops. The possibility of accurately and promptly measuring variations within the olive grove and of correctly managing the collected data, adapting the growing strategy accordingly, can significantly increase the effectiveness of agronomic and phytosanitary interventions. It can reduce the inputs entered into the system and their environmental impact, containing production costs and at the same time increasing the quantity and quality of the final product.

In particular, the use of remote sensing techniques and DSS tools, due to their versatility and their growing application in the olive sector, could substantially contribute to the optimization of agronomic practices in the sector.

In recent years, the evolution of remote sensing techniques, the growing availability of increasingly accurate and reliable technologies, and therefore of more precise and detailed data, favored the increase in Italy the use of such investigation and monitoring systems in numerous sectors, including agriculture. Furthermore, the availability of data acquired by satellites and the possibility of carrying out investigations through the use of drones have also allowed a significant reduction in operating costs, thus further contributing to the proliferation of services that apply this technology to crop management (Nolè et al., 2018).

This is of course an extremely important opportunity also for the olive sector. The application of remote sensing techniques can contribute significantly to the increase in productivity of the olive grove and to the contextual reduction of the environmental impact of agricultural practices. In particular, the main practical applications of these techniques, also integrated with visual and instrumental detection activities carried out on the ground, can support (Bernardi & Zimbalatti, 2013):

(a) Monitoring and investigations aimed at optimizing fertilization and phytosanitary control and interventions.
(b) The choice of the most suitable irrigation method and the amount of water to be supplied according to specific and precise needs.
(c) The assessment of the morphological characteristics of the plants and the consequent planning of the pruning and most appropriate pruning methods.

(d) The planning and organization of the harvesting operations (Fig. 14.5), as well as, the estimate of the yield of the trees and of the consequent production of oil.

The correct application of remote sensing techniques, also in combination with other monitoring and control tools, therefore contributes to defining, at relatively low costs, an extremely detailed information framework, which can support farmers in the management of their olive groves and at the same time facilitate administrators and policymakers in the development of effective and sustainable agricultural policies.

However, the availability of a greater amount of information does not automatically translate into an improvement in production quantity and quality and in the environmental performance. In absence of an adequate interpretative framework, in fact, an excess of data could even result in a stall, with consequent decrease of the system performance. The introduction of the so-called Decision Supporting Systems (DSS) in agriculture can support operators in managing consistent, even heterogeneous data flows. The DSS are in fact software platforms capable of acquiring and processing data from multiple sources through specific algorithms in order to provide indications to technicians or directly to farmers.

Between 2015 and 2017, UNAPROL promoted the creation of a DSS software based on a complex network of integrated models, capable of predicting the needs for agronomic and phytosanitary interventions in olive growing. The system, called Olivo.net®, was created by Horta, a spin-off of the Catholic University of the Sacred Heart of Piacenza, in collaboration with the main Italian universities and experts in

Fig. 14.5 Harvesting of the olives by electric shaker rakes

the olive-oil-producing sector (Anselmi et al., 2017). The creation of the IT platform has engaged the team for 2 years and has seen the collaboration of over 170 technicians of the producers' organizations belonging to UNAPROL, present throughout the national territory [https://www.horta-srl.it/portfolio-item/olivo-net/ Accessed July 2020].

The tool, which can be used today via Internet by farms and technicians, has been designed to: (1) provide technical assistance and additional information to olive growers, to integrate and improve specific skills; (2) organize field monitoring activities; (3) favor the traceability of production from farm to fork; and (4) assess the sustainability of productive processes by applying a set of specific indicators. In particular, the DSS addresses technicians and olive growers providing a decision support system capable of integrating the experiences and knowledge already acquired, and more generally to increase the quality olive supply chain with providing indicators to assess and monitor the performance of the productive process (Anselmi et al., 2017).

14.3 Main Challenges

Currently, the main challenges to the Italian olive sector are to increase the production and to improve the final product quality, while keeping costs under control and reducing the environmental impact of operations. All this in a context of climate change and in a context of land abandonment in many regions especially in southern Italy. A sustainable intensification process is needed, that should be pursued by intervening in every production stage, starting from agronomic operations.

The rational management of soil, trees, and crops, with the adoption of low-impact practices and a correct exploitation of modern technologies can help improving the productive functions of olive agroecosystems while enhancing their capacity to deliver invaluable ecosystem services.

No increase in productivity will be possible without a wider mechanization of the sector, both as regards the tillage and for the olive harvest and pruning. In this regard, on the one hand, the use of efficient and low-consumption machinery must be envisaged, on the other hand, the optimization and rationalization of the harvesting is required in order to limit the consumption of fossil fuels and the associated emissions. The ability of the olive grove to sequester carbon and therefore to counteract climate change must be increased by adopting cultivation practices such as grassing, shredding pruning residues, amending with by-products of the olive-oil supply chain. In this sense, it would be helpful to envisage income supplements for the farmers in the context of the carbon credit market.

Politics also plays its role in favoring the establishment of a generational turnover in companies, olive and non-olive, and to favor the protection of territories, to avoid the abandonment of the countryside and therefore of olive groves, favoring sustainable pipelines and short supply chains that ensure income, also thanks to the possibility of e-commerce.

References

Alfei, B., Pannelli, G., & Ricci, A. (2013). *Olivo: Coltivazione, olio, territorio*. Edagricole-Il Sole24Ore.

Anselmi, A., Caruso, G., Cola, G., Gucci, R., Meriggi, P., Rossi, V., Sandali, P., Scaramozzino, N., & Servili, M. (2017). *Olivo.net, il grande fratello che rivoluziona l'olivicoltura* (Olivo e Olio n. 5/2017). Edagricole.

Bernardi, B., & Zimbalatti, G. (2013). La raccolta in olivicoltura richiede maggiore innovazione tecnologica. *La Rivista di Scienze dell'Alimentazione, 42*(1), 105–110.

Calabrese, G., Perrino, E. V., Ladisa, G., Aly, A., Tesfmichael Solomon, M., Mazdaric, S., Benedetti, A., & Ceglie, F. G. (2015). Short-term effects of different soil management practices on biodiversity and soil quality of Mediterranean ancient olive orchards. *Organic Agriculture, 5*, 209–223. https://doi.org/10.1007/s13165-015-0120-8

Caliandro, A, & Stelluti, M. (2005). Ruolo dell'olivicoltura nella lotta alla desertificazione. Accordo di programma MATT-CNLSD 2005. Available at. https://www.minambiente.it/sites/default/files/archivio/allegati/desertificazione/Ruolo_dellxolivocoltura_nella_lotta_alla_desertificazione.pdf. Accessed July 2020.

Campiglia, E., Mancinelli, R., Paolini, R., & Pandozy, G. (2005). L'inerbimento dell'oliveto con leguminose annuali autoseminanti. ARSIAL – Regione Lazio – Università degli Studi della Tuscia, Roma. Available online at. http://www.agroambientelazio.it/public/doc_file/20140911130480.pdf. Accessed July 2020.

Celano, G., Favia, M., Perretti, B., Xiloyannis, C., & Palese, A. (2011). Una proposta di gestione sostenibile dell'oliveto: una analisi economica, energetica ed ambientale. In: Atti II Convegno Nazionale dell'Olivo e dell'Olio, Firenze Società di Ortoflorofrutticoltura Italiana (SOI), Perugia 21–23 september 2011, vol 10: 378–381.

Costantini, E., & Lorenzetti, R. (2013). Soil degradation processes in the Italian agricultural and forest ecosystems. *Italian Journal of Agronomy, 8*, 233–243.

Gucci, R., & Cantini, C. (2012). *Potatura e forme di allevamento dell'olivo*. Edagricole-New Business Media.

Gucci, R., Caruso, G., Sifola, M., Pellegrini, S., Pagliai, M., & Vignozzi, N. (2011). Gestione sostenibile di oliveti intensivi con l'inerbimento permanente. *Rivista di Frutticoltura e Ortofloricoltura, 12*, 38–42.

Nolè, G., Pilogallo, A., Lanorte, A., & Santis, F. (2018). Remote sensing techniques in olive-growing: A review. *Current Investigations in Agriculture and Current Research, 2*, 205–208. https://doi.org/10.32474/CIACR.2018.02.000138

Panagos, P., Borrelli, P., Poesen, J., Ballabio, C., Lugato, E., Meusburger, K., Montanarella, L., & Alewell, C. (2015). The new assessment of soil loss by water erosion in Europe. *Environmental Science & Policy, 54*, 438–447. https://doi.org/10.1016/j.envsci.2015.08.012

Proietti, P., & Regni, L. (2019). Handbook for a sustainable management of the olive groves. Available at https://olive4climate.eu/wp-content/uploads/Olive4Climate-Handbook-_ENG.pdf. Accessed July 2020.

Proietti, S., Sdringola, P., Regni, L., Evangelisti, N., Brunori, A., Ilarioni, L., Nasini, L., & Proietti, P. (2017). Extra Virgin Olive oil as carbon negative product: Experimental analysis and validation of results. *Journal of Cleaner Production, 166*, 550–562. https://doi.org/10.1016/j.jclepro.2017.07.230

Simeone, V., Tucci, M., Vitti, R., Calabrese, J., & Perri, E. (2013). Linee guida per la gestione degli oliveti secolari. *La rivista di scienza dell'alimentazione, 1*, 135–136.

Vignozzi, N., Agnelli, A., Platinetti, M., Pellegrini, S., Simoncini, S., Pagliai, M., Xiloyannis, C., Celano, G., & Palese, A. (2013). Olivicoltura sostenibile e gestione del suolo. *La Rivista di Scienza dell'Alimentazione, 42*, 91–96.

Chapter 15
Pest and Diseases and Main Control Strategies in Italian Olive Groves

Annunziato Scaramozzino, Giorgio Pace, Alessandro Triantafyllidis, and Sara Petrucci

Abstract In Italy, the trend toward the adoption of protocols for control of diseases that have a low environmental impact has been promoted since 2014 through the PAN, National Plan for sustainable use of Defense Products. The hardest challenge at the moment is to bring together sustainability and fight against *Xylella fastidiosa*, the pathogen that has been undermining olive growth since 2013. For organic producers, *Bactrocera oleae* is the main issue to tackle, different control strategies are applied to address the problem.

15.1 Main Pathogens and Pests in Italian Olive Groves

Italian olive groves, from a phytosanitary point of view, are relatively stable agroecosystems, overall characterized by a limited complex of harmful pests and diseases.

This is a consequence of a sum of factors, such as the wide availability of different olive cultivars, well adapted to local paleoclimatic conditions, and, in the case of insect pests, the high presence of natural enemies. Therefore, the phytosanitary defense of the olive tree, unlike what happens in the case of other fruit trees, in Italy is generally aimed against few adversities.

Currently, the two main phytosanitary threats that Italian olive growing has to deal with are an insect, *Bactrocera oleae*, and a bacterium, *Xylella fastidiosa*.

The most important olive pests and diseases currently affecting Italian olive groves are mentioned in this subchapter with more in-depth information on the

A. Scaramozzino (✉) · G. Pace
UNAPROL Consorzio Olivicolo Italiano, Roma, Italy
e-mail: a.scaramozzino@unaprol.it; g.pace@unaprol.it

A. Triantafyllidis · S. Petrucci
AIAB Associazione Italiana Agricoltura Biologica, Genova, Italy
e-mail: a.trianta@tin.it; info@sarapetrucci.it

J. Muñoz-Rojas, R. García-Ruiz (eds.), *The Olive Landscapes of the Mediterranean*, Landscape Series 36,
https://doi.org/10.1007/978-3-031-57956-1_15

matter of *Bactrocera oleae* and *Xylella fastidiosa*. A specific sub-paragraph will be dedicated to the latter one due to the severity of the threats it is currently posing.

As for pests, among the 50 most common species of interest in Italy, only *Bactrocera oleae*, *Prays oleae*, and *Saissetia oleae* continue to be a threat of greater or lesser importance for the olive sector and can affect the production. In younger orchards, olive growers could also need to control the phytophagous that infest the vegetation, especially *Palpita unionalis* and, less often, *Dasineura oleae* and *Otiorrhynchus cribicollis*.

15.1.1 Bactrocera oleae

As for defense, *Bactrocera oleae* represents largely the most harmful pest to the Italian olive trees. The insect is distributed in all Italian olive areas, except in some located at higher altitudes where low temperatures do not allow the development of the fly. Production losses attributed to its attacks in recent years (e.g., 2014, 2016, 2019 in Toscana, and 2018 in Puglia) have reached high levels, even higher than 50%.

The measures of control and their effectiveness have a major impact not only on the amount of produced olive oil but also on its quality as it can influence chemical, organoleptic and toxicological characteristics (Gucci et al., 2020). As a key pest, *Bactrocera oleae* requires for its containment continuous field monitoring and specific treatments to be repeated every year.

Organic farming control techniques are generally based on early harvesting and timely milling together with mass trapping and repellent/oviposition deterrent sprays by copper products and clays.

In conventional and integrated farming, the most common control techniques rely on spraying of chemical insecticides. The high environmental impact of this approach is currently mitigated by the increasingly widespread application of forecast models based on field monitoring activities and on the application of intervention thresholds. Also, decision support systems (DSS) based on semi-automatic pest monitoring have been developed for managing the *Bactrocera oleae*, thus allowing precise treatments to be conducted through location-aware systems.

The recent measure on the nonrenewal of approval of the active substance dimethoate (European Commission Implementing Regulation 2019/1090 of June 26, 2019) is posing a major issue to Italian nonorganic olive growers: that compound, in fact, has been used for decades as the main synthesis molecule for the larvicidal control of *Bactrocera oleae* in conventional and integrated farming.

The revocation of the dimethoate (apart from the exemption granted for the 2020 campaign) obliges the olive growers to an uncertain change of strategy due to the current lack of equally effective larvicide alternatives. Consequently, the future strategy for the control of the insect will likely rely increasingly on the use of adulticidal means, already used in organic olive growing. In addition to this, the possibility of conservative biological control, obtained by increasing the role of useful indigenous insects, could become a valid alternative but perhaps limited to traditional olive groves, with low planting density, a strong floristic complexity, and large and diversified insect populations.

In organic olive cultivation for the control of the fly, the use of kaolin is increasingly widespread, useful for the plant also to overcome phases of stress due to water scarcity and summer heatwaves, phenomena that climate change is exacerbating. Among the actual bioinsecticides with an adulticidal action, Spinosad shows a certain effectiveness as a product of microbiological origin that acts above all by ingestion but also by contact, for a maximum number of 5 or 6 treatments per year.

15.1.2 Xylella fastidiosa

In addition to *Bactrocera oleae* and the other parasites mentioned above, new concerns have also emerged regarding the control of other insects, not for the direct damage they can cause, but as potential vectors of *Xylella fastidiosa*, one of the most serious diseases that today threaten Italian olive growing. It is about three insects belonging to the order Hemiptera, of which the ability to transmit the phytopathogenic bacterium *Xylella fastidiosa* has been demonstrated: *Philaenus spumarius*, *Philaenus italosignus,* and *Neophilaenus campestris*. Among the three, *P. spumarius* appears to be the main vector of the dangerous bacterium. (Alfei et al., 2013; Simeone et al., 2013), the causative agent of the olive quick decline syndrome (OQDS).

As for diseases, olive trees, such as other woody perennial crops, can be affected by several systemic pathogens, including bacteria, fungi, and vector-borne viruses. In Italy, until a few years ago, just a handful of them could represent real threats to the production and vitality of the olive plants. Among those, we mention the bacterium *Pseudomonas savastanoi* causing olive knot, the deuteromycete fungus *Spilocaea oleaginea*, an olive peacock spot or olive leaf spot agent and *Verticillium dahliae* which causes verticillosis, a vascular pathology currently expanding following the spread of irrigation and more intensive cultivation systems. Other diseases that are widespread in Italian olive groves, but which generally cause less concern are ultimately *Colletotrichum gloeosporioides,* causing olive anthracnose, and *Mycocentrospora cladosporioides,* agent of cercospora leaf spot.

Situation has changed since 2013 due to the first detection of *Xylella fastidiosa* subsp. *pauca* strain st53 in the south of the Italian peninsula. For the first time, a bacterium-caused illness posed a more serious threat to Italian olive growers than that represented by the *Bactrocera oleae*.

Xylella fastidiosa is a plant pathogen with a broad range of host plants. It is transmitted by insects that feed on the xylem fluid of plants. It is a quarantine pest in the European Union (EU), which means suspected sightings must be reported immediately to the plant health authorities. Since it was first detected in Europe near Gallipoli (Puglia, Italy) in 2013, *Xylella fastidiosa* has spread to other zones of Puglia and caused a rapid decline in olive tree plantations. The epidemic is advancing at a speed of about 10 km a year (Kottelenberg et al., 2021) and, currently, it covers an area equal to 40% of the regional surface of Puglia, threatening more than 25 million olive trees. Varietal genetic resistance and control of vector insects are currently the only means of contrasting the bacterium *Xylella fastidiosa* as, to date, there is no cure capable of healing an infected plant.

15.1.3 Prays oleae

The Olive Moth (*Prays oleae*) is a lepidopteran that turns three generations per year and that in some areas can cause extensive damage. In organic olive growing, action must be taken on the first two generations, borne, respectively, by leaves and flowers, by *Bacillus thuringiensis*, which would be ineffective against the third generation, that loaded with olives, since the insect in that case is inside the fruit and *B. thuringiensis* is not a systemic product.

15.1.4 Otiorrhynchus cribricollis

Weevils (*Otiorrhynchus cribricollis*) cause deep erosions in the leaf flap, recognizable by the typical semicircular shape. A control method that avoids chemical treatments consists of wrapping sections of trunk with spongy synthetic fibers, which have the function of hindering the ascent of adults.

15.1.5 Saissetia oleae

For the sustainable control of *Saissetia oleae*, also called olive scale, the best prevention is given by regular pruning, which makes the trees open and bright. Treatments with white mineral oils are also allowed in organic farming and can be repeated after 15–20 days.

15.1.6 Olive Kot and Peacock-Eye

Olive knot is a disease caused by the bacterium *Pseudomonas savastanoi* (previously known as *P. syringae sub. Savastanoi*), while the peacock-eye, or olive leaf spot, is caused by a fungus, *Spilocaea oleaginea*. The former shows well-noticeable tumor formations, the latter is evident with circular spots on the foliage, but both are favored by excesses of nitrogenous fertilizer. The first preventive measure in cultivation is therefore a balanced management of fertilizers, together with plants with large silks and regular pruning. Copper treatments are also allowed in organic olive cultivation and are generally relevant when conducted with appropriate timing.

15.2 The National Action Plan for the Sustainable Use of Plant Protection Products

The phytosanitary defense strategies of olive crops in Italy must be contextualized in the context of the PAN (National Action Plan for the Sustainable Use of Plant Protection Products) (MIPAAF, 2014).

Directive 2009/128 /EC, implemented in Italy with the legislative decree of 14 August 2012, n. 150, has in fact established a "framework for Community action for the sustainable use of pesticides." For the implementation of this directive, National Action Plans (PAN) have been defined to establish the objectives, measures, times, and indicators for the reduction of risks and impacts deriving from the use of plant protection products. The Action Plan promotes practices of use of the most sustainable plant protection products and provides indications to reduce the impact of plant protection products in agricultural areas, nonagricultural areas, and protected natural areas.

Among the most important novelties of the PAN in Italy stands out the introduction in 2014 of the compulsory integrated defense of crops which includes: a) the application of techniques for the prevention and monitoring of infestations, infections, and weeds; b) the use of biological means for control of pest; c) the use of appropriate cultivation practices; and d) the use of plant protection products with the lowest risk to human and the environment health among those available for the same purpose.

The objectives of the PAN can be summarized as follows:

- reduce the risks and impacts of plant protection products on human health, the environment, and biodiversity;
- promote the application of the principles of integrated defense and alternative approaches or techniques to the use of plant protection products;
- promote and incentivize certified integrated production and organic agriculture;
- protect users of plant protection products and the population concerned;
- protect consumers;
- safeguard the aquatic environment and drinking water;
- conserve biodiversity and protect ecosystems.

For the assessment of the progress made in reducing the risks and impacts deriving from the use of plant protection products on human health, the environment, and biodiversity, a set of 15 indicators was identified in 2015, the data relating to which can be consulted on an *ad hoc* website (https://indicatori-pan-fitosanitari.isprambiente.it) [Accessed February 2021].

Analysis of the data shows a gradual trend for the olive sector to increase the surface areas that adhere to voluntary certification systems that involve the use of production techniques with reduced environmental impact, i.e., integrated production and organic agriculture. However, these areas still represent a very small share of the total olive oil UAA.

The elaborations published by ISPRA (Istituto Superiore per la Protezione e la Ricerca Ambientale - Higher Institute for Environmental Protection and Research) in fact highlight a positive trend of Italian organic sector, which also involves the olive sector. Between 2017 and 2018, both the number of certified operators and the area cultivated with the organic method increased, which at the end of 2018 amounted to 238,129 ha, corresponding to about 22% of the country's total olive area.

A similar upward trend is also observed in relation to the olive UAA adhering to the National Integrated Production Quality System (SQNPI), equal to 17,388 ha (just under 2% of the national olive UAA) at the end of 2018, with a growth of about 5% compared to the previous year.

The Integrated Production Standard implements the directive on the sustainable use of pesticides (Dir 2009/128/CE) and defines the agri-food production system that uses all means for defense of agricultural production from adversity, aimed at minimizing the use of synthetic chemicals and rationalizing fertilization, in compliance with ecological, economic and toxicological principles (MIPAAF, 2015).

The PAN has promoted activities related to the correct use of plant protection products, such as training courses and prescriptions for users, distributors and consultants, control of sprayer equipment, establishment of systems of territorial phytopathological monitoring and alerting, promotion of services for the correct treatment and disposal of packaging and remnants of plant protection products (Rossi et al., 2019).

15.3 The Xylella Case: The Most Serious Phytosanitary Challenge of Italian Olive Growing

The path toward the adoption of more sustainable and environmentally friendly phytosanitary control techniques and protocols finds in Italy today an obstacle not to be underestimated in the spread of *Xylella fastidiosa*, a very serious threat to the olive growing of the country and—potentially—of the whole olive world. For this reason, it was considered appropriate to dedicate a specific study to this phenomenon here.

In 2013, in Puglia, the region where more than half of Italian olive production is concentrated, a serious outbreak of *Xylella fastidiosa* subsp. *pauca* was discovered, an extremely aggressive pathogen that causes olive quick decline syndrome, OQDS (Saponari et al., 2017). The main symptoms of this pathology consist in the withering of leaves and branches and, subsequently, in the death of the tree (Colella et al., 2018; Saponari et al., 2018; Scortichini, 2020).

The desiccations begin to manifest and aggravate in already infected plants, usually in the spring-summer period, and are caused by the occlusion of the xylematic vessels which occurs as a result of the multiplication of the pathogen and the consequent formation of bacterial biofilm capable of preventing the normal flow of raw sap from the root system to the crown.

Being a quarantine bacterium according to the European Plant Protection Organization (EPPO), on the basis of the indications deriving from the European regulations governing the management of this category of pathogens, as a means of remediating the infected area the eradication of the bacterium was proposed; it was quickly understood that the conditions of the epidemic would make this solution difficult to implement, above all because *Xylella fastidiosa* subsp. *pauca* lives not only in the olive tree but also in numerous other cultivated and spontaneous plants and is transmitted by insect vectors, the "spittlebugs," very prolific and widespread (Cavalieri & Porcelli, 2017). The insects belonging to the order Hemiptera, of which the ability to transmit the phytopathogenic bacterium *Xylella fastidiosa* has been demonstrated are: *Philaenus spumarius*, *Philaenus italosignus*, and *Neophilaenus campestris*.

Due to the initial extent of the outbreaks and the rapid spread of infections, eradication measures were soon replaced by containment measures including constant monitoring on the borders of the contaminated area, removal of infected trees, and mandatory control of the vector insect.

However, the implementation of containment measures also encountered serious difficulties especially at the beginning, including public reluctance to accept control measures, poor cooperation with stakeholders, disinformation by some media, and lack of concrete responses from some authorities. This scenario delayed and limited containment efforts and allowed the bacterium to continue its rapid spread across multiple areas of the region. On a research level, the European Commission and regional authorities support several programs aimed at finding effective methods to mitigate and contain the impact of *X. fastidiosa* on the olive tree, the host most affected by this epidemic. Preliminary tests for the presence of resistance in some olive cultivars, namely, Leccino and FS17 (Boscia et al., 2017), represent a promising approach currently being studied for the development of long-term management strategies.

As reported by the official documentation of the Phytosanitary Service of the Puglia Region, to date there is no method for treating an infected plant and there are no registered and authorized products that treat *Xylella fastidiosa*. The fight against *Xylella*, pursuant to European Decision 789/2015 as amended, is carried out through the elimination of sources of inoculation (infected plants) with eradication/containment actions and control of the vector.

Infected plants identified through capillary monitoring must be immediately eradicated. With regard to the vector, which is a polyphagous and ubiquitous insect, one must intervene with mechanical and chemical interventions aimed at reducing the population, both toward the juvenile stages and adults. Insecticide treatments are allowed only on olive trees. Organic olive farms can comply with the obligation of adulticidal plant protection treatments while remaining in the conversion regime without however being able to certify the product as organic.

Contrast measures currently adopted for the containment of *X. fastidiosa* in Puglia

- Monitoring of containment, buffer, and free areas.
- Immediate killing of plants that have become infected in the containment area.
- Immediate killing of infected plants and host plants in the 100 m buffer in a buffer or undamaged area.
- Control (mechanical and chemical) of the vector.
- Control and limitations to nursery activities.
- Control of the movement of the "specified" plants.
- Prohibition of planting host plants within the infected area with derogation for the resistant olive varieties Leccino and FS17.
- Ban on the planting of host species in the "containment area."

Diffusion of the Pathogen Puglia is the first and most important Italian diffusion area of *X. fastidiosa. Xylella fastidiosa* subsp. *pauca* genotype ST53 is present in the region, and to date 34 host plants have been identified among cultivated and wild plants: *Acacia saligna* (Labill.) Wendl., *Amaranthus spp., Asparagus acutifolius* L., *Catharanthus, Chenopodium album* L., *Cistus creticus* L., *Dimorphotheca fruticosa* (L.) Less., *Dodonaea viscosa* (L.) Jacq., *Eremophila maculata* P.J.Muell., *Erigeron sumatrensis* Retz., *Erigeron bonariensis* L., *Euphorbia terracina* L., *Euphorbia chamaesyce* L., *Grevillea juniperina* R.Br., *Heliotropium europaeum* L., *Hebe spp., Laurus nobilis* L., *Lavandula angustifolia* Mill., *Lavandula stoechas* L., *Myrtus communis* L., *Myoporum insulare* R.Br., *Nerium oleander* L., *Olea europaea* L., *Pelargonium x fragrans, Phillyrea latifolia* L., *Polygala myrtifolia* L., *Prunus avium* (L.) L., *Prunus dulcis* (Mill.) D.A.Webb, *Rhamnus alaternus* L., *Rosmarinus officinalis* L., *Spartium junceum* L., *Vinca, Westringia fruticosa* (Willd.) Druce, *Westringia glabra* R.Br. The olive tree is the most affected species.

Pursuant to EU Decision 789/2015 as amended by EU decision 927/2018, the demarcated area defined in Puglia (Regione Puglia, 2020) with the executive act n ° 3 of 16/01/2019 (BURP n.8 of 24.01.2019) consists of:

Infected area: it includes the entire provinces of Lecce, Brindisi, and part of Taranto, it is the area where the bacterium is located, and it is not possible to eradicate it. In this area, the decision does not set any obligation to eliminate infected plants.

Containment area: it is the 20 km belt of the infected area adjacent to the buffer zone in which the monitoring must be carried out and containment measures must be implemented through the grubbing up of the plants that have become infected and the fight against the vector.

Buffer zone: it is the 10 km wide belt that surrounds the infected area. The buffer zone is an unscathed area in which monitoring must be carried out and in the event of an outbreak being found, "eradication measures" must be applied which consist of the elimination of the infected plant and all the plants of the host species, regardless of their state of health, present within 100 m. The fight against the vector must also be carried out. Updates are published on the institutional website: www.emergenzaxylella.it [accessed February 2021].

References

Alfei, B., Pannelli, G., & Ricci, A. (2013). *Olivo: Coltivazione, olio, territorio.* Edagricole-IlSole24Ore.

Boscia, D., Altamura, G., Ciniero, A., Di Carolo, M., Dongiovanni, C., Fumarola, G., Giampetruzzi, A., Greco, P., La Notte, P., Loconsole, G., Manni, F., Melcarne, G., Montilon, V., Morelli, M., Murrone, N., Palmisano, F., Pollastro, P., Potere, O., Roseti, V., Saldarelli, P., Saponari, A., Saponari, M., Savino, V., Silletti, M. R., Specchia, F., Susca, L., Tauro, D., Tavano, D., Venerito, P., Zicca, S., & Martelli, G. (2017). Resistenza a *Xylella fastidiosa* in diverse cultivar di olivo. *L'Informatore Agrario, 11*(2017), 59–63. https://doi.org/10.5281/zenodo.495708

Cavalieri, V., & Porcelli, F. (2017). Main insect vectors of *Xylella fastidiosa* in Italy and worldwide. In A. M. D'Onghia, S. Brunel, & F. Valentini (Eds.), *Xylella fastidiosa & the Olive Quick Decline Syndrome (OQDS). A serious worldwide challenge for the safeguard of olive trees* (Série A Mediterranean Seminars, N° 121, Options Méditerranéennes) (172 pp). IAM Bari: CIHEAM (Centre International de Hautes Etudes Agronomiques Méditerranéennes).

Colella, C., Carradore, R., & Cerroni, A. (2018). Problem setting and problem solving in the case of olive quick decline syndrome in Apulia, Italy: A sociological approach. *Phytopathology, 109*, 187–199. https://doi.org/10.1094/PHYTO-07-18-0247-FI

Gucci, R., Adamo, P., Baldoni, L., Benalia, S., Bernardi, B., Falsone, G., Gerinara, G., Petacchi, R., Priori, S., Proietti, P., Regni, L., Schena, L., Servili, M., Vignozzi, N., & Zimbalatti, G. (2020). Intensificazione sostenibile per la filiera olivicolo-olearia: approfondimenti sulle filiere di interesse per l'ambiente mediterraneo. In L. Cocolin & M. Tagliavini (Eds.), *Atti del XVII Convegno AISSA - Buone pratiche di intensificazione sostenibile - Strumento per lo sviluppo del sistema agroalimentare italiano*. Associazione Italiana delle Società Scientifiche Agrarie. ISBN: 978-88-945925-0-4.

Kottelenberg, D., Hemerik, L., Saponari, M., & van der Werf, W. (2021). Shape and rate of movement of the invasion front of *Xylella fastidiosa* spp. *pauca* in Puglia. *Scientific Reports, 11*, 1061. https://doi.org/10.1038/s41598-020-79279-x

MIPAAF Ministero delle Politiche Agricole Alimentari e Forestali. (2014). Piano di azione nazionale per l'uso sostenibile dei prodotti fitosanitari. Decreto 22 gennaio 2014. Available at https://www.minambiente.it/sites/default/files/archivio/normativa/dim_22_01_2014.pdf. Accessed 17 July 2020.

MIPAAF Ministero delle Politiche Agricole Alimentari e Forestali. (2015). Manuale di difesa integrata. Guida per l'applicazione dei principi generali della difesa integrata obbligatoria definiti dall'allegato III della direttiva 2009/128/C. Available at https://www.minambiente.it/sites/default/files/archivio/allegati/vari/manuale_difesa_integrata_2015.pdf. Accessed 17 July 2020.

Regione Puglia. (2020). Determinazione del Dirigente Sezione Osservatorio Fitosanitario 2 novembre 2020, n. 136 – Reg. (UE) 2020/1201 – DGR 548/2020 – Aggiornamento delle aree delimitate alla *Xylella fastidiosa* sottospecie *Pauca* ST53. Bollettino Ufficiale della Regione Puglia n. 157.

Rossi, V., Sperandio, G., Caffi, T., Simonetto, A., & Gilioli, G. (2019). Critical success factors for the adoption of decision tools in IPM. *Agronomy, 9*, 710. https://doi.org/10.3390/agronomy9110710

Saponari, M., Boscia, D., Altamura, G., Loconsole, G., Zicca, S., D'Attoma, G., Morelli, M., Palmisano, F., Saponari, A., Tavano, D., Savino, V. N., Dongiovanni, C., & Martelli, G. P. (2017). Isolation and pathogenicity of *Xylella fastidiosa* associated to the olive quick decline syndrome in southern Italy. *Scientific Reports, 7*, 17723. https://doi.org/10.1038/s41598-017-17957-z

Saponari, M., Giampetruzzi, A., Loconsole, G., Boscia, D., & Saldarelli, P. (2018). *Xylella fastidiosa* in Olive in Apulia: Where we stand. *Phytopathology, 109*, 175–186. https://doi.org/10.1094/PHYTO-08-18-0319-FI

Scortichini, M. (2020). *Xylella*, nuovi protocolli di convivenza negli oliveti pugliesi. Frutticoltura n.2, pp. 52–54.

Simeone, V., Tucci, M., Vitti, R., Calabrese, J., & Perri, E. (2013). Linee guida per la gestione degli oliveti secolari. *La rivista di scienza dell'alimentazione, 1*, 135–136.

Chapter 16
Challenges for Sustainable Water Resource Management in Italian Olive Groves: Irrigation and Salinity

Nelson Marmiroli, Caterina Agrimonti, Elena Maestri, and Sara Petrucci

Abstract Requirements for water in olive growing are highly differentiated due to the extreme heterogeneity, comparing the Regions in which olive is cultivated. Climate, slopes, and soil are factors affecting evapotranspiration and water availability. Irrigation approaches are also determined by the development status of the plant. Irrigation with salty water is possible based on the cultivars. Subirrigation and controlled deficit irrigation are probably the latest, more sustainable approaches.

16.1 Olive Cultivation and Water Needs

The Italian territory spans different regions with peculiar climatic features, and the olive cultivars grown in each region have adapted to these in the course of their previous history. However, recent years have seen changes in rainfall levels, in temperatures, and in the occurrence of extreme events. These changes are among the main challenges facing olive growing nowadays.

The olive tree, *Olea europaea*, is endemic of the Mediterranean area and has evolved by adapting to environmental conditions with a common occurrence of water stress, with summer drought periods alternating with autumn-winter precipitations. Olive growing is in fact widespread outside of Europe in areas with Mediterranean-like features, such as California, Chile, and South Africa. The resilience of the tree to water stress is due to anatomical and physiological features of the leaves and root system, vascular tissues, and cells; however, the different

N. Marmiroli (✉) · C. Agrimonti · E. Maestri
Interdepartmental Center for Energy and Environment (CIDEA), University of Parma, Parma, Italy
e-mail: nelson.marmiroli@unipr.it; caterina.agrimonti@unipr.it; elena.maestri@unipr.it

S. Petrucci
AIAB Associazione Italiana Agricoltura Biologica, Genova, Italy
e-mail: info@sarapetrucci.it

J. Muñoz-Rojas, R. García-Ruiz (eds.), *The Olive Landscapes of the Mediterranean*, Landscape Series 36,
https://doi.org/10.1007/978-3-031-57956-1_16

cultivars show variability in these attributes, and this affects their adaptability (Alfieri et al., 2019). It is reported that olive trees can tolerate levels of water potential much lower than those tolerated by other trees, as low as −3 MPa (Brito et al., 2019). Due to this natural behavior, olive growing often does not require irrigation: just 14% of olive groves in Italy are irrigated (Fig. 16.1), an area of about 165,000 ha. Olive groves account for 6% of the Italian agricultural area which is irrigated (Masia et al., 2018). Nonetheless, lack of water can jeopardize both the productivity and the quality of olives and oil, and modern techniques are now available to assist growers, as previously mentioned (see Chap. 14).

Irrigation is in fact considered beneficial to traditional olive groves, but essential to high-density groves. The benefits attributed to irrigation are: (i) early beginning of production; (ii) decrease in alternation; (iii) increase in yield per tree; (iv) increase in fruit dimension; (v) increase in the proportion of pulp; and (vi) increase in oil yield per tree. Considering the quality of olive oil, irrigation affects the concentration of phenolic compounds but does not significantly affect the fatty acid composition (Marra et al., 2016). A historical analysis spanning 14 years has shown a positive correlation between precipitation and olive production in different cultivars and in different areas in Central and Southern Italy (Orlandi et al., 2020). Recently, a new simulation model has been proposed for predicting the productivity of olive groves taking into account water stress and climate parameters (López-Bernal et al., 2018): the model, called OliveCan, has been applied to olive groves in Tuscany and has been evidenced that water provision is the main factor affecting yield (Mairech et al., 2020).

Fig. 16.1 Examples of olive groves without irrigation in Region Lazio. (Photos by Elena Maestri)

Gucci (2012) found that irrigation doubled the productivity in intensive olive groves in Tuscany, and increased it by about 40–60% in the Southern Regions of Italy.

The effects of irrigation on olive fruit productivity are mostly evident in those areas with annual rainfall below 800 mm, or in years with particularly long dry periods. In Italy, these areas correspond to Southern regions and the islands, Sicilia and Sardegna, where the summer drought lasts over 3 months and annual rainfall is below 800 mm.

16.2 Main Challenges of Irrigation in Olive Cultivation

In face of the new challenges due to climate changes, water consumption must be optimized to avoid wasting a precious resource (Masia et al., 2018). It must be highlighted that an excess of water might have detrimental effects, such as abundant development of vegetative parts and lower resistance to winter temperatures (Proietti & Regni, 2019). It has been estimated that intensive olive groves with 200–400 trees/ha need 1000–3500 m^3/ha/year, whereas super-intensive groves require 2500 m^3/ha/year. However, it is predicted that water loss due to evapotranspiration in olive groves will increase by 8% until 2050 in the Mediterranean area, and correspondingly irrigation requirements will also increase by about 18% (Tanasijevic et al., 2014). According to Orlandi et al. (2020), all scenarios for climate changes forecast a decrease in olive production in the coming years, partly due to an increase in aridity across the Mediterranean.

Water quantity is relevant, but also the timing of irrigation is critical. The most important phases are the flowering and fruit setting stage in spring, the growth of fruits, and the production of oil components in summer.

Leaf transpiration in olive trees can be as high as 0.08 mL/m^2/s. In summer during the day transpiration always exceeds the water uptake by roots, and therefore the plant loses water from the tissues. In the night, the situation is reversed. Additionally, in traditional olive groves, evaporation from the soil surface can represent 50% of the whole water consumption, adding to the water shortage.

The best and most sustainable solution to address water management takes into consideration the real needs of the plants, in order to estimate the deficit and provide the exact quantity of water needed. This can be obtained by measuring the water status of soil and plant, and evapotranspiration.

Measuring the water status of soils requires extensive sampling, or else the positioning of sensors. It is affected by the soil properties, and it is limited to the most superficial horizons. Heterogeneity in soil is often a peculiar feature in Italian olive groves, and this complicates the evaluation of the water status. In a study comparing the efficacy for different irrigation regimes in the Calabria Region, it was evidenced how measurements of water status are beneficial to support choices about the frequency and quantity of irrigation (Capra & Scicolone, 2018).

Measuring the water status of plants can be done at leaf level or for the whole plant. Evapotranspiration is measured in mm of water dispersed to the atmosphere from soil or plants and can be estimated also through indirect methods based on the meteorological conditions, particularly with the Penman-FAO equation, the Penman-Monteith equation or the Hargreaves formula (Allen et al., 1998). Detailed calculation procedures are reported in Gucci (2012) and in the handbook on olive growing published by the International Olive Council (2007).

Irrigation can be based on Controlled Water Deficit, which means to decrease water supply in those phases that are less sensitive, increasing the efficiency in water use in those periods when it is needed. The purpose is to avoid water stress in the period when cells need to grow, during flowering and fruit setting. This irrigation principle is very effective in areas with problems in water supply. Moreover, it has been shown to increase the quality of oil with a higher content of polyphenols (Gucci et al., 2019; Servili et al., 2007).

In this application, called regulated deficit irrigation (RDI), olive groves will receive a quantity of water that is lower (about 70%) than the requirements; for example, 1600 m^3/ha in a super-intensive olive grove, instead of 2500 m^3/ha. Deficit periods must be alternated to recovery periods. Also, water deficit cannot be applied to young plants.

With deficit irrigation, there is better control of the vegetative growth of the crown and a more balanced development of the root system, which tends to deepen and expand laterally. An extensive root system offers advantages both from the point of view of the absorption of mineral elements and water and in terms of anchoring the tree. Regardless of the volume and the periods of restriction, the adoption of irrigation in controlled deficit requires:

- High uniformity of water distribution.
- A clear knowledge and a periodic evaluation of the water condition of the trees.
- Timely intervention when necessary.

Water-saving scenarios have been formulated for the Puglia region, the main producer of olive oil in Italy, as a component of decision support systems for a more sustainable use of water (Daccache et al., 2015). The application of the OliveCan model, specifically developed to evaluate the impact of water stress, has shown a beneficial effect of deficit irrigation on yields, which becomes more evident in regions with high evapotranspiration or low rainfall levels, since it increases the water use efficiency (Mairech et al., 2020).

The application of deficit irrigation can be made more sustainable and effective by introducing biochar, the product of biomass pyrolysis, as an amendment. A recent literature review (Singh et al., 2019) reports that biochar, with high porosity and surface area, shows beneficial effects on water use efficiency in several crops, as well as other positive impacts on nutrient retention, soil aeration, and increased microbial activity. Research is still needed to show in which way biochar affects soil water content and availability to the plant (Atkinson, 2018).

In order to provide water in a tightly controlled manner, it is necessary to use methods for precise irrigation, such as drip irrigation, microirrigation, and sub-irrigation, which bring the water directly to the root system. These systems increase water use efficiency because they minimize the losses due to evaporation or presence of other plants. Also, they decrease runoff and leaching and are compatible with fertirrigation. However, it is necessary to filter the water before using it, to avoid clogging of the systems.

The drip systems can be positioned on the soil or above the soil, with different advantages and disadvantages. Mostly, they can interfere with mechanized procedures. Microspray systems are also possible and can be more effective in traditional olive groves. As a general rule, olive groves in sandy soils with high permeability will need a higher number of watering heads, whereas clayey soils require a lower number.

Intensive groves will have 1–2 drip lines per each row with a distance of 60–120 cm between heads; super-intensive groves need a distance of 40 cm; traditional groves require heads near each tree (Proietti & Regni, 2019).

Sub-irrigation, subsurface drip irrigation, instead makes use of water-carrying systems below the surface, with a further reduction in evaporation and no interference with mechanical operations. The installation is more complex, the initial costs are higher, but the performance is more effective. This is rapidly gaining favor in Puglia, Sardegna, and other regions, even in the North of Italy (Caruso et al., 2017).

New developments in the use of remote sensing could be promising. Estimation of Leaf Area Index and canopy volume have been performed with Unmanned Aerial Vehicles (Caruso et al., 2019; Torres-Sánchez et al., 2015). A Crop Water Stress Index can be derived from aerial thermal images, as has been shown in Spanish super-intensive olive groves (Egea et al., 2017). An extensive review of tools using multi-spectral image data to assess the crop water status and use has shown that, together with meteorological data, these can be integrated into decision support systems for end-users, advisory, and extension services (Calera et al., 2017).

16.3 Water Salinity and Possible Uses

An advantage of localized irrigation systems consists of the possible use of moderately saline water. Use of water with salts can be important to exploit wastewater in areas where clean water is scarce (Segal et al., 2011). It is of course mandatory to check for any phytotoxicity effects from pollutants or chemical compounds. The amount of nutrients provided with these waters must also be factored in the computations regarding fertilization. Spreading olive mill wastewater is seen as a sustainable solution for disposing of a problematic byproduct, and these can also contribute salts and other compounds (Zipori et al., 2020). The hazards for contamination of aquatic ecosystems must be factored in the decisions.

The problem with salinity in water is a challenge for agriculture worldwide. In Italy, this concerns mostly the territories near the coasts in the mainland (Toscana, Lazio, Puglia, Basilicata, Campania) and in the islands (Sardegna, Sicilia) (Costantini et al., 2004; Righini et al., 2002; https://esdac.jrc.ec.europa.eu/content/

saline-and-sodic-soils-european-union). In these areas, the withdrawal of water for irrigation and other uses leads to infiltration of seawater.

Salinity can harm olive plants and also the soil structure and properties. Salt usually consists of Na^+ and Cl^- ions, but Ca^{2+}, Mg^{2+}, and SO_4^{2-} can also be an issue. The main effect of salt is to decrease the water potential and the gradient needed for plants to take up water. Olive is reported to tolerate water with conductivity up to 5 dS/m with low-stress response (Chartzoulakis, 2005), but this also depends on the cultivar: Frantoio is particularly resistant, whereas Leccino is considered to be sensitive (Tattini et al., 1992).

Irrigation with saline water therefore is applied with the following provisions: use of sufficient water to facilitate leaching of salts; drainage of water to eliminate the leached salts; use of chemicals to increase solubility of salts and improve their leaching.

Interesting considerations can be made regarding the practice of intercropping and use of cover crops in olive groves. These can be beneficial under several aspects, including counteraction against erosion and runoff, but the additional plants will also add competition for water and nutrients. Therefore, a permanent cover can be advisable where irrigation is present; if the area is arid, then the presence of other plants must be temporary and limited to the rainy season (Zipori et al., 2020).

The use of mulching along the rows, in the first years since the planting of young olive groves, can help in saving water. If this is made with plastic materials such as black polyethylene, it offers, among the advantages, a decrease in water losses by evapotranspiration. Due to the high costs for the purchase, installation, and maintenance, as well as the problems of environmental impact consequent to their disposal, this application is to be carefully evaluated. In addition, the visual impact of plastic sheeting can be unpleasant in prestigious landscape contexts. The case of green plant materials for mulching is different, when they derive from the mowing of the turf and are conveyed with special equipment on the row of trees. Their main function in this case would be precisely to limit the evaporation of water during the spring-summer period and to bring new organic substance, with the substantial advantage of not causing additional costs (Loreti, 2012).

16.4 Main Challenges to Sustainable Olive Growing

In conclusion, we can highlight some important factors that make irrigation an important component of sustainability in olive groves in Italy.

- Climate change models predict a reduction of olive groves productivity, and irrigation can be an important adaptation measure to maintain productivity.
- Climate change will decrease the amount of water available in the Mediterranean area and the water quality, and a sustainable use of water will require a decrease in irrigation.

- In the framework of circular economy, irrigation with reclaimed wastewater can become a possibility, due to the tolerance to salinity of olive trees.
- Considering the relevance of water in some phenological stages of olive production, new irrigation regimes must be verified for their impact on oil quality and yield.
- Irrigation is a cost impacting on economical sustainability, and decision support systems are needed to assist the grower in the choices.
- Precision farming and models will become a critical factor in managing water supply to olive groves in the face of climate changes.
- Amendments which can improve soil water retention and increase the efficiency in water use will be a beneficial addition.

References

Alfieri, S. M., Riccardi, M., Menenti, M., Basile, A., Bonfante, A., & De Lorenzi, F. (2019). Adaptability of global olive cultivars to water availability under future Mediterranean climate. *Mitigation and Adaptation Strategies for Global Change, 24*, 435–466. https://doi.org/10.1007/s11027-018-9820-1

Allen, R. G., Pereira, L. S., Raes, D., & Smith, M. (1998). *Crop evapotranspiration. Guidelines for computing crop water requirements* (Irrigation and drainage paper 56). FAO.

Atkinson, C. J. (2018). How good is the evidence that soil-applied biochar improves water-holding capacity? *Soil Use and Management, 34*, 177–186. https://doi.org/10.1111/sum.12413

Brito, C., Dinis, L.-T., Moutinho-Pereira, J., & Correla, C. M. (2019). Drought stress effects and olive tree acclimation under a changing climate. *Plants, 8*(7), 232. https://doi.org/10.3390/plants8070232

Calera, A., Campos, I., Osann, A., D'Urso, G., & Menenti, M. (2017). Remote sensing for crop water management: From ET modelling to services for the end users. *Sensors, 17*, 1104. https://doi.org/10.3390/s17051104

Capra, A., & Scicolone, B. (2018). Irrigation scheduling optimisation in olive groves. *Journal of Experimental Agriculture International, 28*(4), 1–19. https://doi.org/10.9734/JEAI/2018/44582

Caruso, G., Gucci, R., Sifola, M. I., Selvaggini, R., Urbani, S., Esposto, S., Taticchi, A., & Servili, M. (2017). Irrigation and fruit canopy position modify oil quality of olive trees (cv. Frantoio). *Journal of the Science of Food and Agriculture, 97*, 3530–3539.

Caruso, G., Zarco-Tejada, P. J., González-Dugo, V., Moriondo, M., Tozzini, L., Palai, G., Rallo, G., Hornero, A., Primicerio, J., & Gucci, R. (2019). High-resolution imagery acquired from an unmanned platform to estimate biophysical and geometrical parameters of olive trees under different irrigation regimes. *PLoS One, 14*(1), e0210804. https://doi.org/10.1371/journal.pone.0210804

Chartzoulakis, K. S. (2005). Salinity and olive: Growth, salt tolerance, photosynthesis and yield. *Agricultural Water Management, 78*, 108–121. https://doi.org/10.1016/j.agwat.2005.04.025

Costantini, E., Urbano, F., & L'Abate, G. (2004). *Soil regions of italy*. CRA-ISSDS. Available at http://www.soilmaps.it/ita/downloads.html. Accessed 17 July 2020

Daccache, A., d'Agostino, D., Lamaddalena, N., & El Chami, D. (2015). A decision tool for sustainable agricultural policies: The case of water saving scenarios for Apulia region (southern Italy). *Water Policy, 18*, 126–142. https://doi.org/10.2166/wp.2015.050

Egea, G., Padilla-Diaz, C. M., Martinez-Guanter, J., Fernandez, J. E., & Perez-Ruiz, M. (2017). Assessing a crop water stress index derived from aerial thermal imaging and infrared thermometry in super-high density olive orchards. *Agricultural Water Management, 187*, 210–221. https://doi.org/10.1016/j.agwat.2017.03.030

Gucci, R. (2012). *Irrigazione. Volume IX Collana divulgativa dell'Accademia*. Accademia Nazionale dell'Olivo e dell'Olio. ISSN 2281–4930.

Gucci, R., Caruso, G., Gennai, C., Esposto, S., Urbani, S., & Servili, M. (2019). Fruit growth, yield and oil quality changes induced by deficit irrigation at different stages of olive fruit development. *Agricultural Water Management, 212*, 88–98. https://doi.org/10.1016/j.agwat.2018.08.022

International Olive Council. (2007). *Production techniques in olive growing*. International Olive Council. ISBN 978-84-931663-6-6.

López-Bernal, Á., Morales., A, García-Tejera, O., Testi, L., Orgaz, F., De Melo-Abreu, J. P., & Villalobos, F. J. (2018). OliveCan: A process-based model of development, growth and yield of olive orchards. *Front. Plant Science 9*, 632. https://doi.org/10.3389/fpls.2018.00632

Loreti, F. (2012). *La gestione del suolo. Volume VIII Collana divulgativa dell'Accademia*. Accademia Nazionale dell'Olivo e dell'Olio. ISSN 2281-4930.

Mairech, H., Lopez-Bernal, A., Moriondo, M., Dibari, C., Regni, L., Proietti, P., Villaloobos, F. J., & Testi, L. (2020). Is new olive farming sustainable? A spatial comparison of productive and environmental performances between traditional and new olive orchards with the model OliveCan. *Agricultural Systems, 181*, 102816. https://doi.org/10.1016/j.agsy.2020.102816

Marra, F. P., Marino, G., Marchese, A., & Caruso, T. (2016). Effects of different irrigation regimes on a super-high-density olive grove cv. "Arbequina": Vegetative growth, productivity and polyphenol content of the oil. *Irrigation Science, 34*, 313–325.

Masia, S., Sušnik, J., Marras, S., Mereu, S., Spano, D., & Trabucco, A. (2018). Assessment of irrigated agriculture vulnerability under climate change in southern Italy. *Water, 10*, 209. https://doi.org/10.3390/w10020209

Orlandi, F., Rojo, J., Picornell, A., Oteros, J., Perez-Badia, R., & Fornaciari, M. (2020). Impact of climate change on olive crop production in Italy. *Atmosphere, 11*, 595. https://doi.org/10.3390/atmos11060595

Proietti, P., & Regni, L. (2019). Handbook for a sustainable management of the olive groves. Available at https://olive4climate.eu/wp-content/uploads/Olive4Climate-Handbook-_ENG.pdf. Accessed 25 September 2024.

Righini, G., Costantini, E., & Sulli, L. (2002). La banca dati delle regioni pedologiche italiane. *Bollettino della Società Italiana Scienza del Suolo*. Available at http://www.soilmaps.it/download/csi-soil_regions.pdf. Accessed 17 July 2020

Segal, E., Dag, A., Ben-Gal, A., Zipori, I., Erel, R., Suryano, S., & Yermiyahu, U. (2011). Olive orchard irrigation with reclaimed wastewater: Agronomic and environmental considerations. *Agriculture, Ecosystems and Environment, 140*, 454–461. https://doi.org/10.1016/j.agee.2011.01.009

Servili, M., Esposto, S., Lodolini, E., Selvaggini, R., Taticchi, A., Urbani, S., Montedoro, G., Serravalle, M., & Gucci, R. (2007). Irrigation effects on quality, phenolic composition, and selected volatiles of virgin olive oils Cv. Leccino. *Journal of Agricultural and Food Chemistry, 55*, 6609–6618. https://doi.org/10.1021/jf070599n

Singh, M., Kaur Saini, R., & Singh, S. (2019). Potential of integrating biochar and deficit irrigation strategies for sustaining vegetable production in water-limited regions: A review. *HortScience, 54*, 1872–1878. https://doi.org/10.21273/HORTSCI14271-19

Tanasijevic, L., Todorovic, M., Pereira, L. S., Pizzigalli, C., & Lionello, P. (2014). Impacts of climate change on olive crop evapotranspiration and irrigation requirements in the Mediterranean region. *Agricultural Water Management, 144*, 54–68. https://doi.org/10.1016/j.agwat.2014.05.019

Tattini, M., Bertoni, P., & Caselli, S. (1992). Genotypic responses of olive plants to sodium chloride. *Journal of Plant Nutrition, 15*, 1467–1485.

Torres-Sánchez, J., López-Granados, F., Serrano, N., Arquero, O., & Peña, J. M. (2015). High-throughput 3-D monitoring of agricultural-tree plantations with unmanned aerial vehicle (UAV) technology. *PLoS One, 10*(6), e0130479. https://doi.org/10.1371/journal.pone.0130479

Zipori, I., Erel, R., Yermiyahu, U., Ben-Gal, A., & Dag, A. (2020). Sustainable management of olive orchard nutrition: A review. *Agriculture, 10*, 11. https://doi.org/10.3390/agriculture10010011

Chapter 17
Management Practices Boosting Ecosystem Services and Life Cycle Assessment of Italian Olive Groves

Elena Maestri, Nelson Marmiroli, Alessandro Triantafyllidis, Giacomo Francesco, Bruno Francsco Nicolò, Nathalie Iofrida, and Anna Irene De Luca

Abstract The main ecosystem services identifiable for traditional olive groves are listed and analyzed according to the standard definitions (MEA, Ecosystems and human well-being: a framework for assessment. Island Press, Washington, DC, 2005). The possible ecosystem services obtainable with new and improved management practices will be discussed, with the aim of assisting in the choice of new agronomic practices for sustainability. Although being so widespread and having a relevant impact on the economy and the market, olive growing is associated with several adverse effects on the environment that cause resource depletion, land degradation, pollutant emissions, and waste generation. The impacts of olive cultivation and olive oil production may vary significantly according to the practices and techniques applied, therefore, appropriate analysis techniques to assess those impacts are of utmost importance. It is of fundamental importance the choice of evaluation method to approach toward the ecosystem services knowledge. LCA (Life Cycle Assessment) is one of the most widely applied environmental assessment methods in recent years, which allows to characterize and quantify the impacts in terms of specific indicators (e.g., global warming, human toxicity, eco-toxicity, and water depletion potential) and/or end-point indicators (human health, ecosystem, and resources) taking into

E. Maestri · N. Marmiroli
Interdepartmental Center for Energy and Environment (CIDEA), University of Parma, Parma, Italy
e-mail: elena.maestri@unipr.it; nelson.marmiroli@unipr.it

A. Triantafyllidis
AIAB Associazione Italiana Agricoltura Biologica, Genoa, Italy
e-mail: a.trianta@tin.it

G. Falcone · B. F. Nicolò · N. Iofrida · A. I. De Luca (✉)
Department of Agriculture (AGRARIA), Mediterranean University of Reggio Calabria, Reggio Calabria, Italy
e-mail: giacomo.falcone@unirc.it; bruno.nicolo@unirc.it; Nathalie.iofrida@unirc.it; anna.deluca@unirc.it

J. Muñoz-Rojas, R. García-Ruiz (eds.), *The Olive Landscapes of the Mediterranean*, Landscape Series 36,
https://doi.org/10.1007/978-3-031-57956-1_17

account all phases of the life cycle of a product or service. The methodology is standardized and normed by ISO 14040-14044 (ISO 14040:2006 life cycle assessment – principles and framework. International Organization for Standardization, Geneva, 2006a; ISO 14044:2006 life cycle assessment – requirements and guidelines. International Organization for Standardization, Geneva, 2006b). The strength of this methodology mainly stays in the possibility of analyzing all life cycle phases, detecting burden shifts and hotspots, and taking into account background and foreground systems. Many of these indicators can make a useful contribution to the assessment of the impacts on ecosystem services and the Italian olive landscape, even if there is not yet a specific framework or a consolidated approach to assess ecosystem services with life cycle methods. In this paragraph, a short review about the contribution of the Life Cycle Assessment (LCA) approach applied to Italian olive growing and oil production is carried out to investigate the environmental implications of traditional and new sustainable practices of olive-orchards in Italy, also in terms of ecosystem services.

17.1 Introduction

Agricultural cultivation is an activity that contributes to the increase in the depletion and deterioration of biotic and abiotic resources and consequently leading to environmental impacts and reduction of ecosystem services (ES). However, thanks to its multifunctional role, agricultural activities can also have positive effects on ES. According to Millennium Ecosystem Assessment (MEA), more than 60% of ES is being degraded or used unsustainably as a consequence of actions taken to increase the supply of other services, such as food (MEA, 2005).

Olive growing has had a great expansion over the centuries, contributing to shape most of the Italian landscape: indeed, today, olive trees are a strongly characterizing element of the Mediterranean areas, particularly in marginal areas. In Italy, the olive cultivation is present in almost all regions thanks to its genetic variability, which allowed adapting to many environmental conditions, creating a heterogeneous Italian landscape (Agnoletti, 2013).

Olive cultivation is now diffused from the plains to the hilly and mountainous areas, where most of the historic Italian olive groves are concentrated. Therefore, olive trees are present in mixed cultivation systems, in forest formation, in large monocultures, and in terraced systems that have generated a high historical-cultural value of the Italian landscape (Fig. 17.1).

17.2 The Ecosystem Services of Olive Growing

In 2005, the MEA classified ES into four functional groups, as synthetized in Table 17.1.

For many of these services, the economic value is not accounted for in the market and, as a result, they are overused or polluted. Olive growing, as well as any other

Fig. 17.1 Heterogeneity of olive growing in Italy. Top left, Calabria region. Top right, Toscana region. Bottom left, Liguria region. Bottom right, Puglia region. Different tree typology and soil management practices are evidenced

Table 17.1 Ecosystem services

Functional group	Services provided	Examples
Support	Services necessary for the production of all other ecosystem services	Soil formation, nutrient cycle, and primary biomass production
Regulation	Benefits obtained from the regulation of ecosystem processes	Climate, water regime, pathogens
Supply	Products obtained from ecosystems	Food, pure water, fibers, fuel, medicines
Culture	All the nonmaterial benefits obtained from ecosystems	Spiritual, ethical, recreational, aesthetic, social relations

Source: Our elaboration from MEA (2005)

agricultural activity, has an impact on natural resources; therefore, useful practices for sustainable management are of utmost importance to limit pollution and environmental damages. According to Demestihas et al. (2017), ecological functions modified by agricultural practices can provide six ecosystem services, i.e., fruit production, climate regulation, soil nitrogen availability, water regulation, pest and disease control, and pollination.

Olive orchards can sequester annually from 2.4 to 12.5 tons C/ha, especially in young trees than in older ones (Demestihas et al., 2017). Furthermore, pruning, which is intensively used in olive orchards, helps increase photosynthesis rates and, so, carbon sequestration (Demestihas et al., 2017). Italian olive growing, if properly managed, can have a multifunctional and sustainable role concerning ecosystem services, such as contributing to food supply, while protecting the environment and shaping the landscape. Sustainable olive growing practices can assure the conservation and fertility of the soil, land stability, and a better soil–water balance and are able to maintain high levels of biodiversity and agro-ecosystem complexity through integration with connected agro-activities and with the rural landscape (Agnoletti, 2013; Parras-Alcántara et al., 2016). Figure 17.2 provides some examples of the relationship between olive growing and the ecosystem services on which it can have possible impacts.

Burkhard and Müller (2015) applied, on different case studies, an approach named "ES matrix" for the assessment of different land-cover types' capacities to support ecosystem functions, to supply multiple ES, and to identify demands for ES. Results of this study showed a maximum and high relevant capacity of land cover with olive groves in providing cultural services category, especially for recreation and tourism, cultural heritage, and cultural diversity; high capacity also resulted for provisioning of services category to the functions of wood fuel and crops. Medium capacity was shown in the ecological integrity to the functions of exergy capture, storage capacity, biotic water flows, and heterogeneity and in the regulating services category to the function of pest and disease control.

Conventional olive cultivation techniques can produce effects on soil properties and their functions by generating degradation processes of the soil organic carbon (SOC) stock. Nevertheless, the adoption of sustainable field practices (e.g., no

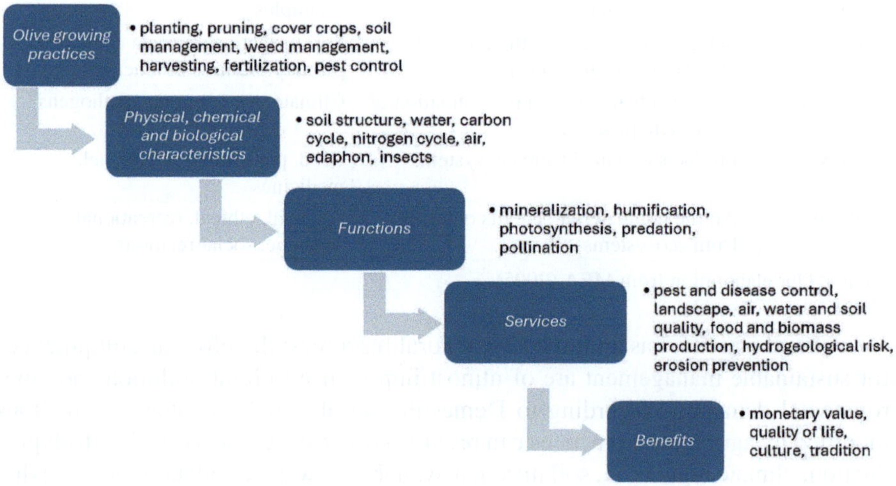

Fig. 17.2 Cause-effect chain between olive growing and ecosystem services (ES). (Haines-Young & Potschin, 2018; Parras-Alcántara et al., 2016; Demestihas et al., 2017; Rugani et al., 2019)

tillage, increased carbon input, recycling of pruning residues and cover crops) could increase SOC stocks (Montanaro et al., 2017).

Regarding the effects of olive cultivation on soil properties, Brunori et al. (2018) have adopted a landscape ecology approach and by means of soil bio-chemical indicators have analyzed the impact of terraced olive agro-ecosystems on the provision of soil-related ES. This study conducted in southern Lazio (Italy) demonstrates that poly-cultural agro-ecosystems can promote the maintenance of soil fertility. In particular, soil management affected positively SOC in olive groves, leading to a high carbon stabilization rate.

Green mulching and intercropping with perennial crops in olive orchards have been successfully adopted to prevent soil erosion and soil degradation and to increase biodiversity (Paris et al., 2019). The use of cover crops has been recognized as a practice that contributes to soil fertility and biodiversity for microflora and microinvertebrates, also by increasing water percolation and organic matter supply (Proietti & Regni, 2019). The presence of animals or livestock in the olive groves is not common in Italy. Disadvantages of this practice concern soil compacting and excessive feeding on olive leaves. However, it could be a system for supplying additional organic matter to the soil (Proietti & Regni, 2019). For the same purpose, also chipped pruning residues or composted pomace are seen as a way for increasing soil fertility through organic matter. However, an excess of pomace could be detrimental to soil fauna or microbial activity (Zipori et al., 2020).

The scientific community is looking for ever better tools for ES evaluation, from various points of view. In last decades, the Life Cycle Assessment (LCA) has proved to be a reliable and widely accepted tool to quantify environmental loads or benefits in several context. Recently, numerous approaches have attempted to incorporate ES in the characterization step of LCA in order to determine one or more impact categories with related indicators (Alejandre et al., 2019; Blanco et al., 2018; Othoniel et al., 2016). In this context, the evaluation of environmental performance by means sustainability tools, as LCA, can be useful as predictive method for planning strategies to avoid the risk of degradation of ES, which could lead to a reduction in human well-being. For instance, the distribution of fertilizers and phytosanitary products on agricultural productions could affect ES so the development and implementation of specific models or assumptions to estimate the fate of these inputs, allows the LCA practitioner to support the design of sustainable agricultural processes from a multifunctional point of view.

17.3 Life Cycle Assessment (LCA) Application to Italian Olive Oil Systems and for Their Ecosystem Services

A rich international scientific literature has been working in recent years to evaluate the environmental performance generated from the olive growing activity, and most studies applied LCA to olive cultivation also in Italian regions to investigate the environmental burden caused from each one of the main steps within the life cycle in olive supply chain, from extraction of raw materials to its end of life. However, there are still very few studies that effectively and convincingly integrate LCA with ES's contexts.

The methodological framework of LCA was standardized in the UNI EN ISO 14040:2006 (ISO, 2006a) and UNI EN ISO 14044:2006 (ISO, 2006b) by the International Organization for Standardization (ISO). The first norm described the principles and framework to perform an LCA study, while the second illustrated the guidelines and the fundamental steps of the analysis.

LCA is a quantitative methodology used to assess all the related and interdependent phases of a product's life cycle, in order to account for its associated potential environmental impacts. Generally, in the LCA application, the life cycle is modeled by distinguishing three main phases of a product/process life: the upstream processes, as production and extraction of the raw materials; the core processes, as processing and assembly of product; the downstream processes, as distribution, use, and end of life of the product.

General framework of LCA analysis was structured in four consecutive steps linked to each another through an iterative approach that allows the revision of previous steps, if necessary (ISO, 2006a, b): goal and scope definition; Life cycle inventory; Life Cycle Impact Assessment; Interpretation of results.

Olive growing has been widely analyzed through LCA methodology, as demonstrated in the scientific literature. Environmental impact analysis with the LCA method is the most widely used evaluation method in the olive oil sector, with the main purpose of comparing alternative production scenarios. The first LCA study carried out in the olive oil sector (Nicoletti & Notarnicola, 2000) aimed to compare olive growing systems conducted in dry and irrigation, together with different olive oil extraction techniques. A decade later, Salomone and Ioppolo (2012) and Salomone et al. (2010) analyzed eight different scenarios including different combinations of production practices, oil extraction methods, and oil waste treatment: the results of the analysis indicated a high environmental load in both for conventional scenarios and for some sub-processes such as fertilization, pesticide use and combustion of spent pomace, and for the sub-processes of composting wet olive pomace with manure in the field compared to the composting of oil wastewater with wet pomace with composting machines, and finally a significant positive contribution, in terms of environmental credits, associated with the use of co-products as fertilizers or fuels.

Pattara et al. (2016) analyzed five case studies in Abruzzo using the Carbon Footprint method, which mainly aimed at quantifying greenhouse gas emissions related to olive cultivation and olive oil production and identifying those responsible for these emissions (excluding the planting and distribution phase of olive trees). Compared to the existing literature, the latter research developed a more detailed collection of data on companies and selected case studies to represent the current production situation at international level in terms of technology and size.

Five liters of extra virgin olive oil, with primary and secondary packaging, was chosen as a functional unit. The results showed that agriculture counts for CO2eq emissions from 3.34 to 7.74 kg (mainly due to treatments with fertilizers and pesticides), followed by the packaging process in the industrial phase for CO_{2eq} emissions from 1.13 to 3.20 kg (for which glass bottles represent the largest load). The study revealed that a realistic reduction in greenhouse gas emissions should be based on the efficient use of pesticides and fertilizers.

The most recent research about olive growing environmental impacts has been published by Stillitano et al. (2019), Espadas-Aldana et al. (2019), and Maffia et al. (2020). The study by Stillitano et al. (2019) showed that olive milling with the innovative plant resulted in olive oil with a significant increase in quality, although the extraction yield was significantly higher when using conventional technology. In terms of environmental results, an average growth of 4.5% of the impacts in all categories was reached. Espadas-Aldana et al. (2019) conducted a review about scientific studies on LCA applied to the olive growing sector: the literature analyzed converged on an unequivocal environmental hotspot, i.e., the agricultural phase; which represents the most impactful phase of the olive oil life cycle, due in particular to fertilization, pesticides treatment, and irrigation. Waste management and distribution also appear to represent a crucial issue. Maffia et al. (2020) evaluated the environmental impacts of the production of olive oil in the Campania region by comparing six olive oil production systems (two organic-certified, two integrated, and two organic-hobbyists). The authors found that the production of 1 L of oil caused CO_2eq emissions on average from 0.22 up to 3.39 kg. The least impacting cultivation method was the hobby one, which did not use pesticides and carried out organic fertilization. On average, packaging represented 60% of total impacts; so, the materials and formats choice by farms have strong effects on the environmental sustainability of oil production.

The inclusion of the evaluation of ecosystem services into the LCA began quite recently and still remains an open issue due to the many limitations of approaches and assessment methods (Raymundo Pavan & Ometto, 2018). In a recent review conducted by Othoniel et al. (2016) they are highlighted the incompleteness and the limitation of the LCA models used to incorporate ES assessment.

As argued by Bruel et al. (2016), a few approaches have already tried to account for ES in LCA, by creating specific characterization factors to assess the ES in the LCIA phase.

Alejandre et al. (2019) recently, proposed a systematic approach to incorporate the evaluations of environmental performance in terms of ES into Life Cycle Impact Assessment (LCIA) methods. Of the different LCIA methods, ReCiPe2016 was the one with the most relations between impact categories and ES. Five mid-point impact categories of ReCiPe2016 were related to ES. In particular, Climate change and stratospheric ozone depletion categories were directly linked with Regulation and Maintenance services; Water use was directly linked with the water provisioning categories; and Mineral resource scarcity and Fossil resource scarcity were related to Abiotic Provision (Alejandre et al., 2019).

Afterward, next to a summarization of 31 ecosystem services groups ecosystem based on CICES V5.1 classification, authors selected services that are considered pertinent for environmental LCAs, by obtaining 15 impact categories that form the optimal coverage of ecosystem services, 11 of which are still entirely lacking in LCAs.

Concluding, LCA is a relatively young methodology, born in the energy and industrial field, soon found an application in the agri-food sector where many critical issues arise from the difficulty of modeling biological processes that inevitably influence the life cycle of a product (De Luca et al., 2015; Gulisano et al., 2018). In particular, agricultural production is always an attempt by humans to govern natural phenomena of which; however, they have only partial control. The LCA practitioner community is spending a lot of energy on understanding these phenomena to create holistic environmental impact assessment models that take into account the interaction between humans, the agro-ecosystem and the natural environment (Alejandre et al., 2019). Some LCIA methods contain the basis for the analysis of ES, however, the scientific community must strive to identify factors that characterize impacts on ecosystem services and make them operational.

17.4 Conclusions

The possible ecosystem services obtainable from olive growing, according to MEA (2005) standard definitions, are long been the subject of great interest especially in order to understand how new and improved management practices can improve their provisioning and then also the overall sustainability of olive territories. As can easily guessed, olive growing, depending on their management (traditional, intensive, super-intensive, organic, conventional, and so on), can be useful to provide several typologies of functions, like as: services necessary for the production of all other ecosystem services; benefits obtained from the regulation of ecosystem processes; products obtained from ecosystems; all the nonmaterial benefits obtained from ecosystems. On the other hand, from an environmental point of view, the impacts of olive cultivation and olive oil production may vary significantly according to the practices and techniques applied, therefore, appropriate analysis techniques to assess those impacts are of utmost importance. In this sense, therefore, it is of fundamental importance the choice of evaluation method to approach toward the ecosystem services knowledge. LCA (Life Cycle Assessment) is one of the most widely applied environmental assessment methods in recent years, which allows to characterize and quantify the impacts in terms of specific indicators (e.g., global warming, human toxicity, eco-toxicity, and water depletion potential) and/or endpoint indicators (human health, ecosystem, and resources) taking into account all phases of the life cycle of a product or service. The strength of this methodology mainly stays in the possibility of analyzing all life cycle phases, detecting burden shifts and hotspots, and taking into account background and foreground systems. Many of these indicators can make a useful contribution to the assessment of the impacts on ecosystem services and the Italian olive landscape, even if still today there is not a specific framework or a consolidated approach to assess ecosystem services with life cycle methods.

References

Agnoletti, M. (Ed.). (2013). *Italian historical rural landscapes*. Springer.

Alejandre, E. M., van Bodegom, P. M., & Guinée, J. B. (2019). Towards an optimal coverage of ecosystem services in LCA. *Journal of Cleaner Production, 231*, 714–722.

Blanco, C. F., Marques, A., & van Bodegom, P. M. (2018). An integrated framework to assess impacts on ecosystem services in LCA demonstrated by a case study of mining in Chile. *Ecosystem Services, 30*, 211–219.

Bruel, A., Troussier, N., Guillaume, B., & Sirina, N. (2016). Considering ecosystem services in life cycle assessment to evaluate environmental externalities. *Procedia CIRP, 48*, 382–387. https://doi.org/10.1016/j.procir.2016.03.143

Brunori, E., Salvati, L., Antogiovanni, A., & Biasi, R. (2018). Worrying about 'vertical landscapes': Terraced olive groves and ecosystem services in marginal land in Central Italy. *Sustainability, 10*, 1164. https://doi.org/10.3390/su10041164

Burkhard, B., & Müller, F. (2015). Ascertainment and assessment of ES – Indicators and quantification approaches. In K. Grunewald & O. Bastian (Eds.), *Ecosystem services – Concept, methods and case studies* (pp. 75–143). Springer.

De Luca, A. I., Falcone, G., Iofrida, N., Stillitano, T., Strano, A., & Gulisano, G. (2015). Life cycle methodologies to improve Agri-food systems sustainability. *Rivista di Studi sulla Sostenibilità, 1*, 135–150. https://doi.org/10.3280/RISS2015-001009

Demestihas, C., Plénet, D., Génard, M., et al. (2017). Ecosystem services in orchards. A review. *Agronomy for Sustainable Development, 37*, 12.

Espadas-Aldana, G., Vialle, C., Belaud, J.-P., Vaca-Garcia, C., & Sablayrolles, C. (2019). Analysis and trends for life cycle assessment of olive oil production. *Sustainable Production and Consumption, 19*, 216–230. https://doi.org/10.1016/j.spc.2019.04.003

Gulisano, G., Strano, A., De Luca, A. I., Falcone, G., Iofrida, N., & Stillitano, T. (2018). Evaluating the environmental, economic, and social sustainability of agro-food systems through life cycle approaches. In C. M. Galanakis (Ed.), *Sustainable food systems from agriculture to industry improving production and processing*. Academic Press Inc, Elsevier Science. https://doi.org/10.1016/B978-0-12-811935-8.00004-4

Haines-Young, R., & Potschin, M. B. (2018). *Common International Classification of Ecosystem Services (CICES) V5.1 and guidance on the application of the revised structure*. Available from www.cices.eu. Accessed 18 July 2020.

ISO International Organization for Standardization. (2006a). *ISO 14040:2006 life cycle assessment – Principles and framework*. International Organization for Standardization.

ISO International Organization for Standardization. (2006b). *ISO 14044:2006 Life cycle assessment – Requirements and guidelines*. International Organization for Standardization.

Maffia, A., Pergola, M., Palese, A. M., & Celano, G. (2020). Environmental impact assessment of organic vs. integrated olive-oil systems in Mediterranean context. *Agronomy, 10*, 416. https://doi.org/10.3390/agronomy10030416

Millennium Ecosystem Assessment (MEA). (2005). *Ecosystems and human well-being: A framework for assessment*. Island Press.

Montanaro, G., Xiloyannis, C., Nuzzo, V., & Dichio, B. (2017). Orchard management, soil organic carbon and ecosystem services in Mediterranean fruit tree crops. *Scientia Horticulturae, 217*, 92–101. https://doi.org/10.1016/j.scienta.2017.01.012

Nicoletti, G. M., & Notarnicola, B. (2000). L'impatto ambientale della filiera dell'olio di oliva: pratiche agricole e tecniche di estrazione. In: *Atti del XIX Congresso Nazionale di Merceologia, "La sfida per il terzo millennio: tecnologia, innovazione, qualità e ambiente"*, Sassari, 27–29 September 2000.

Othoniel, B., Rugani, B., Heijungs, R., Benetto, E., & Withagen, C. (2016). Assessment of life cycle impacts on ecosystem services: Promise, problems, and prospects. *Environmental Science & Technology, 50*, 1077–1092. https://doi.org/10.1021/acs.est.5b03706

Paris, P., Camilli, F., Rosati, A., Mantino, A., Mezzalira, G., Dalla Valle, C., Franca, A., Seddaiu, G., Pisanelli, A., Lauteri, M., Brunori, A., Re, G. A., Sanna, F., Ragaglini, G., Mele, M., Ferrario, V., & Burgess, P. J. (2019). What is the future for agroforestry in Italy? *Agroforestry Systems, 93*, 2243–2256. https://doi.org/10.1007/s10457-019-00346-y

Parras-Alcántara, L., Lozano-García, B., Keesstra, S., Cerdà, A., & Brevik, E. C. (2016). Long-term effects of soil management on ecosystem services and soil loss estimation in olive grove top soils. *Science of the Total Environment, 571*, 498–506. https://doi.org/10.1016/j.scitotenv.2016.07.016

Pattara, C., Salomone, R., & Cichelli, A. (2016). Carbon footprint of extra virgin olive oil: A comparative and driver analysis of different production processes in Centre Italy. *Journal of Cleaner Production, 127*, 533–547.

Proietti, P., & Regni, L. (2019). *Handbook for a sustainable management of the olive groves*. Available at https://olive4climate.eu/wp-content/uploads/Olive4Climate-Handbook-_ENG.pdf. Accessed 18 July 2020.

Raymundo Pavan, A. L., & Ometto, A. R. (2018). Ecosystem services in life cycle assessment: A novel conceptual framework for soil. *Science of the Total Environment, 643*, 1337–1347.

Rugani, B., Maia de Souza, D., Weidema, B. P., Bare, J., Bakshi, B., Grann, B., & Verones, F. (2019). Towards integrating the ecosystem services cascade framework within the life cycle assessment (LCA) cause-effect methodology. *Science of the Total Environment, 690*, 1284–1298.

Salomone, R., & Ioppolo, G. (2012). Environmental impacts of olive oil production: A life cycle assessment case study in the province of Messina. *Journal of Cleaner Production, 28*, 88–100.

Salomone, R., Ioppolo, G., & Battaglia, G. (2010). *Environmental impacts of olive oil production: A life cycle assessment case study in the province of Messina*. Paper presented at the VII international conference on life cycle assessment in the agri-food sector, Bari, Italy.

Stillitano, T., Falcone, G., De Luca, A. I., Piga, A., Conte, P., Strano, A., & Gulisano, G. (2019). A life cycle perspective to assess the environmental and economic impacts of innovative technologies in extra virgin olive oil extraction. *Food, 8*(6), 209. https://doi.org/10.3390/foods8060209

Zipori, I., Erel, R., Yermiyahu, U., Ben-Gal, A., & Dag, A. (2020). Sustainable management of olive orchard nutrition: A review. *Agriculture, 10*, 11. https://doi.org/10.3390/agriculture10010011

Chapter 18
Market Challenges in Italian Olive Groves

Annunziato Scaramozzino, Giorgio Pace, Teodora Stillitiano, and Anna Irene De Luca

Abstract Despite the Italian olive oil production covers around 15% of the world total, ranking second after Spanish, demonstrating that Italian olive growing shows huge potentiality at the same time, it is suffering from a lack of competitiveness due to ancient structural problems: fragmentation, lack of strategic planning, high costs of production and, strong reliance on import. One problem above all, the market price of olive oil in Italy, and especially at the origin, often fails to adequately remunerate the production costs and therefore to ensure a fair income to companies that are often faced with a market in which the willingness of most consumers to pay for quality extra virgin olive oil is still scarce.

This limit is strongly accentuated not only by the structure of the olive industry but also by the marketing policies in place in the large-scale retail trade, where olive oil is often treated as a commodity and used as a "coquette" product, creating confusion within the category, not facilitating a conscious choice and favoring low-quality oils of foreign origin, especially from Spain. The "mass" extra virgin olive oil sector has the leading market in Spain and the average price of Italian "mass" extra virgin olive oil is, however, higher than the Iberian one, therefore less competitive. Focus on enhancing olive oil quality attributes, such as biodiversity, certified productions, and high competence of operators represent the strong points on which the market challenges can be taken.

A. Scaramozzino (✉) · G. Pace
UNAPROL Consorzio Olivicolo Italiano, Rome, Italy
e-mail: a.scaramozzino@unaprol.it; g.pace@unaprol.it

T. Stillitiano · A. I. De Luca
Department of Agriculture, Mediterranean University of Reggio Calabria, Reggio Calabria, Italy
e-mail: teodora.stillitano@unirc.it; anna.deluca@unirc.it

J. Muñoz-Rojas, R. García-Ruiz (eds.), *The Olive Landscapes of the Mediterranean*, Landscape Series 36,
https://doi.org/10.1007/978-3-031-57956-1_18

18.1 Main Olive Oil Flows in Italy

The production structure of the Italian olive sector, according to the latest studies published by ISMEA (2019), is characterized by a utilized area of 1.07 million ha, with a total number of 825 thousand farms, more than 4 thousand active mills, an estimated production turnover of around 1.2 billion euro/year, equal to 3% of the total for Agriculture and an industrial turnover of approximately 3.3 billion euro/year, equal to 2.4% of the total national agri-food industry turnover.

The reduced average farm area, which despite the concentration of olive farms in progress in recent years, is equal to just 1.3 ha, and the high number of mills (in Spain the number of mills oscillates between 1600 and 1700) indicates an extreme fragmentation of Italian olive oil production, where to date only 37% of companies can be competitive on the market. Italian olive oil production, with an annual average of around 320,000 tons, covers around 15% of the world's total, ranking second after Spanish with an average production of 45%.

The country is also the world's leading importer of olive oil with an annual average of 535,000 ton imported mainly from Spain, Greece, and Tunisia. This huge flow of olive oil is necessary both to satisfy the high internal demand, which amounted annually to about 500,000 tons on average and to fuel an important export flow. This flow places Italy in the second position among exporters with around 330,000 tons per year (Stillitano et al., 2016).

The strong Italian dependence on foreign products derives to a large extent from the structure of the Italian packaging industry, but it should be emphasized that, as shown by the above figures, national production would still not be sufficient to satisfy internal consumption.

The purchase of large quantities of the product abroad is mainly driven by the need of the national processing industry to reach certain volumes of olive oil which is imported in bulk and which is used to obtain blends, which in turn are bottled and exported or distributed on the domestic market. A significant part of the national agricultural and industrial production framework, finding itself competing with large flows of low-priced import oils, focuses on medium-low production, characterized by a limited economic enhancement of the product. This production, nevertheless, satisfies a large part of the population that still recognizes themselves—as we will explain further below—in a product of medium-low quality and low price (Stillitano et al., 2016).

The United States ranks first among the main foreign buyers of Italian olive oil, followed by Germany and France, but in recent years direct exports to the USA and Germany have tended to shrink. In the international context, Italy plays, in any case, a leading role in foreign trade, also for the structure of the bottling industry which always makes it an almost obligatory stage of international trade.

Currently, it is worth mentioning, that olive oil represents only 4–5% of world fat consumption, but its diffusion grows slowly but steadily, and this implies a good margin of market growth, especially in countries where there is still no tradition of consumption.

However, this growth risks benefiting above all those countries whose production is more competitive in terms of price compared to olive oils produced in Italy. In fact, "Made in Italy" oil is characterized by a higher cost than that of the main international competitors, due to the higher production costs that distinguish the Italian olive production chain (Stillitano et al., 2019).

Concerning domestic demand, it is observed that the consumption preferences of Italians are decidedly oriented toward extra virgin olive oils, but—as anticipated above—the trend is to consume above all low-quality oil at low prices, this is also due to the continuation of the economic crisis, and recently by the COVID-19 pandemics. This causes serious problems at the market level for the Italian olive production chain, since the level of production costs in Italy is everywhere structurally high compared to that of the competing countries.

The market price, especially at the origin, often fails to adequately remunerate the production costs (estimated in a range of 3.4/8.5 euros/kg depending on the areas) and therefore to ensure a fair income to Italian olive companies that are often faced with a market in which the willingness of most consumers to pay for quality extra virgin olive oil is still scarce.

This "cultural" limit is strongly accentuated not only by the structure of the olive industry but also by the marketing policies in place in the large-scale retail trade, where olive oil is often treated as a commodity and used as a "coquette" product, creating confusion within the category, not facilitating a conscious choice and favoring low-quality oils of foreign origin, especially from Spain. The "mass" extra virgin olive oil sector has the leading market in Spain and the average price of Italian "mass" extra virgin olive oil is, however, higher than the Iberian one, therefore less competitive.

18.2 Opportunities for Italian Quality and Certified EVOOs

In recent years, however, alongside the general trend of the internal market described above, there is also the emergence, in a minority, but growing portion of consumers, of interest in superior quality extra virgin olive oils, which give guarantees regarding food safety, organoleptic characteristics and origin. This new consumption model, which also implies a greater propensity to spend, is also starting to determine more structured reactions from many small and medium-sized enterprises. Those companies are starting to implement production differentiation processes, to obtain a better positioning and conquest of new market segments (denomination, organic, monovarietal oils, etc.) (D'Adamo et al., 2019).

From this point of view, an element of essential strength of Italian olive growing, as already illustrated in Chap. 13, is made up of a unique varietal heritage, consisting of over 500 cultivars or varieties of olive trees. This great diversity, combined with the high heterogeneity of the olive growing areas, allows Italy to boast 46 oils with PDO or PGI certification of origin, a type of products with a high propensity to export (the share destined for foreign markets for some denominations ranges from 50% to 70% (ISMEA, 2019).

The sector of Designation of Origin oils, which most of all indicate the immense varietal and territorial heritage of Italy, despite the high number of awards, however, currently represents a segment in the country that is not yet sufficiently valued. Indeed, certified production represents just 2–3% of the total oil produced in terms of quantity. This percentage rises to 6% if analyzed in terms of value, a still very small share, especially if we consider the fact that over 75% of the national total of oils certified as PDO/PGI is absorbed by only 5 of the 46 Denominations.

It is, therefore, a growing sector, but far too little developed compared to its potential, the latter destined to expand also in light of the recent introduction in the country (Financial Law 2020) of rules governing the so-called "oil tourism" (see Chap. 11), i.e., all those activities aimed at getting to know olive oil, carried out directly at the production sites. These activities consist mainly of visits to places of culture, production or display of tools useful for the cultivation of the olive tree, tasting and marketing of the company's production of olive oil in combination with other foods, in educational and recreational initiatives in the areas of cultivation and production. The new discipline represents a further strengthening of food and wine tourism, a strategic asset for the promotion of Made in Italy and of the territories through the enhancement of identity products of the Italian agro-food heritage. This will especially benefit olive oil companies, which, by producing oils with a certified origin, place the link with the territory at the heart of their marketing strategy. Also, in this, the recent COVID-19 pandemics has deeply modified the tourism patterns in the country.

18.3 SWOT Analysis of Italian Olive Growing

Today, Italian olive growing has enormous growth potential but suffers overall from a lack of competitiveness due to its structural problems. This represents a lost precious opportunity for economic development and—for many olive-growing areas—the lack of simple economic sustainability. A missed chance, especially in light of the worldwide growth in consumption of olive oil, of the high appeal enjoyed by Made in Italy food and the simultaneous spread of the Mediterranean Diet in the world (ISMEA, 2013, 2019; MIPAAF, 2016; Sansone et al., 2015). Italy plays a leading role in the World scenario, being first in the consumption and imports of olive oil, and second in production and export. To be maintained, this role requires a strategic vision and governance that aims at integrating the whole production chain while emphasizing the role of olive oil as a "cultural" product. The CAP reform and the new 2023–2027 programming could be an opportunity to thoroughly analyze the Italian olive sector and to develop an overall strategy oriented to global sustainability: economic, environmental, and social (ISMEA, 2021).

The Italian olive sector is highly differentiated, especially as regards the agricultural phase, where different production models and different approaches to the market coexist. On the one hand, there are professional and strongly market-oriented olive firms, on the other hand, there are companies with a less entrepreneurial approach, more oriented toward self-consumption or direct sales. Italy has many strengths, starting with its huge reserve of varietal and pedo-climatic biodiversity that make it possible to produce oils with unique aromatic profiles, widely recognized as premium ones. At the same time, the small size of the farms, the low profitability, and the widespread presence of unproductive olive groves represent worrying elements of weakness in the sector. As for the transformation, the widespread presence of mills in the olive-growing areas, on the one hand, promotes quality, on the other hand, makes it difficult to achieve economic sustainability and to produce the critical masses of product required by the bottling industry. In some Italian areas, the integration of the entire production chain, from the field to bottling, and the realization of economies of scale are leading to high levels of profitability, but also to the employment and social development of the territories concerned (ISMEA, 2021).

In summary, the main Strengths, Weaknesses, Opportunities, and Threats related to the Italian olive sector can be identified as follows (ISMEA, 2019):

Strengths

- Presence of important areas suitable for the cultivation of olive trees both in terms of quantity and product quality.
- High potential to differentiate the productions, by variety, production methods, origin, etc.
- The high environmental, landscape, historical, cultural, and anthropological value of the olive groves and the olive growing activity enhances their potential in terms of tourist attractiveness (oil tourism).
- Territorial extension of the crop and important contribution in terms of employment.
- Increasing attention to quality products (PDO/PGI, organic, new product classes such as 100% Italian, High quality, etc.)
- Medium-high know-how, even if not distributed evenly in all the appropriate areas.
- Widespread location of the oil mills in suitable areas, with a greater guarantee of timely processing and the average quality and professionalism of the Italian oil mills.
- A large presence of olive oil product chains was traced (around 400 with 8000 farms involved).

Weaknesses

- Lack of adequate centralized strategic planning in the olive sector.
- Fragmentation of the production structure (small farm size) and diffusion of olive growing in difficult areas: poor land mobility. The very high unit value of the land.
- Main presence of traditional groves, often old, and limited diffusion of mechanization and irrigation.

- Delay in the implementation of technological innovations and missed investments.
- Large and growing interannual fluctuations in production in qualitative and quantitative terms with relative commercial repercussions.
- Generational turnover is still extremely limited.
- The high cost of labor and bureaucracy.
- Widespread presence especially in certain areas of non-entrepreneurial olive growing.
- Little use of EU funds for rural development.
- High production costs and lower profitability, especially in the case of medium-small companies.
- A little attitude of the Italian operators in the supply chain to make a system.
- Still little incisive role of producer organizations in the concentration of the offer, in the enhancement of the product and the strengthening of the contractual capacity of the production phase toward the distribution and industrial phases.
- Delay in the application of the results of scientific research.
- Threat formed by the continuous diffusion of Xylella fastidiosa in Puglia.
- High dependence of the industry on imports is also due to the excessive internal production alternation.
- Excessive dependence of the internal market on the rules dictated by the large distribution operators.
- Little diffusion among consumers and restaurateurs of a true "culture" of quality oil.
- High "promotional pressure" and use of extra virgin olive oil as a coquettish product in stores and trivialization of the product (still considered by many to be a commodity).

Opportunities

- Greater stabilization of farmers' income with decoupled payments.
- Cultivation techniques allow the stabilization of olive production, reducing the phenomenon of the alternation.
- New CAP programming.
- Renewed awareness of the role of olive growing in the protection and planning of the territory.
- Possibility of restructuring the olive groves especially for their multifunctional environmental value and of investing in new —at least partially— mechanizable plants.
- Development of short production chains also through agreements between operators.
- Increasing consumer sensitivity toward certified quality products.
- Increasing consumers' attention toward the national and regional origins of olive oil.
- Production of larger volumes of quality Italian products.

- Slow but progressive increase in international consumption and greater opportunities for product segmentation.
- Driven promotion of extra virgin olive oil as a characterizing part of the Mediterranean diet and the Made in Italy agri-food sector.
- EU programs with resources to support the promotion of oil consumption.
- High appreciation of the health characteristics of the product.
- Fair growth margins in the ho.re.ca sector.

Threats

- Abandonment of olive groves, especially nonprofessional or marginal ones, and the consequent erosion of the olive growing area.
- Loss of opportunities for smallholders due to difficulties in accessing rural development funds and credit.
- Growing international competition on production costs and quality.
- Further loss of income.
- Leveling to international prices.
- Growth in investments for the production of quality oils in competing Countries.
- Growing international competition on production costs.
- Aggressive policies of competing Countries in the acquisition of control of Italian brand companies.
- Consumption contraction due to the economic and financial crisis.
- The concentration of decision-making power in the hands of a few final distributors.

18.4 Conclusions

Despite the quality of olive oil production and its role in terms of quantities in the international market—that continues to be very significant, although it is lately threatened by other countries production—the Italian olive oil sector is suffering more and more structural and commercial difficulties. Especially for mass products, Italy fails to be competitive and, probably that is why it should focus principally on the high-quality olive oil market. On the other hand, also from the demand side, Italian consumers hold the highest share affecting the international market. Given the complexity and diversity of national agricultural production, the sector needs new drives that suggest innovative and quality-oriented to improve competitiveness, while safeguarding the distinctive role of olive growing for the socio-economic development and in general for the sustainability of many Italian rural areas.

References

D'Adamo, I., Falcone, P. M., & Gastaldi, M. (2019). Price analysis of extra virgin olive oil. *British Food Journal, 121*, 1899–1911. https://doi.org/10.1108/BFJ-03-2019-0186

ISMEA. (2013). Piano di settore olivicolo oleario. *Indagine sui costi di produzione delle olive da olio.* http://www.ismea.it. Accessed 17 July 2020.

ISMEA. (2019). *Scheda di settore 'olio di oliva'*. Available at http://www.ismeamercati.it. Accessed 17 July 2020.

ISMEA. (2021). *La competitività della filiera olivicola: Analisi della redditività e fattori determinanti. Rete Rurale Nazionale 2014–2020*. MIPAAF Ministero delle Politiche Agricole Alimentari e Forestali. https://www.reterurale.it/flex/cm/pages/ServeAttachment.php/L/IT/D/3%252Fe%252Fc%252FD.641f6f625e97c6f2b6fd/P/BLOB%3AID%3D22376/E/pdf. Accessed July 2020.

MIPAAF Ministero delle Politiche Agricole Alimentari e Forestali. (2016). *Piano di settore olivicolo oleario.* https://www.politicheagricole.it/flex/cm/pages/ServeBLOB.php/L/IT/IDPagina/1984. Accessed 17 July 2020.

Sansone, L., Carroni, A. M., Fedrizzi, M., Santilli, E., Pipitone, G., Pagano, M., Guerrieri, M., Puri, D., Zaffina, F., Salis, M., Ruda, P., Vatrano, T. P., Bazzoffi, P., & Zelasco, S. (2015). Evaluation of environmental and economic effectiveness of the cross compliance 4.3 standard 'maintenance of olive groves and vineyards in good vegetative conditions'. It. *Journal of Agronomy, 10*(1). https://doi.org/10.4081/ija.2015.691

Stillitano, T., De Luca, A., Falcone, G., Spada, E., & Gulisano, G. (2016). Economic profitability assessment of Mediterranean olive growing systems. *Bulgarian Journal Of Agricultural Science, 22*, 517–526.

Stillitano, T., Falcone, G., Nicolò, B. F., Di Girolamo, C., Gulisano, G., & De Luca, A. I. (2019). Technical efficiency assessment of intensive and traditional olive farms in southern Italy. *AGRIS on-line Papers in Economics and Informatics, 11*(2), 81–93. https://doi.org/10.7160/aol.2019.110208

Part IV
Olive Landscapes Across the Mediterranean: Regional Heterogeneity and Case Studies: Olive Groves and Landscapes in Greece

Chapter 19
A Historical and Social Insight of Olive Groves and Landscapes in Greece

Kostas Chartzoulakis

Abstract Olive trees are native to the coastal areas of the eastern Mediterranean Basin and Western Asia, as proven by fossil Olea pollens. Fossilized leaves of olive 37,000 years old, found in the paleosols of the volcanic Greek island of Santorini, prove that Greece is one of the homelands of olive trees. Archaeological data and historical findings confirm that during the Minoan period (3000–1000 BC) olive cultivation and olive oil trading were widespread in Crete. Olives and olive oil appear steadily over time in all aspects of Greek life, both individual and collective: religion, art, social and political events, sports, and of course, gastronomy.

The main characteristics of olive culture in Greece are: (a) small average farm sizes (less than two hectares), (b) the utilization of hilly and mountainous areas, and (c) the need for irrigation in southern regions of the country. Olive trees grows in poor, rocky, or fertile soils, at elevations up to 1000 m with annual rainfall ranging from 200 to 1000 mm. Since more than 70% of olive-growing land is placed on hilly or mountainous areas, there are severe problems of soil erosion (30% of total land), when no proper management is practiced.

Olive growing in Greece has a multi-functional use including agricultural (oil and table olive), environmental (protect from erosion, mitigate climate change, etc.), and social (keep farmers in marginal and hilly areas) functions. More than 550,000 families cultivate olive trees and for many of them olive products are the sole income of the family.

Greece today is the third olive oil producer country in the world with an average production 320,000–350,000 tons per year with an area covered with olives around 970,000 ha. Table olive production in Greece fluctuates between 200,000 and 250,000 tons per year. The olive growing area occupies more than 21% of the total cultivated land of the country while 29% of Greek farms are specialized in olives.

K. Chartzoulakis (✉)
Hellenic Agricultural Organization—ELGO-Demeter/NAGREF—Institute for Olive Tree & Subtropical Plants, Chania (Crete), Greece
e-mail: Chartzoulakis@elgo.gr

J. Muñoz-Rojas, R. García-Ruiz (eds.), *The Olive Landscapes of the Mediterranean*, Landscape Series 36,
https://doi.org/10.1007/978-3-031-57956-1_19

The family run small businesses that form the backbone of Greek olive production are in a better position to meet the requirements of organic olive oil production.

19.1 Introduction

Greece is located in south-east Europe, at the south edge of Balkan Peninsula, and occupies 131,957 km^2, of which one-fifth consists of islands in the Aegean, Ionian, and Mediterranean seas (Fig. 19.1). The climate is typical Mediterranean with long, hot and dry summers and mild winters. Precipitation varies greatly with the average approaching 850 mm/year; western Greece receives the majority of rainfall, more than 1500 mm/year, while eastern Greece, along with the islands of Aegean Sea and

Fig. 19.1 Olive growing areas in Greece. In this map, the names and location are indicated for the main urban nodes acting as core centres across the olive production network in Greece. Whe considered at a National level, their strategic distribution across the country becomes clear, along with their location in relation to the main bio-physical regions in Greece

Crete, have considerably smaller rainfalls, although overall rainfall is not uniformly distributed throughout the year.

Greece is predominantly an agricultural country, although less than one-third of its area is cultivated. The main tree crops are olives (970,000 ha), citrus (58,400 ha), deciduous trees (78,100 ha), and other fruits (78,200 ha). Herbaceous crops (cereals, root crops, industrial varieties, vegetables, forage plants, etc.) cover a total area of 2,175,100 ha, vines (126,100 ha) and other crops (17,100 ha), while fallow land covers 453,400 ha (National Statistical Service of Greece).

19.2 Historical Background of Olive Cultivation

Olive (*Olea europaea* L.) is native to the coastal areas of the eastern Mediterranean Basin and Western Asia, and it is now widely spread to nearby countries. It is estimated that the olive tree cultivation began more than 7000 years ago in the E Mediterranean. However, fossil Olea pollens have been found in Macedonia, Greece, and other places around the E Mediterranean, indicating that this genus is an original element of the Mediterranean flora. Fossilized leaves of Olea, found in the paleosols of the volcanic Greek island of Santorini (Fig. 19.2) and the island of Nisyros, have been dated to about 37–50,000 years old (Lacroix, 1896), a fact that proves that Greece is one of the homelands of the olive tree.

According to P. Anagnostopoulos, in a statement he made at the Athens Academy in 1951 based on findings from excavations, Crete is the homeland of the domesticated olive. This statement is strengthened by the fact that the name of "olive" is Greek. As far back as 3000 BC, olives were grown commercially in Crete; they even may have been the source of the wealth of the Minoan civilization (Faure, 1973). In the palace of King Minos at Malia, storerooms have been found, which could hold approximately 10,000 hectolitres of olive oil. G. and E. Sakellarakis's excavations in Fourni, Archanes, Crete ensure that during the Middle-Minoan period, the cultivation of the olive tree was one of the inhabitants' most fundamental activities.

To reinforce the cultural relevance of olives in green history, it can be noted how in Homer's Odyssey, Odysseus crawls beneath two shoots of an olive that grows from a single stock and in the Iliad, (XVII.53ff) there is a metaphoric description of an isolated olive tree in the mountains, by a spring. Olive wood was one of those used to fashion the most primitive Greek cult figures, called *xoana*, referring to their wooden material; subsequently, they were reverently preserved for centuries.

It was purely a matter of local pride that the classic Athenians claimed that the olive grew firstly in their city-state. In an archaic Athenian foundation myth, Athena—an Olympian God—won the sponsorship of Attica from Poseidon via a gift in the form of an olive. Though, according to the fourth century BC botanist, Theophrastus, olive trees ordinarily attained an age of about 200 years old, he mentions that the very old olive tree of Athena still grew on the Acropolis. Furthermore, an olive tree in western Athens, named "*Platon's Olive Tree*", was rumored to be a remnant of the grove within which Platon's Academy was situated, which would

Fig. 19.2 Fossilized leaves of Olea

date to approximately 2400 years b.p. Meanwhile, an olive tree in Crete has been claimed to be over 3000 years old. The prosperity of classical Athens during the Golden Ages was attributed, among others, to the high production of olive oil (Foxhall, 2007).

When the Romans republic occupied Greece after the year 146 BC, olive oil production continued and Romans were able to learn the secrets of cultivation. During Byzantine times (IV–XIV C AD) things did not change much. The production of the olive oil in Greek territories remained significant because of the commercial opportunities open by the vast size of the Empire. The Empire itself embraced half of the olive oil productive areas in the known world and the product was exported throughout the world.

From this point, and lasting until today, Greece became the world's one of the most important exporters of high-quality olive oil. Olive trees have become a national symbol for Greeks, lasting until today.

19.3 Area Under Olive Cultivation

Olive culture in Greece represents nowadays 21–23% of the total of cultivated area (about 60% of the tree crops) with 970,000 ha nationally, embracing all bigger farms but also small orchards and scattered trees (840,000 ha for oil and 120,000 ha for table olives) with more than 160,000,000 olive trees currently under production.

In 2023, Greece is the third olive-oil producer country in the world with an average production of 320,000–350,000 tons per year, while 29% of Greek farms are engaged in olive growing. Its production has more than doubled along the last 40 years; moving from 180,000 tons in the 1970s to 400,000 tons in 2010. This increase in production is attributed to factors, including (a) new modern plantations, mainly established during the 1970s and 1980s, (b) lately, intensification and mechanization as well as the expansion in the use of external inputs and irrigation, (c) improvements in olive cultivation and oil extraction technologies, (d) more favorable net income for olive growers comparing to other crops, driven by both the availability of multiple CAP subsidies and high olive oil prices, and (e) the lack of capability for too many other crops due to agro-climatic and socioeconomic conditions.

The distribution of olive cultivation among regions is very uneven, although more than 58% of the total olive trees are grown in semi-mountainous and mountainous areas. Southern Greece (Crete and Peloponnese) produces about 50% of the Greek olive oil, while other important regions include the Ionian Islands, Continental Greece, and the northern Aegean Islands.

Greece is one of the richest countries in the world in terms of autochthonous olive cultivars. About 44 native cultivars had already been recorded by 1985, although recent studies have enlarged this number to more than 100 (Kostelenos, 2010).

Tree densities vary from 80 to 120 trees per ha, noted in traditional extensive plots, to 200–360 plants or more in intensive new groves. Tree density and planting patterns depend partly on local traditions but water availability is also a determining factor. In recent years, drip irrigation has been introduced on a large scale, even in traditional plantations on hilly areas of Greece.

The main characteristics of olive culture in Greece are (a) the small average farm size (less than 2 ha), (b) spread mainly across hilly and mountainous areas, and (c) the need for irrigation, mainly in southern regions of the country. Olive trees grow in poor, rocky, or fertile soils, at elevation up to 900 m with annual rainfall ranging between 200 and 1000 mm.

Since more than 70% of olive groves are placed in hilly or mountainous areas (Fig. 19.3), there are severe problems of soil erosion (30% of total land), when no proper soil conservation practices are implemented.

Fig. 19.3 Olive growing in hilly areas of central Greece

19.4 Impact on the Economy and the Environment

Olive growing in Greece has traditionally been considered as a multi-functional land use system, embracing agricultural (oil and table olive production), environmental (protect from erosion, as a forest that is not susceptible to fire, etc.), and social (keep farmers in marginal and hilly areas) functions. About 450,000 families cultivate olive trees and for many of them, the income from olive oil and table olives is the sole or main source of income. Olive oil consumption in Greece is the highest in the world, reaching 12.8 kg per capita for 2014, compared to 11.3 kg in Spain and 10.5 kg in Italy (oliveoilmarket.eu).

Annual production of olive oil has remained relatively stable in recent years, fluctuating from 300,000 to 350,000 tons per year, depending on the climatic conditions and more than 85% is classified as extra virgin. There are 27 Protected Designation of Origin (PDO) and 10 Protected Geographical Indications (PGI) of Greek Olive Oils. Many of these oils have won gold or silver prizes in international contests like Mario Solinas (IOC), LA International EVOO (Los Angeles, USA), NY International Olive Oil Competition (New York, USA), BIOL International (Italy), Great Taste Gold (UK), Der Feinschmecker (Germany), AVPA (Paris, France), ITQI (Brussels). According to Greek Statistical Authority and the Greek Olive Oil Exporters Association (SEVITEL) along 2018, Greece exported 164.329 tons of olive oil, of which 46,737 tons were labeled.

Table olive production in Greece fluctuates between 180,000 and 220,000 tons per year, representing 8% of the world production and 15–20% of that in the

European Union. There are more than 64,000 table olive farmers and more than 100 companies engaged in the processing, packaging, marketing and exporting. Black olives of varieties "Kalamata" and "Konservolia" represent more than 60% of the production. There are 10 Protected Designation of Origin (PDO) for table olives.

Currently, the main problems of olive growing in Greece include: (a) the low prices of olive oil, (b) the ineffective organization of standardization and marketing, (c) the high cost of harvesting due to lack of mechanization, (d) the small size of olive groves (80% are smaller than 5 ha and only 0.8% > than 20 ha), frequently scattered across diverse plots, and (e) the relatively old age of trees.

Despite these problems, Greece has some innate advantages for the production of organic olive products (Spirou, 2019). Olive trees are perfectly adjusted to the Greek environment and thanks to largely adequate climate and soil conditions, the olive tree bears yields even in poor years and given proper care, harvests may become abundant. At the same time, the model of organic production is in line with the fractured small-scale production model still prevalent in Greece. The family run small businesses that form the backbone of Greek olive production are in a better position to meet the needs of the organic rather than conventional olive production, which is best fit to a larger scale of operation. Furthermore, organic olive oil farmers can also explore and capture the advantages offered by a great variety of terroirs, which bring character and identity that turn olive oil and table olives unique and well positioned to claim a higher price on the shelf.

Organic olive-growing output in Greece has grown from 22,000 tons to 132,000 tons along the past 5 years, accounting for an average of 8.7% of the country's total olives production in that same period. It is estimated that there is strong potential for growth in sales of the domestic and international markets, given that the current turnover is at a very low level, which leaves room for further development. Moreover, conditions are in place for new entrants into the business of organic olive oil to succeed (Spirou, 2019). In order to secure a foothold, small enterprises need to gear their sales toward export markets even more so than traders of conventional olive oil.

Olive oil production, considering the whole chain from farm management practices to oil extraction at the olive mill, is associated with a variety of positive and negative environmental impacts on soil, water, biodiversity, and landscape.

A positive effect of olive trees in hilly and mountainous areas is the prevention of soil erosion (Metzidakis & Chartzoulakis, 2001), but only under the right management conditions. To avoid erosion, terracing is the common practice for soil conservation, and seasonal plowing, if applicable, is done perpendicularly to the hill slopes. In cases that no dry-stone wall was constructed between terraces (Fig. 19.4), wild vegetation and bushes are left to grow in the separating edges, so preserving them from erosion.

The most common practice is to grow or allow growing herbaceous cover crops during the rainy season (autumn to winter), where more than 70% of the total annual rainfall occurs. This allows for vegetation to take advantage of soil moisture available, reducing run-off, which prevents soil erosion. Cover crops are grazed or cut early in the spring, before the moisture competition arises (Fig. 19.5). Erosion has

Fig. 19.4 Dry-stone wall in terraces with olives

Fig. 19.5 Grazed olive orchard by sheeps

been noticed in some areas due to inadequate protection measures, especially where intensive cultivation has minimized vegetation.

Olive growing also plays an important role in mitigating the effects of climate change through the positive role in atmospheric CO_2 fixation (Mantonaro et al., 2018). There is also a positive environmental impact from the production of compost by using olive by-products, including pomace cake, leaves, pruning residues, etc., for soil amendments and/or organic fertilizer. Olive groves also provide an agricultural ecosystem to native insect and bird species.

However, the intensification of olive cultivation, mainly in plains has raised environmental concerns including those derived from the over-application of chemical pesticides and fertilizers and soil compacting through the use of mechanized tilling. Furthermore, a negative environmental impact of olive growing, especially after the replacement of traditional olive mills with three-phase centrifugal extraction plants, is the production of huge amounts of wastewater (OMW). These wastes are characterized by high pollution loads (mainly organic) and result in negative impacts on the environment (mainly to surface or groundwater). Although many management practices have been tested at mill level (aerobic, anaerobic, use of membranes, composting, etc.), their application in most cases is limited due to the high investment or running cost and the technological knowledge required. In the late 1980s of past century, evaporation ponds were developed as an alternative for the disposal of OMW. However, despite of the low cost of this method, the odors produced cause serious problems in villages and touristic areas near the ponds, especially during summer. Controlled application of pre-treated row OMW on the soil of olive orchards, an environmentally acceptable and economically feasible management, has been applied since 2017, mainly in small-medium, family based olive mills (Chartzoulakis et al., 2010).

19.5 Olive Tradition and Culture

In Greece, olive tree is considered a symbol of geography, peace, and prosperity, but not only. Rather, it is an integral part of Greek culture. More than a mere natural feature of the landscape, the olive tree frequently becomes the landscape itself, even defining its geography. The importance of the olive in shapingthe relationship of the Greek people to their land and the great significance the olive fruit had since prehistoric times for agricultural production and for the economy in general has been highlighted over the years, along with the multiple symbolic values that accompany the olive in religious worship, art, and culture.

Olives and olive oil appear steadily over time in all aspects of Greek life, both individual and collective: religion, art, social and political events, sports and of course, gastronomy. Every aspect of Greek lifestyle, meets in the presence of abundant olive trees and fruits.

In ancient Greece, olive trees were acknowledged for their multiple benefits and usefulness, as shown by angiography and wall paintings in Crete and mainland

Fig. 19.6 Painting with olive harvesting in ancient Greece

Greece (Fig. 19.6). According to the Greek mythology, Hercules planted an olive tree, after completing his famous 12 labors. The olive is a favorite subject in Minoan Art. Olive trees, olive branches, and olive blooms are depicted in many wall paintings and plasters, found at the palace of Knossos (1600–1400 BC) "The dance in the Sacred Grove" and other scenes with olive foliage, blooming branch, branches, and modeled olives. During this period, olive branches were considered a symbol of peace, wisdom, honorable distinction, and victory. At the first Olympic Games, in 776 BC, ancient Greeks crowned winners with a wreath (*kotinos*) made of branches cut off from a single wild olive tree, *Kallistephanos*, in the temple of Hera in Ancient Olympia, a tradition that was brought back during the Athens 2004 Olympic games.

The Athenians represented Athena with an olive wreath on her helmet and an amphora with olive oil or an olive branch. One of the seven wonders of the ancient world, the gold statue of Zeus at Olympia, was crowned with olive branches. At that time, Thales of Miletus used his astronomical observations to forecast an excellent olive yield for 596 BC while Democritus studied the relationship between fruitful olive yields and the positions of the stars.

Olive oil had many uses in ancient Greece, differing according to social status. Poor people did not consume olive oil while the rich were allowed to use olive oil in cooking, for cleaning their bodies and for lighting. The ancient Greeks applied olive oil on their bodies and hair as a smear matter of grooming and good health (Fig. 19.7). Moreover, a popular use of olive oil in ancient Greece was for oiling athletes' bodies before exercise in the gymnasium and at games. Furthermore, at the Panathenea games (600 BC), the winner's prize was a decorated amphora, full of olive oil.

The olive tree is the holy tree of all religions, from the days of idolatry up to the times when the monotheistic religions sprang up around the Mediterranean,

Fig. 19.7 Use of olive oil by ancient Greeks

including especially the Jew and Christian faiths. In Greek mythology, the goddess Athena gave to the citizens of Athens an olive tree to win Neptune in a contest that was held among these two Gods, and elected the patron of the city. The Orthodox Church, which still holds official status in Greece, uses olive oil for its faithful, to cure and sanctify the soul and the body, and as a weapon in the field of battle against evil and ammunition for eternal life, a tradition that remains in rituals. For the holy Baptism mystery, which marks the start of the path that will familiarise men with God, among the material elements of the sacrament are water and oil.

Ancient Greeks believed that olive oil was beneficial to human health and recommended it to heal multiple afflictions, such as dermatological problems, lacerations, and burns, as gynecological ailment, emethic, to cure ear infections and for birth control, etc. According to the code established by the father of medicine, Hippocrates, olive oil was held to be beneficial for over 60 therapeutic uses. When medicine was not enough to save the patient, olive oil was used in laying out the dead. Women washed their bodies and anointed them with olive oil or scented oils. Nowadays, medical science has proven all these beneficial effects of olive oil and olive products and many medicines or health supplements and cosmetics such as soaps, hand creams, body lotions, shampoos, nourishing and moisture creams, are still based on olive products.

Homer and Pindarus, Aescylous, and Sophocles have devoted unique lines to the olive tree, hymns to its existence, and its multiple values. The olive tree is also a protagonist in modern Greel poetry and prose. Palamas, Ritsos, Sikelianos, Polemis, Seferis, and Elytis do not fail to mention the olive tree "with the wrinkles of our parents." There are also many proverbs which refer to olive cultivation that stand for various meanings reminding us of how profound its bonds are with the life of the Greek people, especially in relations with peasants and in the countryside.

19.6 Olive and Touristic Activities

The connection of olives to the Greek culture is shown in the multiple museums dedicated to olive cultivation. The purpose of these museums is to preserve and highlight the aspects of olive cultivation in relation to the residents of each region, the way they cultivated olive trees, produced olive oil, and also the technological progress in olive oil production. Olive museums are currently located in the islands of Lesvos, Corfu, Kephalonia, Thasos, and Andros. In the mainland of Greece, there are olive museums in Sparta (www.oliveoilmuseums.gr), Serres, Pylio. In the island of Crete, there are two olive museums. The first is located in Kapsaliana in Rethimno and the second in Vouves, Chania (www.olivemuseumvouves.com), near to the ancient olive tree of Vouves (Fig. 19.8) that is estimated to be more than 3.000 years old. Furthermore, since 1998 the "Olive Tree Routes" (www.olivetreeroute.gr) have been taking place with cultural itineraries examining the olive tree as a "bridge" from Messinia, and also linking Greece to the other Mediterranean olive-growing countries.

There are many popular festivals that take place in olive growing areas where the visitors are able to come into contact with olive traditions, products and producers from across the country, comprising multiple actions such as corporate presentations, contests, tastings, and high-quality products produced by the olive industry. More often the festivals are accompanied by traditional folklore music orchestras that sing songs related to olive cultivation. Furthermore, a new trend regarding the touristic development has been created in the past few years, showcasing fusion of mild touristic activity and agricultural activities, whereby tourists can be hosted in

Fig. 19.8 The oldest olive tree in Vouves, Crete Greece

villages that have kept their agricultural traditions, and where guests can participate in the cultivation, harvesting and management of olive trees, and production of olive oil.

References

Chartzoulakis, K., Psarras, G., Moutsopoulou, M., & Stefanoudaki, E. (2010). Application of olive mill wastewater to a Cretan olive orchard: Effects on soil properties, plant performance and the environment. *Agriculture, Ecosystems & Environment, 138*(3–4), 293–298.

Faure, P. (1973). *La vie quotidienne en Crète au temps de Minos Hachette, Paris*. http://epp.eurostat.ec.europa.eu/cache/ITY_OFFPUB/KS-SF-09-096/EN/KS-SF-09-096-EN.PDF

Foxhall, L. (2007). *Olive cultivation in ancient Greece. Seeking the ancient economy* (pp. xvii, 294). Oxford University Press. ISBN 978-0-19-815288-0.

Kostelenos, G. (2010). *Elements of olive culture. History, description and geographical distribution of olive cultivars in Greece* (p. 430) ISBN: 978-960-93-2479-3.

Lacroix, A. (1896). Les minéraux néogènes des scories plombeuses athéniennes du Laurium (Grèce). *Comptes Rendus de l'Académie des Sciences de Paris, 123*, 955–998.

Mantonaro, G., Nuzzo, V., Xiloyiannis, C., & Dichio, B. (2018). Climate change mitigation and adaptation in agriculture: The case of olive. *Journal of Water and Climate Change, 9*(4), jwc2018023.

Metzidakis, I., & Chartzoulakis, K. (2001). Soil conservation in olive orchards. A case study in Crete, Greece. *Proceedings of the international workshop on 'land and water use for sustainable olive growing' C.O.I.*, Naples, Italy, pp. 289–295.

Spirou, S. (2019, January 31). Rise of organic olive oil in Greece. *The Olive Oil Times*. https://www.oliveoiltimes.com/world/the-rise-of-organic-olive-oil-in-greece/66787

Chapter 20
Pedoclimatic and Landscape Conditions of Greek Olive Groves

Georgios Koubouris and Georgios Psarras

Abstract Greece belongs to the temperate continental climate zone of the Northern Hemisphere. The continental part of the country represents about 80% of total land while the remaining 20% is distributed among 6.000 islands. Soil is commonly dry and rocky and only 20% is considered arable. Low content in organic matter is also a common limitation of soils in Greece. Land is mainly sloping and mountainous and therefore prone to soil erosion. The climate is typical Mediterranean with mild rainy winter, relatively warm and dry summer, and extensive sunshine period throughout most of the year. The average annual temperature can be lower than 8 °C in some Northern mountainous areas while it can be higher than 20 °C in other areas. In the same rationale, average annual precipitation is as low as 300 mm in Cyclades islands and as high as 2000 mm in Pindos mountain. The climate in the major part of the country is suitable for this crop, favoring constant expansion even nowadays that available land is scarce and mainly exploited for touristic and urban uses. Indeed, 14% of agricultural land and 75% of the land devoted to fruit tree growing, respectively, is covered by olive groves with approximately 150 million olive trees.

Keywords Climate · Topography · Temperature · Rainfall · Sloping land

20.1 Olive Growing Distribution and Climate

Determined mainly by climatic limitations and especially frost, olive cultivation is mostly concentrated in the southern half of the mainland, along coastal areas and in most of the large islands. Crete and Peloponnese in Southern Greece are the main

G. Koubouris (✉) · G. Psarras
ELGO Dimitra – Institute of Olive Tree, Subtropical Crops and Viticulture, Chania-Crete, Greece
e-mail: koubouris@elgo.iosv.gr; psarras@elgo.iosv.gr

J. Muñoz-Rojas, R. García-Ruiz (eds.), *The Olive Landscapes of the Mediterranean*, Landscape Series 36,
https://doi.org/10.1007/978-3-031-57956-1_20

producing regions providing over 60% of the national olive oil, although significant production of table olives is taking place in areas of Central and Northern Greece. Therefore, olive trees are cultivated in all Regions of Greece, occasionally under marginal climatic conditions. This is mainly achieved by the significant genotypic variability, resulting in growing cultivars that are well adapted to the climatic challenges of each area.

In terms of climate, Greece belongs to the temperate continental climate zone of the Northern Hemisphere. The continental part of the country represents about 80% of total land while the remaining 20% is distributed among 6.000 islands. The climate in Greece is typical Mediterranean with mild rainy winter, relatively warm and dry summer, and extensive sunshine period throughout most of the year, which exceeds a mean value of 3100 h year^{-1} in areas of Crete and Rhodes islands. The most dominant climate type is hot-summer Mediterranean climate (Csa according to Köppen-Geiger) observed mainly in coastal areas, including some of the most important olive-growing areas of Greece. Significantly smaller portions of land are characterized by warm-summer Mediterranean climate (Csb) and humid subtropical climate (Cfa). The complexity of landscape effects on climate is complemented by the incidence of oceanic climate (Cfb) in cold mountainous areas and cold semi-arid climates (BSk) mainly in Northern Greece Climate-data.org. (2020).

Although mean annual temperature can be lower than 8 °C in some Northern mountainous areas (noncultivated land), the typical mean values in the olive growing areas range between 15 °C (Northern Greece) and 20 °C (Crete). In general, a notable precipitation gradient exists, with mean annual precipitation being reduced from West to East and from North to South. Therefore, olive trees growing in the Ionian Sea islands receive a mean annual precipitation of more than 1000 mm. On the other hand, olive tree growing in small islands in the Central Aegean Sea is not favored by the low annual precipitation (less than 300 mm in Cyclades) and limited irrigation water availability.

Vegetative growth of olive trees starts around February to March and ceases in October to November while flowering occurs in April to June and fruit maturation is observed from October to December, in most cases. Typical climatic challenges that may affect the yield on an annual basis include:(i) mild winters (insufficient chilling for flower differentiation), (ii) long dry periods (e.g., February to October), (iii) early heat waves or nontypical rainy conditions during flowering. According to the Climate Quality Index, most of the areas in Greece are characterized as moderately suitable (Fig. 20.1). Indeed, over 70% of the Greek territory is characterized by moderate climate quality, while less than 22% is characterized by high climate quality (Karamesouti et al., 2018).

Air temperature extremes (<0 °C or > 30 °C) and precipitation deficit (since most olive groves were rainfed) have been the main factors determining the geographical spread of olive cultivation. The optimum monthly mean temperatures for its cultivation are centered on 7 °C in winter and 25 °C in summer (Moriondo et al., 2013). Temperature below 0 may cause damage in plant tissues (Larcher, 2000).

Therefore, olive varieties with low frost tolerance such as Koroneiki are rarely found in Northern Greece due to frequent incidence of heavy winters. Furthermore,

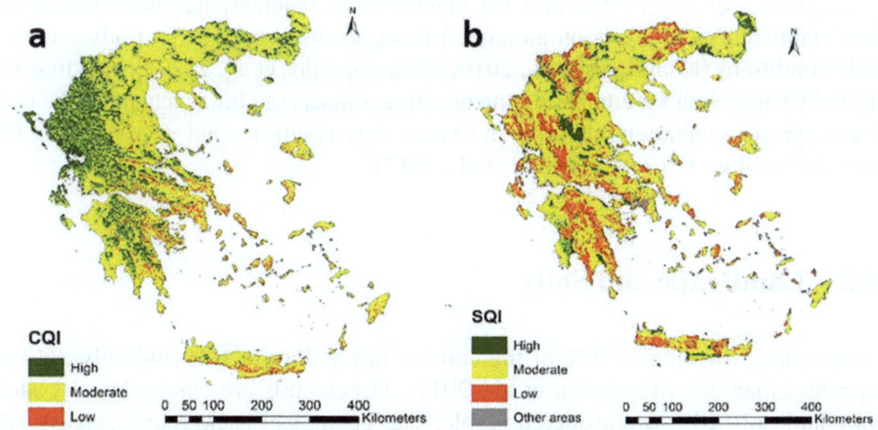

Fig. 20.1 Climate Quality Index (CQI) (**a**), and Soil Quality Index (SQI) (**b**). (Source: Karamesouti et al., 2018)

the olive tree may be more sensitive at certain phenological stages. For example, during flower development, high temperatures (30–40 °C) may diminish pollen germination (Koubouris et al., 2009, 2015a, b), and low temperature may lead to anomalies in fruit set (Koubouris et al., 2010a). Temperature levels can also cause earlier or later development of flowers from year to year (Koubouris et al., 2010b). In addition, mild low-temperature levels during winter are necessary for olive trees to differentiate flowers and produce fruit (Koubouris et al., 2019). It is therefore recommended to avoid planting of varieties with high chilling demands such as Chondrolia Chalkidikis in areas with warm winters such as most of the plain areas of Southern Greece.

The impacts of climate change are expected to be related with changes in the frequency of extreme climate events than to changes in mean conditions (Giannakopoulos et al., 2009). In the same study, the number of days per year without precipitation are expected to increase affecting water deficit in crops while the number of days with night frost is projected to decrease in marginal areas of Northern Greece reducing the risk for sensitive crops (Giannakopoulos et al., 2009).

Signs of this expected impact of climate change have been recorded during the past decade, when olive growing has been affected in consecutive years by nontypical climatic extreme incidents at different seasons. Such examples include a heat wave during May in 2013, which resulted in flower damage and the lowest olive production in the last 30 years, extensive fruit drop caused by storms during 2017 and 2018 and lately, during 2019, a dramatic loss of yield and olive oil quality deterioration, linked to high air temperature and relative humidity during autumn which favored fruit infestation by pests and diseases.

Consecutive years of low yields during the present decade resulted in a significant drop in mean annual olive oil production from 20,430 Mg in the period 1998–2008 to 12,652 Mg in the period 2009–2018.

Adaptation of sustainable orchard management practices has been linked to more stabilized yield levels on an annual basis, even during years of adverse climatic conditions (Kourgialas et al., 2019; Michalopoulos et al., 2020), providing an alternative approach for combating the negative impacts of climate change. The use of appropriate adaptation measures in Greece may result in a net yearly benefit of about 162 million € (Georgopoulou et al., 2017).

20.2 Landscape and Soils

It is estimated that about 50% of the country has a slope > 10% and only 36% a slope less than 5% (Yassoglou et al., 2017). Greek soils are commonly dry and rocky and only 20% is considered arable. The geological background consists of limestone, marls, and volcanic rocks (Yassoglou et al., 2017). Among this limited arable land, about 14% in total, or 75% of the land devoted to fruit tree growing is covered by olive groves totaling approximately 150 million olive trees. Traditionally, olive tree cultivation is mostly concentrated in sloping land, in shallow and stony soils, where, practically, no alternative crops could be cultivated on an economically viable basis. Although the expansion of olive growing during the second half of the twentieth century, along with the adaptation of intensive management systems, lead to cultivation in more fertile soils of lower slope, the limited area of large plains in the Greek landscape is still commonly occupied by other crops. According to Karamesouti et al. (2018), moderate soil quality characterizes over 70% of the country, while more than 20% of the soils are of low quality. Based on CORINE 2011 land cover data, about 60% of olive trees are located in elevations up to 200 m above sea level, while about 40% of the olive-cultivated land occupies the semi-mountainous zone (200–800 m). In terms of slope, this is translated to about 60% of olive groves being in sloping or mountainous land, facing serious management difficulties and increased production cost. The remaining 40% of olive trees grown in areas of reduced slope are threatened by land use change for urban uses and restrictions in pest management due to proximity with houses. The above-mentioned landscape features of olive growing in Greece indicate that appropriate orchard management is a necessity for reducing soil erosion and desertification risks in olive orchards.

In terms of soil chemical and physical properties, alkaline soils are quite common in the olive tree growing zone in Greece (Yassoglou et al., 2017). Major Greek cultivars are well adapted in growing at high soil pH (>8.0), while special management (liming) is needed in order to be fully productive at acidic soils (pH < 6.0). Organic matter is generally low, as in most soils around the Mediterranean (EIP-AGRI Focus Group, 2015). Olive growing in coarse-textured soils is not common. Typically, medium-textured soils are used, while soils with a clay percentage higher than 30% are also common in certain areas (Yassoglou et al., 2017).

References

Climate-data.org. (2020). *Climate classifications of Greece*. https://en.climate-data.org/europe/greece-173/. Accessed on 29 June 2020.

EIP-AGRI Focus Group. (2015). Soil organic matter in Mediterranean regions. *Final Report*. https://ec.europa.eu/eip/agriculture/sites/agri-eip/files/eip-agri_fg_soil_organic_matter_final_report_2015_en_0.pdf

Georgopoulou, E., Mirasgedis, S., Sarafidis, Y., et al. (2017). Climate change impacts and adaptation options for the Greek agriculture in 2021–2050: A monetary assessment. *Climate Risk Management, 16*, 164–182.

Giannakopoulos, C., Kostopoulou, E., Varotsos, K. et al. (2009). *Climate change impacts in Greece in the near future*. https://www.bankofgreece.gr/RelatedDocuments/WWF_Climate_change_impacts_in_Greece_in_the_near_future.pdf

Karamesouti, M., Panagos, P., & Kosmas, C. (2018). Model-based spatio-temporal analysis of land desertification risk in Greece. *Catena, 167*, 266–275.

Koubouris, G. C., Metzidakis, I. T., & Vasilakakis, M. D. (2009). Impact of temperature on olive (Olea europaea L.) pollen performance in relation to relative humidity and genotype. *Environmental and Experimental Botany, 67*, 209–214.

Koubouris, G. C., Metzidakis, I. T., & Vasilakakis, M. D. (2010a). Influence of cross pollination on the development of parthenocarpic -shotberries- olive (Olea europaea L.) fruits. *Experimental Agriculture, 46*, 67–76.

Koubouris, G. C., Metzidakis, I. T., & Vasilakakis, M. D. (2010b). Phenological, morphological and functional indicators of genetic variability and their implication in the sexual reproductive system of Olea europaea L. (Oleaceae). *Scientia Horticulturae, 123*, 547–550.

Koubouris, G. C., Tzortzakis, N., Kourgialas, N. N., et al. (2015a). Growth, photosynthesis and pollen performance in saline water treated olive plants under high temperature. *International Journal of Plant Biology, 6*, 28–32.

Koubouris, G. C., Kavroulakis, N., Metzidakis, I. T., et al. (2015b). Ultraviolet-B radiation or heat cause changes in photosynthesis, antioxidant enzyme activities and pollen performance in olive tree. *Photosynthetica, 53*, 279–287.

Koubouris, G., Limperaki, I., Darioti, M., et al. (2019). Effects of various winter chilling regimes on flowering quality indicators of Greek olive cultivars. *Biologia Plantarum, 63*, 504–510.

Kourgialas, N. N., Morianou, G., Psarras, G., et al. (2019). *Improving water use efficiency in irrigated fields: The LIFE+ AGROCLIMAWATER case study*. IX International Symposium on Irrigation of Horticultural Crops, 17–20 June 2019, Matera—Italy. Book of Abstracts, p. 40.

Larcher, W. (2000). Temperature stress and survival ability of Mediterranean sclerophyllous plants. *Plant Biosystems, 134*, 279–295.

Michalopoulos, G., Kasapi, K. A., Koubouris, G., et al. (2020). Adaptation of Mediterranean olive groves to climate change through sustainable cultivation practices. *Climate, 8*(4), 54.

Moriondo, M., Trombi, G., Ferrise, R., et al. (2013). Olive trees as bio-indicators of climate evolution in the Mediterranean Basin. *Global Ecology and Biogeography, 22*, 818–833.

Yassoglou, N., Tsadilas, C., & Kosmas, C. (2017). In A. E. Hartemink (Ed.), *The soils of Greece, world soils book series*. Springer. https://doi.org/10.1007/978-3-319-53334-6

Chapter 21
Varieties and Olive Grove Production Systems

Georgios Koubouris and Vasileios Stournaras

Abstract Olive is the most important tree crop in Greece as in many countries of the Mediterranean basin. There is a significant number (80–100) of Greek olive varieties cultivated in Greece. 'Koroneiki', 'Mastoidis', 'Megareitiki' and 'Lianolia Kerkyras' are the main varieties for olive oil production. 'Koroneiki' is the main variety cultivated for olive oil production with fruit of around 20–25% olive oil content and olive oil of excellent quality. 'Mastoidis' is a variety with one of the highest fruit olive oil contents (25–30%) depending on the date of harvest. 'Amfissis', 'Kalamon' and 'Chondrolia Chalkidikis' are the main varieties for table olives that are exported in many countries worldwide. 'Amfissis' is mainly cultivated in Central Greece, 'Kalamon' in Western Greece and 'Chondrolia Chalkidikis' in the area of Chalkidiki. Traditional, semi-intensive and intensive are the main plantation systems. Super high-density system has also been implemented, in small-scale, in some regions of Greece during the last years. Replacement of local varieties by 'Koroneiki' poses a risk for biodiversity. High production cost in mountainous areas threatens the already fragile socio-economic life of inhabitants and often leads to abandonment. Adoption of new technologies, organization in cooperative schemes and training of farmers should be intensified.

Keywords Olive varieties · Genetic resources · Olive oil · Table olives · Production systems · Biodiversity

---

G. Koubouris (✉)
Hellenic Agricultural Organization DIMITRA—Institute of Olive Tree, Subtropical Crops and Viticulture, Chania-Crete, Greece
e-mail: koubouris@elgo.gr

V. Stournaras
Hellenic Agricultural Organization DIMITRA—Institute of Olive Tree, Subtropical Crops and Viticulture, Kalamata, Greece
e-mail: vstournaras.elgo@gmail.com

J. Muñoz-Rojas, R. García-Ruiz (eds.), *The Olive Landscapes of the Mediterranean*, Landscape Series 36,
https://doi.org/10.1007/978-3-031-57956-1_21

21.1 Olive Varieties

Olive tree (*Olea europaea* L.), which probably originates from the Middle East (Zohary & Hopf, 1994), has been cultivated for thousands of years and includes many varieties, not fully characterized, creating problems in their classification (Barranco et al., 2000). The problem becomes more acute because, very often, the same cultivars at various locations are known under different names (Koubouris et al., 2019). During the twentieth century, structured protocols were developed for morphological description of olive varieties by the International Olive Council through the RESGEN project (Barranco et al., 2000) and by the International Union for the Protection of New Varieties of Plants (UPOV, 2011) with information regarding tree, shoot, leaf, fruit and endocarp traits allowing for the development of the 'identity card' of each variety and the publication of official plant variety catalogues (European Commission, 2020). Some of the criteria for classification of olive cultivars are the fruit characteristics, such as shape, colour, size and flesh, pit ratio, percentage of oil, shape of endocarp, as well as leaf size and inflorescence characteristics (Georgiadou et al., 2019; Koubouris et al., 2010, 2018, 2019). The most important Greek olive varieties are presented in Table 21.1 and Fig. 21.1.

Table 21.1 List of important Greek olive varieties

Variety	Synonyms or other common names	Purpose (olive oil, table olive)	Region of cultivation in Greece
Adramytini	Adramytiki Lesvou, Aivaliotiki, Fragolia	Olive oil	Aegean islands (Lesvos, Chios), Evia
Amfissis	Voliotiki, Stylidos, Agriniou, Artas, Konservolia	Table olive, olive oil	Central and Western Greece, Thessaly
Chondrolia Chalkidikis		Table olive	Macedonia Northern Greece
Kalamon	Kalamatiani, Tsigkeli, Aetonychi, Kalamata	Table olive	Central and Western Greece, Peloponnese, Crete
Koroneiki	Psilolia, Vatsiki, Kritikia	Olive oil	Peloponnese, Crete, continental areas and islands of Greece
Koutsourelia	Patrini	Olive oil	Peloponnese
Lianolia Kerkyras	Prevezana, Pargas	Olive oil	Western Greece continental areas and Ionian islands (Corfu)
Mastoidis	Athinolia, Tsounati	Olive oil	Crete, Peloponnese, Central Greece
Megareitiki	Perachoritiki, Vododitiki	Olive oil, table olive	Central Greece (Megara Attikis)
Valanolia	Mytilinia, Kolovi, Valana	Olive oil, table olive	Aegean islands (Lesvos)

Fig. 21.1 Main Greek olive varieties

'Koroneiki' is the main variety cultivated for the production of olive oil with fruit of around 20–25% olive oil content and olive oil of excellent quality. It is mainly cultivated in Peloponnese and Crete with some genetic variability (Avramidou et al., 2020) and is expanding in other areas of the world too. Due to the size and the formation of the tree and its suitability for semi-mechanized harvest, 'Koroneiki' is often used in modern intensive plantations with impressive typical fruit yields of between 7.5 and 15 Mg ha^{-1} under irrigated intensive management. It is frost sensitive and thus cultivated in Southern Greece and mainly in plain areas. 'Mastoidis' is a variety with one of the highest fruit oil content fluctuating around 25–30%, depending on the date of harvest. It is mainly cultivated in mountainous areas of Crete, Peloponnese and Central Greece due to its frost tolerance. Productivity is medium and alternate, while olive oil quality is excellent. However, in some areas, farmers still apply a traditional harvesting practice of collecting naturally fallen overripe fruit which results in degraded olive oil. During the recent years, young farmers modernize olive grove management and achieve olive oil of superior organoleptic properties from 'Mastoidis'. 'Megareitiki' is mainly cultivated in the area of Megara Attikis. The fruit has average an weight of 3.5–4.5 g and olive oil content of 22–25%. This cultivar requires pollination since it is partially self-sterile, has medium tolerance to cold and salinity and high tolerance to drought (Therios, 2009). 'Lianolia Kerkyras' is cultivated in Western Greece continental areas and islands and mainly in the island Kerkyra (Corfu). The fruit is small, with high olive oil content (20–22%) and excellent quality. It is partially self-fertile and has stable and high productivity and high tolerance to drought, wind and cold (Therios, 2009).

'Amfissis', 'Kalamon' and 'Chondrolia Chalkidikis' are the main varieties for table olives that are exported in many countries worldwide. 'Amfissis' has high and alternate productivity and large fruit of medium olive oil content. It is a hardy variety that adapts to several environments. It is mainly cultivated in Central and Western Greece. 'Kalamon' has the natural tendency of high and alternate productivity; however, under management for table olives, fruit load is regulated to achieve moderate yields every year with large fruit size. Its cultivation started from Southern Peloponnese, but nowadays rapid expansion is observed to many areas all over Greece and mainly in Central and Western Greece. 'Kalamon' is moderately resistant to cold and resistant to olive knot. Apart from its main use for table olives, it gives also olive oil of excellent quality. 'Chondrolia Chalkidikis' is also a top table olive variety with large fruit and is grown in the peninsula of Chalkidiki in intensive plantations. It is managed for green table olives. It is resistant to drought; however, regular irrigation is needed for high-quality table olives. It is also resistant to cold.

21.2 Olive Grove Production Systems

Through the time, systematic olive orchards replaced the spontaneous trees, and great progress was realized in the cultivation techniques. Nowadays, olive is one of the most important crops in Greece, covering about 950,000 ha (Greek Ministry of Rural Development and Food, 2020a). The total number of olive trees in Greece (2017/2018) is estimated about 166 million (ELSTAT, 2020), and the mean number of tree density is calculated about 175 trees per hectare.

Traditional, semi-intensive and intensive are the main plantation systems in olive cultivation according to agro-ecological, technological and socio-economic criteria.

Traditional plantation system is characterized by low-density plantations (<100 trees ha^{-1}) of old trees (>50 years) often irregularly planted in steep or moderately steep slopes (Duarte et al., 2008) (Fig. 21.2a). Low inputs and extensive management without mechanization, which characterize this system in Greece, result in relatively low yields that in the case of traditional olive grove of Crete fluctuate around 1250 kg ha^{-1}. One of the main strengths of this system is its high biodiversity, whereas one of the weaknesses is the low profitability that often leads to abandonment. These plantations are located in hard-to-reach areas and commonly managed by ageing farmers. Due to the low input of chemicals, the transition of this system to organic farming to achieve higher product prices is considered easy without drastic modifications. However, the age and the lack of agricultural education by farmers are serious obstacles towards this evolution. These plantations have high socio-economic and environmental importance and should be supported by rural policy measures.

Semi-intensive plantation system is characterized by olive groves of variable tree density, fluctuating between 100 and 150 trees ha^{-1} (Xiloyannis et al., 2008). These orchards consist of well-formed and productive trees of about 30 years old or more, with increased inputs compared to the traditional system but usually no irrigation. Typically, yields range between 2500 and 3750 kg ha^{-1} in Crete.

Intensive plantation system includes olive groves with a high tree density (about 250 trees ha^{-1}) of productive trees with yield of up to 5000 kg ha^{-1}, significantly higher than organic or traditional ones (Metzidakis et al., 2008) (Fig. 21.2b). High yield and quality of the obtained olive oil together with the partial mechanization are the main strengths of intensive plantations, while soil erosion promoted by tillage, pollution when improper use of agrochemicals and low biodiversity due to intensive pest management are the main weaknesses. In the last 15 years, super high-density system (tree density more than 1000 trees $ha^{-1)}$ has been implemented, in small-scale, in some Prefectures of Greece, such as Ilia, Lakonia, Thessalia and Aetoloakarnania (Fig. 21.2c). They covered an area of approximately 500 ha, and the main cultivated varieties are 'Koroneiki' and 'Arbequina'.

Organic plantation system is characterized by olive groves without a clear profile (tree density, age, distances) since they may result from any olive grove after the modification of the orchard management (Gómez et al., 2008). Yield levels also vary significantly due to the variable levels of organic material inputs as a means of fertilization or soil amendment. This system is very labour demanding. The high

Fig. 21.2 Olive grove production systems in Greece

biodiversity, soil organic matter content, product quality and price are the strengths of organic plantations, while the decreased yield, the high labour demand and the difficulties in pest management are the weaknesses. Organic olive groves covered about 35,000 ha (2018) (about 3.7% of the total land cultivated by olive trees), and transitional to organic olive groves covered about 15,000 ha (Greek Ministry of Rural Development and Food, 2018). Integrated olive production has recently gained a lot of attention, and more than five certification agencies are activated in organic and integrated production revealing the tendency for further future development.

There are 19 brands of olive oil and 10 of table olives characterized as protected designation of origin (PDO) and 12 brands of olive oil and 1 of table olive characterized as protected geographical indication (PGI) (Greek Ministry of Rural Development and Food, 2020b). Replacement of local varieties grown in traditional olive groves by high-density 'Koroneiki' plantations poses a risk of biodiversity decline. Massive establishment of Greek varieties such as 'Kalamon' in non--E.U. countries with lower production cost has caused serious changes in the supply-demand balance and reduced profitability for Greek farmers. High production cost in mountainous areas threatens the already fragile socio-economic life of inhabitants and often leads to abandonment not only of the olive groves but of rural areas once and for all, in favour of starting a new life in big cities. Adoption of new technologies, organization in cooperative schemes and training of farmers should be intensified for Greek olive growing to withstand international competition.

References

Avramidou, E. V., Koubouris, G. C., Petrakis, P. V., et al. (2020). Classification binary trees with SSR allelic sizes: Combining regression trees with genetic molecular data in order to characterize genetic diversity between cultivars of *Olea europaea* L. *Agronomy, 10*, 1662.

Barranco, D., Cimato, A., Fiorino, P., et al. (2000). *World catalogue of olive varieties* (p. 360). International Olive Oil Council. http://www.internationaloliveoil.org/store/view/79-world-catalogue-of-olive-varieties-english

Duarte, F., Jones, N., & Fleskens, L. (2008). Traditional olive orchards on sloping land: Sustainability or abandonment? *Journal of Environmental Management, 89*, 86–98.

ELSTAT (Hellenic Statistical Authority). (2020). *Statistical data for 2017/2018*. www.statistics.gr

European Commission. (2020). https://ec.europa.eu/frumatis/. Accessed 18/09/2024.

Georgiadou, E. C., Koubouris, G., Goulas, V., et al. (2019). Genotype-dependent regulation of vitamin E biosynthesis in olive fruits as revealed through metabolic and transcriptional profiles. *Plant Biology*. https://doi.org/10.1111/plb.12950

Gómez, J. A., Amato, M., Celano, G., et al. (2008). Organic olive orchards on sloping land: More than a specialty niche production system? *Journal of Environmental Management, 89*, 99–109.

Greek Ministry of Rural Development and Food. (2018). *Statistical data on organic cultivation*. http://www.minagric.gr/images/stories/docs/agrotis/Biologika/statist_fitikis2018.pdf

Greek Ministry of Rural Development and Food. (2020a). *Statistical data on tree cultivation*. http://www.minagric.gr/index.php/el/pinakas-3-dendrodeis-monimes-kalliergeies/routedownload/statistical-works-data/pinakas-3-dendrodeis-monimes-kalliergeies

Greek Ministry of Rural Development and Food. (2020b). *PDO and PGI olive products*. http://www.minagric.gr/index.php/el/for-farmer-2/2012-02-02-07-52-07/ellinikaproionta/1182-elaioladaelies-2

Koubouris, G. C., Metzidakis, I. T., & Vasilakakis, M. D. (2010). Phenological, morphological and functional indicators of genetic variability and their implication in the sexual reproductive system of Olea europaea L. (Oleaceae). *Scientia Horticulturae, 123*, 547–550.

Koubouris, G., Bouranis, D., Vogiatzis, E., et al. (2018). Leaf area estimation by considering leaf dimensions in olive tree. *Scientia Horticulturae, 240*, 440–445.

Koubouris, G. C., Avramidou, E. V., Metzidakis, I. T., et al. (2019). Phylogenetic and evolutionary applications of analyzing endocarp morphological characters by classification binary tree and leaves by SSR markers for the characterization of olive germplasm. *Tree Genetics & Genomes*. https://doi.org/10.1007/s11295-019-1322-0

Metzidakis, I., Martinez-Vilela, A., Castro Nieto, G., et al. (2008). Intensive olive orchards on sloping land: Good water and pest management are essential. *Journal of Environmental Management, 89*, 120–128.

Therios, I. N. (2009). *Olives: Crop production science in horticulture* (Vol. 18). CABI.

UPOV – International Union for the Protection of New Varieties of Plants. (2011). *Technical guideline for the conduct of tests for distinctness, homogeneity, and stability in olive, UPOV code: OLEAA_EUR, Olea europaea TG/099/4*. UPOV.

Xiloyannis, C., Martinez Raya, A., Kosmas, C., et al. (2008). Semi-intensive olive orchards on sloping land: Requiring good land husbandry for future development. *Journal of Environmental Management, 89*, 110–119.

Zohary, D., & Hopf, M. (1994). *Olive: Olea europaea. Domestication of plants in the Old World* (pp. 137–143). Clarendon Press.

Chapter 22
Soil Management Practices, Fertilization, Pruning and Harvesting of Greek Olive Groves

Georgios Psarras, Georgios Koubouris, and Vasileios Stournaras

Abstract During the past, frequent soil tillage and herbicide spraying were increasingly employed in olive groves in Greece contributing in soil erosion. In the recent decade, a different rural movement has evolved in Greece aiming at sustainably growing olive trees and preventing soil erosion. Fertilizing is still applied empirically in most olive orchards in Greece. Boron deficiency is quite common, followed by K and N deficiencies. On the contrary, phosphorus is usually found in excess in soils, mainly due to the extensive use of composite fertilizers. In olive groves destined for olive oil production, most farmers apply pruning every 2–3 years, with extensive use of handheld machined chainsaws at an indicative cost of 4–7 € tree^{-1}. On the other hand, trees grown for producing table olives are pruned annually. Harvesting is also differentiated between production systems at an indicative cost of 0.15–0.20 € kg^{-1} fruit. In the case of table olives, fruit is harvested manually by workers, and this is a very costly practice. When olives are grown for olive oil, the dominant practice is to use handheld machined harvesting sticks at significantly higher cost compared to over-the-row harvesters or trunk vibrators employed in other countries.

Keywords Soil management · Fertilization · Pruning · Harvesting · Olive groves

---

G. Psarras (✉) · G. Koubouris
Hellenic Agricultural Organization 'DIMITRA'—Institute of Olive Tree, Subtropical Crops and Viticulture, Chania-Crete, Greece
e-mail: psarras@elgo.gr; koubouris@elgo.gr

V. Stournaras
Hellenic Agricultural Organization 'DIMITRA'—Institute of Olive Tree, Subtropical Crops and Viticulture, Kalamata, Greece
e-mail: vstournaras.elgo@gmail.com

J. Muñoz-Rojas, R. García-Ruiz (eds.), *The Olive Landscapes of the Mediterranean*, Landscape Series 36,
https://doi.org/10.1007/978-3-031-57956-1_22

22.1 Soil Management

Olive orchard management in Greece is usually applied under traditional or intensive cropping systems. The super-high-density (SHD) cropping system has only occasionally been applied, and its impact on olive tree growing in Greece is practically minimal. Soil management plays a crucial role in the conservation or degradation of soil and water resources, and its optimization would contribute to increasing efficiency of resources, improving soil fertility as well as plant nutrition and preventing environmental disorders. In Greece, the soil management strategies are more related to the intensity of the cropping system and less to the geographical location of the olive grove. In traditional olive orchards soil management is usually minimal, especially in areas of high slope. In some of these areas, weed control by herbicides is the only intervention applied, occasionally linked to the need of facilitating harvesting, i.e. need for a clean soil surface for permanent laying of harvesting nets. On the contrary, under intensive cropping system, both the frequent soil tillage (Fig. 22.1a) and herbicide spraying were increasingly employed in olive groves in Greece in line with the trend for intensification of agricultural systems during the past decades (Gómez et al., 2008; Metzidakis et al., 2008). Since the majority of olive growing, even under intensive cultivation, is realized in sloping land, soil erosion became an important limitation of soil fertility and a serious threat for the environment—on-site effects (Kourgialas et al., 2016)—as well as on nearby infrastructures, -off-site effects (Kourgialas et al., 2017a). Soil loss by water erosion in important olive growing areas in Southern Greece (Peloponnese, Crete and Aegean islands) ranges from 3.68 to 7.83 tonne ha^{-1} year^{-1} (Kourgialas, 2021). Assessing desertification risk for Crete, one of the most important olive growing areas of Greece, Morianou et al. (2018) highlighted the importance of factors like land slope, shallow soil profiles and intensive grazing in scoring moderate to high soil and management sensitivities, in a significant part of the agricultural land of the island. Overgrazing is a common exogenous factor that could affect the incorporation of sustainable orchard management practices. In a recent study, grazing resulted in failure to achieve successful introduction of cover crops in 3 out of 10 olive

Fig. 22.1 Intensive soil tillage (**a**) employed during the past decades following intensification of olive grove management resulted in serious land degradation, while a more sustainable approach (**b**) with cover crops has been gaining ground mainly during the past 5 years. Photos taken in Messinia

orchards of a pilot area in Eastern Crete (LIFE AgroClimaWater project, unpublished data, www.lifeagroclimawater.eu). In SHD olive groves, weed management is applied through herbicides and weed mowers. In some cases that SHD plantations have been established on heavy soils or in fields previously grown with cotton or tobacco, tree losses due to soil-borne pathogens (e.g. *Verticillium* wilt) and delays in harvesting due to poor drainage of rainwater have been observed.

During the last decade, a different rural movement has evolved in Greece aiming at sustainably growing olive trees (Michalopoulos et al., 2020; Kavvadias & Koubouris, 2019; Koubouris, 2018). This more ecologically friendly mindset resulted in increased application of non-disturbing soil management practices (Fig. 22.1b). In general, weed mowing has replaced soil tillage in several areas, at least for farmers following the integrated management principles. If coupled with mulching and establishment of cover crops, this management strategy may further contribute to sustainable soil management (Bechara et al., 2018; Kavvadias et al., 2018; Kourgialas et al., 2017b).

22.2 Fertilization

Fertilizing is still applied empirically in most olive orchards in Greece. The only exception are orchards under integrated management, but even in this case, soil, instead of leaf, analyses are periodically used. Fertilizer use has been reduced, as compared to the past decades, but this is due to a general effort for reduction of production cost and not based upon the actual nutritional requirements of olive trees. For the same reason, composite fertilizers are still used extensively, leading to an unbalanced nutrition of olive orchards, with excess of soil availability of P being quite common in Greek olive orchards.

The above-mentioned facts are characteristic for most olive-growing areas in Greece. Indicatively, they are illustrated in Fig. 22.2, through a survey of the nutritional status of commercial olive orchards in the area of Chania, Crete (Psarras et al., 2016), during the past decade. Occurrence of P deficiency is quite rare, but despite that, farmers kept adding P through composite fertilizers on a frequent basis. The percentage of deficiencies in key nutrients for olive growing, like N, K and B, was observed in more than 40% of the orchards. However, for N and K, this was mostly due to significantly higher percentages of deficiency in rainfed orchards, indicating the impact of low soil water content on sufficient nutrient uptake in Southern Greece (Psarras et al., 2016). Another significant finding was that Zn deficiency at 80% of the orchards was related to high P availability in the soil, indicating problems of unbalanced nutrition, as it has also been reported by Jimenez-Moreno and Fernandez-Escobar (2016). As compared to a similar survey in the same area during the 1990s (Loupassaki et al., 2005), the percentage of orchards deficient in key nutrients for olive growing like N, K and B has been increased significantly, highlighting the reduced use of fertilizers.

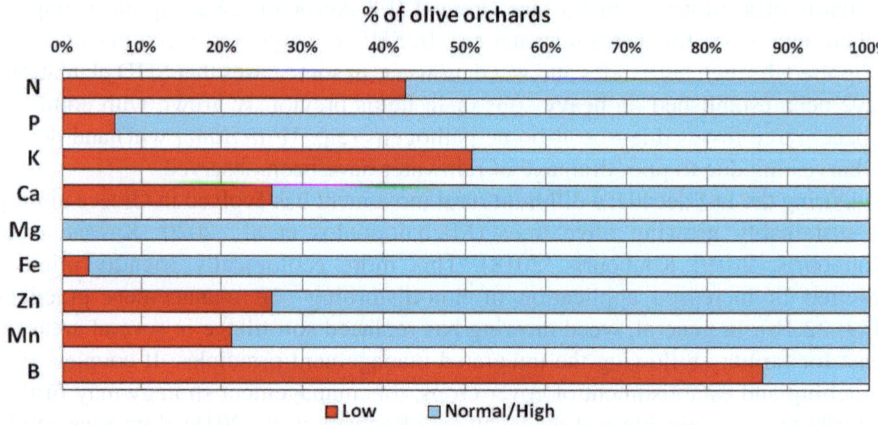

Fig. 22.2 Occurrence of nutrient deficiencies (as determined by low concentration in leaves) as a percentage of total number of olive orchards, during a nutritional status survey in the Prefecture of Chania, Crete. (Psarras et al., 2016)

Typically, olive tree fertilizing in Greece is focused on application of N, P and K. Boron is the only micronutrient usually applied, although quite often in the form of composite fertilizers that also contain B and less frequently in the form of borax. In general, under intensive cropping, fertilizing is applied annually, especially for cultivars like 'Koroneiki' which, under proper management, may give relatively good yields in consecutive years. In traditional olive orchards, fertilizing could be either minimal or quite often applied on a biennial cycle, being higher during the 'on' year.

Traditionally, the main application of chemical fertilizers in olive orchards is taking place during the raining season (winter time). Foliar application of nutrients (mainly B and in a lesser degree N) and bio-regulators has been increased during the last decade. On the contrary, fertigation is practically not included in the fertilizing schedule of irrigated olive orchards, with limited use taking place only in the case of table olives. The increased frequency in occurrence of non-typical climatic phenomena during winter time in recent years (e.g. long dry periods or extremely high frequency and amount of precipitation) has increased the use of slow- or controlled-release fertilizers, as a way to reduce N losses. Although similar results could be achieved by the application of organic materials or composts (Michalopoulos et al., 2020), their use in conventional olive groves is relatively limited due to non-sufficient availability and high cost per nutrient unit in most olive-growing areas.

Fig. 22.3 In the case of olive oil production (**a**), a free tree shape is commonly aimed through pruning as shown here in cv. Koroneiki, while when destined for table olives (**b**), trees are trained in detail, and canopies are strictly structured as depicted for cv. Chalkidikis. Photos taken in Chania and Chalkidiki, respectively

22.3 Pruning

Pruning is considered as one of the most drastic olive tree management practices to regulate vegetative growth and fruit production as well as to adjust tree size to the available resources, mainly water, nutrients and space. In the case of olive groves destined for the production of olive oil, farmers apply pruning every year or once every 2–5 years depending on the region and the tree canopy (Fig. 22.3a). They usually employ handheld machined chainsaws for fast and less detailed tree formation at an indicative cost of 4–7 € tree^{-1} including management (recycling or burning) of pruning residue. On the other hand, trees grown for producing table olives are pruned every year or even twice a year for fruit thinning, by pruning scissors or small saws resulting in a very sophisticated shaping of vegetation (Fig. 22.3b). There is also the case of traditional olive groves characterized by naturally growing trees exceeding 10–15 m in height and seldom pruned (i.e. once every 5 years) at indicative cost of 10–15 € tree^{-1}, including management (recycling or burning) of pruning residue. Large branches are considered as a by-product of pruning, and they are sold for firewood at 70–140 € m^{-3} or used by farmers for heating. In the case of super-high-density olive groves, a combination of mechanical and manual pruning is applied contributing to significant cost reduction. However, failure to comply with the sophisticated pruning plan required may cause reduced yields and pest problems due to excessively dense vegetation.

22.4 Harvesting

Harvesting is also differentiated between orchards aiming at olive oil or table olive production. When olive trees are grown for olive oil, the dominant practice is to use handheld machined harvesting sticks (Fig. 22.4a). Compared to the traditional use

Fig. 22.4 The dominant techniques for olive harvesting in Greece: (**a**) use of handheld mechanized sticks in the case of olive oil production, (**b**) manual harvest in the case of table olives, (**c**) manual harvest after natural fruit drop on permanently laid plastic net on the ground in traditional olive groves in mountainous areas and (**d**) over the row harvester in super-high-density orchard representing less than 1% of the national acreage. (Photos (a–c) taken in Chania and (d) taken in Thessalia)

of harvesting sticks without an external energy source, significant reduction of cost has been achieved, though it still remains quite high in Greece representing up to 50% of the production cost, at an indicative level of 0.15–0.20 € kg^{-1} fruit. In the case of table olives, fruit is harvested manually by workers, at an indicative cost of 0.1–0.2 € kg^{-1} fruit when harvesting green olives and varying between 0.2 and 0.4 € kg^{-1} fruit when harvesting black olives, since not all fruits of a tree are at the same maturity stage, and workers have to select the olives that are suitable for harvest (Fig. 22.4b). Portable branch vibrators are employed in some cases with promising results. However, trunk vibrators have not found wide application so far in Greece. A plausible cause is the high diversity of local olive cultivars grown and the lack of adaptation of this harvesting practice to their tree pruning, fruit removal force and harmless trunk vibration. Another obstacle is related to the landscape, since most olive groves are located in sloping land, which may be considered risky for accidents in case of tractor use.

In a less frequent case, representing less than 10% of total acreage, in traditional olive groves growing in sloping land, farmers cover the soil through a plastic net and wait for natural fruit drop to harvest manually many times a year, commonly every

1–2 weeks from October to March at an indicative cost of 0.05 € kg^{-1} fruit (Fig. 22.4c). This harvesting practice, which is currently in recession, results in olive oil of very low quality and price. During the recent years, young farmers modernize olive grove management and achieve olive oils of superior organoleptic properties in this type of traditional olive groves. Finally, in less than 1% of the national acreage planted under a super-high-density system, the specialized tractors are employed for harvesting, achieving the lowest cost among olive production systems applied in Greece, at about 0.03 € kg^{-1} (Fig. 22.4d).

References

Bechara, E., Papafilippaki, A., Doupis, G., Sofo, A., & Koubouris, G. (2018). Nutrient dynamics, soil properties and microbiological aspects in an irrigated olive orchard managed with five different management systems involving soil tillage, cover crops and compost. *Journal of Water and Climate Change, 9*(4). https://doi.org/10.2166/wcc.2018.082

Gómez, J. A., Amato, M., Celano, G., & Koubouris, G. (2008). Organic olive orchards on sloping land: More than a specialty niche production system? *Journal of Environmental Management, 89*, 99–109.

Jimenez-Moreno, M. J., & Fernandez-Escobar, R. (2016). Response of young olive plants (*Olea europaea*) to phosphorus application. *Hortscience, 51*(9), 1167–1170.

Kavvadias, V., & Koubouris, G. (2019). Sustainable soil management practices in olive groves. In D. G. Panpatte & Y. K. Jhala (Eds.), *Soil fertility Management for Sustainable Development* (pp. 167–188). Springer.

Kavvadias, V., Papadopoulou, M., Vavoulidou, E., Theocharopoulos, S., Koubouris, G., Psarras, G., Manolaraki, C., Giakoumaki, G., & Vasiliadis, A. (2018). Effect of sustainable management of olive tree residues on soil fertility in irrigated and rain-fed olive orchards. *Journal of Water and Climate Change, 9*(4). https://doi.org/10.2166/wcc.2018.143

Koubouris, G. (2018). Editorial – Advances in prediction, monitoring and mitigation of climate change effects on water resources and good agricultural practices for crop adaptation to environmental stresses. *Journal of Water and Climate Change, 9*(4), 631–632.

Kourgialas, N. N. (2021). A critical review of water resources in Greece: The key role of agricultural adaptation to climate-water effects. *Science of the Total Environment, 775*. https://doi.org/10.1016/j.scitotenv.2021.145857

Kourgialas, N. N., Koubouris, G. C., Karatzas, G. P., & Metzidakis, I. (2016). Assessing water erosion in Mediterranean tree crops using GIS techniques and field measurements: The effect of climate change. *Natural Hazards, 83*, 65–81.

Kourgialas, N. N., Karatzas, G. P., & Koubouris, G. C. (2017a). A GIS policy approach for assessing the effect of fertilizers on the quality of drinking and irrigation water and wellhead protection zones (Crete, Greece). *Journal of Environmental Management, 189*, 150–159.

Kourgialas, N. N., Doupis, G., Psarras, G., Sergentani, C., Digalaki, N., & Koubouris, G. C. (2017b). Soil management and compost effects on salinity and seasonal water storage in a Mediterranean drought-affected olive tree area. *Desalination and Water Treatment, 99*, 1–7.

Loupassaki, M., Digalaki, N., Psarras, G., Chartzoulakis, K., & Androulakis, I. (2005). Survey of fertilizing practices and nutritional status of olive orchards in Chania Prefecture. *Proceedings of the 22nd conference hellenic horticultural society*, Vol. 12, pp. 331–334 (in Greek).

Metzidakis, I., Martinez-Vilela, A., Castro Nieto, G., & Basso, B. (2008). Intensive olive orchards on sloping land: Good water and pest management are essential. *Journal of Environmental Management, 89*(2), 120–128.

Michalopoulos, G., Kasapi, K. A., Koubouris, G., Psarras, G., Arampatzis, G., Hatzigiannakis, E., Kavvadias, V., Xiloyannis, C., Montanaro, G., Malliaraki, S., Angelaki, A., Manolaraki, C., Giakoumaki, G., Reppas, S., Kourgialas, N., & Kokkinos, G. (2020). Adaptation of Mediterranean olive groves to climate change through sustainable cultivation practices. *Climate, 8*(4). https://doi.org/10.3390/cli8040054

Morianou, G. G., Kourgialas, N. N., Psarras, G., & Koubouris, G. C. (2018). Mapping sensitivity to desertification in Crete (Greece), the risk for agricultural areas. *Journal of Water and Climate Change, 9*(4), 691–702.

Psarras, G., Digalaki, N., Darioti, M., & Bertaki, M (2016) Survey of nutritional status in olive orchards of Chania Prefecture during a year with a long dry season. *Proceedings of the 27th conference hellenic horticultural society*, Vo. 17A, pp. 269–273. (in Greek).

Chapter 23
Pest and Diseases and Main Control Strategies of Greek Olive Groves

Kiki Varikou, Panagiotis Skouras, and Emmanouil Markakis

Abstract Olive crop in Greece is attacked by diverse pest species. The olive fruit fly (OFF), *Bactrocera oleae* (Gmelin) (Diptera; Tephritidae) is the major insect pest affecting olive plantations of Greece, resulting in a decrease of the olive yield and lower quality of the extracted olive oil, which might lead to a reduction of the economic profitability. The degree of infestation is high; about more than 90% of orchards can be infested by the OFF in absence of treatment and under optimum climate conditions for the development of the insect. *Bactrocera oleae* control in Greece is mainly based on insecticides particularly organophosphates, pyrethroids and spinosyns as bait applications (3–5) as well as cover sprays (1–3) and rarely mass trapping. However, pyrethroid resistance has been developed in populations of *B. oleae* from Greece. Other olive pests that could cause damage of major economic importance from time to time include olive moth *Prays oleae* (Bernard) or the black scale, *Saissetia oleae* (Olivier), which their control is based only on cover sprays with registered insecticides.

Moreover, olive trees are commonly affected by a wide range of diseases that may occasionally cause severe yield losses and tree death in major olive-growing regions in Greece. Among them, fungal diseases such as *Verticillium* wilt, olive leaf spot, *Cercospora* leaf spot, olive anthracnose, olive fruit rot and the bacterial olive knot disease are the most predominant, whereas others like esca or other wood

K. Varikou
Laboratory of Entomology, Institute of Olive Tree, Subtropical Crops and Viticulture, Hellenic Agricultural Organization—DEMETER, Chania, Crete, Greece
e-mail: varikou@elgo.iosv.gr

P. Skouras
Laboratory of Agricultural Entomology and Zoology, Department of Agriculture, University of the Peloponnese, Kalamata, Greece
e-mail: skouras@elgo.iosv.gr

Emmanouil Markakis (✉)
Department of Agriculture, School of Agricultural Sciences, Hellenic Mediterranean University, Stavromenos, Heraklion, Crete, Greece
e-mail: markmanos@hmu.gr

J. Muñoz-Rojas, R. García-Ruiz (eds.), *The Olive Landscapes of the Mediterranean*, Landscape Series 36,
https://doi.org/10.1007/978-3-031-57956-1_23

decay-associated diseases are considered to be emerging diseases for the Greek as well as for the Mediterranean olive industry. In most cases, complete eradication of such diseases is practically unfeasible. However, when olive diseases are early identified and recommended treatments properly applied, they can be controlled efficiently. Herein, we provide a brief description of dominant pests and diseases in terms of their impact, incidence (mainly for pathogens) and distribution across the major olive-producing areas in Greece. Moreover, we present the basic management practices that are commonly suggested and applied by the Greek olive growers.

23.1 Major Pests in Greece and Their Damages

Greece is among the leading olive oil-producing countries, along with Spain and Italy. The recorded olive pests in Greece exceed 70 (Anagnou-Veroniki et al., 2008). The olive fruit fly (OFF) [*Bactrocera* (*Dacus*) *oleae* (Gmelin) (Diptera; Tephritidae)] is the most destructive olive pest with its larvae consuming olive mesocarp and causing both quantitative and qualitative injuries to the produced olive oil. It has been reported that the OFF may infest more than 90% of olive fruit in untreated orchards (Chaniotakis, 1994). Besides OFF, a number of occasional pests damage olive tree production to various extents, depending on weather conditions or other circumstances such as improper plant protection product applications or inadequate management of the beneficial entomofauna of the olive farm (Fig. 23.1). Insect species belonging to the above category include (i) the olive moth [*Prays oleae* (Bernard), Lepidoptera: Yponomeutidae] whose larvae of carpophagous generation feed on the olive mesocarp of mainly small fruit cultivars; (ii) the nymphs of black scale [*Saissetia oleae* (Olivier), Homoptera: Coccidae], olive scale [*Parlatoria oleae* (Colvee), Homoptera: Diaspididae] or oleander scale [*Aspidiotus nerii* Bouche, Homoptera: Diaspididae] stuck on olive plant tissues and ingest its fluid reducing the photosynthesis, causing leaf loss, excreting honey dew at which sooty mould is developed (*S. oleae*); (iii) the females of scolytids insects such as *Hylesinus oleiperda* (Danthoine) and *Phloeotribus scarabaeoides* (Bernard) (Coleoptera: Scolytidae) excavate breeding or feeding tunnels causing shoot weakening, dieback or wind breakage; (iv) adults of *Rhynhites cribripennis* Desbrochers (Coleoptera: Attelabidae) feed on olive fruits, while its larvae complete its development within olive kernel causing fruit drops and significant yield losses (Perdikis et al., 2009); (v) *Closterotomus (=Calocoris) trivialis)* Costa (Hemiptera: Miridae) is a polyphagous species attacking mainly weeds as well as early olive inflorescence under the absence of its main hosts, reducing significant fruit setting and consequently olive oil yield (Perdikis et al., 2009; Varikou & Birouraki, 2014); and finally (vi) females of *Resseliella oleisuga* (Targionni-Torzzetti) (Diptera: Cecidomyiidae) lay their eggs in bark crevices, and the developed larvae cause cortical necrosis and twig drying (Delrio & Lentini, 2017).

Fig. 23.1 Olive pests (adults) and their symptoms

23.2 Main Pests Control Techniques

In Greece current control methods for all pests include mainly 1–2 cover sprays with organophosphorus or pyrethroid insecticides. Unfortunately, chemical sprays are applied not only in cases that pest population or infestation exceeds the economic thresholds but also as a preventative method for pests that can be easily suppressed by beneficial species and other cultural practices.

The protection of the olive fruit production against the olive fruit fly, with the exception *B. oleae*, is funded by the Greek Agriculturist Ministry ('Collective control against the olive fruit fly') since 1937, with an estimated annual cost of approximately 20 million euros. This scheme consists of monitoring the population (traps, olive samplings and agriculturists under contract) and (3–5) bait sprays (addition of a food attractant into the insecticide solution and applied as a spot treatment to the half trees of the orchard) from the ground applications depending on the year, variety and region of the country (Varikou & Birouraki, 2014). However, despite the potential of the method to control the pest, its application is characterized by a series of drawbacks and limitations which have remained practically unchanged for decades, without any adoption of technological solutions for its improvement (Fouskitakis et al., 2017). Moreover, growers apply 1–3 cover sprays (mixing two active ingredients) additionally to (3–5) bait sprays in order to protect their olive production and maximize their financial benefits.

Organophosphates (OPs) were initially used to control OFF (mainly in bait sprays), followed by the more recent introduction of pyrethroids, spinosad and other compounds (Vontas et al., 2001). Vontas et al. (2001) reported up to tenfold of OP resistance in a laboratory-selected strain and a field population. Later, Skouras et al. (2007) found much higher levels of OP resistance in *B. oleae* field populations from Crete and mainland Greece (up to 64-fold compared with the laboratory strain). Pyrethroid resistance was found in *B. oleae*, in alpha-cypermethrin of up to 55-fold in field populations from south Greece and Crete. On the other hand, low levels of spinosad resistance were detected in field populations of *B. oleae* from Greece (Kampouraki et al., 2018).

The presence of toxic residues in olive oil is another important issue that has been reported by several researchers (Amvrazi & Albanis, 2009), as a consequence of the non-sustainable use of these insecticides. The criteria of the International Olive Oil Council (IOOC) concerning olive oil unfortunately do not include chemical residues of the applied insecticides. An extra virgin oil can probably include at least two registered active ingredients according to the Greek local practice. Producing extra virgin oil free from residues is of major importance when the mean olive oil consumption per Greek is approximately estimated at 20 kg (Kiritsakis, 1989).

Current approaches for a more environmentally friendly management of herbivorous insect species according to the sustainable use of insecticides (promoted by the European Commission 2009/128/CE) as well as the progressively loss of key active ingredients (e.g. dimethoate) contribute to find an alternative to chemical methods or to improve the current methods in order to produce olive oil free of residues. This is an urgent demand nowadays, since olive pest outbreaks occur more often now than previously, possibly due to the impact of climate change; an olive fly outbreak was recorded in Peloponnese in autumn 2018 and in 2019 in Crete due to an abnormal increase of the mean daily temperature not more than 2° during the harvest period.

23.3 Major Diseases in Greece and Their Damages

Main pathogens in Greek olive groves include:

1. *Verticillium* wilt (VW) caused by the fungus *Verticillium dahliae* Kleb. is the most important soil-borne disease affecting olive in Greece and worldwide (Jimenez-Diaz et al., 2012). The disease causes severe losses in orchards mainly

in central and northern Greece. Surveys revealed 2–3% average disease incidence and 1% tree mortality in such areas. However, recent records showed up to 50% disease incidence in young orchards planted in fields previously cultivated with cotton (Jimenez-Diaz et al., 2012). VW is rarely observed in southern Greece possibly due to the higher temperature and the extensive use of 'Koroneiki', one of the most resistant olive cultivars to this particular disease. The fungus is capable of surviving in the soil by means of microsclerotia for up to 20 years in an absence of susceptible plant host, whereas no effective chemical control treatment is currently known. Thus, disease management is especially difficult and relies primarily on preventive measures. The exploitation of resistant cultivars (e.g. Kalamon, Koroneiki) and/or rootstocks is considered as the most efficient control measure in practice, whereas the use of pathogen-free propagated material is of vital importance (Antoniou et al., 2008; Jimenez-Diaz et al., 2012; Markakis et al., 2009). Co-cultivation of olive tree with other susceptible crops or planting new orchards in fields with VW history is strongly discouraged. Soil solarization and biocontrol agent implementation are also efficient practices in disease controlling in Greece (Markakis et al., 2016; Tjamos et al., 1991).

2. Olive leaf spot (OLS) or peacock spot caused by the fungus *Spilocaea oleagina* (syn. *Cycloconium oleagineum*) is the most important leaf spot disease in Greece. Infection is favoured by high humidity and low-intermediate temperature conditions (6–12 °C), whereas pathogen's activity is eliminated during warm and dry conditions (Viruega et al., 2013). Therefore, OLS is a major problem in areas with high relative humidity such as in western, central and northern regions of the country. Strong olive tree bloom fails to develop in severely affected trees; thus crop production is substantially reduced. In areas with disease-favourable conditions (e.g. Corfu, western Greece), the disease can lead to complete production loss in the absence of effective control treatments. OLS is controlled by a combination of cultural measures such as proper pruning and balanced nitrogen fertilization along with applications of copper-based and/or organic fungicides (e.g. dodine, pyraclostrobin, trifloxystrobin, kresoxim-methyl and tebuconazole) at the correct time (Trapero & Blanco, 2010).
3. Cercospora leaf spot (CLS) has been recorded since 1973 (Pappas, 1993) and is widely distributed in most of the olive-growing regions of the country (e.g. in Arta, Fthiotida, Evoia, Preveza and Messinia prefectures and in Crete). CLS is caused by the fungus *Pseudocercospora cladosporioides*, affecting olive leaves and fruits. The disease may occasionally cause severe defoliation, leading to reduced vegetative growth and fruit production. Besides, evident disease symptoms include more or less extended lesions on olives fruits, leading in rot and fruit drop. Therefore, CLS affects negatively table olives and olive oil quantity and quality. Climatic conditions across the Mediterranean suggest that fungal infection may occur in autumn, winter and spring and favoured by rainfalls that facilitate conidial spread. Besides copper treatments, no other fungicides are officially available for controlling CLS in Greece. However, treatments with

pyraclostrobin + boscalid is Spain were effective under field conditions, even without the application of supplementary copper (Romero et al., 2020).

4. Olive anthracnose (OA) is the main disease of olive fruit worldwide. Until recently, *Colletotricum gloeosporiodides* was thought to be the only responsible species for OA in Greece. The disease was first observed in Corfu island in 1920, and it is currently considered endemic in the coasts of the Epirus region, in Paxoi and Lesvos islands and in the region of Chalkidiki. In January 2015, severe symptoms of mummified fruits were observed in olive groves in the Aitoloakarnania region (west-central Greece), in the table olive cultivar 'Kalamon'. Continuous surveys in spring and autumn of the same year revealed severe symptoms in inflorescences and fruits in orchards of cv 'Koroneiki' in the island of Zakynthos (West Greece), resembling that of anthracnose. Besides olive fruits and inflorescences, the organs affected also included leaves, stalks, shoots and twigs. Since then, autumn and winter heavy rainfalls resulted in extensive spread of anthracnose in West Greece and Peloponnese in 2016, causing severe losses. Based on morphological and molecular analyses, the causal agent of the epidemic was identified as *C. acutatum* (Iliadi et al., 2018) for the first time in Greece. In 2018 growing season, severe OA epidemics also occurred in Messinia and Ilia prefectures, affecting olive oil quality (increased acidity and peroxide number) and quantity. OA can result in destructive epidemics in favourable conditions (severe rainfalls, intermediate temperature and attacks by *B. oleae*), such as those existing commonly in several regions in Greece, and cause up to 100% yield losses (Cacciola et al., 2012). Chemical control in Greece relies on copper-based and/or other chemical fungicides (e.g. azoxystrobin, difenoconazole, dodine, pyraclostrobin), while correct timing of application (even from the flowering stage) is crucial for successful disease management.
5. Olive fruit rot (OFR) or Dalmatian disease caused by *Botryosphaeria dothidea* is another very common disease of olive in Mediterranean countries, including Greece, and is directly linked with the olive fruit fly *B. oleae* (Latinovic et al., 2013). However, OFR has a comparatively minor role compared to the above-mentioned diseases. OFR affects olive fruits exclusively causing typical spots, rot, shrinkage and mummification on fruits. Wounds made by *B. oleae*'s oviposition sites are readily invaded by the pathogen, promoting disease development. Therefore, disease management primarily relies on the efficient control of the olive fly, while copper-containing fungicides are indicated for pathogen's control.
6. Olive knot (OK) caused by the bacterium *Pseudomonas savastanoi* pv. *savastanoi* is among the most relevant diseases affecting olive cultivation worldwide and has been observed since ancient times. OK is widespread all over Greece with variable disease indices among areas and cultivars (Panagopoulos, 1993). Although olive trees are rarely killed, OK is capable of reducing tree vitality and causing twig and branch dieback. Bacterial cells are spread by wind and rain and may infect any wound, including leaf, blossom and peduncle scars, pruning wounds, frost or hail injuries, etc. Conventionally, OK management in Greece is carried out by applying copper-based formulations. Moreover, removing severely

affected tissues, avoiding harvesting and pruning in moist conditions and pruning tool disinfestation along with the use of pathogen-free propagating material are suggested (Panagopoulos, 1993).

7. Wood decay (WD) or Esca caused primarily by *Fomitiporia mediterranea* is considered to be an emerging disease for olive trees in Greece (Markakis et al., 2017; Paplomatas et al., 2006a, b). In the last decades, a significant outbreak of WD has been observed in olive orchards throughout Greece, and *F. mediterranea* is consistently identified as the causal agent of the disease in routine diagnostic tests (Markakis et al., 2017; Paplomatas et al., 2006a, b). In particular, regions with high disease incidence are Messinia and Arkadia prefectures in Peloponnese (south-central Greece), Evros prefecture in Thrace (northeastern Greece) (Fig. 23.2) and Crete (southern Greece). The increased frequency of WD in olive tree has been attributed to several factors involving pathogen dispersion and adaption from other hosts to olive (Markakis et al., 2017) and alterations in conducting cultivation practices (Paplomatas et al., 2006b). Olive trees affected by *F. mediterranea* exhibit premature aging, tree decline and reduced yields. The pathogen cannot be currently controlled by chemicals. To manage WD, farmers are suggested to remove and destroy the infected wood and carpophores on diseased trees together with an additional segment of visibly healthy tissue proximal to the lesions. Coating of the large pruning wounds with protective paste, labelling of the diseased trees and pruning them separately from the healthy ones in the next cultivation period are additional prevention practices in the field. Moreover, disinfestation of the pruning tools and the use of resistant cultivars (e.g. Kalamon and Koroneiki) are also suggested (Markakis et al., 2019; Paplomatas et al., 2006b).

Fig. 23.2 Wood decay (Esca) symptoms in olive trees in Thrace (northeastern Greece)

23.4 Conclusions

Conventional management of a typical Greek olive farm includes the control of diverse pest/disease complexes that can be endemic and widespread or of occasional and local significance. In the last decades, due to outbreaks of more or lesser-known olive-associated pests and diseases, the non-proper use of the pesticides, the climate change phenomena and the progressively reduction of major active ingredients, there is a global need for effective and contemporary control strategies. In Greece, this is mostly translated to *remote pest and disease monitoring coupled with combined and improved control methods at the single-orchard level and with respect to the agroecosystem.* The proper control method based on the on-time and accurate pest/disease monitoring can also lead to reduced use of chemicals according to the principles of viable and sustainable farming necessitated by an integrated pest management and control framework.

A case scenario for the OFF could be the application of mass trapping during all the year, application of low volume and attractive bait spraying solution and covering the 'susceptible' olive trees with a registered kaolin clay. At the end, copper and copper nanoparticles can play an important role in sustainable agriculture due to their both antimicrobial and pesticidal properties.

References

Amvrazi, E. G., & Albanis, T. A. (2009). Pesticide residue assessment in different types of olive oil and preliminary exposure assessment of Greek consumers to the pesticide residues detected. *Food Chemistry, 113*(1), 253–261.

Anagnou-Veroniki, M., Papaioannou-Souliotis, P., Karanastasi, E., & Giannopolitis, C. N. (2008). New records of plant pests and weeds in Greece, 1990–2007. *Hellenic Plant Protection Journal, 1*(2), 55–78.

Antoniou, P. P., Markakis, E. A., Tjamos, S. E., Paplomatas, E. J., & Tjamos, E. C. (2008). Novel methodologies in selecting and screening olive varieties and rootstocks for resistance to *Verticillium dahliae*. *European Journal of Plant Pathology, 122*, 549–560.

Cacciola, S. O., Faedda, R., Sinatra, F., Agosteo, G. E., Schena, L., Frisullo, S., & Magnano di San Lio, G. (2012). Olive anthracnose. *Journal of Plant Pathology, 94*, 29–44.

Chaniotakis, N. (1994). Protection of olive orchards in organic farming. In: *Proceeding of national conference of organic farming*, Kalamata, Greece.

Delrio, G., & Lentini, A. (2017). Pest management in organic olive. In V. Vacante & S. Kreiter (Eds.), *Handbook of pest management in organic farming* (pp. 218–235). CABI.

Fouskitakis, G. N., Doitsidis, L., Varikou, K. N., Rigakis, I. I., Sarantopoulos, I., Mpirouraki, A., & Maniadaki, K. (2017). Integrated pest control for the olive-fruit fly: remote pest monitoring and optimized bait-sprays. In *HAICTA*, pp. 721–733.

Iliadi, M. K., Tjamos, E. C., Antoniou, P. P., & Tsitsigiannis, D. I. (2018). First report of *Colletotrichum acutatum* causing anthracnose on olives in Greece. *Plant Disease, 102*, 820.

Jimenez-Diaz, R. M., Cirulli, M., Bubici, G., Jimenez-Gasco, M. M., Antoniou, P. P., & Tjamos, E. C. (2012). Verticillium wilt, a major threat to olive production: Current status and future prospects for its management. *Plant Disease, 96*, 304–329.

Kampouraki, A., Stavrakaki, M., Karataraki, A., Katsikogiannis, G., Pitika, E., Varikou, K., Vlachaki, A., Chrysargyris, A., Malandraki, E., Sidiropoulos, N., Paraskevopoulos, A., Gkilpathi, D., Roditakis, E., & Vontas, J. (2018). Recent evolution and operational impact of insecticide resistance in olive fruit fly *Bactrocera oleae* populations from Greece. *Journal of Pest Science, 91*, 1429–1439.

Kiritsakis, A. (1989). Tocopherols. In: *The Olive Oil* (pp. 61–64). Agricultural Cooperative Editions.

Latinovic, J., Mazzaglia, A., Latinovic, N., Ivanovic, M., & Gleason, M. L. (2013). Resistance of olive cultivars to *Botryosphaeria dothidea*, causal agent of olive fruit rot in Montenegro. *Crop Protection, 48*, 35–40.

Markakis, E. A., Tjamos, S. E., Antoniou, P. P., Paplomatas, E. J., & Tjamos, E. C. (2009). Symptom development, pathogen isolation and real time QPCR quantification as important factors for evaluating resistance of olive cultivars to *Verticillium* pathotypes. *European Journal of Plant Pathology, 124*, 603–611.

Markakis, E. A., Tjamos, S. E., Antoniou, P. P., Paplomatas, E. J., & Tjamos, E. C. (2016). Biological control of Verticillium wilt of olive by *Paenibacillus alvei*, strain K165. *BioControl, 61*, 293–303.

Markakis, E. A., Kavroulakis, N., Ntougias, S., Koubouris, G. C., Sergentani, C. K., & Ligoxigakis, E. K. (2017). Characterization of fungi associated with wood decay of tree species and grapevine in Greece. *Plant Disease, 101*, 1929–1940.

Markakis, E. A., Ligoxigakis, E. K., Roussos, P. A., Sergentani, C. K., Kavroulakis, N., Roditakis, E. N., & Koubouris, G. C. (2019). Differential susceptibility responses of Greek olive cultivars to *Fomitiporia mediterranea*. *European Journal of Plant Pathology, 153*, 1055–1066.

Panagopoulos, C. G. (1993). Olive knot disease in Greece. *EPPO Bulletin, 23*, 417–422.

Paplomatas, E. J., Elena, K., Tsopelas, P., Tzima, A., Paraskevopoulos, A., & Papanikolaou, A. (2006a). Infection of olive trees with the fungus *Fomitiporia punctata* (*Phellinus punctatus*). *Phytopathologia Mediterranea, 45*, 69.

Paplomatas, E. J., Paraskevopoulos, A., Tsopelas, P., Elena, K., & Malandraki, I. (2006b). Study of the epidemiology of esca disease on olive. *Phytopathologia Mediterranea, 45*, 166–167.

Pappas, A. C. (1993). *Mycocentrospora cladosporioides* on olive in Greece. *EPPO Bulletin, 23*, 405–409.

Perdikis, D., Garantonakis, N., Kitsis, P., Giatropoulos, A., Paraskevopoulos, A., Cassis, G., & Panagakis, S. (2009). Studies on the damage potential of *Closterotomus triviallis* and *Aphanosoma italicum* on olive fruit setting. *Bulletin of Insectology, 62*, 215–219.

Romero, J., Ávila, A., Agustí-Brisach, C., Roca, L. F., & Trapero, A. (2020). Evaluation of fungicides and management strategies against Cercospora leaf spot of olive caused by *Pseudocercospora cladosporioides*. *Agronomy, 10*, 271. https://doi.org/10.3390/agronomy10020271

Skouras, P. J., Margaritopoulos, J. T., Seraphides, N. A., Ioannides, I. M., Kakani, E. G., Mathiopoulos, K. D., & Tsitsipis, J. A. (2007). Organophosphate resistance in olive fruit fly *Bactrocera oleae* populations in Greece and Cyprus. *Pest Management Science, 63*, 42–48.

Tjamos, E. C., Biris, D. A., & Paplomatas, E. J. (1991). Recovery of olive trees from Verticillium wilt after individual application of soil solarization in established olive orchards. *Plant Disease, 75*, 557–562.

Trapero, A., & Blanco, M. A. (2010). Diseases. In D. Barranco, R. Fernández-Escobar, & R. Rallo (Eds.), *Olive growing* (pp. 521–578). Junta de Andalucía/Mundi-Prensa/RIRDC/AOA.

Varikou, K., & Birouraki, A. (2014). Life history of *Closterotomus (Calocoris) trivialis* (Costa) (Heteroptera: Miridae) in olive and citrus orchards in Crete. *Crop Protection, 59*, 14–21.

Viruega, J. R., Moral, J., Roca, L. F., Navarro, N., & Trapero, A. (2013). *Spilocaea oleagina* in olive groves of southern Spain: Survival, inoculum production, and dispersal. *Plant Disease, 97*, 1549–1556.

Vontas, J. G., Cosmidis, N., Loukas, M., Tsakas, S., Hejazi, M. J., Ayoutanti, A., & Hemingway, J. (2001). Altered acetylcholinesterase confers organophosphate resistance in the olive fruit fly *Bactrocera oleae*. *Pesticide Biochemistry and Physiology, 71*, 124–132.

Chapter 24
Challenges for Sustainable Water Resource Management in Greek Olive Groves: Irrigation and Salinity

Nektarios N. Kourgialas

Abstract Irrigated agriculture is the world's largest water user. It has been estimated that in order to satisfy food demand during the next 30 years, world food production must increase by 70%. Considering that most of the agriculture in the world is rainfed, an important strategy to achieve the above objective is the transition from rainfed to irrigated agriculture. Concerning water supply in Greece, the volume of available natural water resource is about 14,500 hm^3, and water demand per year is estimated at 8243 hm^3, of which 83% is used for irrigation. About 45% of the total cultivated area of Greece is irrigated, and groundwater is the main source of its supply. In Greece, sprinkler and drip irrigation are the main irrigation systems used for olive cultivation, while about 30% of the total olive cultivated areas are irrigated. One of the most common water management problems encountered in coastal olive cultivation areas of Greece is saline irrigation caused mainly by intensive agriculture activities (over-pumping of aquifers). In this chapter, the current status and future priorities regarding the effective irrigation for olive cultivation were identified. Also, quantities of irrigation needed for olive groves at regional units of Greece are suggested.

Keywords Water supply in Greece · Effective irrigation · Saline water · Precision agriculture

N. N. Kourgialas (✉)
Hellenic Agricultural Organization (ELGO Dimitra), Institute of Olive Tree, Subtropical Crops and Viticulture (IOSV), Water Resources-Irrigation & Env. Geoinformatics Laboratory, Chania, Greece
e-mail: kourgialas@elgo.gr

J. Muñoz-Rojas, R. García-Ruiz (eds.), *The Olive Landscapes of the Mediterranean*, Landscape Series 36,
https://doi.org/10.1007/978-3-031-57956-1_24

24.1 Effective Irrigation for Olive Trees: The Case of Greece

Irrigated agriculture is the world's largest water consumer, while at the same time water resources are under increasing pressure from rapidly growing demands and climate change. Considering that most of the agriculture in the world is rainfed, an important strategy to increase food production is the transition from rainfed to irrigated agriculture. In the Mediterranean basin, Greece is considered one of the most important olive-producing countries (Puppo et al., 2019). Specifically, olive (*Olea europaea* L.) is the main crop in Greece covering more than 75% of the total cultivation area, while there has been a marked increase in olive cultivation areas over dry land in recent years. About 45% of the total cultivated area of Greece is irrigated, and groundwater is the main source of water supply (Kourgialas et al., 2019). The volume of available water resources in Greece is about 14,500 hm^3, and the water demand per year is estimated at 8243 hm^3, of which 83% is used for crop irrigation. In Greece, sprinkler and drip irrigation are the main irrigation systems used for olive cultivation, while about 30% of the total olive cultivated areas are irrigated (Eurostat, 2018).

Olives are grown in many parts of southern, central and northern Greece to produce olive oil and table olives (Fig. 24.1). In the world ranking, Greece ranks third in olive oil production, while in table olives production it ranks sixth. The significant spread of olive cultivation in Greece is due, among other factors, to the characteristic that the olive is a drought-resistant tree and its cultivation is appropriate for the climatic and soil conditions of Greece, while it can be productive even when not irrigated. This does not mean that it is not affected by the lack of water, which has adverse effects on flowering, fruiting, fruit development and the level of production in Greece; olive trees are cultivated in both lowland and semi-mountainous areas. In these zones, rainfall occurs mainly in autumn and winter, resulting in a lack of water in the spring-summer period where high temperatures can lead to a significant reduction in productivity (less than 1–2 kg of oil per tree).

Proper olive irrigation has a positive effect on flowering, improving fruit set and fruit development while helping to reduce alternate bearing. Olive trees are particularly sensitive to lack of water during the flowering stage (April–May), and therefore watering during that period should begin early so that trees do not start flowering which leads to a significant yield reduction. The critical stages of the olive tree on an annual basis in terms of its water needs are (a) just before and until flowering (March–May) (during this period, irrigation is necessary to have better flower growth, better fruit set and shoot growth); (b) from fruit set to pit hardening (May–June), a period of intense fruit growth; and (c) from oiling to fruit harvesting (August–September or until the first rains of autumn) in order to increase the oil content and have a greatest final growth of the fruit flesh (cell expansion). Irrigation is necessary for both table olives and olive oil varieties. However, irrigation is more necessary for the table olive, because the aim is to achieve high-quality large fruits. Table olives should be irrigated especially during the third stage (cell expansion) of fruit growth to increase their size (Kourgialas, 2017).

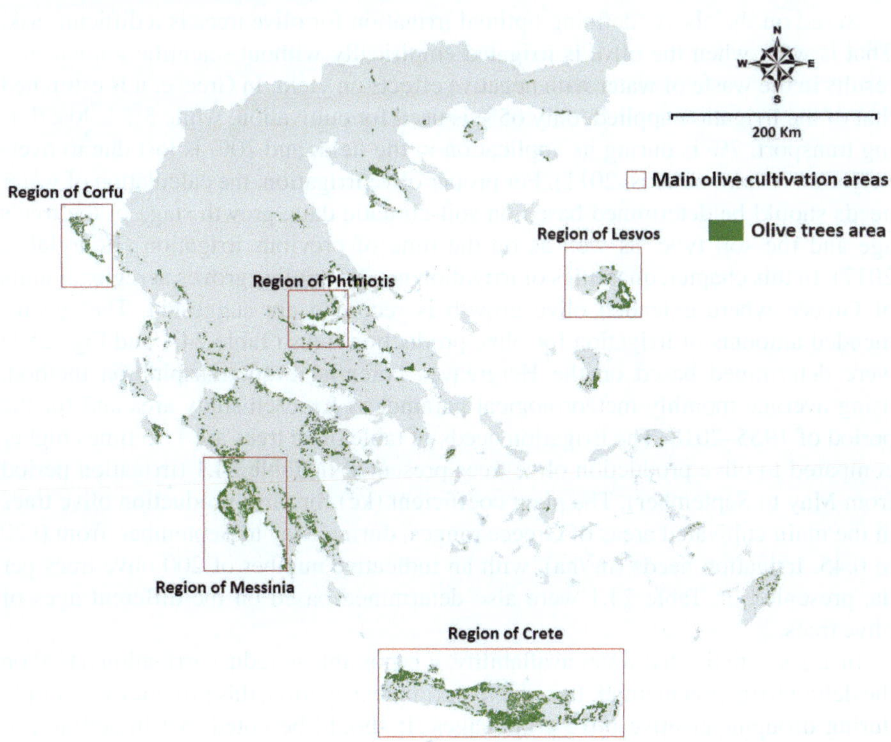

Fig. 24.1 Olive's spatial distribution and the main olive cultivation areas in Greece

Excessive irrigation, on the other hand, can cause significant problems such as weed growth and disease. The olive tree is sensitive to excessive moisture and does not behave well in very moist soils. These soils do not allow proper ventilation, and so the roots suffocate, and the tree can die. Trees grown in moist-saturated soils are more susceptible to changing weather conditions and soil diseases, such as *Phytophthora* and *Verticillium* wilt. Also, when watering is excessive, blastomania is observed (excessive growth of shoots and minimal fruiting), while the quality of the olive oil is reduced. In spring (eye differentiation, flowering/vegetation) excessive irrigation can lead to the leaching of nitrogen and other nutrients, which are necessary for good fruit set. In summer as well as in fall, excessive irrigation can delay the maturation of the oils, lead to softening of the table olives, and increase the damage from *Bactrocera (Dacus) oleae* (olive fruit fly). On the other hand, a late irrigation can make the new vegetation more sensitive to winter frosts. Regarding the olive oil quality characteristics, the more water applied, the more likely it is that the oil will have lower polyphenol content and stability. In fact, the water content of the fruit can influence the amount of polyphenols remaining in the oil after processing. Moderate irrigation trees produce oils that are fruitier, with a balanced ratio of bitterness and pungency (Panagou, 2015; Kourgialas, 2017).

Based on the above, defining optimal irrigation for olive trees is a difficult task. That is why, when the olive is irrigated empirically without scientific guidance, it results in the waste of water with negative effects on yield. In Greece, it is estimated that of the irrigation applied, only 65% is used for cultivation, while 8% is lost during transport, 7% is during its application in the field, and 20% is lost due to over-irrigation (Chartzoulakis, 2011). For proper olive irrigation, the calculation of water needs should be determined based on soil-climatic data, growth stage of the tree's age and the soil type, as well as on the time of previous irrigation (Kourgialas, 2017). In this chapter, quantities of irrigation needs for olive groves at regional units of Greece where extended olive growth is recorded are suggested. The recommended amounts of irrigation for olive production trees (Table 24.1 and Fig. 24.1) were determined based on the Hargreaves potential evapotranspiration method, using average monthly meteorological parameters for each study area and for the period of 1955–2018. The irrigation needs of table olive trees are 1.75 times higher compared to olive production olive trees presented in Table 24.1 (irrigation period from May to September). The plant coefficient (kc) for olive production olive trees in the main cultivated areas of Greece ranges, during May to September, from 0.20 to 0.45. Irrigation needs (m^3/ha), with an indicative number of 200 olive trees per ha, presented on Table 24.1 were also determined based on the different ages of olive trees.

In areas with limited water availability, it is possible to reduce irrigation based on the deficit irrigation method. In order to optimize irrigation, this provides water only during drought-sensitive olive crop stages. It should be noted that in addition to proper irrigation of olive trees, special attention should be paid to proper cultivation practices, which play a key role (a) in increasing and/or maintaining water storage in the soil and (b) in reducing evapotranspiration losses, especially during summer in areas with water shortage (Kourgialas et al., 2019).

Table 24.1 Suggested mean May–September irrigation needs for olive groves of different plantation ages in the main cultivated areas of Greece

Main regional units of olive cultivation in Greece	Irrigation needs (m^3/ha) for mature trees >10 years old	Irrigation needs (m^3/ha) for trees 4–10 years old	Irrigation needs (m^3/ha) for trees <4 years old
Crete	2090	1460	840
Lesvos	2180	1520	870
Messinia	2020	1410	810
Phthiotis	2060	1440	820
Corfu	1090	760	430

24.2 Olive Trees Saline Irrigation Management: Current Status and Future Priorities for Greece

Due to the scarcity of fresh water, one of the most common problems encountered in coastal agricultural areas is seawater intrusion caused mainly by intensive agriculture activities (over-pumping of aquifers). Also, increased demand for olive oil production and the fact that olive is relatively tolerant to many environmental stress-causing factors have led to the increased irrigation in arid regions where the cultivation was previously rainfed. In addition, rainfed olive trees, especially in arid islands, must now be irrigated to survive to the adverse effects of climate change. In these areas, it is a common practice to utilize saline groundwater to irrigate olive orchards. In these dry areas, the water for irrigation can also be obtained through the recycling of wastewater, characterized by relatively high concentrations of dissolved salts. These approaches exacerbate the problem of low yield in olive cultivation, due to the fact that salts decrease the osmotic potential of the soil water, making water and nutrient uptake by crop roots more difficult (Minhas et al., 2020).

Regarding the tolerance to saline water of the most common olive varieties in Greece, it is important to point out that 'Koroneiki' (small size olive cultivar) is the most common variety for olive production. It is resistant to water stress and satisfyingly tolerant to water salinity (irrigation water with ECw equal to 6.5 dS/m could lead to a more than 25% olive yield reduction, compared to freshwater application). The 'Megareitiki', a medium size olive cultivar, can reduce its yield at levels more than 25% when irrigated water has ECw equal to 5.5 dS/m, compared to freshwater irrigation. On the other hand, large-sized fruit olive cultivars such as 'Chondrolia Chalkidikis' are much more sensitive to water salinity; they can lose their yield in a degree higher than 25% when irrigated water has values of ECw equal to 3.5 dS/m, compared to freshwater irrigation (Kourgialas et al., 2017; Arambatzis et al., 2018).

Based on the current irrigation practices as well as international literature, the determination of the precise volume of saline water to irrigate olive trees remains questionable (Ben-Gal et al., 2017). Specifically, salinity is not a common factor in the calculation of the actual crop evapotranspiration, and this omission is likely to lead to overestimation of the actual crop evapotranspiration (applied irrigation dose) (Minhas et al., 2020; Ben-Gal et al., 2008). On the contrary, strategies for handling salinity usually aim at preventing the build-up of salts in the root zone to levels that limit the root water uptake, minimizing the damaging effects of salinity on crop transpiration and soil evaporation for an optimal crop growth. Also, leaching requirements for maximizing yield, dependent from the climate and soil type, are largely determined by the salinity of the irrigation water (Dudley et al., 2008). Irrigation under saline conditions must aim at meeting both actual crop evapotranspiration needs and leaching requirements, in order to maintain a favourable salt balance in the root zone.

The solutions of water shortage and salinity intrusion problems can be in the use of treated water and/or the improvement of water efficiency (maintenance of drip irrigation systems and using effective filters or the means to control emitter clogging under saline water drip irrigation such as adding acids or applying magnetic fields to irrigation water). Generally, the reuse of treated water for irrigation should be analysed based on the applicable legislation. In Greece, reusing treated water is not broadly accepted due to legislation gaps. Specifically, in Greece, the applicability of reusing wastewater in agriculture is limited by the obstacles rising from the strict legislation in Greece and the EU, as well as the low public acceptance due to financial, environmental and public health issues. To overcome these difficulties, the EU proposed guidelines for wastewater reuse for crop irrigation, defining minimum monitoring needs and requirements. Especially for olive cultivation in Greece, some extra actions by government agencies to overcome the reluctance of farmers to apply treated water for irrigation could be raising awareness about the benefits for farmers and local authorities from the applicability of this practice from an environmental and circular economy point of view (Kourgialas, 2021). On the contrary, other alternatives for saving water have been studied, ensuring olive yield at the same time, and are of high priority, namely, the development of useful tools (e.g. mathematical models that incorporate both fresh and saline water, under different agronomic practices) and the implementation of proper agricultural practices. Recent progress in the field of wireless sensor technology has led to the adoption of precision agriculture which permits agronomists to design irrigation plans, both with freshwater and saline irrigation. New technological approaches in precision agriculture for olive orchards include monitoring soil water content in the root zone automatically using digital tensiometers and determining crop water requirements in online settings. Also, multiple sensors for measuring soil moisture and salinity are recommended for accurate monitoring of the whole soil profile.

Considering that the main olive cultivated areas of Greece are climate change 'hotspots', farmers will have to adapt to the changing climate, which indicates changes in management practices. It is of the utmost significance that policymakers in Greece prioritize adaptation measures which are in line with the increase of water use efficiency in olive cultivation. So, based on the above-mentioned techniques, it will be possible to design more effective irrigation schemes, optimizing the water application at the 'right' time and in the 'right' amount and rate, to avoid excess run-off, water logging or deep percolation. In fact, it is important not to forget that agricultural research focuses on the optimal olive yield and is of high importance for the Greek economy.

References

Arambatzis, G., Panagopoulos, A., Pisinaras, B., Xatzigiannakis, E. (2018). *Agricultural water use—Climate change*. (Book in Greek). Hellenic Agricultural Organization. ISBN: 978-618-83573-0-3

Ben-Gal, A., Ityel, E., Dudley, L., Cohen, S., Yermiyahu, U., Presnov, E., Zigmond, L., & Shani, U. (2008). Effect of irrigation water salinity on transpiration and on leaching requirements: A case study for bell peppers. *Agricultural Water Management*. https://doi.org/10.1016/j.agwat.2007.12.008

Ben-Gal, A., Beiersdorf, I., Yermiyahu, U., Soda, N., Presnov, E., Zipori, I., Ramirez Crisostomo, R., & Dag, A. (2017). Response of young bearing olive trees to irrigation-induced salinity. *Irrigation Science*. https://doi.org/10.1007/s00271-016-0525-5

Chartzoulakis, K. S. (2011). The use of saline water for irrigation of olives: Effects on growth, physiology, yield and oil quality. *Acta Horticulturae*. https://doi.org/10.17660/ActaHortic.2011.888.10

Dudley, L. M., Ben-Gal, A., & Shani, U. (2008). Influence of plant, soil, and water on the leaching fraction. *Vadose Zone Journal*. https://doi.org/10.2136/vzj2007.0103

Eurostat. (2018). *Statistics explained*. http://ec.europa.eu/eurostat/statisticsexplained/. 14 Sept 2018.

Kourgialas, N. (2017). Effective olive irrigation in various areas of the Greece territory. *Agriculture-Livestock, 29*(7), 31–34. (in Greek).

Kourgialas, N. N. (2021). A critical review of water resources in Greece: The key role of agricultural adaptation to climate-water effects. *Science of the Total Environment*. https://doi.org/10.1016/j.scitotenv.2021.145857

Kourgialas, N. N., Doupis, G., Psarras, G., Sergentani, C., Digalaki, N., & Koubouris, G. C. (2017). Soil management and compost effects on salinity and seasonal water storage in a Mediterranean drought-affected olive tree area. *Desalination and Water Treatment*. https://doi.org/10.5004/dwt.2017.21548

Kourgialas, N., Koubouris, G., & Dokou, Z. (2019). Optimal irrigation planning for addressing current or future water scarcity in Mediterranean tree crops. *Science of the Total Environment*. https://doi.org/10.1016/j.scitotenv.2018.11.118

Minhas, P. S., Ramos, T. B., Ben-Gal, A., & Pereira, L. S. (2020). Coping with salinity in irrigated agriculture: Crop evapotranspiration and water management issues. *Agricultural Water Management*. https://doi.org/10.1016/j.agwat.2019.105832

Panagou, E. (2015). *Sectoral study—Olive oil & table olives. Final Report* (in Greek).

Puppo, L., Garcíac, C., Bautistad, E., Hunsakerd, D. J., Berettae, A., & Girona, J. (2019). Seasonal basal crop coefficient pattern of young non-bearing olive trees grown in drainage lysimeters in a temperate sub-humid climate. *Agricultural Water Management*. https://doi.org/10.1016/j.agwat.2019.105732

Chapter 25
Management Practices Boosting Ecosystem and Landscape Functions and Services in Greek Olive Groves

Georgios Koubouris and Nektarios N. Kourgialas

Abstract Agriculture has been shaping the landscape of Greece for thousands of years. Following a dynamic change in the twentieth century by expansion of olive groves, landscape seems relatively stable during the past two decades. During the recent decade, minimum tillage practice has gained wide adoption by olive farmers all over Greece in the framework of modern farming aiming at sustainably producing food. Soil mulching on the olive grove ground, combining spontaneous or to a lesser extent cover crops residues and trimmed tree pruning residue as well as composted olive mill pomace and manure, is a practice which is currently booming with positive effects on plant, bird, insect and soil microbe diversity. In most mountainous areas, a mixed economic model consisting of olive growing, sheep farming and other minor activities is common. This socio-economic model plays a fundamental role in the resilience of these populations against economic crisis and climatic extremes and should be safeguarded by long-term and continuous policy planning and implementation. Agricultural practices when planned as a holistic olive grove management system, taking into account all the elements of the agroecosystem and society, are easier to apply, produce multiplicative effects and are more sustainable in the long term.

Keywords Good agricultural practices · Carbon sequestration · Greenhouse gases emissions · Circular economy

G. Koubouris (✉) · N. N. Kourgialas
ELGO Dimitra—Institute of Olive Tree, Subtropical Crops and Viticulture, Chania-Crete, Greece
e-mail: koubouris@elgo.gr; kourgialas@elgo.gr

J. Muñoz-Rojas, R. García-Ruiz (eds.), *The Olive Landscapes of the Mediterranean*, Landscape Series 36,
https://doi.org/10.1007/978-3-031-57956-1_25

25.1 Ecosystem Services

Ecosystem services (ES) are defined as "the benefits that humans derive from nature" (Millennium Ecosystem Assessment, 2005). Ecosystem services are divided in provisioning services (products obtained from ecosystems), regulating services (benefits obtained from regulation of ecosystem processes), cultural services (non-material benefits obtained from ecosystems) and supporting services (services necessary for the production of all other ecosystem services) (Millennium Ecosystem Assessment, 2005). In order to integrate ES in designing agri-environmental policy, both the demand for and supply of ES from agricultural environments should be determined (Tienhaara et al., 2020). In Greece, a set of 40 mapping and assessment of ecosystems and their services (MAES) indicators was drafted, distributed in six general indicator groups, i.e. biodiversity, environmental quality, food, material and energy, forestry, recreation and water resources in the frame of the LIFE-IP 4 NATURA project (Kokkoris et al., 2020).

25.2 Soil and Water Functions

Agriculture has been shaping the landscape of Greece for thousands of years. There are historical reports about soil ploughing by animals and irrigation through river channels in ancient Greek history and mythology. Landscape changes sometimes are smooth and harmonic while in certain instances cause environmental degradation. Greece is prone to desertification, and it is often experiencing intense droughts (Tsesmelis et al., 2019). In addition, soil erosion is the main threat in agricultural Mediterranean areas (Barão et al., 2019). Agricultural modernization during the previous 50 years did not uniformly and necessarily lead to land degradation (Detsis et al., 2017). Similarly, land abandonment due to low profitability not always has positive environmental impacts and may lead to deteriorating or improving conditions of plant cover depending on the soil and climate characteristics of the area (Kosmas et al., 2000).

During the first period of the recent green revolution (1970–1985), the traditional practices did not deteriorate soil resources (Kosmas et al., 2016). Land management is considered positive also during the period 1985–2000 for most of the country, while during 2000–2015 the mechanization and the intensification of agriculture impose even higher pressure to soil resources (Karamesouti et al., 2018). During the intensification of agriculture in Greece, soil tillage was the main driver of land degradation. Olive farmers repeatedly tilled the soil during the year to minimize water and nutrient losses by weeds and to achieve the view of a "tidy" olive grove.

Only during the recent decade, minimum tillage practice has gained wide adoption by olive farmers all over Greece in the framework of modern farming aiming at sustainably producing food (Michalopoulos et al., 2020; Arampatzis et al., 2018; Kavvadias et al., 2018). The use of soil rotators has been reduced, and weed control is mainly achieved by non-soil-disturbing methods such as tractor-mounted flail lawn mower or spraying herbicides (Kavvadias & Koubouris, 2019). This is especially determinant for soil conservation, taking into account that Greece consists mainly of sloping land. The ground of the olive grove remains covered by weeds during the winter, which is the rainy period in Greece, and trees are dormant (Fig. 25.1a). Therefore, no competition between weeds and olive trees for water and nutrients occurs, and soil is protected from soil erosion and run-off. At the end of winter, weeds are either mowed or sprayed by herbicides, and their growth is arrested throughout the dry period of late spring to early autumn. An additional benefit derived from this practice is the soil organic matter build-up. Soil organic matter is linked to many provisioning (e.g. yield, biomass production), regulating (e.g. preventing soil erosion, water regeneration, storage of air carbon dioxide) and supporting (e.g. plant nutrients, water) ecosystem services (Montanaro et al., 2017).

Soil mulching by either weed or cover crops established in a lesser extent is supported by spreading on the olive grove ground the trimmed pruning residue as well as composted olive mill by-products and manure (Fig. 25.1b). This dense soil cover by organic materials provides an efficient protection of soil from erosion and increased water storage through slowing water movement on soil surface, increasing water infiltration in deeper soil layers and decreasing evaporation from the soil surface. Some off-site ecosystem services provided are mitigation of flood damage, of sedimentation of water bodies and of saltwater intrusion into groundwater (Montanaro et al., 2017). Last but not least, these good agricultural management practices contribute to landscape conservation, which is also a valuable ecosystem service.

25.3 Management in Greece Which Boosts Biodiversity

In a study (Bechara et al., 2018) in Crete in which olive groves were cultivated with leguminous cover crops, composted olive mill pomace, trimmed pruning residue and minimum tillage, there were significant positive effects on many biological indicators of soil quality, such as mycorrhizal colonization of spontaneous weed flora, actinobacteria, *Azotobacter* and proteolytic bacteria, indicating clearly the favourable effects of these management practices on soil biology and ecosystem complexity. Plant cover composition in olive grove soil also affects biodiversity in olive groves (Fig. 25.1c). Indeed, patches with sown plant mixtures attracted higher numbers of pollinating Hymenoptera as well as parasitoid insects against olive pests, compared to native vegetation, in a typical dry Mediterranean area in Southern Greece (Karamaouna et al., 2019). For terrestrial isopods, an important component of biodiversity, species richness increased with increasing air humidity, woody plant species richness and soil organic matter in organic olive groves, while in conventional olive groves, it decreased possibly due to herbicide and industrial fertilizer application (Solomou et al., 2019). Organic olive groves in Central Greece had

Fig. 25.1 Elements that provide ecosystem services in olive groves of Greece. (**a**) Soil protection from erosion and rainfall water storage are achieved by natural vegetation cover and stone terraces, (**b**) carbon is stored by recycling pruning residue and olive mill by-products, (**c**) biodiversity is protected by maintaining uncultivated margins, minimizing weed control during wintertime and establishing cover crops, (**d**) landscape complexity by mixed crop systems and supplemental activities such as animal husbandry and beekeeping provide a multitude of ecosystem services, (**e**) agritourism bridges two previously competitive economic sectors, (**f**) last but not least olive oil and table olives of excellent quality are produced

significantly greater values of plant cover, species richness and Shannon and Simpson indexes than the conventional ones (Solomou et al., 2013).

Earthworm biomass and density were significantly higher in organic olive groves than in conventional olive groves in Central Greece (Solomou et al., 2012). Also, biomass of earthworm was also higher by 20–42% in Italian fields with cover crops in comparison to frequently tilled soil in vineyards and orchards (Paoletti et al., 1998). Soil spreading the liquid waste from three-phase olive mills, under the conditions suggested by the Greek legislation, did not harm earthworms (Chalkia et al., 2020). A study on bird diversity monitoring showed significant differences in the Acoustic Complexity Index (ACI) and Bioacoustic Index (BIO) between organic and conventional management systems in olive groves in Southern Peloponnese, Greece (Myers et al., 2019). This result is possibly related to the use of synthetic fertilizers and pesticides as well as with management of ground vegetation.

25.4 Economic and Social Aspects

The accelerated rates of land degradation during the past century are not only a serious threat for soil resources but also for other environmental and socio-economic structures (Nkonya et al., 2015; Requier-Desjardins et al., 2011). The Mediterranean islands, in particular, are strongly influenced by tourism development, which has increased the demand for land to develop accommodation and entertainment facilities (Vogiatzakis et al., 2008).

Mediterranean landscapes are considered as complex and adaptive systems that provide multiple ecosystem services (Crews et al., 2016). In this context, socio-economic factors are important drivers of landscape change and are essential for expansion and renaissance to occur (Wolpert et al., 2020). Following a dynamic change in the twentieth century by expansion of olive groves, landscape in the olive-growing areas of Greece seems relatively stable during the past two decades with few new plantations and some cases of abandonment and land use change towards urban and tourism facilities (Wolpert et al., 2020).

A recent study in the Ionian Islands, representing satisfactorily the overall situation in Greece, identified that the dominant landscape elements were natural forests and olive orchards, leading large areas to be dominated by excess ecosystem services supply or by similar amounts of both supply and demand (Lorilla et al., 2019). High values of recreation supply were found in mountainous, highly vegetated and, often, remote areas (Lorilla et al., 2019). In most of the mountainous olive-growing areas, a mixed economic model consisting of olive growing, sheep farming, bee-keeping and other minor activities such as fruit tree and herb growing is common (Fig. 25.1d).

This socio-economic model plays a fundamental role in the resilience of these populations against economic crisis and climatic extremes and should be safeguarded by long-term and continuous policy planning and implementation. Otherwise, this fragile socio-ecological model may irreversibly degrade by abandonment and migration of citizens to urban areas with dramatic consequences. Noteworthy, olive growing is carried out by almost 500,000 holdings in Greece, and all these people benefit from either main or supplementary income that is valuable for their well-being.

25.5 Holistic Olive Grove Management

Ecosystem services provided by the olive agroecosystem are multiple and interconnected. For example, recycling olive mill by-products as soil amendment for the olive groves not only contributes to carbon storage, plant nutrition, protection from erosion and biodiversity enhancement but also solves the problem of waste management and prevents pollution of water bodies by accidental leakages of storage tanks and unpleasant waste odours that cause an important annoyance to nearby tourism facilities, not needing to mention the basic ecosystem services provided by agroecosystems such as producing oxygen and storing and cleaning the water in the soil.

Approximately 39% of the terrestrial area in Greece is covered by agricultural areas (Kokkoris et al., 2020) and directly linked to provisional services. In addition, in many areas of Greece where olive growing and tourism are the main economic sectors, remarkable progress has been observed to smoothly combine them by providing olive products for the hotels and organizing agritourism activities (Fig. 25.1e). Last but not least, Greece is among the top producers in the world for olive oil and table olives of excellent quality (Fig. 25.1f), and a significant part of house heating needs in rural areas is covered by olive tree pruning firewood.

Agricultural management practices are easier to apply when planned as a holistic olive grove management system, taking into account all the elements of the agroecosystem as well as the surrounding landscape and society. By doing so, it is produced a multiplicative effect boosting the sustainability in the long term. The majority of natural ecosystems in Greece are at above-adequate condition (Kokkoris et al., 2018); however, specific data about olive groves are limited and should be collected in future studies.

References

Arampatzis, G., Hatzigiannakis, E., Pisinaras, V., et al. (2018). Soil water content and olive tree yield responses to soil management, irrigation, and precipitation in a hilly Mediterranean area. *Journal of Water and Climate Change, 9*(4). https://doi.org/10.2166/wcc.2018.224

Barão, L., Alaoui, A., Ferreira, C., et al. (2019). Assessment of promising agricultural management practices. *Science of the Total Environment, 649*, 610–619.

Bechara, E., Papafilippaki, A., Doupis, G., et al. (2018). Nutrient dynamics, soil properties and microbiological aspects in an irrigated olive orchard managed with five different management systems involving soil tillage, cover crops and compost. *Journal of Water and Climate Change, 9*(4). https://doi.org/10.2166/wcc.2018.082

Chalkia, C., Vavoulidou, E., Koubouris, G., et al. (2020). Spreading raw olive mill wastewater is compatible with the growth and the beneficial functions of the earthworm *Octodrilus complanatus*. *Applied Soil Ecology, 153*, art. no. 103625.

Crews, T. E., Blesh, J., Culman, S. W., et al. (2016). Going where no grains have gone before: From early to mid-succession. *Agriculture, Ecosystems and Environment, 223*, 223–238.

Detsis, V., Briassoulis, H., & Kosmas, C. (2017). The socio-ecological dynamics of human responses in a land degradation-affected region: The Messara Valley (Crete, Greece). *Land, 6*(3), art. no. 45.

Karamaouna, F., Kati, V., Volakakis, N., et al. (2019). Ground cover management with mixtures of flowering plants to enhance insect pollinators and natural enemies of pests in olive groves. *Agriculture, Ecosystems and Environment, 274*, 76–89.

Karamesouti, M., Panagos, P., & Kosmas, C. (2018). Model-based spatio-temporal analysis of land desertification risk in Greece. *Catena, 167*, 266–275.

Kavvadias, V., & Koubouris, G. (2019). Sustainable soil management practices in olive groves. In *Soil fertility management for sustainable development* (pp. 167–188). Springer.

Kavvadias, V., Papadopoulou, M., Vavoulidou, E., et al. (2018). Effect of addition of organic materials and irrigation practices on soil quality in olive groves. *Journal of Water and Climate Change, 9*(4). https://doi.org/10.2166/wcc.2018.090

Kokkoris, I. P., Dimopoulos, P., Xystrakis, F., et al. (2018). National scale ecosystem condition assessment with emphasis on forest types in Greece. *One Ecosystem, 3*, art. no. e25434.

Kokkoris, I. P., Mallinis, G., Bekri, E. S., et al. (2020). National set of MAES indicators in Greece: Ecosystem services and management implications. *Forests, 11*, 595.

Kosmas, C., Gerontidis, S., & Marathianou, M. (2000). The effect of land use change on soils and vegetation over various lithological formations on Lesvos (Greece). *Catena, 40*(1), 51–68.

Kosmas, C., Karamesouti, M., Kounalaki, K., et al. (2016). Land degradation and long-term changes in agro-pastoral systems: An empirical analysis of ecological resilience in Asteroussia – Crete (Greece). *Catena, 147*, 196–204.

Lorilla, R. S., Kalogirou, S., Poirazidis, K., et al. (2019). Identifying spatial mismatches between the supply and demand of ecosystem services to achieve a sustainable management regime in the Ionian Islands (Western Greece). *Land Use Policy, 88*, art. no. 104171.

Michalopoulos, G., Kasapi, K. A., Koubouris, G., et al. (2020). Adaptation of Mediterranean olive groves to climate change through sustainable cultivation practices. *Climate, 8*(4), art. no. 54.

Millennium Ecosystem Assessment. (2005). *Ecosystems and human well-being: Synthesis*. Island Press. https://www.millenniumassessment.org/documents/document.356.aspx.pdf

Montanaro, G., Xiloyannis, C., Nuzzo, V., et al. (2017). Orchard management, soil organic carbon and ecosystem services in Mediterranean fruit tree crops. *Scientia Horticulturae, 217*, 92–101.

Myers, D., Berg, H., & Maneas, G. (2019). Comparing the soundscapes of organic and conventional olive groves: A potential method for bird diversity monitoring. *Ecological Indicators, 103*, 642–649.

Nkonya, E., Mirzabaev, A., & Von Braun, J. (2015). *Economics of land degradation and improvement: A global assessment for sustainable development*. Springer.

Paoletti, M. G., Sommaggio, D., Favretto, M. R., et al. (1998). Earthworms as useful bioindicators of agroecosystem sustainability in orchards and vineyards with different inputs. *Applied Soil Ecology, 10*(1–2), 137–150.

Requier-Desjardins, M., Adhikari, B., & Sperlich, S. (2011). Some notes on the economic assessment of land degradation. *Land Degradation and Development, 22*, 285–298.

Solomou, A. D., Sfougaris, A. I., Vavoulidou, E. M., et al. (2012). The effects of farming practices on earthworm dynamics in olive groves of Central Greece. *Zoology in the Middle East, 58*, 119–126.

Solomou, A. D., Sfougaris, A. I., Kalburtji, K. L., et al. (2013). Effects of organic farming on winter plant composition, cover and diversity in olive grove ecosystems in Central Greece. *Communications in Soil Science and Plant Analysis, 44*(1–4), 312–319.

Solomou, A. D., Sfugaris, A. I., & Sfendourakis, S. (2019). Terrestrial isopods as bioindicators for environmental monitoring in olive groves and natural ecosystems. *Journal of Natural History, 53*(27–28), 1721–1735.

Tienhaara, A., Haltia, E., Pouta, E., et al. (2020). Demand and supply of agricultural ES: Towards benefit-based policy. *European Review of Agricultural Economics, 47*(3), 1223–1249.

Tsesmelis, D. E., Karavitis, C. A., Oikonomou, P. D., et al. (2019). Assessment of the vulnerability to drought and desertification characteristics using the standardized drought vulnerability index (SDVI) and the environmentally sensitive areas index (ESAI). *Resources, 8*(1), art. no. 6.
Vogiatzakis, I., Pungetti, G., & Mannion, A. M. (2008). Introduction to the Mediterranean Island landscapes. In *Mediterranean island landscapes* (pp. 3–14). https://doi.org/10.1007/978-1-4020-5064-0_1
Wolpert, F., Quintas-Soriano, C., & Plieninger, T. (2020). Exploring land-use histories of tree-crop landscapes: A cross-site comparison in the Mediterranean Basin. *Sustainability Science*. https://doi.org/10.1007/s11625-020-00806-w

Chapter 26
Market Challenges in Greek Olive Groves

Vassilis Zampounis and Pavlos Karanikolas

Abstract The Greek olive oil sector faces a number of serious challenges that need to be tackled effectively to upgrade its position and increase the share of value added acquired. Four such inter-related market challenges are identified in this chapter. Firstly is the need for a profound reorganization of the whole sector as a response to the small-scale structure and extreme fragmentation. The strengthening and reorganization of the agricultural cooperatives should be pursued by reducing production costs and enlarging of scale, especially at the processing and distribution stages. Secondly is in establishing marketing standards and beneficial trade rules, the need for a reversal of bulk/branded portions of Greek olive oil in both domestic and export markets, as well as increasing exports and changing the terms of integration of Greek olive oil into global value chains. Thirdly is the formation of a quality strategy for olive oil, having as a focal point the valuation of positive externalities provided by the extensive and semi-extensive Greek cultivation systems, including environmentally friendly practices and preserving local cultivars of olive trees. Finally, several actions are needed to address the lack of coordination and poor governance of the sector as a whole, including the efficient operation of the inter-professional organization and effective collaborative and collective efforts.

V. Zampounis (✉)
Axion Ekdotiki, Athens, Greece
e-mail: zampounis@olivenews.gr

P. Karanikolas
Agricultural University of Athens, Athens, Greece
e-mail: pkaranik@aua.gr

J. Muñoz-Rojas, R. García-Ruiz (eds.), *The Olive Landscapes of the Mediterranean*, Landscape Series 36,
https://doi.org/10.1007/978-3-031-57956-1_26

26.1 Introduction

Olive cultivation is central to the extremely heterogeneous Greek agriculture. From antiquity until today, olive oil is a hallmark of Greek diet and culture. According to official data (IOC, 2021), Greece is the third-largest producer of olive oil in the world, after Spain and Italy. Nevertheless, two trends stand out in recent years: (a) the huge increase in Spanish production, which has far exceeded the other countries and (b) the alternation of Italy, Greece and Tunisia in the second place, depending on the annual fluctuations of production. The Greek olive oil sector faces several serious challenges that need to be tackled effectively to upgrade its position in a rapidly changing international environment in which competition is intense from highly efficient and integrated competitors.

Nowadays, olive groves for olive oil production occupy 18% of the utilized agricultural area (UAA) in Greece. These are mainly cultivated in extensive or semi-extensive farming systems. The average density in olive orchards is 139 trees per hectare (ha), i.e. much lower than 180 trees per ha, which is the threshold of intensive systems (ELSTAT, 2019; IOC, 2015). On the other hand, in terms of the farms' productive specialization, olive cultivation is the most numerous farming type, including 39% of all Greek farms, despite their reduction from 359,404 in 2011 to 233,140 in 2019 (elaborated data of Integrated Administrative Control System [IACS] 2019). They also employ 21.2% of all labour force in Greek Agriculture in 2016 (Eurostat, 2016).

This chapter aims at identifying some key market challenges for the entire Greek olive oil sector, keeping in mind that the primary goal should be to upgrade this sector and increase the share of value added acquired by its actors. In the next section, four market challenges are identified, along with the most recent data, the related problem to be overcome and the actions to be taken.

26.2 The Key Market Challenges of the Greek Olive Oil Sector

26.2.1 A Profound Reorganization of the Whole Sector

Small-scale structure and extreme fragmentation prevail in the whole olive oil sector in Greece. Firstly, 84% of olive oil-producing farms in Greece have a UAA of up to 5 ha (Eurostat, 2016), while, compared to the national average, their mean economic size was one-third in 2019 (€4332 and €12,399, respectively; calculations based on IACS elaborated data). As for the milling sector, there are about 2100 olive mills in Greece (from which 1500 work regularly), with a milling capacity much lower than that of Spanish mills. Furthermore, in an environment of intensified international competition and prevailing low prices, even the largest Greek exporting enterprises are very small compared with the much larger (and fewer) Italian importers. The latter set up an oligopsony and, in most of the time, impose the terms of the transactions in the trade of bulk olive oil.

All the above structural characteristics result in a meagre farm family income for farms specializing in olive oil production, which in 2015–2017 amounted to about 6711 € compared to 10,962 € of all-farms average in Greece (elaborated data from the Farm Accountancy Data Network Database).

In addition, within the context of EU Common Market Organization, during the 2018–2021 time span, 92 operational programmes for olive oil-related actions are implemented in Greece, with a 33.3 million € budget; of these, 90 programmes have been submitted by producer organizations and 2 programmes by associations of producer organizations (MRDF, 2021). The value of olive oil produced by all these organizations amounts to 52 million €, roughly 7% of the national production value. These organizations' membership is 15,951, which corresponds to 6.8% of all operators of the farms specialized in olive oil production. Most of the budget of the above-mentioned operational programmes are devoted to upgrading the quality and competitiveness; about 2% of the budget concerns information dissemination actions, while a negligible amount is allocated to market surveillance actions, on which the commercial policy of each entity should be based. These programmes' contribution to solving chronic problems of the sector is minimal, especially concerning the reduction of bulk sales (see below).

It is well known that the small scale entails high costs of production. But is the cost of production an insurmountable obstacle for the Greek olive oil sector? Table 26.1 shows the production costs data from 57 olive oil farms, derived from 3 field surveys within the 2013/2014–2019/2020 time span. The average area of olive groves in these farms ranges between 6.9 and 8.7 ha. The cash expenses per kg of olive oil range between 1.57 and 1.87 euros, while the depreciation of fixed capital (farm buildings, machinery-equipment, drillings, tree stock) ranges from 0.38 € to 0.49 € per kg. If we add the opportunity cost of the resources owned by farm families, then the 'grand total' cost of production of olive oil increases, ranging from 2.36 € to 4.70 € per kg of olive oil. In all these surveys, the smaller the farm, the higher the costs of production.

Table 26.1 Production costs of olive oil in Greece

Field research	Two-year average	Number of olive farms	Average cultivated area with olive groves (ha)	Cost of production (€ per kg of olive oil)			
				Cash expenses	Depreciation of fixed capital	Opportunity costs of own resources	Total
Lesvos	2014–2016	13	8.0	1.87	0.49	2.34	4.70
East Mani	2013–2015	21	8.7	1.57	0.44	1.25	3.27
Messini	2018–2020	23	6.9	1.60	0.38	0.39	2.36

Source: Agricultural University of Athens, Laboratory of Political Economy and European Integration

On the other hand, it has been documented that several small, medium, large and very large farms have similar or identical costs per kilogramme of olive oil (Karanikolas et al., 2018).

Therefore, in order to overcome the ubiquitous extreme fragmentation and small scale, a series of interventions should aim at improving and increasing the added value of Greek olive oil:

- The strengthening and reorganization of the agricultural cooperatives mentioned above.
- The enlargement of scale and the reduction of production costs, particularly in harvesting, which represents the primary cost factor. It is worth investing in developing a small, light and versatile vibratory harvesting machine capable of being used on steep slopes, uneven surfaces, etc. On the contrary, heavy machines are unsuitable, while high-density planting in Greece is minimal.
- The spreading of the continuous milling (Spanish) system, with the aim of both improving the quality of the oil and reducing its costs of production.
- In pursuing to limit the bulk distribution, the introduction of the packaging of the producer's olive oil in a 5-L container. This will be packaged by the mill and delivered to the producer sealed, indicating on a label all the information about the producer.
- The renewal and rationalization of plant capital through gradual planting of trees, in a programme tailored to meet trade policy targets (selection of variety) and needs for protection against olive diseases (e.g. *Xylella fastidiosa*).
- Improvement of plant protection through a radical change in olive fly control approach, with continuous updating/training for olive producers through agricultural extension, for the adoption of integrated management/good agricultural practices.
- The emphasis on all the cultivation practices that minimize pesticide residues and other contaminants, e.g. mineral oil saturated hydrocarbons (MOSH) and mineral oil aromatic hydrocarbons (MOAH).

26.2.2 Establishing Marketing Standards and Beneficial Trade Rules

The well-known problem of retaining consumers' trust in the image of olive oil as a high-quality product has to be seen in both the domestic and export markets. Firstly, in the domestic market, one of the most astonishing characteristics is the selling of olive oil mostly in bulk (80% of 120,000 tonnes of domestic consumption), while 20% or less is branded oil, distributed through official market channels. Thus, apart from self-consumption and widespread informal networks based on kinship and friends, the largest part is distributed directly by producers to consumers in urban centres and non-olive oil-producing regions of Northern Greece by other intermediaries, often travelling sellers. This is a deep-rooted habit, since most of the consumers are convinced that olive oil bought from 'producers' is 'authentic' in contrast to branded oil, which they believe to be 'processed' and 'with additives' (Zampounis, 2017a).

The negative consequences of not challenging this widespread belief have been detrimental for the entire sector. For instance, this practice has led (i) to loss of added value: an unknown, small but significant proportion is lost as 'oil' which is either degraded or adulterated, ultimately at the expense of consumers, and (ii) to the decline of the total proportion of branded bottled olive oil to such a low level that it is impossible to develop strong businesses that are capable of promoting Greek olive oil in the domestic market, let alone the export market. Therefore, an additional consequence is a widening gap between the producer's price (raw material) and the price on supermarket shelves for the consumer, making olive oil even less competitive against other seed oils. Another source of loss of added value is the mass catering sector (HORECA), due to its non-compliance with national legislation on the mandatory serving of oil at tables in small branded packages etc., a practice that has become widespread in other olive oil-producing countries such as Spain and Italy. For touristic countries, this has very indirect positive effects on product promotion.

Secondly, during the period 2017–2019, Greece has exported on average 155,485 tonnes per year, mostly to Italy (57%), Spain (12%), the rest of EU (18%) and the rest of the world (13%). It has to be noted that three-quarters of all this quantity is exported in bulk (Zampounis, 2017b). The fact that bulk exports are sold on average at 1.5 €/kg (Zampounis, 2017a) less in comparison with small branded containers has led to harsh criticism, e.g. that Greece does not take full advantage of its product. In recent years, after the intense Spanish penetration into the Italian market, Greek exports have become increasingly more difficult, with prices being pushed consistently downward, while quality requirements have become more demanding.

It has to be recalled that during the 1980s and 1990s, full advantage was not taken of subsidies, consumption aid and restitution nor of promotional campaigns. On the other hand, risk-free solutions prevailed for the problem of bulk olive oil, while there was a lack of businesses willing to invest long term in branded products. The two largest private companies (Elais/Unilever and Minerva) focused on the domestic market. At the same time, cooperatives did not concern themselves with bottling or export of certified (PDO/PGI) quality olive oil, with few exceptions, such as the Union of Sitia (in Crete), which ended up at an administrative/managerial/financial impasse. In the early 2000s, the third-tier co-op 'Elaiourgiki' controlled almost 25% of the olive oil market in Greece. In 2010, the four largest cooperatives had a share of 13.5% in this market; however, most of them have gone bankrupt or been acquired by private companies in recent years. On the other hand, 70% of the marketed, labelled olive oil is distributed and sold in retail chains, as opposed to 30% sold in smaller, specialty stores (Iliopoulos, 2012).

It is also worth noting that after the severe crisis that began in 2010, a number of new, dynamic, small businesses were developed by young people with a high level of education, who in some cases undertook a family property. Focusing on olive oil standardization, branding and quality, they have achieved diversification and remarkable export performance. Therefore, there is an urgent need to (i) reverse the bulk/branded portions of Greek olive oil in both domestic and export markets; (ii) increase the extremely low shares (under 4%) of Greek olive oil exports across all international markets; and (iii) change the integration of Greek olive oil into global value chains.

26.2.3 The Quest for a Clear Quality Strategy

Approximately 70% of Greek production is extra virgin olive oil, i.e. superior quality (IOC, 2012). However, only 20% of Greek production reaches the labelling/branding stage, compared with 50% in Spain and 80% in Italy, with the remainder sold in bulk form, including 70% of exports, mainly to Italy for re-export (ICAP, 2012). Additionally, Greece's market share in the world market of branded olive oil decreased from 6% during the 1990s to 4% during the past 5 years.

It has to be noted that 23 Greek olive oils with protected geographical indications (PDO and PGI) have been registered in the official quality system of the EU up to now. On average, 13,283 tonnes per year have been sold from these categories during the period 2016–2018 in both domestic and export markets (ELGO-DEMETRA Organization, 2020). This means that only 5.5% of all olive oil produced in Greece is sold as quality-certified olive oil.

Furthermore, the quality strategy should rely on approaches that go beyond intensification and enlargement of scale. For example, recent research findings (Tzouramani et al., 2020) indicate that the prevailing extensive or semi-extensive Greek olive oil system, conducted by professional farms, performs very well in terms of economic-social-environmental sustainability. Besides, recent research findings indicate that, although multiple strategies with synergistic effects co-exist at the farm/farm household level, specific territorial resources are partially mobilized by olive cultivators' strategies; consequently, olive oil identities are valourized on the market to some extent but less so through positive externalities (Karanikolas et al., 2021).

These findings could support the formation of a quality strategy for olive oil, having as a focal point *the valuation of positive externalities* provided by this system, including environmentally friendly practices, the preservation of local cultivars of olive trees and the localization of the food systems into which olive farms are integrated. As has been documented by an ample number of studies, extensive olive tree systems lag behind intensive systems in terms of yields, economic outcomes and profit; however, in many cases, they provide landscape and habitat diversity, along with multiple benefits for the local communities (Rodríguez Sousa et al., 2019; Russo et al., 2016). Likewise, intensified olive farming is a major cause of one of the critical environmental problems affecting the EU today, i.e. widespread soil erosion and desertification in all southern EU countries (Beaufoy, 2001).

The repositioning of the whole chain towards the promotion of high-quality branded olive oil requires a profound restructuring of the sector to include more vertically integrated production, both upstream in the olive production stage and downstream in the branding stage. The industry also lacks widespread access to distribution channels, which can be costly to develop. Finally, the standardization of quality control, which is vital for promoting premium olive oil, is not supported by the existing fragmented nature of Greek olive oil cooperatives.

26.2.4 Institutional Organization and Governance

Lack of coordination and poor governance, which have been identified as a corollary of the fragmented Greek agri-food chains (Karantininis, 2017), are also key characteristics of the olive oil sector's institutional organization. Concerted efforts should be made to address the weaknesses and gaps in the entire web of organization, networking, cooperation and policy development, between all those who directly or indirectly are involved in and influence the value chain; these include the Ministry of Rural Development and Food and its subsidiary organizations, the professional bodies (from growers/cooperatives, up to packaging plants) and the scientific community. Of paramount importance is *the failure of the olive oil inter-professional organization* to play a key role in coordinating all players and the sector's governance.

An upgraded institutional organization of the sector could act as a catalyst for efforts to pursue integration both forwards and backwards and accomplish economies of scale in processing, distribution and logistics, research and marketing. *Further consolidation and coordination* at all levels of the sector, primarily through efficient collaborative and collective efforts, should be pursued through cooperatives, producer groups and operational programmes. Finally, the establishment of flexible small cooperative schemes, the empowerment of private inter-professional and sector representatives, as well as collaborations or other contractual methods could improve the control over the supply of olive oil (Iliopoulos et al., 2012).

References

Beaufoy, G. (2001). *EU policies for olive farming – Unsustainable on all counts*. WWF and BirdLife International. www.assets.panda.org/downloads/olivefarmingen.pdf. Accessed 14 Dec 2020.

ELGO-DEMETRA Organization. (2020). *Production of greek certified olive oils*. Unpublished Data, Athens.

ELSTAT [Hellenic Statistical Authority]. (2019). *Livestock/crops surveys*. https://www.statistics.gr/en/statistics/agr. Accessed 15 Sep 2020.

Eurostat. (2016). *Farm structure survey*. https://ec.europa.eu/eurostat/data/database. Accessed 7 Jul 2020.

ICAP. (2012). *Olive oil – Kernel oil, sectoral study*. ICAP, Athens.

Iliopoulos C (2012). Support for Farmers' Cooperatives; Country Report Greece. Wageningen UR, Wageningen.

Iliopoulos, C., Giagnocavo, C., Theodorakopoulou, I., & Gerez, S. (2012). Case Study Report; Structure and Strategy of Olive Oil Cooperatives: Comparing Crete, Greece to Andalusia, Spain. Wageningen UR, Wageningen.

IOC [International Olive Council]. (2012). *Country profiles – Greece.* https://www.internationaloliveoil.org/wp-content/uploads/2019/11/ENGLISH_POLICY_GREECE-2012_OK.pdf. Accessed 8 Feb 2021.

IOC [International Olive Council]. (2015). *Olive oil production by country.* http://www.internationaloliveoil.org/estaticos/view/131-world-olive-oil-figures. Accessed 5 Jun 2020.

IOC [International Olive Council]. (2021). *Olive oil figures.* https://www.internationaloliveoil.org/what-we-do/economic-affairs-promotion-unit/#figures. Accessed 10 Feb 2021.

Karanikolas, P., Bebeli, P., & Thanopoulos, R. (2018). Farm economic sustainability and agrobiodiversity: Identifying viable farming alternatives during the economic crisis in Greece. *Journal of Environmental Economics and Policy, 7*(1), 69–84. https://doi.org/10.1080/21606544.2017.1360212

Karanikolas, P., Martinez-Gomez, V., Galli, F., Prosperi, P., Hernández, P., Arnalte-Mur, L., Rivera, M., Goussios, G., Fastelli, L., Oikonomopoulou, E., & Fonseca, A. (2021). Food system integration of olive oil small farms in Southern Europe. *Global Food Security, 28*(3), 100499. https://doi.org/10.1016/j.gfs.2021.100499

Karantininis, K. (2017). *A new paradigm for greek agriculture*. Palgrave Macmillan.

MRDF [Ministry of Rural Development and Food]. (2021). *Operational programs for the olive oil.* Unpublished data, Athens.

Rodríguez Sousa, A. A., Barandica, J. M., Sanz-Cañada, J., & Rescia, A. J. (2019). Application of a dynamic model using agronomic and economic data to evaluate the sustainability of the olive grove landscape of Estepa (Andalusia, Spain). *Landscape Ecology, 34*(7), 1547–1563. https://doi.org/10.1007/s10980-019-00773-3

Russo, C., Cappelletti, G. M., Nicoletti, G. M., Di Noia, A. E., & Michalopoulos, G. (2016). Comparison of Europea olive production systems. *Sustainability, 8*(8), 825. https://doi.org/10.3390/su8080825

Tzouramani, I., Mantziaris, S., & Karanikolas, P. (2020). Assessing sustainability performance at the farm level: Examples from Greek agricultural systems. *Sustainability, 12*(7), 2929. https://doi.org/10.3390/su12072929

Zampounis, V. (2017a). Conclusions about olive policies. In *Encyclopedia of olive oil* (Vol. 38, pp. 676–677). Gaia Epicheirein, Axion Ekdotiki. (in Greek).

Zampounis, V. (2017b). Statistical analysis of olive sector. In *Encyclopedia of olive oil* (Vol. 35, p. 577). Gaia Epicheirein, Axion Ekdotiki. (in Greek).

Part V
Olive Landscapes Across the Mediterranean: Regional Heterogeneity and Case Studies: Olive Groves and Landscapes in Portugal

Chapter 27
Historical and Social Insights into the Olive Groves and Landscapes in Alentejo (Portugal): Trajectories of Change and Sociocultural Perceptions

José Muñoz-Rojas, Nuno Guiomar, Ignacio García Pereda, and Teresa Pinto-Correia

Abstract Olive groves are a keystone land use and crop in Portugal, having historically shaped the ecological, economic and socio-cultural values of Portuguese rural landscapes. They are also a key element for the Mediterranean diet and thus also a public health asset. These land-use systems have over the past century undergone radical transformations, following subsequent stages of decline and expansion. Such changes have been triggered by a combination of global market oscillations, socio-cultural shifts, increased availability of water for irrigation and of agrochemical fertilizers and public policy support including public and private funding. Following up from the latest trends towards intensification, globalization and financialization, a series of public debates are arising with contrasted positions being held. These debates

J. Muñoz-Rojas (✉)
MED-Mediterranean Institute for Agriculture, Environment and Development CHANGE-Associated Lab for Global Change & Sustainability Department of Landscape, Environment and Planning, Universidade de Évora, Valverde - Évora, Portugal
e-mail: jmrojas@uevora.pt

T. Pinto-Correia
MED-Mediterranean Institute of Agriculture, Environment and Development & Laboratório Associado CHANGE, Universidade de Évora, Évora, Portugal

Departamento de Paisagem, Ambiente e Ordenamento, Universidade de Évora, Évora, Portugal
e-mail: mtpc@uevora.pt

N. Guiomar
MED-Mediterranean Institute of Agriculture, Environment and Development & Laboratório Associado CHANGE, Universidade de Évora, Évora, Portugal
e-mail: nnunogui@uevora.pt

I. G. Pereda
Centro Inter-Universitário de História das Ciências e das Tecnologias, Universidade de Lisboa, Lisbon, Portugal
e-mail: igpereda@ciencias.ulisboa.pt

J. Muñoz-Rojas, R. García-Ruiz (eds.), *The Olive Landscapes of the Mediterranean*, Landscape Series 36,
https://doi.org/10.1007/978-3-031-57956-1_27

are largely underpinned by discourses that too frequently misrepresent historical trends and their complexities. It is expected that improved historical and socio-cultural information contributes towards a better-informed debate that can lead the sector towards consensus, social acceptance and, ultimately, increased sustainability standards.

Keywords Historical trends · Social discourses · Olive groves · Agricultural sustainability · Portugal

27.1 Origins and Key Drivers of Olive Grove Expansion and Intensification in Portugal and Alentejo

Olive tree cultivation in the Mediterranean semi-dry landscapes of Portugal was traditionally managed as rainfed land-use systems, dating back to ancient times (Vossen, 2007). In Alentejo (southern Portugal), olive cultivars always have played a critical role in shaping the agricultural economy (Reis, 2014).

Since the opening of the Lisbon Agronomy School, in the 1850s, one of the main objectives of Portuguese agronomists was the enhancement of olive cultivation. The productivity of traditional olive groves is nowadays considered relatively low, whereby yields of 2500–3000 kg/ha were once considered high (Dalla-Bella, 1818; Prego, 1902; Câmara, 1905) in the light of contemporary agronomic standards, whilst the crop costs associated with such farming and cropping systems (harvesting, pruning, etc.) are considered high (Radich, 1996).

One of the historically proposed solutions was the "cultural intensification" of the olive groves, but the interpretation of what intensification means in practice has evolved in time (Gomes & Leitão, 2016). The first use of inorganic fertilizers, already in the twentieth century, meant that former strategies, such as the application of manure, could be abandoned (Infante-Amate, 2012). Agricultural modernization demanded strong state intervention and was associated with profound social, economic and institutional changes, with examples of these being the organization of the 1933 Lisbon "Congrès international d'oléiculture" and the creation of the National Olive Oil Board (*Junta Nacional do Azeite*—JNA) in 1937 (Cidraes, 1939; Langhans, 1949; Galvão, 1952) and the Elvas Oliviculture Station in 1958 (Cidraes, 1965). Policies of autarchy and economic intervention dominated the oil sector up until the second half of the twentieth century. Along this period, the JNA decided to invest large quantities of money to promote Portuguese oil in international markets.

Later on, several agronomists, during the *Estado Novo* years (1933–1974), proposed important changes in management practices, including:

- To better study traditional cultivars useful for innovative breeding programs (Bobone, 1933). In 1970, six olive cultivars were considered the most interesting from an economic point of view (Cidraes, 1970): Galega, Carrasquenha, Redondil, Cordovil de Elvas and Cobrançosa e Blanqueta de Badajoz.
- To upgrade tree densities, to a maximum of about 400 trees/ha. Before 1965, in Italy and Portugal the tree density of traditional olive groves was of 100 trees/ha (Cidraes, 1965), but Cidraes (1938) had already recommended that the spacing between trees could not be less than 10 m.

Fig. 27.1 (**a**) Machine "Olivia" designed for harvesting olives, acorns and almonds, among other fruits, through suction; (**b**) small olive harvesting vibrator designed by Teodoro Gonçalves. Image

- To develop, improve and implement irrigation, which was primarily considered for table olives (Almeida, 1965).
- To use "mechanical olive harvesters" (Fig. 27.1). In the 1960s, a lack of rural workers (or rising labour costs) began to affect Alentejo, mainly driven by the rural exodus to the larger metropolitan areas (Cidraes, 1965). During this period, mechanical harvesting, with trunk shakers and vacuum or suction harvesters, was considered an efficient method to reduce harvesting costs (Gomes & Leitão, 2016). In this period, Italian machines began to be imported and tested in the fields of the agrarian stations as *Estação Agronómica Nacional*, in Oeiras (Almeida, 1965). One of the new trunk shakers was even designed by a Portuguese engineer, Teodoro Gonçalves.
- To support the development of cooperatives of olive farmers. Between 1950 and 1956 there were created 44, 39 of each with an olive press (Debates Parlamentares, 1956).

27.2 Dynamic Trajectories of Change: Is the Recent Boom Unprecedented?

Following the aforementioned intensification of farming systems, the area covered by olive groves rapidly expanded until the early 1960s (Gomes & Leitão, 2016). Nonetheless, an equally rapid decline was subsequently observed (Fig. 27.2). This variation is reflected in the trends of production of olive oil, which shows higher inter-annual variability rates over the various decades of the twentieth century, probably due to alternate bearing in olive trees. In the mid-1980s the olive sector started to become restructured, with the main public intent to increase the area covered by olive groves, and to densify or replace the existing ones through grafting.

In the late 1980s, after Portugal joined the then EEC (now EU) and under Council Regulation (EEC) n.° 3828/85 of 20 December 1985, the National Plan for Oliviculture (Portaria n.° 259/87 of 2 April 1987) was approved. Considering the

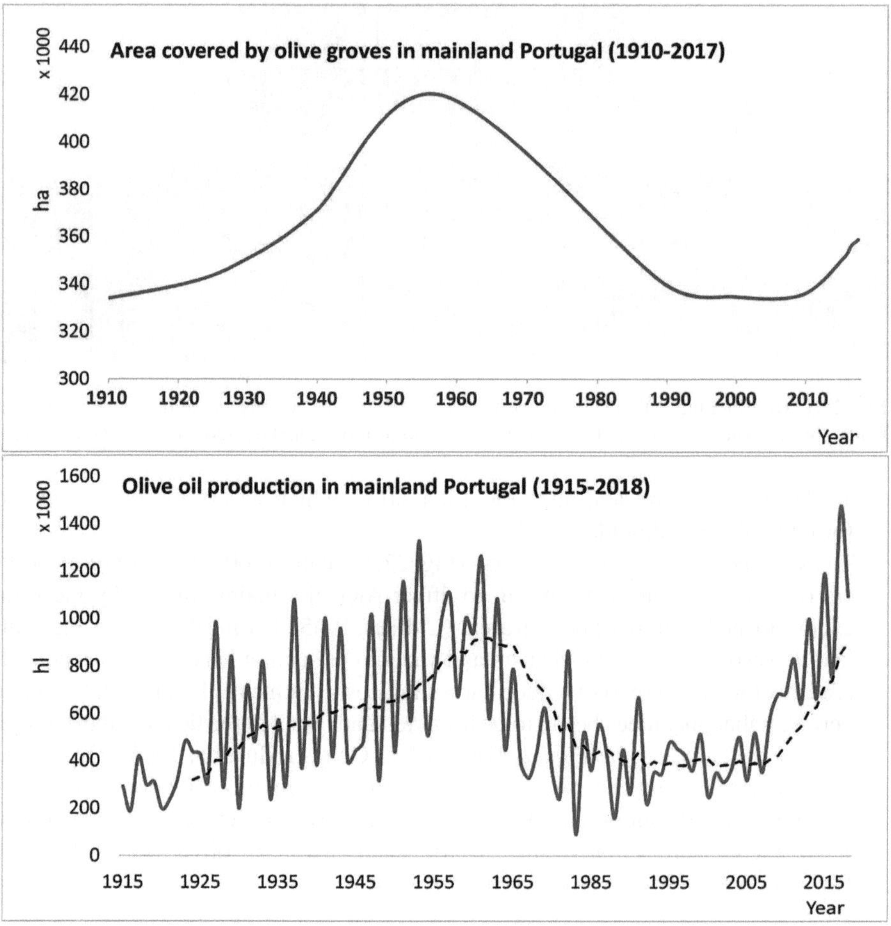

Fig. 27.2 Trends of the area covered by olive groves between 1910 and 2017 (top) and of olive oil production between 1915 and 2018 (bottom) in mainland Portugal. (Sources: Agricultural and Forest Map of 1910, agricultural surveys and census (1945–2016; National Institute for Statistics—INE))

requirements to shift the production mode that were imposed by the gradual mechanization of farming operations for olive harvesting and cleaning, this plan was revised in 1991 (Portaria n. 601/91 of 4 July 1991). However, the area covered by olive groves continued to decline between 1989 and 1999 (about 2%), and the average production of olive oil between the 1980s and 1990s grew by only 8% without substantial changes to be noted in the following 10 years (Reis, 2014).

It was only in the late 1990s that incentives for increasing olive oil production were approved under the reform of the Common Organization of the Market in Oils and Fats (Council Regulation (EC) n. 1638/98 of 20 July 1998). Two years later, the European Commission Decision 2000/406/CE of 9 June 2000 allowed Portugal to expand the area of olive groves by 30,000 ha until 2006. These changes were undertaken due to the inability of the olive sector to react to the increased global olive oil demand, which had resulted in a high external deficit, underpinned by a degree of national self-sufficiency of only 65% along the decade 1993–2002 (GPP, 2007) and of less than 50% at the end of this same period (Reis, 2014).

The low profitability rates of the traditional low-input farming systems and the consequent dependency on subsidies are demonstrated by FADN data (2000–2007), the data from the Portuguese Institute for Statistics concerning the Economic Dimension of farms (INE, 2018) and other studies conducted to analyse olive-farming systems (GPP, 2007). These data were used as the central argument not only for the need increase of the area of cultivated olive groves but also for increasing land-use intensity.

However, it was only after 2005 that significant increases in the area covered by olive groves have been observed (GPP, 2007), resulting from the expansion of 25,000 ha in the following 3 years (Reis et al., 2010), an amount that was then doubled by 2014 clearly exceeding the values set out by the previous European Commission decision. Most of the new olive groves grew in the area of the irrigation project of the Alqueva in the Alentejo region (Fig. 27.3d), considered to this day as the largest project of artificial water storage and crop irrigation in Europe (Silveira et al., 2018). The expansion of the Alqueva irrigation area, which includes a complex network of linkages to other supply and water storage systems, was carried out mainly in Baixo Alentejo (NUTS-III), in itself the driest district in Portugal. According to the data from the CORINE Land Cover 2012, in comparison with the 2006 database and considering only mature olive tree stands (land cover class 223), the areas with olive groves, only in Baixo Alentejo, have increased more than 35,000 ha, corresponding to 38.93% of the total changes observed at national level between 2006 and 2012 and to 60.78% of the total area that has undergone land-use changes in Alentejo. Between 2012 and 2018 the area of olive groves has expanded by a further 16,000 ha.

It is important to note that between the beginning of the last century and this day, today there has been a spatial shift, of the preferential area for olive trees, from the more mesic areas of the central region of mainland Portugal to the more xeric territories of Alentejo (Fig. 27.3a–c).

27.3 Beyond Extent: Production, Productivity and Efficiency

Nowadays, and according to data from the Portuguese Institute for Statistics (hereafter INE), the current area of olive groves for oil production is about 343,557 ha being in area. This turns olive groves onto the main permanent crop in Portugal, with a share of 48%. These changes have had repercussions on the production of table olives and olive oil. The average annual production was, in the period 2000–2007, of 390,493.62 hl of olive oil and 252,247.50 tonnes of olives (INE, 2001–2008). These values have nearly doubled in the following sexennial period

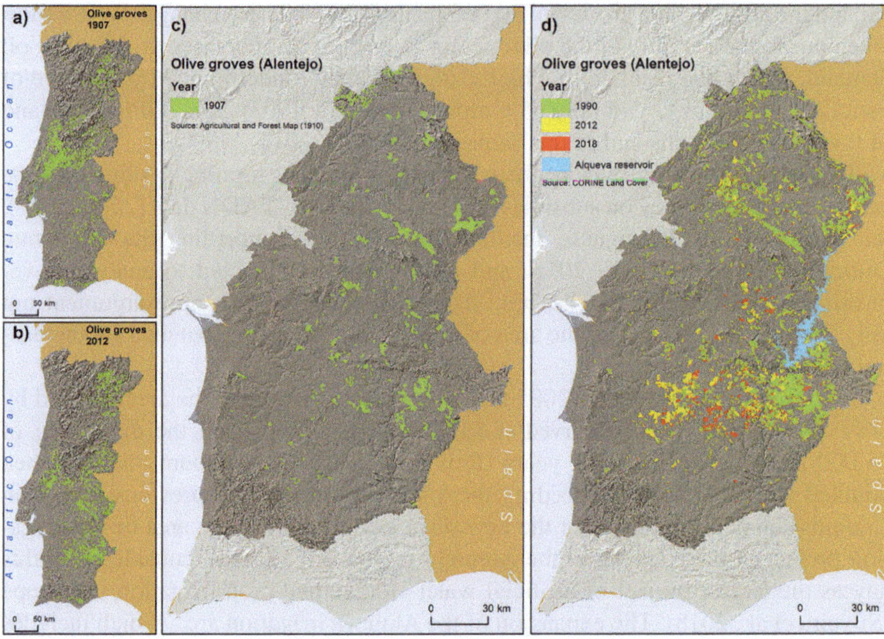

Fig. 27.3 Area covered by olive groves in 1907 in mainland Portugal (**a**) and in the Alentejo region (**c**), in 2012 in mainland Portugal (**b**) and in 1990, 2012 and 2018 in the Alentejo region (**d**). (Sources: Agricultural and Forest Map (1910) and CORINE Land Cover)

(2008–2014), in which there have been recorded mean annual productions of 677,249.14 hl of olive oil and 421,386.42 tonnes of olives (increases of 74% and 67%, respectively) (INE, 2009–2015), with data subsequently expanding even further. As a result of this, Portugal is currently the fourth largest European producer of olive oil, mostly originating in the inland regions of Alentejo and Trás-os-Montes. Currently Alentejo is the region that contributes most to both olive and olive oil production (between 60% and 70%; Fig. 27.4) nationally. In this region, the contribution to the overall national production doubled during the period 2008–2014 in both of the production indicators, when compared to the previous 7 years.

This production growth is not only the result of increased area of olive groves but is also due to a rapid intensification process (Gomes & Leitão, 2016) that has now embraces all production systems, from conventional to organic and even biodynamic. Drip irrigation has been established in intensive orchards characterized by a high tree density (300–400 trees ha^{-1}) and super intensive orchards characterized by a very high tree density (400–1700 trees ha^{-1}) and, in some cases even more, up to 3000 trees ha^{-1} as reported by Vossen (2007). In fact, approximately 88% of the areas installed in the Alentejo and 76% of the latest groves are irrigated (GPP, 2007). Due to the high dependence on irrigation and the increasingly emergent need for a more rational use of water as a scarce resource, the estimation of water requirements and water consumption by super-intensive olive groves, as well as their temporal

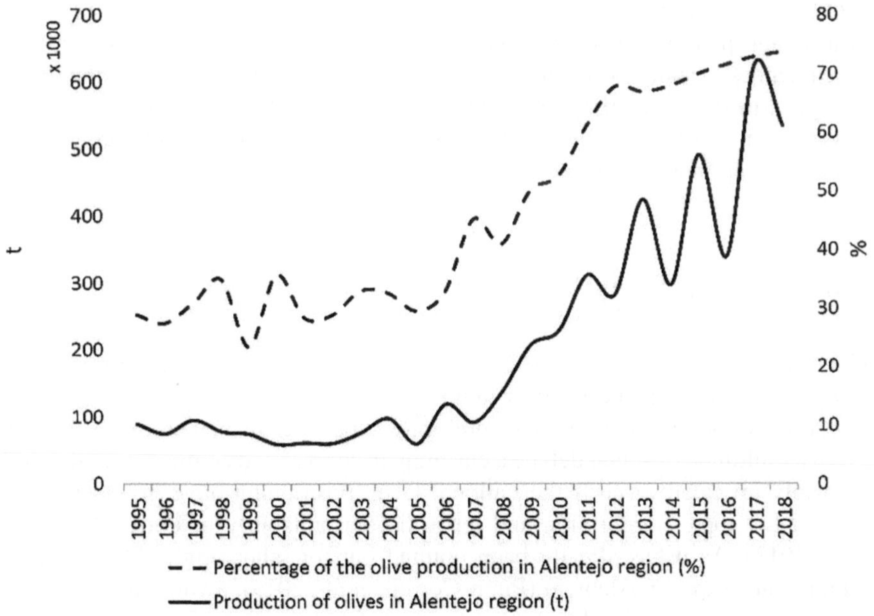

Fig. 27.4 Evolution of the production of olives in the Alentejo region between 1995 and 2018. (Sources: Agricultural surveys and census (1945–2016; National Institute for Statistics—INE))

distribution, has also become a concern involving groups composed of researchers, entrepreneurs and managers of water resources and with increasing frequency social and environmental activists acting at levels from the local to the national. These needs are promoting the development of more innovative methods for water management and nutrient inputs directly related with real-time plant needs (e.g. Patumi et al., 2002; Gómez-Rico et al., 2007; Cameira et al., 2014) or guidelines to reduce soil erosion and loss of organic matter (Santilli et al., 2011).

Intensive production models also allow for the use of continuous mechanical straddle harvesters and the achievement of higher yields within a few years after planting, largely driving the exponential growth of production after 2009 (Fig. 27.4). However, there are few cultivars already well adapted to this system. This is a key reason behind the late expansion of imported varieties in Portugal, such as *Olea europaea* cv. Picual and *Olea europaea* cv. Arbequina, which grow fast, rapidly reaching higher yields (Silveira et al., 2018). In these intensive olive groves, more resistant varieties to pests and diseases can however also be used, considering that suitable phytosanitary treatments are applied. Nevertheless, and as for the irrigation scheme, also the optimization of the mechanization procedures is varietal dependent (Camposeo et al., 2013), reinforcing the need to identify, map and monitor the distribution of different cultivars (Gomes et al., 2020), an issue that is currently undergoing by novel projects funded by the Portuguese Foundation for Science and Technology-FCT (see OLEADAPT-fct).

Other key changes have also taken place that are apparent in the landscape composition and patterns, such as in the farms' structure (Rolo, 2009). As a result of these changes, Alentejo was one of the few regions which registered an increase in average farm size of 6.1 ha/farm in 1999 to 8.3 ha in 2009 (significantly higher than the continental average value of 2.6 ha/farm), resulting from the replacement of the traditional low-input olive groves for new intensive ones.

27.4 Culture, Social Acceptance and Resulting Conflicts

As a result of all of the above-described transformations in the olive grove farming systems in Portugal, and especially in the region of Alentejo, an equally drastic and rapid transformation of traditional rural landscapes has also taken place. This has driven a multitude of social debates, clashing discourses and controversies that have ultimately upscaled towards the national level, positioning olive groves as a key political hot topic for farming and territorial sustainability in Portugal (Silveira et al., 2018). As it has already been pointed out for other fragile Mediterranean farming landscapes in Alentejo (Pinto-Correia et al., 2019), a wide range of social discourses can be identified that are reflective of the diverse sets of mindsets framing the decisions and strategies that drive landscape changes. Such discourses may eventually happen synchronically. This is all of direct relevance for the olive groves of Portugal and Alentejo, which are now placed in a paradoxical situation in which a clear trend towards rapid intensification and expansion in the most productive areas and abandonment in marginal ones are both happening. This is underpinned by a strong public support which persists despite the subsequent increase in the level of risks related to climate change, soil erosion (Rodríguez Sousa et al., 2021) and biodiversity and landscape degradation. In responses to the multiple social-ecological and territorial challenges arising from such complex dynamics, diverse social positionings can be preliminarily detected through their underpinning discourses, including:

- "Green Revolution", which is largely espoused by the larger conventional farmers, trade agents, agrotechnological and financial investment firms, as well as having been largely supported by the public sector. This discourse supports the rapid intensification and financialization of olive groves and olive oil production for purposes of food security, global market competitiveness and national economic growth.
- "Eco-Futures", espoused by younger and smaller farmers and also by few bigger ones and by academics and activists in the field of agro-ecology. This advocates for organic production, regional varieties and shorter market chains. Despite being minor, this is an increasingly growing viewpoint that counts with marginal political support.

– “Think locally”, restricted to local cooperatives and smaller and older producers, advocating shorter value chains, local consumption and higher-quality products with local varieties.

At the moment it seems self-evident that each of these discourses advocates for the simplification in the interpretations of a reality that is complex and where all models of production and trade could, and even should, coexist in the future. This is even if we disregard aspects of greenwashing and other dubious discourses which are also rapidly permeating the sector and region. Whether or not we are able to recognize and foster the wide range of options that exist for olive groves and olive oil in Portugal will ultimately define the future sustainability of the sector. However, unless we better disentangle its past, this shall indeed become a lost cause, with future trajectories of change and sustainability becoming devoid of the necessary empirical baseline that can only be set by better understanding its past.

Acknowledgements This work was funded by the SUSTAINOLIVE project (Grant n° 1822; PRIMA EU Programme) and by FCT-Portugal (UIDB/05183/2020 as members of MED-Universidade de Évora and LA/P/0121/2020 as members of the CHANGE- Associated Lab-Universidade de Évora).

References

Almeida, F. J. (1965). Perspectivas da olivicultura portuguesa. Sep. do Boletim da Junta Nacional do Azeite. Lisboa (Portugal). N°. 71, p 48.

Bobone. (1933). Essai sur la caracterisation des variétés de l’olivier. Étude biometrique. s. 1. imp. s. d. 5 fotogravs. 12 diagrs. 9 esquems. De pags. 85 a 146. 23,8 X 15,7.

Câmara, M. S. (1905). Subsídio para o estudo das variações de oliveiras portuguesas. Typographia La Bécarre, Lisboa.

Cameira, M. R., Pereira, A., Ahuja, L., & Ma, L. (2014). Sustainability and environmental assessment of fertigation in an intensive olive grove under Mediterranean conditions. *Agricultural Water Management, 146*, 346–360.

Camposeo, S., Vivaldi, G. A., & Gattullo, C. E. (2013). Ripening indices and harvesting times of different olive cultivars for continuous harvest. *Scientia Horticulturae, 151*, 1–10.

Cidraes, F. (1938). *O Baixo Alentejo, grande produtor de azeite*. Minerva Comercial.

Cidraes, F. (1939). Métodos de reprodução da oliveira, Direcção-Geral dos Serviços Agricolas, Folha de Divulgação, Série II, 6: 1–21.

Cidraes, F. (1965). A crise actual da olivicultura e as soluções que se impõem (p. 31). Lisboa: Junta Nacional do Azeite. Lisboa (Portugal).

Cidraes, F. (1970). A Colheita Mecânica de Azeitona, Separata do Boletim da Junta Nacional do Azeite n°79, Janeiro – Junho, 1970. Lisboa (Portugal).

Dalla-Bella, J. (1818). Memoria sobre a cultura das oliveiras em Portugal. Typographia da Academia Real das Sciencias, Lisboa (Portugal).

Debates Parlamentares. (1956, April 23).

Galvão, J. M. (1952). Manual do olivicultor. Livraria Castro e Silva, Lisboa (Portugal).

Gomes, R., & Leitão, D. (2016). Deliverable 4.1. Case study. Intensive olive groves (Portugal). Deliverable from the PEGASUS H2020 project (No 633814), p. 17.

Gomes, L., Nobre, T., Sousa, A., Rei, F., & Guiomar, N. (2020). Hyperspectral reflectance as a basis to discriminate olive varieties – A tool for sustainable crop management. *Sustainability, 12*(7), 3059.

Gómez-Rico, A., Salvador, M. D., Moriana, A., Pérez, D., Olmedilla, N., Ribas, F., & Fregapane, G. (2007). Influence of different irrigation strategies in a traditional Cornicabra cv. olive orchard on virgin olive oil composition and quality. *Food Chemistry, 100*, 568–578.

GPP. (2007). Olivicultura: diagnóstico sectorial. Gabinete de Planeamento e Políticas. Ministério da Agricultura do Desenvolvimento Rural e das Pescas, Lisboa. (Portugal).

INE. (1943–2018). Estatísticas agrícolas. Instituto Nacional de Estatística, Lisboa (Portugal).

Infante-Amate, J. (2012). The ecology and history of the Mediterranean olive grove: The Spanish great expansion, 1750–2000. *Rural History, 23*(2), 161–184.

Langhans, F.-P. (1949). Apontamentos para a história do azeite em Portugal. Junta Nacional do Azeite, Lisboa (Portugal).

Patumi, M., D'Andria, R., Marsilio, V., Fontanazza, G., Morelli, G., & Lanza, B. (2002). Olive and olive oil quality after intensive monocone olive growing (Olea europaea L. cv. Kalamata) in different irrigation regimes. *Food Chemistry, 77*, 27–34.

Pinto-Correia, T., Muñoz-Rojas, J., Thorsoe, M. H., & Noe, E. B. (2019). Governance discourses reflecting tensions in a multifunctional land use system in decay; Tradition versus modernity in the Portuguese Montado. *Sustainability, 2019*(11), 3363. https://doi.org/10.3390/su11123363

Prego, J. M. (1902). Olivaes e lagares: Andaluzia, Catalunha, Nice, Bari, Portugal. Typographia da Livraria Ferin, Lisboa (Portuhgal).

Radich, M. C. (1996). Agronomia no Portugal oitocentista: uma discreta desordem. Celta Editora. Lisboa (Portugal).

Reis, P. (2014). O olival em Portugal: dinâmicas, tecnologias e relação com o desenvolvimento rural. Animar, Lisboa (Portugal),

Reis, P., Rolo, J., Calha, I., & Martins, V. (2010). Environmental pressure on intensive olive groves in the Alentejo region (Portugal)–the use of pesticides. 120th EAAE Seminar, Chania (Greece).

Rodríguez Sousa, A. A., Muñoz-Rojas, J., Pinto-Correia, T., Aguilera, P. A., Barandica, J. M., & Rescia, A. J. (2021). A comparative analysis of soil loss tolerance and productivity of the olive groves in the protected designation of origin (PDO) areas Norte Alentejano (Portugal) and Estepa (Andalusia, Spain). *Agronomy, 11*, 665. https://doi.org/10.3390/agronomy11040665

Rolo, J. C. (2009). Uma leitura da base produtiva olivícola do Alentejo. *Revista da Associação Portuguesa de Horticultura, 99*, 1246–1290.

Santilli, E., Lombardo, L., Varlaro, M. E., Nannelli, R., Gagnarli, E., & Bati, C. B. (2011). Effectiveness of the GAEC cross compliance standard maintenance of olive groves in good vegetative condition in avoiding the deterioration of habitats and land abandonment. *Italian Journal of Agronomy, 6*(s1), e15.

Silveira, A., Ferrão, J., Muñoz-Rojas, J., Pinto-Correia, T., Guimarães, M. H., & Schmidt, L. (2018). The sustainability of agricultural intensification in the early 21st century: Insights from the olive oil production in Alentejo (Southern Portugal). In A. Delicado, N. Domingos, & L. de Sousa (Eds.), *Changing societies: Legacies and challenges. Vol. iii. The diverse worlds of sustainability* (pp. 247–275). Imprensa de Ciências Sociais da Universidade de Lisboa.

Vossen, O. (2007). Olive oil: History, production, and characteristics of the world's classic oils. *HortScience, 42*, 1093–1100.

Chapter 28
Pedoclimatic and Landscape Conditions of Olive Groves in Portugal and Alentejo

José Muñoz-Rojas, Nuno Guiomar, Teresa Pinto-Correia, and Sergio Prats

Abstract Olive groves in Alentejo have exponentially increased their extent over the past 30 years through irrigation and technological intensification, including also the use of exogenous breeds and cultivars. This has been facilitated by the implementation of the Alqueva irrigation network. This rapid process of change has permitted overcoming many of the bio-physical limitations of the regional soils and climate. Ultimately, the expansion and intensification of olive groves have focused on the areas with most productive agricultural soils and access to water, with traditional cultivars and farming systems restricted to marginal agricultural areas, now frequently in risk of abandonment. Besides their multiple ecological, social and economic impacts, these rapid changes have also marked their scars upon the traditional rural landscapes. Nonetheless, when looked in further detail, the differentiations between intensive and extensive farming systems, and their sustainability impacts, become fuzzier than originally thought. In such a complex context, the novel concept of "sustainable intensification" might open new opportunities to progress towards more realistic solutions for the future.

J. Muñoz-Rojas (✉)
MED-Mediterranean Institute for Agriculture, Environment and Development CHANGE-Associated Lab for Global Change & Sustainability Department of Landscape, Environment and Planning, Universidade de Évora, Valverde - Évora, Portugal
e-mail: jmrojas@uevora.pt

N. Guiomar · S. Prats
MED-Mediterranean Institute of Agriculture, Environment and Development & Laboratório Associado CHANGE, Universidade de Évora, Évora, Portugal
e-mail: nnunogui@uevora.pt; sergio.prats@uevora.pt

T. Pinto-Correia
MED-Mediterranean Institute of Agriculture, Environment and Development & Laboratório Associado CHANGE, Universidade de Évora, Évora, Portugal

Departamento de Paisagem, Ambiente e Ordenamento, Universidade de Évora, Évora, Portugal
e-mail: mtpc@uevora.pt

J. Muñoz-Rojas, R. García-Ruiz (eds.), *The Olive Landscapes of the Mediterranean*, Landscape Series 36,
https://doi.org/10.1007/978-3-031-57956-1_28

Keywords Soil capability for agriculture · Landscape simplification · Climate change · Sustainable intensification · Olive groves · Alentejo

28.1 Introduction

Alentejo, which covers 1/3 of Portuguese mainland, is the Portuguese region where olive groves have expanded more substantially over the past 30 years (Silveira et al., 2018), with 17.5% in surface growth. Circa 48% of agricultural land and 96% of the land devoted to fruit tree growing in Alentejo are nowadays covered by olive groves, summarizing approximately circa 1.35 million olive trees (Silveira et al., 2018). Vegetative growth in the region starts around February–March and ceases in October–November, while flowering occurs in April–June, and fruit maturation is observed from September to December, in most cases. With temperatures sufficiently mild, and thus not acting as a limiting factor, and water availability having proven as a historical deficiency, over 52% of the Portuguese olive groves can now be found in the Alentejo, largely triggered by the recent process of intensification, technological infrastructure implementation and expansion linked to the Alqueva water reservoir, which is located in the eastern part of the region and which is the largest artificial water body in Europe.

28.2 Climate Conditions: Opportunities and Limitations

Mainland Portugal belongs to the temperate climate zone of the Northern Hemisphere, being located at the interface between the semi-humid oceanic (Csb) and semi-arid Mediterranean (Csa) climates, according with the Köppen-Geiger climate classification (Peel et al., 2007).

Regarding the Mediterranean climate where olive trees grow, across Portugal, it is characterized by average annual temperatures that can be lower than 10 °C in some mountainous areas of the northeast, while they can reach values of 18 °C in the southern plains of Baixo Alentejo. In general, a decreasing rainfall gradient exists from west to east and from north to south, with annual mean values ranging between 350 mm and over 2500 mm and with main records in the Mediterranean regions concentrated in the months between October and May.

The region of Alentejo, occupying approximately 30% of Continental Portugal, is fully embedded in a semi-arid Mediterranean (Csa) climate. The bioclimate of the Alentejo region is Mediterranean rain-stational oceanic (Monteiro-Henriques, 2010). In this region, olive groves are spread over four ombrotypes (lower dry, upper dry, lower subhumid, upper subhumid) and two thermotypes (upper thermomediterranean and upper mesomediterranean). Based on the data provided by Monteiro-Henriques (2010), in the area covered by olive groves in 2018, the mean compensated thermicity index is 346.53 (s.d. = 13.74), the annual ombrothermic index is 3.28 (s.d. = 0.52), the mean annual temperature of the coldest month is 9.35 °C (s.d. = 0.50 °C), and the mean annual temperature of the warmest month is 23.76 (s.d. = 0.63).

The rainfall gradients are somehow fractally distributed, and in Alentejo a sub-gradient between 1200 mm and 300 mm of mean annual rainfall can be found from west to east. However, the temporal distribution of rainfall shows higher inter-annual variability as showed by Carvalho and Lourenço (2014) for the regional capital of Évora, with a mean annual value of 680 mm but with records ranging between 320 and 1080 mm. The mean positive rainfall in the areas covered by olive groves in Alentejo is 625.71 mm (standard deviation = 82.00 mm).

Typical challenges related to climate change that increasingly affect olive yields on an annual basis may include milder than standard winters (Kjellström et al., 2007; Viceto et al., 2019); long dry periods (e.g. February to October; Costa et al., 2012); increased frequency, intensity and severity of droughts (Vicente-Serrano et al., 2014; Jenkins & Warren, 2015; Coll et al., 2017; Paulo et al., 2016); longer and earlier heat waves (Parente et al., 2018; Viceto et al., 2019); non-typical rainy conditions during the flowering season and increases in the occurrence of heavy precipitation (Costa et al., 2012; Espírito-Santo et al., 2014; Nunes et al., 2016); and, lastly, higher than expected frequency of periods with atmospheric instability (Viceto et al., 2017).

According to Valverde et al. (2015) permanent crops well adapted to the Mediterranean climate, such as olive trees and vineyards, show less variation in irrigation requirements in face of adverse climate change scenarios. The results obtained by Fraga et al. (2020) pointed out to the negative impacts of changes in the temperatures and rainfall records affecting olive yields in Alentejo in rainfed olive groves, with the adoption of irrigation becoming a key possible adaptation measure, although now largely brought into question, to maintain productivity. The eco-physiological response of olive trees to irrigation is largely positive (Palese et al., 2010; Jiménez-Herrera et al., 2019; García et al., 2020; Gonçalves et al., 2020), allowing the stabilization of its inter-annual productivity which is a marked characteristic of olive trees. However, the optimization of the irrigation schemes is variety dependent, given the differences in the phenology of olive cultivars, and these changes may produce effects not only in the productivity (*stricto* sensu) but also in flowering, fruiting, fruit ripening and also in fruit retention and detachment forces limiting mechanical harvest efficiency (Camposeo et al., 2013).

It is also relevant to consider that the availability and quality of water for irrigation (either through groundwater abstraction or from surface reservoirs) are already shifting as a result of climate change and also of the expansion and intensification of olive groves. This is especially problematic in a region that progressively increases the area of crops that are dependent on extra availability of water.

Thus, under the current intensification trends, water availability is a critical factor, and the growing dependence on this scarce resource raises the level of threat foreseen in the most common business-as-usual scenarios for the future (Costa, 2011).

Although this is partly tackled through the increasing technological control of key production factors, especially water, which is largely linked to the process of intensification, the future resilience against a likely scenario of climate change and desertification is increasingly raising doubts regarding the resilience of currently prevailing models of soil and water use (Fig. 28.1).

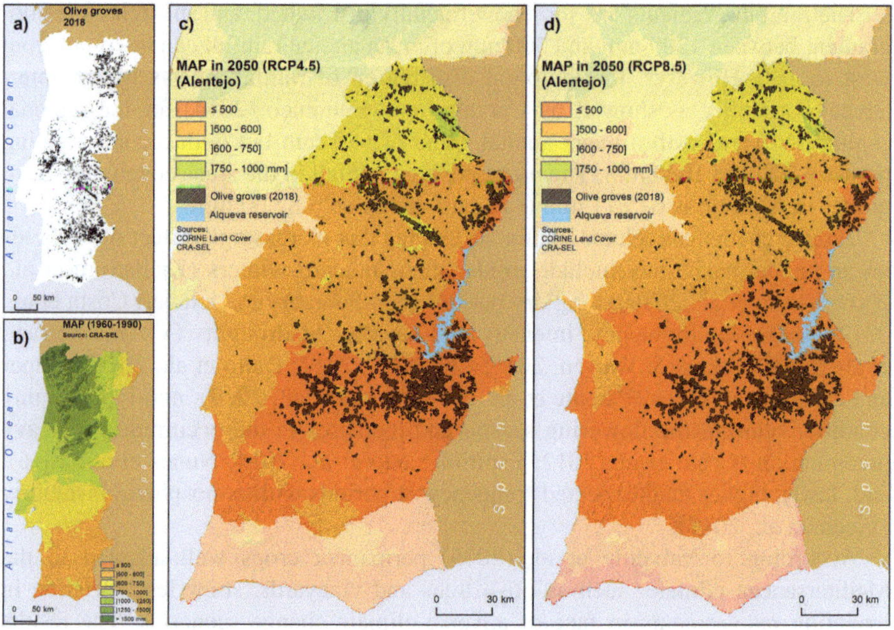

Fig. 28.1 Distribution of olive groves (**a**) and mean annual rainfall records between 1960 and 1990 (**b**) in mainland Portugal and estimated mean annual rainfall (MAP) in 2050 in Alentejo, considering two likely scenarios of climate change: RCP 4.5 (**c**) and RCP 8.5 (**d**). (Source: Climate data have been generated with the ClimateEU v4.63 software package, available at http://tinyurl.com/ClimateEU, based on methodology described by Wang et al., 2016)

28.3 Soil Types and Implications for Olive Grove Farming and Productivity

Soil classes in Portugal differ largely across biogeographic and ecological regions. According to the soil map of Portugal at 1:1.000.000 scale (SROA, 1971), Cambisols, Lithosols and Rankers prevail in the mountainous and humid regions of the north of the country, while Luvisols, Planosols and at time Podzols, Vertisols and also Lithosols are spread across the plains, hills and valleys of Ribatejo and Alentejo regions, in central and southern Portugal, respectively. Overall, most of the Portuguese soils are poor in organic matter. According to Carvalho and Lourenço (2014), 70.4% of the Portuguese soils have contents of less than 1% of organic matter. Low phosphorus content, a determining factor for plant nutrition, is a further limitation of Portuguese soils of relevance. Most of the samples analysed by Arrobas and Coutinho (2002) showed levels of total phosphorus below 1000 mg.kg^{-1}. Such low values imply the need to add fertilizers to turn them into fertile agricultural soils. Approximately 70.2% of the Portuguese soils have moderate cation exchange capacity (between 10 and 20 meq. per 100 g of soil), but only 4.2% exceed this level (Carvalho & Lourenço, 2014).

Most of the soils nationally are acid, with 82.9% having a pH lower than 5.5. Under such conditions, either aluminium (Foy, 1984) or manganese (Foy et al., 1978; Carvalho et al., 2015) toxicity problems may easily occur. Nearly 49.61% of the soils have low levels of permeability, with only 21.21% of the Portuguese soils being more than 50-cm thick, thus also being commonly shallow and rocky (about 45.22% of soils have medium texture, and 40.12% of them have coarse texture according to the spatial data provided by the EPIC-WebGIS data infrastructure). In such context, the most productive ones are generally restricted to the sedimentary terraces of the main rivers.

The main Portuguese olive cultivars are well adapted to such poor soils, with low soil organic matter, as is also the case of most soils around the Mediterranean, which are also largely characterized by also steeper slopes that too frequently result in great rates of soil erosion (Van Walleghem et al., 2011). The cultivation and productivity of olive groves in poorer soils are made possible by their significant genotypic variability, resulting in the traditional breeding of cultivars that are well adapted to the climatic challenges of each area. However, more recent olive tree cultivation, largely intensive and super-intensive, is concentrated in slightly sloping and mainly plain slopes, with richer soils (Fig. 28.2c).

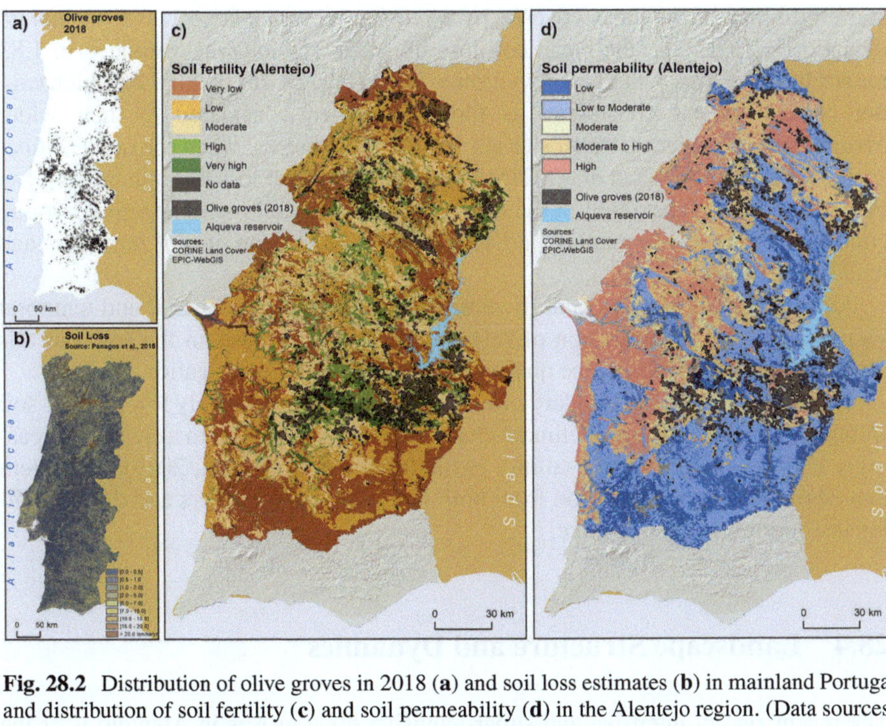

Fig. 28.2 Distribution of olive groves in 2018 (**a**) and soil loss estimates (**b**) in mainland Portugal and distribution of soil fertility (**c**) and soil permeability (**d**) in the Alentejo region. (Data sources: Panagos et al., 2015; CORINE Land Cover 2018; EPIC-Web GIS http://epic-webgis-portugal.isa.ulisboa.pt/; the soil classes were reclassified in soil fertility classes based on CEEM, 1996)

Table 28.1 Percentage of the area covered by olive groves in Alentejo in 2008 and 2018 (calculated upon spatial data from the CORINE LC database in 2008 and 2018), considered by the soil fertility classes defined in the classification by CEEM (1996)

Classes of soil fertility	Alentejo region	Olive groves in 1990		Olive groves in 2018	
		a	b	a	b
Very low	31.01	7.61	3.04	5.78	2.51
Low	30.21	17.76	1.24	14.37	1.30
Moderate	22.40	35.19	3.23	35.56	3.51
High	11.27	31.94	8.65	32.91	11.70
Very high	5.12	7.50	15.60	11.38	21.52

[a]Percentage in relation of total area covered by olive groves
[b]Percentage in relation of the total area covered by each soil fertility class

In 1990, 74.63% of olive cultivars were located on soils with moderate to very high fertility rates, with this percentage increasing to 79.85% when considering the impacts of the latest intensive and super-intensive olive groves (Table 28.1). As a result, olive groves currently occupy more than 20% of the most fertile soils in the Alentejo region (Table 28.1).

Among the soil risks associated with olive groves, there are two that should be highlighted in this context: water erosion and salinization/sodification. According to the datasets produced by Panagos et al. for European regions (2014, 2015), mean soil erodibility in the areas covered by olive groves in Alentejo is 0.33 (standard deviation = 1.63), and the mean soil loss by water erosion was estimated in 1.83 tonnes ha-1 per year (standard deviation = 1.12) (Fig. 28.2b), neither of which cannot be nonetheless considered as too high, when considered in their wider Mediterranean context (Panagos et al., 2015). Nonetheless, when looked at a finer scale (Rodríguez Sousa et al., 2021), the potential for soil loss through erosion can be considered as substantial in the olive groves of Alentejo with negative farming conditions, such as eliminating herbaceous cover vegetation, use of fertilizers and inadequate slopes.

Annual rainfall patterns, and its recent changes, affect the spatial and temporal variability in the accumulation of salinity excesses and crusts in the soil, but this process depends mainly on the quality of the water used for irrigation.

The current irrigation standard practices seem to have relatively low risks of soil salinization/sodification, but climate change scenarios point out to a clearly increasingly high level risks of soil salinity in the future (Ramos et al., 2019), a problem that demands a strong attention from both farmers, land managers and the key policy institutions and actors alike.

28.4 Landscape Structure and Dynamics

The recent rise in intensive and super-intensive olive groves in Alentejo is at the core driver of shifting regional agricultural landscapes (Silveira et al., 2018). Olive oil produced under intensive and super-intensive models is frequently a commodity and associated farming systems on which foreign and private financial capital investment, rapid intensification and strong technological inputs are strongly driving the rapid transformation of the traditional regional landscapes that were shaped by extensive and smaller-scale olive groves, Montado silvo-pastoral systems, vineyards and extensive cereals (Muñoz-Rojas & Pinto-Correia, 2019). This is a transition process that is undoubtedly affecting the structure of the supply chain and power relationships in the production of food commodities in this region of Portugal more strongly than in any other areas of the country (Silveira et al., 2018).

Specifically, the main concentration of intensive (and super-intensive) olive groves happens in the area leaning towards the northern edge of the lower Alentejo sub-region, around the district capital of Beja (Fig. 28.3). This is an area characterized by plain slopes with productive soils and extensively cultivated landscapes that were traditionally occupied by rainfed crops, mainly cereals (Pelúcio-Pimenta, 2014).

It is important to acknowledge that, in addition to the availability of water arising from the construction of the Alqueva reservoir (the dam was completed in 2002), other key bio-physical and ecological drivers exist that have clearly determined the magnitude and spatial distribution of the recent expansion of intensive and super-intensive olive groves in the region (Neves et al., 2013; Reis, 2014). These drivers include the location of lower slopes and related rich agricultural soils, the strict prohibition for new olive groves to substitute other traditional land-uses and landscape elements that are effectively protected by law (e.g. Holm Oaks and Cork

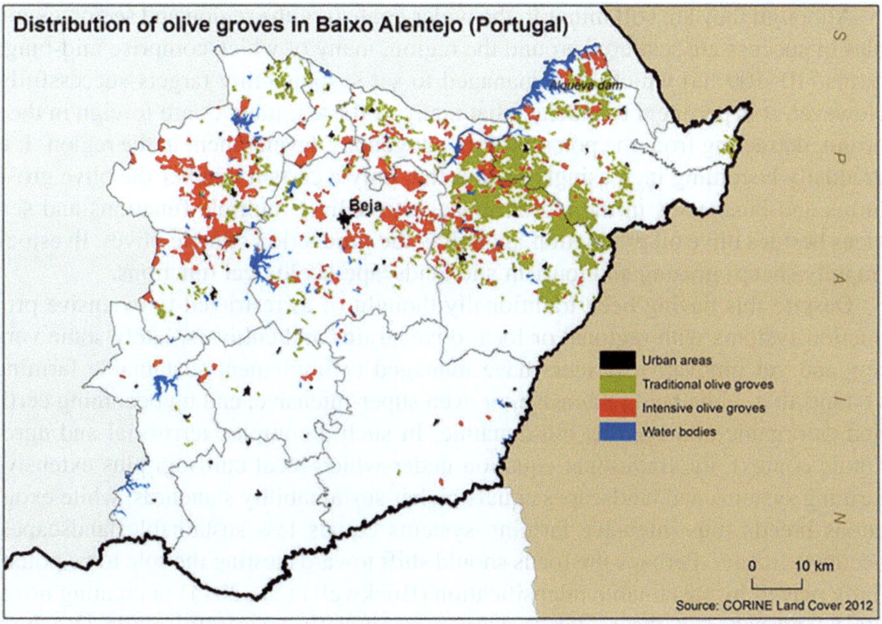

Fig. 28.3 Spatial distribution of olive groves, distinguishing between intensive and traditional olive groves in the lower Alentejo sub-region of Portugal. (Based on data by produced by co-author Nuno Guiomar)

Oaks) and the concentration of intensive and super-intensive groves on certain strains and cultivars of olive trees, such as Arbequina and Picual, both original from NE Spain.

Despite of the recent expansion of intensive and super-intensive of olive groves in the region, a complex landscape mosaic of traditional (extensive) and intensive olive groves is still found occupying large areas, especially in the northern edge of Alentejo, with poorer soils and more rugged terrain.

In addition, nature and landscape conservation designations are frequently indicated as key territorial factors influencing changes in farming systems, in this case by limiting the expansion of intensive olive groves. In Alentejo, these (e.g. the Parque Natural do Vale do Guadiana, to the east of the region and neighbouring Andalusia, in Spain) are located in areas with the poorest soils and bio-physical conditions for intensive agriculture. Consequently, this is also where the main opportunities may be identified for adding value to traditional extensive crops, such as olives, a target that could be achieved by implementing novel rural development strategies and action plans underpinned by what has been generically termed as "territorial agriculture" (Cairol et al., 2009).

We also find examples of alternative funding schemes that are intended for maintaining traditional olive groves and their associated landscapes. These are still largely underdeveloped in the region but include green capital investment funds of international origin, payment for ecosystem services schemes and innovation projects (e.g. SmartFarmers) funded under associations for the protection of rural heritage and landscape such as ADPM, operating in Mértola (http://adpm.pt/).

Although they are still minor in the wider context of the region and sector, examples of success are scattered around the region, many of which comprise mid-range farms (10–100 ha) which have managed to set sustainability targets successfully. However, it is pertinent to indicate that many of these initiatives are foreign in their origin, detracting from the potential for endogenous development in the region. It is gradually becoming increasingly clearer that only a certain number the olive grove farms and businesses in the region manage to deliver multiple functions and services besides olive oil production, including the production of table olives, livestock (mainly sheep) grazing and tourism and landscape-ecological functions.

Despite this having been traditionally thought of as restricted to extensive production systems with regional or local olive strains and cultivars, lately some very few and yet innovative farmers have managed to implement sustainable farming systems that while being intensive, or even super-intensive, end up becoming certified as organic and even as bio-dynamic. In such a complex territorial and agronomic context, the traditional equation under which local cultivars plus extensive farming systems and landscapes equals higher sustainability standards, while exogenous breeds plus intensive farming systems equals less sustainable landscapes, becomes fuzzier. Perhaps the focus should shift towards testing the role to be potentially played by sustainable intensification (Buckwell et al., 2014) in creating novel olive landscapes in the region that are economically successful while also help tackle the imminent threats of climate change, soil degradation, landscape simplification and biodiversity loss.

Acknowledgements This work was funded by the SUSTAINOLIVE project (Grant n° 1822; PRIMA EU Programme) and by FCT-Portugal (UIDB/05183/2020 as members of MED-Universidade de Évora and LA/P/0121/2020 as members of the CHANGE- Associated Lab-Universidade de Évora).

References

Arrobas, M., & Coutinho, J. (2002). Caracterização do fósforo em solos de Portugal. Revista das Ciências Agrárias, pp. 109–122.

Buckwell, A., Heissenhuber, A., & Blum, W. (2014). *Sustainable intensification of european agriculture*. A review sponsored by the RISE Foundation. IEEP, pp. 98. Available at: https://www.risefoundation.eu/images/files/2014/2014_%20SI_RISE_FULL_EN.pdf

Cairol, D., Coudel, E., Knickel, K., Caron, P., & Kroger, M. (2009). Multifunctionality of agriculture and rural areas as reflected in policies: The importance and relevance of the territorial view. *Journal of Environmental Policy and Planning, 11–4*, 269–289.

Camposeo, S., Vivaldi, G. A., & Gattullo, C. E. (2013). Ripening indices and harvesting times of different olive cultivars for continuous harvest. *Scientia Horticulturae, 151*, 1–10.

Carvalho, M., & Lourenço, E. (2014). Conservation agriculture–a Portuguese case study. *Journal of Agronomy and Crop Science, 200*(5), 317–324.

Carvalho, M., Goss, M., & Teixeira, D. (2015). Manganese toxicity in Portuguese Cambisols derived from granite: Causes, limitations of soil analyses and possible solutions. *Revista de Ciências Agrárias, 38*, 518–527.

CEEM. (1996). A região de Évora numa perspectiva de ecologia da paisagem. Final report of the STRIDE-Amb. 12 research Project, Universidade de Évora (Portugal).

Coll, J. R., Aguilar, E., & Ashcroft, L. (2017). Drought variability and change across the Iberian Peninsula. *Theoretical and Applied Climatology, 130*(3–4), 901–916.

Costa, A. C. (2011). Local patterns and trends of the standard precipitation index in southern Portugal (1940–1999). *Advances in Geosciences, 30*, 11–16.

Costa, A. C., Santos, J. A., & Pinto, J. G. (2012). Climate change scenarios for precipitation extremes in Portugal. *Theoretical and Applied Climatology, 108*(1–2), 217–234.

Espírito-Santo, F., Ramos, A. M., de Lima, M. I. P., & Trigo, R. M. (2014). Seasonal changes in daily precipitation extremes in mainland Portugal from 1941 to 2007. *Regional Environmental Change, 14*(5), 1765–1788.

Foy, C. D. (1984). Physiological effects of hydrogen, aluminum, and manganese toxicities in acid soils. *Soil Acidity and Liming, 2*, 57–97.

Foy, C. D., Chaney, R. T., & White, M. C. (1978). The physiology of metal toxicity in plants. *Annual Review of Plant Physiology, 29*(1), 511–566.

Fraga, H., Pinto, J. G., & Santos, J. A. (2020). Olive tree irrigation as a climate change adaptation measure in Alentejo, Portugal. *Agricultural Water Management, 237*, 106193.

García, J. M., Hueso, A., & Gómez-del-Campo, M. (2020). Deficit irrigation during the oil synthesis period a ects olive oil quality in high-density orchards (cv. Arbequina). *Agricultural Water Management, 230*, 105858.

Gonçalves, A., Silva, E., Brito, C., Martins, S., Pinto, L., Dinis, L. T., Luzio, A., Martins-Gomes, C., Fernandes-Silva, A., Ribeiro, C., et al. (2020). Olive tree physiology and chemical composition of fruits are modulated by different deficit irrigation strategies. *Journal of the Science of Food and Agriculture, 100*, 682–694.

Jenkins, K., & Warren, R. (2015). Quantifying the impact of climate change on drought regimes using the standardised precipitation index. *Theoretical and Applied Climatology, 120*(1–2), 41–54.

Jiménez-Herrera, R., Pacheco-López, B., & Peragón, J. (2019). Water stress. Irrigation and concentrations of pentacyclic triterpenes and phenols in Olea europaea L Cv. picual olive trees. *Antioxidants, 8*, 294.

Kjellström, E., Bärring, L., Jacob, D., Jones, R., Lenderink, G., & Schär, C. (2007). Modelling daily temperature extremes: Recent climate and future changes over Europe. *Climatic Change, 81*(1), 249–265.

Monteiro-Henriques, T. (2010). Landscape and phytosociology of the Paiva River's hydrographical basin. PhD dissertation, Instituto Superior de Agronomia, Universidade Técnica de Lisboa.

Muñoz-Rojas, J., & Pinto-Correia, T. (2019). Deliverable 2.2. SUFISA Portuguese National Report, pp. 203. Available at: https://www.sufisa.eu/wp-content/uploads/2018/09/D_2.2-Portugal-National-Report.pdf

Neves, B., Miranda Pires, I., & Roxo, M. J. (2013). Culturas intensivas e superintensivas e a susceptibilidade à Desertificação: o caso do olival no Alentejo. IX Congresso da Geografia Portuguesa, Évora, 28–30 November 2013. Seminário do olival de regadio-IRIRCAMPO.

Nunes, A. N., Lourenço, L., Vieira, A., & Bento-Gonçalves, A. (2016). Precipitation and erosivity in southern Portugal: Seasonal variability and trends (1950–2008). *Land Degradation & Development, 27*(2), 211–222.

Palese, A. M., Nuzzo, V., Favati, F., Pietrafesa, A., Celano, G., & Xiloyannis, C. (2010). Effects of water deficit on the vegetative response. Yield and oil quality of olive trees (Olea europaea L. cv Coratina) grown under intensive cultivation. *Scientia Horticulturae, 125*, 222–229.

Panagos, P., Meusburger, K., Ballabio, C., Borrelli, P., & Alewell, C. (2014). Soil erodibility in Europe: A high-resolution dataset based on LUCAS. *Science of the Total Environment, 479*, 189–200.

Panagos, P., Borrelli, P., Poesen, J., Ballabio, C., Lugato, E., Meusburger, K., Montanarella, L., & Alewell, C. (2015). The new assessment of soil loss by water erosion in Europe. *Environmental Science & Policy, 54*, 438–447.

Parente, J., Pereira, M. G., Amraoui, M., & Fischer, E. M. (2018). Heat waves in Portugal: Current regime, changes in future climate and impacts on extreme wildfires. *Science of the Total Environment, 631*, 534–549.

Paulo, A., Martins, D., & Pereira, L. S. (2016). Influence of precipitation changes on the SPI and related drought severity. An analysis using long-term data series. *Water Resources Management, 30*(15), 5737–5757.

Peel, M. C., Finlayson, B. L., & McMahon, T. A. (2007). Updated world map of the Köppen-Geiger climate classification.

Pelúcio Pimenta, S. (2014). A agricultura da região Alentejo nos ultimos 25 anos, e perspectivas do quadro da PAC post 2013. Tesina de Mestrado en Zootecnia, Universidade de Évora (Portugal), pp. 146.

Ramos, T. B., Darouich, H., Šimůnek, J., Gonçalves, M. C., & Martins, J. C. (2019). Soil salinization in very high-density olive orchards grown in southern Portugal: Current risks and possible trends. *Agricultural Water Management, 217*, 265–281.

Reis, P. (2014). O olival em Portugal. Dinâmicas, tecnologias e relação com o desenvolvimento rural. Instituto Nacional de Investigação Agrária e Veterinária, Portugal, pp. 40.

Rodríguez Sousa, A. A., Muñoz-Rojas, J., Pinto-Correia, T., Aguilera, P. A., Barandica, J. M., & Rescia, A. J. (2021). A comparative analysis of soil loss tolerance and productivity of the olive groves in the protected designation of origin (PDO) areas Norte Alentejano (Portugal) and Estepa (Andalusia, Spain). *Agronomy, 11*, 665. https://doi.org/10.3390/agronomy11040665

Silveira, A., Ferrão, J., Muñoz-Rojas, J., Pinto-Correia, T., Guimarães, M. H., & Schmidt, L. (2018). The sustainability of agricultural intensification in the early 21st century: Insights from the olive oil production in Alentejo (Southern Portugal). In A. Delicado, N. Domingos, & L. de Sousa (Eds.), *In changing societies: Legacies and challenges. Vol. iii. The diverse worlds of sustainability* (pp. 247–275). Imprensa de Ciências Sociais. https://doi.org/10.31447/ics9789726715054.10

SROA. (1971). Carta dos Solos de Portugal, 1:1.000.000. Serviço de Reconhecimento e Ordenamento Agrário, Secretaria de Estado da Agricultura, Lisboa (Portugal).

Valverde, P., Serralheiro, R., de Carvalho, M., Maia, R., Oliveira, B., & Ramos, V. (2015). Climate change impacts on irrigated agriculture in the Guadiana river basin (Portugal). *Agricultural Water Management, 152*, 17–30.

Van Walleghem, T., Infante-Amate, J., González de Molina, M., Soto Fernández, D., & Gómez, J. A. (2011). Quantifying the effect of historical soil management on soil erosion rates in Mediterranean olive orchards. *Agriculture, Ecosystems & Environment, 142*(3–4), 341–351.

Vicente-Serrano, S. M., Lopez-Moreno, J. I., Beguería, S., Lorenzo-Lacruz, J., Sanchez-Lorenzo, A., García-Ruiz, J. M., et al. (2014). Evidence of increasing drought severity caused by temperature rise in southern Europe. *Environmental Research Letters, 9*(4), 044001.

Viceto, C., Marta-Almeida, M., & Rocha, A. (2017). Future climate change of stability indices for the Iberian Peninsula. *International Journal of Climatology, 37*(12), 4390–4408.

Viceto, C., Cardoso Pereira, S., & Rocha, A. (2019). Climate change projections of extreme temperatures for the iberian peninsula. *Atmosphere, 10*(5), 229.

Wang, T., Hamann, A., Spittlehouse, D. L., & Carroll, C. (2016). Locally downscaled and spatially customizable climate data for historical and future periods for North America. *PLoS One, 11*, e0156720.

Chapter 29
Olive Growing Systems, Varieties and Olive Oil Production Systems

Raquel Garcia, Maria João Cabrita, and António Bento Dias

Abstract Portuguese olive groves suffered a great setback in the decades of 1950s–1960s of the twentieth century with the rural exodus. Portugal's entry into the EEC/EU in 1986 created conditions to the reconversion of existing olive groves, enabling the plantation of new orchards with 200–300 trees per hectare and the use of varieties from other regions.

At the beginning of the twenty-first century, the construction of the Alqueva Dam in the Alentejo region led to an increase in the area of irrigated olives groves. The hedge olive grove concept with continuous mechanized harvesting and the availability of water for irrigation contributed to the increase of this new growing system, reaching about 50,000 ha in two decades.

Furthermore, olive oil production systems are also the focus of this section. Indeed, olive oil production encompasses, previously, the collection of olives and their washing followed by an extraction process. This step, comprising crushing/malaxing with or without water and finally centrifuging/decanting processes, is typically feasible using two types of systems: a traditional discontinuous press process and a continuous three-phase centrifugal or two-phase centrifugal extraction systems. Thus, a descriptive summary of the different olive oil extraction systems will be presented covering their advantages and drawbacks. Also, recent approaches and innovations introduced in the olive oil extraction chains, in order to improve the

R. Garcia (✉) · M. J. Cabrita
MED—Mediterranean Institute for Agriculture, Environment and Development & CHANGE - Global Change and Sustainability Institute, Departamento de Fitotecnia, Escola de Ciências e Tecnologia, Universidade de Évora, Évora, Portugal
e-mail: raquelg@uevora.pt; mjbc@uevora.pt

A. B. Dias
MED—Mediterranean Institute for Agriculture, Environment and Development & CHANGE—Global Change and Sustainability Institute, Universidade de Évora, Pólo da Mitra, Universidade de Évora, Évora, Portugal

Departamento de Engenharia Rural Escola de Ciências e Tecnologia, Universidade de Évora, Évora, Portugal
e-mail: adias@uevora.pt

J. Muñoz-Rojas, R. García-Ruiz (eds.), *The Olive Landscapes of the Mediterranean*, Landscape Series 36,
https://doi.org/10.1007/978-3-031-57956-1_29

working efficiency of the production system and to obtain high-quality extra virgin olive oils (EVOOs), will be highlighted. In Portugal, a change in the sector of olive oil has undergone reflecting in the modernization of the mills to increase production capacity, offsetting the increase in production and increasing consumption.

Keywords Olive growing systems · Traditional rainfed groves · High-density groves · Hedge groves · Olive oil · Crushing · Malaxation · Centrifugation · Two-phase and three-phase systems

29.1 Portuguese Olive Growing Systems

In Portugal, the better climate regions for olive production are located in the eastern areas from the north until the south of the country, near the border with Spain (Fig. 29.1). According to INE (2019), the olive grove area in Portugal was 377,200 ha, occupying 197,600 ha in the Alentejo region.

Like other regions of the Mediterranean basin, in the Alentejo region exists traces of olive cultivation since antiquity, including olive trees with more than 2000 years (Bohm, 2013).

There has been an evolution of olive cultivation since antiquity. Initially, this would be based on olive trees scattered across the soil surface, that Barroso et al. (2013) named traditional scattered olive groves (Fig. 29.2).

In this olive growing system, the distance between trees is not the same, and in some cases the plots have rocky outcrops that make mechanization of the cultural

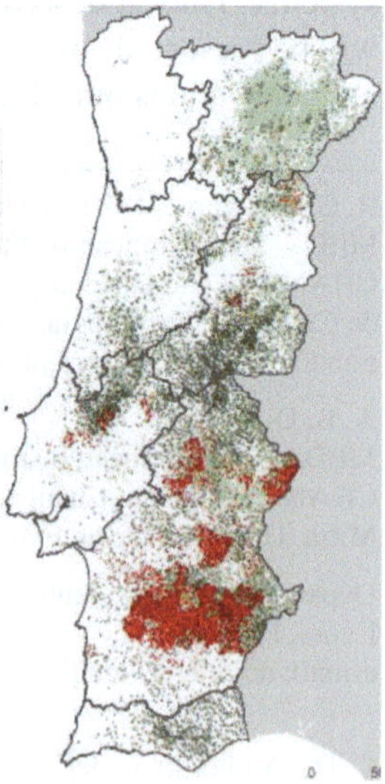

Fig. 29.1 Distribution of olive groves in Portugal. (Adapted by António B. Dias from INE, 2019)
Legend:
10 ha of olive grove with:
–40 to 60 tree per ha;
–61 to 100 tree per ha;
–101 to 300 tree per ha;
–301 to 700 tree per ha;
–701 to 1500 tree per ha;
– more than 1500 tree per ha;

Fig. 29.2 Traditional scattered rainfed olive orchard (Santarém region). (Photo by António B. Dias)

practices very difficult. This will have contributed to the abandonment, although it is still possible to find this kind of olive orchards. From this kind of olive orchard there was an evolution to the traditional aligned olive orchard (Fig. 29.3) with trees distributed on the ground according to a certain geometry.

There are olive groves planted in a square array as well as in an equilateral triangle (or quincunx). The more usual planting density was 100 trees per hectare although there were also less density plantations, from 70 trees per hectare (12 m × 12 m) to 50 trees tees per hectare (14 m × 14 m). This low density, according to Figueira (2010), allowed the association with autumn-winter cereals (Fig. 29.3) and was due to the low and irregular rainfall characteristic of the so called Margem Esquerda do Guadiana—*Left Bank Guadiana river* (Moura and Serpa municipalities) (Fig. 29.4).

The traditional aligned olive orchard was rainfed and the variety most used was Galega vulgar variety. Galvão (1952) refers that in the area of the "Left Bank Guadiana river" in addition to Galega, the most used varieties were Cordovil of Serpa and Verdeal Alentejana. Varieties like Bical, Maçanilha, Carrasquenha, Galego grado, Gama and Ocal could appear with some trees in the orchards (Galvão, 1952).

According to Figueira (2010), based in Galvão (1952) in the area of Elvas to Campo Maior, in addition to Galega vulgar, varieties such as Conserva of Elvas,

Fig. 29.3 Traditional aligned rainfed olive orchard (Elvas region). (Photo by António B. Dias)

Carrasquenha and Blanqueta of Elvas exist. This may be associated with the tradition of this area in the production of table olives, which normally requires more graded varieties. The installation in Elvas at the end of the 1950s of the twentieth century of the Oliviculture Station has also contributed to a greater diversity of the varieties used.

According to Figueira (2010), from the 1960s onwards, there was a setback in the area and productivity of the olive groves associated with the rural exodus that increased since that date.

In 1979, the area of olive orchards in the Alentejo region (INE, 1979 *cit. in* Figueira, 2010) was about 158,000 ha of traditional rainfed olive orchards.

Portugal's entry into the EEC/EU in 1986 created conditions for the intensification of olive groves. Several programs were implemented, which initially included:

– Grubbing up of olive groves located in less apt soils.
– Reconversion of existing olive groves, using densification in low-density olive groves.
– Grafting for varietal reconversion and plating of new olive groves.

These measures were decisive to change the slump in which the sector found itself, leading to the appearance of new olive groves with more graded olive varieties and with higher planting density.

Fig. 29.4 Traditional aligned rainfed olive orchard in association with cereals (Left Bank Guadiana river). (Photo by António B. Dias)

High-density olive groves (Fig. 29.5) were planted, initially with about 200 trees per hectare (7 m × 7 m) that evolved in following programmes to higher densities, from 286 trees per hectare (7 m × 5 m) to 400 trees per hectare (7 m × 3.5 m).

This increase in planting density was associated with the authorization to install drip irrigation systems in these new orchards.

In terms of varieties, the following ones should be used (Figueira, 2010): Blanqueta of Elvas, Conserva of Elvas, Redondil, Carrasquenha, Verdeal Alentejana and Cordovil of Serpa, but other varieties could also be used with favourable recommendation of the Ministry of Agriculture.

This allows the use of varieties from other regions, namely, the varieties Picual (Spain, Andalusia) and Cobrançosa (Trás-os-Montes, North Portugal).

During this phase of change in the olive sector, to ensure the typicality of the olive oils produced in the Alentejo region (Fig. 29.6) the following Protected Denomination of Origins (PDOs) were created: "PDO Moura olive oils", "PDO North Alentejo olive oils" and "PDO Alentejo interior olive oils".

At the beginning of the twenty-first century, super high-density or hedge olive groves (Fig. 29.7) appeared with an array of 3.75 m × 1.35 m (1975 trees per hectare) and using the Catalan Arbequina variety. The fully mechanized harvesting and the achievement of high production from the third year of planting will have been decisive for the rapid acceptance of this system.

Fig. 29.5 High-density irrigated olive orchard (Alqueva dam region). (Photo by António B. Dias)

This planting system requires dwarf varieties with early bearing, and as a consequence the most used varieties are Arbequina and Arbosana. There are also plantations of Koroneiki and Sikitita (variety obtained by the Spanish breeding program of University of Córdoba and IFAPA). The large Spanish nurseries have more olive varieties in their catalogues, although their use is limited.

In Portugal, some experimental fields were installed to evaluate the behaviour of some Portuguese varieties in this planting system. The results showed lower production levels of the Portuguese varieties used than the ones registered by Arbequina variety (Dias et al., 2018).

Nowadays the three types of olive groves coexist with the traditional olive groves located in areas where irrigation is not possible. In irrigated areas, there is great pressure to replace high-density olive orchards by hedge (super high-density) olive groves due to the fully mechanized harvesting of the latter.

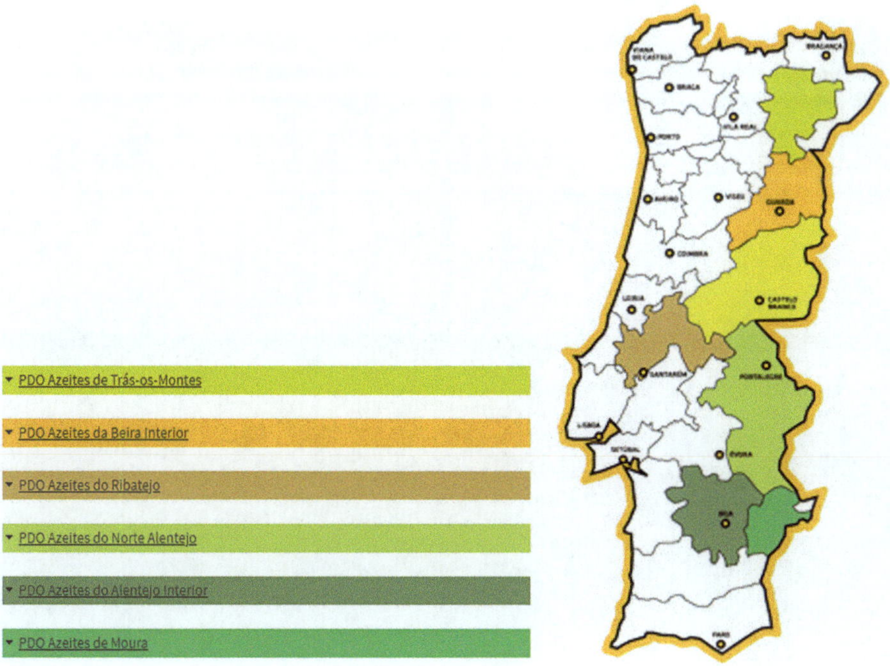

Fig. 29.6 Portuguese protected denomination of olive oils according to own interpretation by the author of this chapter, Prof. António Dias

29.2 Olive Oil Production Systems

Virgin olive oils (VOOs) are obtained from the fruit of the olive tree by mechanical or other physical means, being allowed only in this process the performance of treatments like washing, decantation, centrifugation or filtration (Regulation EU no. 1308/2013).

The production of VOO initiates with the olive's harvesting. Commonly, the appropriate harvesting time is a compromise between harvesting efficiency and final oil quality. After collecting olives, they are washed to remove dirt, leaves and twigs. The time between harvest and processing should be less than 24 h in order to avoid significant changes in the oil sensory profile of healthy and optimally ripe olives and not to compromise the highest-grade olive oils. The traditional olive oil processing is initiated by the crush of the of olives affording a paste, as depicted in Kapellakis et al. (2008). The purpose of the crushing is to facilitate the release of the oil from the vacuoles of mesocarp cells in the olive fruit. In the oldest mills, large granite stones were used to crush olives constituting the traditional olive oil processing.

Further, the malaxation process occurs involving the mixing of the paste, normally during 20–45 min in order to allow small oil droplets to combine into bigger ones. During this step, the olive enzymes produce aromas and flavours, playing a

Fig. 29.7 Hedge or super high-density irrigated olive orchard. First hedge olive orchard planted in Portugal (Santarém region). (Photo by António B. Dias)

crucial role on the development of several attributes that contribute to the organoleptic properties of VOOs. In fact, malaxation is a key step in the whole olive oil extraction process since, if well conducted, it affords optimal oil extraction and good values of antioxidants and flavours in VOO. Although longer mixing times enhance yield, it could result in an increase of oxidation processes, decreasing shelf life and quality of olive oils. In the traditional olive oil processing, the next step encompasses the spread of the paste on fibre disks, which are stacked on top of each other and then pressed in a hydraulic press. In this procedure, pressure is applied to the disks, compacting the solid phase and percolating the liquid phases (oil and vegetation water).

Vegetation water is an aqueous phase coming basically from the water content of the fruit, which is a water-brown watery residue liquid with a high organic matter content. To increase the speed of percolation, water can be used to run down the sides of the discs. Next, olive oil is separated by decantation or centrifugation (Kapellakis et al. 2008). Basically, the traditional VOO extraction method consists of three main processes: crushing, malaxation and centrifugation. The traditional olive oil production system is a "discontinuous" system due to the necessity to proceed according to "loads" or sequential pressing cycles (Yuksel Aydar, 2019; Kapellakis et al. 2008).

Traditional olive oil processing shows an adequate grinding of the olives since the ancient grindstones enable to break up the drupe's pulp while only slightly touching the nut and the skin, hampering the release of oxidation enzymes present in these organs. Other advantages could be encountered comprising the reduced added water that minimizes the decrease of polyphenol contents from olive oil, which is highly advantageous owing to the benefits of their consumption for human health. Moreover, the production of a residue called pomace during the olive oil processing, which possess a low amount of water, enables its easier management. However, some drawbacks could be ascribed to this traditional process, like difficulties on cleanness, the non-continuous feature of the overall process with waiting periods that expose the olive paste to oxygen and light, the requirement of laborious manual procedures and a relatively longer production time from harvest until pressing operation. Owing to all those limitations, the traditional extraction system has been gradually falling out of use.

In the last years, improvements on olive oil processing have occurred, prompted by the need to develop more efficient, continuous and optimized extraction processes. A modern olive mill based on centrifugation process is well depicted in Kapellakis et al. (2008).

Nowadays, the implementation of modern olive mills that are semi- or fully automated progressively substituted the old crushers made of granite crushers by metal crushers. They are composed by a stainless steel body and crusher with rotation at high speed. The efficiency of the crushing step is crucial to guarantee the taste and aroma of olive oil and the yield of the extraction process. Modern milling is very gently hampering the overheating of the paste, which is highly advantageous, since it minimizes the tendency to oxidation. Indeed, at any stage of the oil processing, the temperature must not exceed 27 °C.

In modern olive oil processing, the use of metal crushers could contribute to lower yield owing to the formation of emulsions, but it could be overcome by malaxation operation since it promotes the coalescence of droplets produced by the crushing of the olives. Indeed, the malaxation process could be considered as the bottleneck step in the overall extraction process hindering the development of a full and continuous process. Next, olive paste malaxation occurs by means of stirring the olive paste slowly in malaxers. This equipment has two or three cylindrical vats in tandem possessing each one a rotating helix with several wings, being the vats made of stainless steel with double walls for circulation of heated water. The paste is mixed at slow speeds during 20–75 min, being the mixing time dependent of paste characteristics. Modern malaxers have a horizontal design and are also equipped with an atmosphere-controlled system by means of an inert gas which reduces the oxidation, producing higher-quality VOOs. Indeed, the variables of malaxation step—temperature and malaxation time—are automatically controlled and are correlated with extraction yields (Jolayemi et al., 2016). As abovementioned, rising the temperature of the paste and the mixing time enhances the yield. Nevertheless, the temperature of the paste should not be higher than 27 °C in order to avoid degradations, namely, changes in the oil colour, increases of acidity and alterations in the profile of volatile compounds which promotes the appearance of negative sensory attributes.

In fact, the combination of an inert atmosphere with temperature below 30 °C during the malaxation step is of vital relevance, affording VOOs with highest sensory and nutritional quality. Further, the olive paste constituted mainly by a liquid fraction—olive oil and olive mill wastewater (OMWW)—and a solid fraction will be subjected to an extraction process, frequently based on centrifugation (Fig. 29.8).

In fact, this is the most used operation in modern olive oil mills owing to the higher quality of the VOOs produced and moreover the lower production costs associated to the process. It encompasses a separation of olive oil from the other constituents (water and solids) and represents a continuous process in which the separation is carried out inside a decanter. In this structure occurs an efficient separation between the solid fraction (olive pomace) to one end of the decanter centrifuge, while the liquid phase (olive oil and water) is moved to the other end.

In this procedure, the paste is transferred from the bottom of the malaxing vats into the decanter centrifuge using a pump, being added lukewarm water to increase the fluidity of the paste, and further pumped into three-phase centrifuge enabling the separation of the liquid and solid phases. The volume of added water varies being dependent of the olive cultivar and the physical characteristics of the harvested olives. Despite the enhancement in olive oil yield, the increased production of OMWW with low biodegradation potential represents a negative environmental impact.

Additionally, three-phase decanting results in loss of olive oil polyphenolics due to the higher quantity of water used (Fig. 29.8). Aiming to circumvent these disadvantages, several technological improvements lead to the introduction in modern olive mills of a separation system called as two-phase decanters that do not need the addition of water, being the olive paste separated in two phases—olive oil and wet pomace—as illustrated in Fig. 29.8 (Jiménez Marquez et al., 1995). In this case, the waste produced increases, but the loss of phenolic compounds is minimized with a noticeable influence in the bitterness perception of the VOOs produced with this system (Kalogeropoulos et al., 2014; Caponio et al., 2014).

A final centrifugation process is mandatory in order to remove vegetation water and small solids from VOOs, which is carried out in vertical centrifuges (Guerrini et al., 2016; Giovacchino et al., 2001).

From the comparison between the three- and two-phase systems arises also the following considerations:

- The design of two-phase decanters is simpler than the three-phase decanters, influencing the purchase price of the two equipment.
- The oil yield obtained by means of two-phase system is relatively higher owing to the retention of the oil in the solid fraction.
- The processing capacity of the two-phase decanter is higher than that of the three-phase system and has a lesser cost of operation. The main olive processing method used in many countries of the Mediterranean area, including Portugal, is the continuous centrifugation system, called a two-phase system. Indeed, this

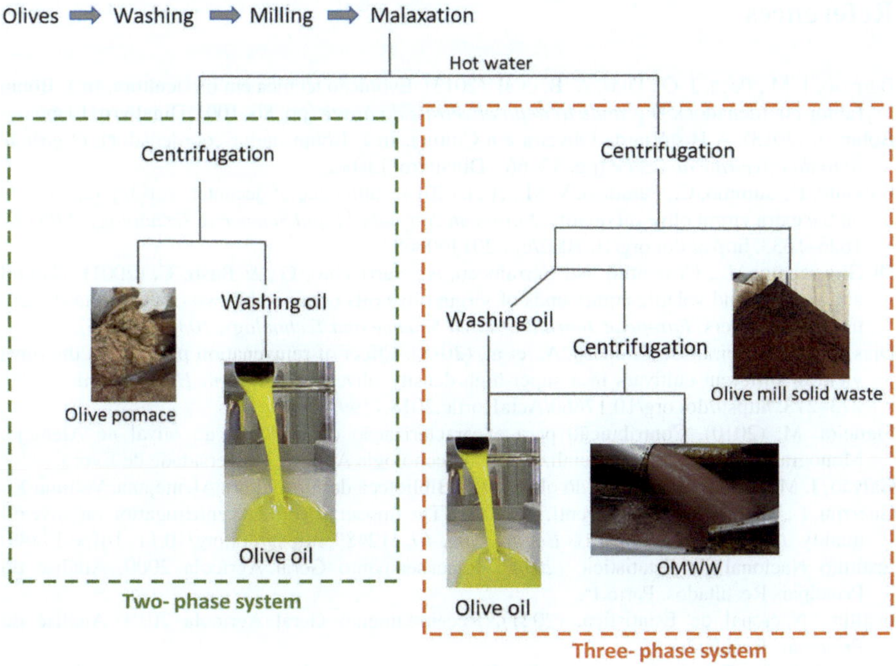

Fig. 29.8 Flow diagram of the continuous two- and three- phase olive oil production systems

technological system leads to a better quality of the olive oil, which is obtained with lower environmental impact. Also, other factors could be ascribed, like lower investment and operating costs, the latter due to reduced consumption of water and energy. In overall, the process is more sustainable from an environmental point of view, as well as from that of recovery and added value.

After processing, the VOO produced is stored in sealed stainless steel tanks and is frequently filtered prior to bottling. Most frequently, coloured glass is the choice for olive oil packaging owing to its properties, namely, blocks the UV light, is impermeable to oxygen and is inert. The bottling with inert gas extends the VOO shelf life and nowadays is a common practice.

In summary, olive oil production systems experienced great technological advances in recent years driven by the needs to improve automation into the olive oil production systems with minor environmental impacts. Likewise, the quality of the VOO in terms of organoleptic and nutritional properties and how those characteristics are influenced by the production systems are also crucial issues that should be present on the design of more advanced technologies to produce VOOs.

Acknowledgements This work was funded by FCT-Portugal (UIDB/05183/2020 as members of MED-Universidade de Évora and LA/P/0121/2020 as members of the CHANGE- Associated Lab-Universidade de Évora).

References

Barroso, J. M., Peça, J. O., Dias, A. B. et al. (2013). Evolução técnica em olivicultura. In J. Böhm (autor-coordenador), *O grande livro da oliveira e do azeite* (pp. 86–100). Dinalivro, Lisboa.

Bohm, J. (2013). A História da Oliveira em Cultura. In J. Böhm (autor-coordenador), *O grande livro da oliveira e do azeite* (pp. 52–66). Dinalivro, Lisboa.

Caponio, F., Summo, C., Paradiso, V. M., et al. (2014). Influence of decanter working parameters on the extra virgin olive oil quality. *European Journal of Lipid Science & Technology, 116*(12), 1626–1633. https://doi.org/10.1002/ejlt.201400068

Di Giovacchino, L., Costantini, N., Serraiocco, A., Surricchio, G., & Basti, C. (2001). Natural antioxidants and volatile compounds of virgin olive oils obtained by two or three-phases centrifugal decanters. *European Journal of Lipid Science and Technology, 103*, 279–285.

Dias, A. B., Figueira, M., Pinheiro, A., et al. (2018). Effect of rejuvenation pruning on the olive yield of different cultivars in a super-high-density olive orchard. *Acta Horticulturae, 1199*, 273–278. https://doi.org/10.17660/ActaHortic.2018.1199.42

Figueira, M. (2010). Contribuição para a caracterização da evolução do olival no Alentejo. Monografia do Curso de Especialização em Tecnologia Agrária, Universidade de Évora.

Galvão, J. M. M. (1952). Manual do olivicultor. Biblioteca de Agricultura Alentejana. Volume X.

Guerrini, L., Pantani, O. L., & Parenti, A. (2016). The impact of vertical centrifugation on olive oil quality. *Journal of Food Process Engineering, 40*, e12489. https://doi.org/10.1111/jfpe.12489

Instituto Nacional de Estatistica. (2011). Recenseamento Geral Agrícola 2009. Análise do Principais Resultados. Parte 1ª.

Instituto Nacional de Estatistica. (2021). Recenseamento Geral Agrícola 2019. Análise do Principais Resultados.

Jolayemi, O. S., Tokatli, F., & Ozen, B. (2016). Effects of malaxation temperature and harvest time on the chemical characteristics of olive oils. *Food Chemistry, 211*, 776–783. https://doi.org/10.1016/j.foodchem.2016.05.134

Kalogeropoulos, N., Kaliora, A. C., Artemiou, A., et al. (2014). Composition, volatile profiles and functional properties of virgin olive oils produced by two-phase vs three-phase centrifugal decanters. *LWT: Food Science and Technology, 58*(1), 272–279. https://doi.org/10.1016/j.lwt.2014.02.052

Kapellakis, I. E., Tsagarakis, K. P., & Crowther, J. C. (2008). Olive oil history, production and by-product management. *Reviews in Environmental Science and Biotechnology, 7*, 1–26. https://doi.org/10.1007/s11157-007-9120-9

Marquez, A. J., Fernández, M. H., & Ojeda, M. U. (1995). Extraction of virgin olive oil by two-phase continuous system. Influence of different variables of the process on certain parameters related to oil quality. *Grasas y Aceites, 46*(4–5), 299–303. https://doi.org/10.3989/gya.1995.v46.i4-5.942

Regulation EU n° 1308/2013.

Yüksel Aydar, A. (2019). Emerging extraction technologies in olive oil production. In *Technological innovation in the olive oil production chain*. IntechOpen https://doi.org/10.5772/intechopen.81390.

Chapter 30
Soil Management Practices, Fertilization, Pruning and Harvesting of Olive Groves

R. M. A. Machado and A. B. Dias

Abstract The cultural practices used in olive grove depend on cultivation method: traditional olive grove or modern olive grove. This chapter will analyse the practices of soil management, fertilization, pruning and harvesting methods in different cultivation methods. In traditional olive groves, producers tend to not invest significantly in the inputs as the groves do not provide great economic returns. In the majority of the cases, these producers only harvest the olives. On the other hand, modern (high density and hedge) olives groves drip irrigated with high productivities are made under intensive management making systematic use of pesticides, fertilizers and pruning. Currently, the producers are incorporating fertilization practices to minimize the impacts on soil, water and air. Biostimulant application is increasing, which contributes to improving soil fertility and plant resilience and reducing inorganic nutrients application and soil and water salinization. Lastly, it will be analysed the pruning and harvesting methods used. Manual pruning with chainsaw is the common practice used by growers. Mechanical pruning tends to be more applied in hedge groves due to the high work rate registered in the control of canopy height ("topping"). Trunk shaking and collecting canvas are the harvesting solutions used in traditional and high-density olive groves, while the over-the-row harvesting machine use is generalized in hedge groves.

R. M. A. Machado (✉)
MED—Mediterranean Institute for Agriculture, Environment and Development & CHANGE—Global Change and Sustainability Institute, Universidade de Évora, Pólo da Mitra, Universidade de Évora, Évora, Portugal

Departamento de Fitotecnia, Escola de Ciências e Tecnologia, Universidade de Évora, Évora, Portugal
e-mail: rmam@uevora.pt

A. B. Dias
MED—Mediterranean Institute for Agriculture, Environment and Development & CHANGE—Global Change and Sustainability Institute, Universidade de Évora, Pólo da Mitra, Universidade de Évora, Évora, Portugal

Departamento de Engenharia Rural Escola de Ciências e Tecnologia, Universidade de Évora, Évora, Portugal
e-mail: adias@uevora.pt

J. Muñoz-Rojas, R. García-Ruiz (eds.), *The Olive Landscapes of the Mediterranean*, Landscape Series 36,
https://doi.org/10.1007/978-3-031-57956-1_30

Keywords Fertigation · Foliar fertilization · Mechanical pruning · Harvest solutions · Sustainable intensification

30.1 Soil Management and Fertilization

30.1.1 Traditional Olive Orchards

In the traditional olive groves, the most common density is 100 trees/ha being these plots also used for other agricultural activities, such as grazing, or the inclusion of secondary crops (Duarte et al., 2005). For many years the weed control to prevent them from competing with the olive trees for moisture and nutrients was achieved by using disc harrows or cultivators, which contributed to soil erosion (Barroso et al., 2013). Currently weed control is made by sheep and/or by mechanical cutting, or it is not made. In the major part of traditional olive groves are not applied inorganic fertilizers, and some of them have been converted to organic farming to increase the income due to organic production subsidy. Overall, when fertilization is made, it consists of the application of complex fertilizers (e.g. 20–10–10, 20–20–20, etc.), broadcast on the soil, during the winter, since the movement of nutrients to tree roots is highly dependent of the rainfall events. Some producers, due to a high diffusion by the commercial agents and observation of the positive effects on plant growth, apply nutrients by foliar fertilization.

30.1.2 High-Density and Hedge Olives Groves

30.1.2.1 Soil Management

For crop establishment, it is usually carried out a deep tillage (to improve root growth, soil aeration, etc.), by subsoiler or by a deep ploughing. To incorporate fertilizers and materials to correct the soil pH is carried out a harrowing of the soil. After olive tree planting, the soil of the majority of olive orchards is not disturbed. Weed control is made by the combination of herbicide application and of the weed cutting (Fig. 30.1). In the olive tree rows, in the range corresponding to the canopy, weed control is done through the application of herbicide.

A tractor-mounted under-tree spray boom is used for herbicide application. The number of applications will depend on the climatic conditions of the year. In dry years an application with "residual" herbicide at the beginning of autumn and a total herbicide at the beginning of spring are enough. Weed control in the inter-row is made with a tractor-mounted rotary shredder. Cover cropping is increasing in the modern olive groves. Currently, there is an effort to improve the quality of the herbaceous vegetation, sowing mixtures of legumes and grasses. Cover cropping accompanied by no-tillage practices reduces the negative impacts of the use of

Fig. 30.1 Cover cropping and herbicide as a soil management practice. (Photo by António Bento Dias)

herbicides and increases soil fertility, carbon sequestration and biodiversity of the olive grove ecosystem (e.g. increases soil microbial activity and diversity, etc.), among others.

30.1.2.2 Fertilization

The modern olive groves are characterized by high productivity due to irrigation (drip irrigated: a single drip line or one drip line on either side of the crop row), high tree density and use of high yielding cultivars and early entry into production. The increase in productivity causes a higher demand for nutrients that are supplied through an annual fertilization programme that usually takes into account soil and water analyses, the expected yield, nutrient uptake by 1000 kg of fruit, seasonal nutrient requirements of olive trees, foliar analyses, results of fertilizer experiments and nutrient recycling. Soil and water analyses are carried out prior to crop establishment and during the life of the orchard.

Soil analysis is useful for diagnosing problems due to salts, pH and nutrient imbalances and for assessing soil nutrient availability. In many cases, nutrient soil availability is not considered in the annual fertilization plan. Irrigation water analyses are useful for diagnosing problems (e.g. high salinity and HCO_3- content, etc.) and for assessing nutrient content. When the concentrations of nutrients (e.g., NO_3-, Ca^{2+}, etc.) in the irrigation water are significant, they are deducted from the amounts

Table 30.1 Critical nutrient levels (dry weight basis) to July olive leaf samples (Fernández-Escobar, 2004)

Element	Deficient	Adequate	Toxic
N (%)	1.4	1.5–2	
P (%)	0.05	0.1–0.3	
K (%)	0.4	>0,8	
Ca (%)	0.3	>1	
Mg (%)	0.08	>0.1	
Mn (ppm)		>20	
Zn (ppm)		>10	
Cu (ppm)		>4	
B (ppm)	14	19–150	185
Na (%)			>0.02
Cl (%)			>0.5

to be applied during fertigation. In general, the nutrient uptake by 1000 kg of olives is around 15, 4, 25 and 3 kg of N, P_2O_5, K_2O and MgO, respectively.

Foliar analysis provides an indication of tree nutritional status and is useful for detecting toxicities. The nutritional status of a plant is assessed by comparing the nutrient concentration in the plant tissue sample to a pre-established critical range, below which nutrients are considered deficient, and above which they are considered excessive (Table 30.1). There are two recommended sampling times for collecting leaf samples: in July (during endocarp hardening) and in the winter. However, the nutritional status of trees is typically evaluated only in July. The critical or sufficiency ranges of nutrient concentrations for olive leaf samples collected in July are presented in Table 30.1.

The nutrients to satisfy the crop nutritional requirements are applied mainly through the fertigation, complemented by foliar fertilization. In some cases, during the end of autumn or winter, organic substances and/or mineral fertilizers are broadcast on the soil or localized in bands in depth at the side of the tree lines. The organic matter from pruning by-products that are left on the soil inter-rows and from cover crops in addition to improving the soil is also a source of N and other nutrients (difficult of quantifying).

Fertigation

Fertigation increases the water and nutrient use efficiency, nutrient availability and the yield (Machado & Serralheiro, 2020), and it may contribute to reduce alternate bearing. Fertigation allows the adjustment of nutrient supply to plant requirements and the timing of application, and the concentration of fertilizers is easily controlled. Fertilizers are generally fractionated according to the seasonal nutrient requirements of olive trees. During the irrigation period, the N, P and K nutrients are applied continuously, weekly or biweekly through a trickle *irrigation system.*

Usually, nitrogen is provided in a greater proportion in the spring–summer period (March–July), a time when there is a greater demand for this nutrient as a consequence of the great vegetative growth and the fruit set and initial fruit growth. Phosphorus is applied in practically equal amounts throughout the irrigation period. Potassium is provided in a greater proportion from the endocarp hardening until the autumn, to supply the high sink demand of the olives. In addition to those nutrients, the following are also applied by fertigation: Mg, Ca and B and more and more biostimulants (e.g. humic substances, seaweed extract, silicon, mycorrhizas, etc.), to increase soil fertility, productivity and resistance to biotic and abiotic factors.

Foliar Fertilization

Foliar fertilization is a common practice that contributes to rising yield and fruit quality, and it may reduce alternate bearing. Olive leaves absorb foliar fertilizers well, except iron. Foliar fertilization is the fastest and most effective method to supplement and enrich plant nutrition and to correct nutritional deficiencies. Overall, foliar fertilization is made at peak nutrient requirement periods, in the spring (during vegetative growth and of the inflorescences) and at the end of the summer (fruit growth period). However, it is also made in other states of growth. The application of B in spring and K at the end of the summer is widespread, but other macro- and micronutrients and biostimulants are also applied.

30.2 Pruning and Harvesting

30.2.1 Pruning Practices

30.2.1.1 Traditional Olive Groves

In the traditional olive orchards in the Alentejo region were used two pruning systems (Dias, 2006): the "high vase" and the "low bowl", commonly called "arreia".

The "high vase" pruning system (Galvão, 1952 *cit. in* Dias, 2006) was used in the Baixo Alentejo area, especially in the left bank of the Guadiana River. In this system the main branches were maintained, the secondary and tertiary branches inside the canopy were removed, and all the dead and bad positioned shoots inside the canopy were eliminated to increase air and light exposure (Fig. 30.2a). This system was used at a time with great labour availability but fell into disuse.

The "low bowl" pruning system was more diffused in the Alto Alentejo area. This system left the trees virtually devoid of fruitful branches in the pruning year (Fig. 30.2b). As a consequence, the trees respond to pruning emitting several vigorous branches which do not produce that year or the next. To restore balance between fruiting and vegetative growth, it is necessary to carry out an intermediate pruning between two successive "low bowl" pruning for thinning wood branches, which led

Fig. 30.2 "High vase" pruning system (**a**); "arena" in traditional olive growing at the AltoAlentejo area (**b**). (Photos by António Bento Dias)

to additional costs. Although Ramos et al. (1995) reported the convenience of making more rational pruning, the "low" bowl" system continues to be used extensively.

In the Alentejo region, specialized labour for manual pruning was becoming scarce, and the evaluation of mechanical pruning as an alternative to manual pruning began in 1997. It consists in the use of a disc saw pruning machine mounted on the front loader of an agricultural tractor (Peça et al., 2002). The cutting bar could be positioned at different height, with possible cuts at angles ranging from horizontal ("topping") (Fig. 30.3) to vertical ("hedging").

The evaluation mentioned above contributed for the definition of a pruning strategy (Dias, 2006) to reduce pruning costs and maintain olive yield (Fig. 30.4). It consists of performing a horizontal cut of the olive canopy ("topping") at intervals of time that can stretch up to 7 years.

After the second "topping", a manual complement should be performed within the next 2 years. Then, after a new period of 5–6 years, a "topping" intervention is performed, restarting the routine. Therefore, there is only a single manual intervention (complement) over a 15-year period (Dias et al., 2014).

Despite the advantages of mechanical pruning ("topping"), in traditional olive groves, growers still do the severe pruning "arreia". This may be due to:

- A lot of small olive groves.
- Olive groves planted in hillside areas that do not allow the use of pruning machines.
- Inadequate diffusion of the use of mechanical pruning, in particular the pruning strategy mentioned above.
- Little receptivity of olive growers to new production technologies, possibly due to old age.

Fig. 30.3 "Topping" at the uppermost part of olive canopy. (Photo by António Bento Dias)

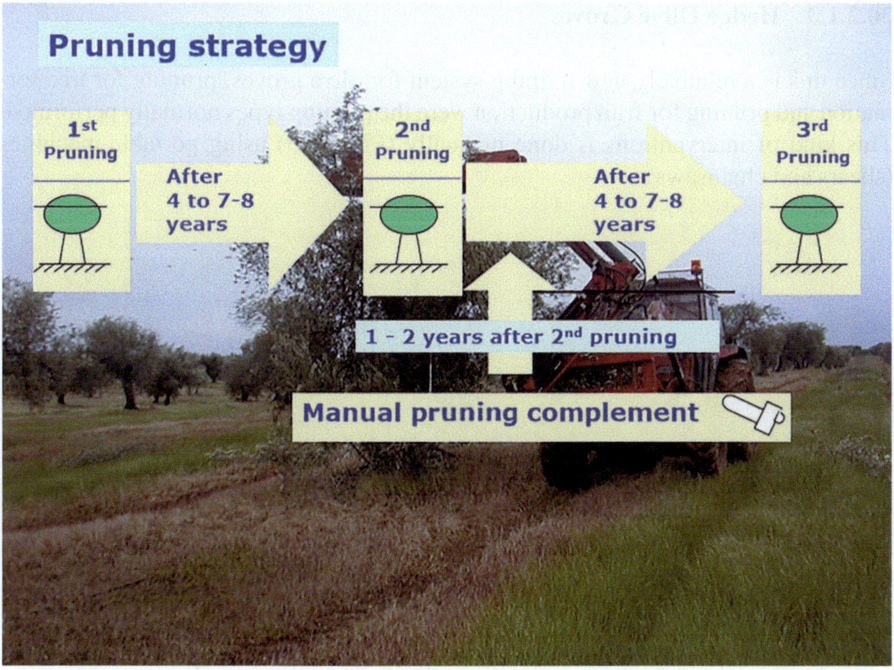

Fig. 30.4 Pruning strategy for traditional olive orchards. (Dias, 2006)

30.2.1.2 High-Density Olive Groves

In high-density olive groves, the type of pruning used is conditioned by its age (Pastor & Humanes, 1998; Gucci & Cantini, 2001). The following pruning interventions are for:

- *Tree formation* applied in young trees to ensure the formation of the tree structure; therefore, pruning interventions should be limited to correct the canopy and eliminate shoots poorly positioned to minimize the unproductive phase.
- *Fruit production* applied in orchards at full production ensuring the balance between fruiting and vegetative growth.
- *Renewal or rejuvenation pruning* applied to olive orchards with decreased vegetative growth and decreased yields; as a consequence pruning should be more intense.

The canopy volume of irrigated high-density olive orchards in full production could reach up to 15,000 m^3/ha (Pastor & Humanes, 1998), and, in consequence, the frequency of pruning intervention could be greater than in traditional olive orchards.

The knowledge obtained with the use of mechanical pruning in traditional olive orchards was also applied in high-density orchards. As expected, the obtained results allowed to reduce pruning costs and the time needed to prune, with maintenance of olive yield (Dias et al., 2012, 2018a, b). As pruning tests should be long-term experiments, evaluation of the use of mechanical pruning is still ongoing.

30.2.1.3 Hedge Olive Groves

Since this is a relatively new training system for olive groves, pruning for tree formation and pruning for fruit production were the pruning types normally performed. This kind of interventions is done manually (Fig. 30.5) using portable machines (shears and chainsaws).

Fig. 30.5 Manual pruning with a chainsaw in a hedge olive orchard. (Photo by António Bento Dias)

In the first hedge olive groves planted in Portugal, the objective was the formation of the trees with a central axis, which require the placement of a tutor, usually bamboo. In this phase pruning interventions are scarce, limiting to control the appearance of undesirable shoots in the trunk, in order to leave it without branches from the ground up to about 0.50 m height.

In the first years of olive production, the pruning intervention should be scarce and limited, to eliminate only badly inserted shoots or small branches. Since it is a training system that aims to form a hedge, its size must be adequate to the dimensions of the harvesting machine. As consequence, all woody branches that exceed the limits of the machine must be eliminated, both in height and in width. These pruning interventions are selective and are performed annually using manual portable machines (assisted manual pruning).

The need to prune every year, the high number of plants per hectare and the existing olive groves area mean that the labour that is needed to carry out this cultural practice is high.

The high work rate of the disc saw pruning machines has contributed to an increasing use to control of the hedge dimension, mainly in height, with "topping" in the uppermost part of the canopy (Fig. 30.6).

The "topping" of the hedge allows an increase of 50% in the work rate of manual pruning of the remaining part (side faces) in comparison to exclusive manual pruning (Dias et al., 2011).

Fig. 30.6 Summer "topping" in an olive hedge grove. (Photo by António Bento Dias)

Table 30.2 Graphic depiction and interpretation by A. Dias of the strategy for hedging proposed by Rius and Lacarte (2015), whereby half the area of each olive grove is treated on a half-year basis

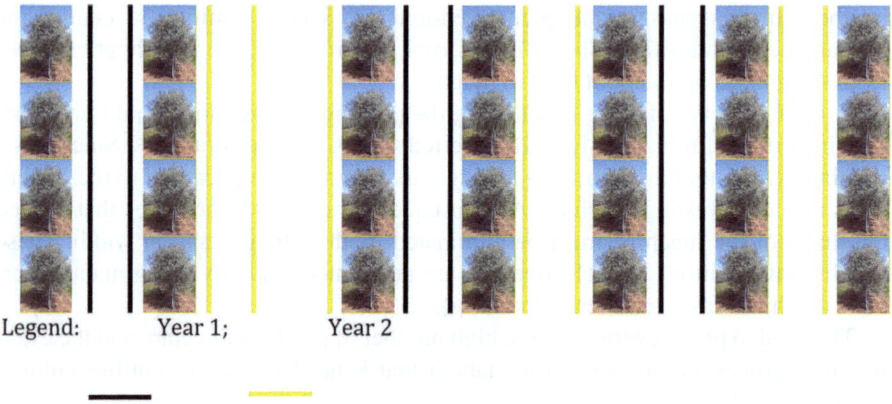

(Adapted by Dias, A.B. from Rius and Lacarte, 2015)

Mechanical pruning also standardizes the size of the hedge and makes it adequate it to the harvesting machine. Although "hedging" can contribute to reducing labour, a long-term evaluation of the influence on olive yield is needed in order to balance advantages and disadvantages.

It should be noted that in some companies with large olive grove areas, hedging is a common practice, following the strategy (Table 30.2) proposed by Rius and Lacarte (2015).

The necessity to reduce manual interventions in the hedge formation phase has contributed to the emergence of the "smart tree" concept.

The formation of the central axis is no longer privileged. The adoption of a disorganized but quite branched hedge seeking to achieve the desired canopy volume as soon as possible is preferred.

The hedge formation begins in the nursery with a horizontal cut at the top of the young plant.

In the field (Fig. 30.7), when plants reach a certain height, canopy cuts are made, either on top or in side faces, in order to control the growth in height and promote the branching inside the canopy.

It should also be noted that mechanical pruning ("topping" and "hedging" both sides of canopy) followed by a manual complement intervention (Dias et al., 2018a, b) can also be an alternative for the regeneration of SHD olive orchards with excessive canopy development by:

- Reducing tree density by half, after removing alternate tree rows.
- Cutting trees near the ground to encourage regrowth from which new trees will be formed.
- Removing all tree branches, keeping only the central leader (Tous et al., 2012).
- Cutting the central leader at 5–10 cm from the bottom (Hidalgo et al., 2012).

Fig. 30.7 Smart tree training system. (Photo by António Bento Dias)

The use of the mechanized pruning solution mentioned above could recover olive production in the second year after pruning. The knowledge obtained on several pruning trials made in different olive orchards shows that mechanical pruning could be used as the main solution for olive pruning, which requires manual pruning complement. The frequency and intensity of manual pruning complement need more experimental evaluation.

Although the experimental results show the advantage in the use of pruning machines, its use by olive growers is not widespread. Most of the pruning machines sold are intended for olive hedge groves (personal information obtained from industry).

30.2.2 Harvesting Solutions

30.2.2.1 Traditional Olive Orchards

Manual harvesting has been used for centuries. At the beginning of the twentieth century, olives were detached by hand from the tree by men using stairs and collected from the ground by women (Fig. 30.8).

Fig. 30.8 Manual olive harvesting at Herdade dos Machados, Moura (Alentejo, Portugal). The photograph is a personal property of António Bento Dias. (The author of the photograph is anonymous, and thus no copyright is implied for reproduction)

Fig. 30.9 Self-propelled trunk shaker imported from the USA in 1973. (Photo by António Bento Dias)

Later, the harvesting method changed: the detachment began to be done by men with wood poles, and the olives were collected in canvas manually placed under the tree canopy. This procedure is still used in small traditional olive groves.

The decrease in the availability of labour was crucial for the mechanization of olive harvesting. In Portugal, there were two key moments: the import from California, in 1973, of two self-propelled shakers by Eng. José Franco Falcão to be used in his own farm in the Alentejo region (Fig. 30.9) and the import from Italy in the decade of 1970s of a shaker with an inverted umbrella by Eng. Camilo Mendonça to be used in the Trás-os-Montes region (Barroso et al., 2013).

Nowadays in traditional olive orchards, growers use two kinds of harvesting solutions: tractor-mounted trunk shaker or small portable equipment.

In larger olive farms, olive harvesting is done by tractor-mounted trunk shakers complemented by manual beating will poles and canvas manually placed under the trees (Fig. 30.10a) to catch the olives.

An alternative harvesting solution with less labour requirements and cheaper cost (Peça et al., 2000) is the trunk shaker with inverted umbrella (Fig. 30.10b).

The diffusion of portable olive harvesting equipment (Fig. 30.11) enabled small olive growers to have solutions to improve olive detachment. This equipment could be divided into two groups: portable combs and branch shakers.

Fig. 30.10 Trunk shaking with complementary manual beating (**a**); tractor-mounted trunk shaker with inverted umbrella (**b**). (Photos by António Bento Dias)

Fig. 30.11 Olive harvesting with handheld machines. (Photo by António Bento Dias)

30.2.2.2 High-Density Groves

Olive harvest in intensive orchards is usually performed by a tractor-mounted trunk shaker and a canvas manually placed on the ground under the tree. The canvas is unloaded into a smaller storage canvas left on the ground which is loaded into a trailer by a rear-mounted hydraulically crane (Fig. 30.12b).

The inverted umbrellas linked to the trunk shaker have limited use since trees are very closely spaced to allow the umbrella to open (Peça et al., 2014).

The use of a winch-type canvas collector mounted in the front loader of a tractor (Fig. 30.12a) contributes to reducing the number of workers necessary to move the canvas.

In order to reduce the high level of strain put on workers, a mechanized rolling canvas prototype (Fig. 30.13) was developed (Peça et al., 2008). Although less demanding in labour strength, it did not make any improvement in working capacity, and it requires more tractors.

Today there is technology available for canopy shaking allowing continuous mechanical harvesting using the over-the-row principle. However, the over-the-row principle requires an adequacy of the canopy volume to the harvester dimension.

Large over-the-row olive harvesters (Fig. 30.14), mainly of South American and Australian design, are too heavy and expensive, hardly suitable to the difficult wet soil conditions encountered in the Mediterranean countries. The side-row concept is the obvious alternative (Peça et al., 2014).

Fig. 30.12 Winch-type canvas collector (**a**); trailer + rear-mounted hydraulically crane (**b**). (Photos by António Bento Dias)

Fig. 30.13 Rolling canvas prototype. (Photo by António Bento Dias)

Fig. 30.14 Large over-the-row harvester. (Photo by António Bento Dias)

Fig. 30.15 The side-row continuous canopy shaking fruit harvester. (Photo by António Bento Dias)

The SRCH or "side-row continuous canopy shaking fruit harvester" project (2009–2012) was set to develop a prototype (Fig. 30.15) to detach fruits from the tree branches and collect and transport the fruits to a temporary storage bag (Peça et al., 2014), which has been intensively tested ever since.

30.2.2.3 Hedge Groves

This concept of the olive grove is based on super high-density planting, training the trees as a hedge and harvesting done by machines based on the over-the-row principle.

The over-the-row harvesting machine (Fig. 30.16) detaches, collects and storages the olives temporarily. When the storage bins are full, it is required to discharge the fruit into a loader to make the transport from the field to the olive mill.

The use of these machines requires adequacy of hedge dimension to the machine, as mentioned in the pruning chapter.

Fig. 30.16 Over-the-row harvesting machine in a hedge grove. (photo by António Bento Dias)

Acknowledgements This work was funded by FCT-Portugal (UIDB/05183/2020 as members of MED-Universidade de Évora and LA/P/0121/2020 as members of the CHANGE- Associated Lab-Universidade de Évora).

References

Barroso, J. M., Peça, J. O., Dias, A. B., Pinheiro, A. C., & Santos, F. L. (2013). Evolução técnica em olivicultura. In: O Grande Livro da Oliveira e do Azeite. Ed. J. Bohm. Portugal Oleícola. Lisboa, Dinalivro.

Dias, António B. (2006). A mecanização da poda do olival. Contribuição da máquina de podar de discos. Tese de Doutoramento. Universidade de Évora.

Dias, A. B., Pinheiro, A., & Peça, J. O. (2011). Resultado preliminar da aplicação da poda mecânica num olival da variedade Arbequina conduzido em sebe. Actas Portuguesas de Horticultura, n° 14, Associação Portuguesa de Horticultura, pp. 65–73.

Dias, A. B., Falcão, J. M., Pinheiro, A., & Peça, J. O. (2012). A utilização da máquina de podar de discos num olival de 400 árvores por ha. Actas Portuguesas de Horticultura, n° 21, Associação Portuguesa de Horticultura, pp. 107–114.

Dias, A. B., Pinheiro, A., & Peça, J. O. (2014). Fifteen-year evaluation of the influence of mechanical pruning on olive yield. *Acta Horticulturae, 1057*, 335–340. https://doi.org/10.17660/ActaHortic.2014.1057.39

Dias, A. B., Figueira, M., Pinheiro, A., & Peça, J. O. (2018a). Effect of rejuvenation pruning on the olive yield of different cultivars in a super-high-density olive orchard. *Acta Horticulturae, 1199*, 273–278. https://doi.org/10.17660/ActaHortic.2018.1199.42

Dias, A. B., Pimentel, R. S., Pinheiro, A., & Peça, J. O. (2018b). Resultados da avaliação da poda com máquina de discos num olival intensivo da variedade Galega. Actas de Horticultura n° 73, Sociedad Española Ciencias Horticolas, pp. 194–201.

Duarte, F., Mansinho, M., Barreira, M., Jones, N., Lúcio, C., & Nunes, A. (2005). Overview of Olive Production Cost Analysis for Trás-os-Montes SMOPS. In L. Fleskens (ed.), "OLIVERO" Working Paper Series – N° 3 Overview of production costs for Sloping and Mountainous Olive Plantation Systems (SMOPS) under different circumstances (pp. 75–91). Wageningen: Luuk Fleskens (Ed.).

Fernández-Escobar. (2004). Fertilization. In El cultivo del olivo, D. Barranco, R. L. Fernández-Escobar (eds.), Rallo Ediciones Mundi-Prensa (Madrid), and Junta de Andalucía (Sevilla). 6th Edition.

Gucci, R., & Cantini, C. (2001). *Potatura e Forme di Allevamento dell'Olivo*. Edagricole—Edizioni Agricole.

Hidalgo, J., Vega, V., & Hidalgo, J. C. (2012). Experiencias sobre la renovación de plantaciones en seto en el cultivo del olivar. *Vida Rural, 341*, 42–48.

Machado, R. M. A., & Serralheiro, R. (2020). Salt stress alleviation through fertilization in fruit crops. In A. K. Srivastava & C. Hu (Eds.), *Fruit crops diagnosis and management of nutrient constraints* (pp. 502–505). Elsevier. https://doi.org/10.1016/B978-0-12-818732-6.00033-2

Pastor, M., & Humanes, J. (1998). *A poda del olivo – moderna olivicultura*, 3ª edición (corregida y actualizada), Editorial Agrícola Española S.A., Madrid.

Peça, J. O., Almeida, A., Pinheiro, A. C., Dias, A. B., Reynolds de Souza, D., Santos, L., & Lopes, J. (2000). Manual da colheita mecanizada de azeitona. "Projecto PAMAF 2072 – Sistemas de Colheita Mecânica de Azeitona", Edição Fundação Luís de Molina.

Peça, J. O., Dias, A. B., Pinheiro, A. C., Santos, L., Morais, N., Pereira, A. G., & Reynolds de Souza, D. (2002). Mechanical pruning of olive trees as an alternative to manual pruning. *Acta Horticulturae, 586*, 295–299. https://doi.org/10.17660/ActaHortic.2002.586.57

Peça, J. O., Almeida, A., Pinheiro, A. C., Dias, A. B., Santos, L., Lopes, J., Gomes, J. A., & Reynolds de Souza, D. (2008). Mechanical harvesting of 400 trees per hectare olive orchards based on a rolling canvas prototype. *Acta Horticulturae, 791*, 363–367. https://doi.org/10.17660/ActaHortic.2008.791.52

Peça, J. O., Dias, A. B., Pinheiro, A., Cardoso, V., Reynolds de Souza, D., & Falcão, J. M. (2014). Side-row continuous canopy shaking harvester for intensive olive orchards. *Acta Horticulturae, 1057*, 391–397. https://doi.org/10.17660/ActaHortic.2014.1057.48

Ramos, A. M. S., Santos, L. M. S. S., & Candeias, M. M. J. (1995). A poda da oliveira, *Agroforum*, N°s 8 e 9, Ano 4, 39–44.

Rius, X., & Lacarte, J. M. (2015). La revolución del Olivar. El cultivo en seto. 2ª Edicion, ampliada y revisada, COMOGRAFIC SA, Barcelona (Spain).

Tous, J., Romero, A., Hermoso, J. F., Msallem, M., & Larbi, A. (2012). Olive orchard design and mechanization: Present and future. In *VII international symposium on olive growing 1057* (pp. 231–246). https://doi.org/10.17660/ActaHortic.2014.1057.27.

Chapter 31
The Phytosanitary Protection of the Portuguese Olive Grove: Current and Future Perspectives

Fernando Trindade Rei

Abstract This chapter describes the situation of the olive phytosanitary protection in Portugal, one of the most important crops in the country. Having two main regions where the olive grove stands out, we proceed to the description of the main olive enemies associated with phytophagous arthropods, fungi and bacteria. The differences associated with each of the two most relevant olive regions are also described. The methodologies currently used to limit the main enemies of the culture are presented, as well as new alternative methods to chemical control, recently studied and developed in Portugal. The influence of predicted climate changes for the Iberian Peninsula is also discussed for the main pest (*Bactrocera oleae*) and olive disease (*Colletotrichum* sp.) control.

Keywords *Olea europaea* · Olive fly · Olive anthracnose · Integrated pest management (IPM) · Massive trapping · Dimethoate resistance · Microorganism symbiosis · Antimicrobial peptides (AMP) · Endophytic fungi · Climate change

31.1 Introduction

Although the olive culture is ubiquitous in all regions of the country, Portugal has two main olive groves regions: one located to the south, in the Alentejo region, comprising 50% of the total olive area (INE, 2019; EDIA, 2019), and the other in the interior north of the country, in the region of Trás-os-Montes. There are differences between these two regions, both in terms of olive farm structure and size. In the south, the medium to high-scale olive groves are predominant, where professional management is becoming the rule. This contrasts with the small farms with

F. T. Rei (✉)
Entomolab/MED – Mediterranean Institute for Agriculture, Environment and Development & CHANGE – Global Change and Sustainability Institute, Departamento de Fitotecnia, Escola de Ciências e Tecnologia, Universidade de Évora, Pólo da Mitra, Évora, Portugal
e-mail: frei@uevora.pt

J. Muñoz-Rojas, R. García-Ruiz (eds.), *The Olive Landscapes of the Mediterranean*, Landscape Series 36,
https://doi.org/10.1007/978-3-031-57956-1_31

familiar management that prevail in the northern region. With the availability of the irrigation perimeter associated to the Alqueva reservoir, the Alentejo region has been experiencing a high increase in intensive and super-intensive olive groves, in detriment of the traditional low-density rainfed orchards which have been prevalent in the past years (EDIA, 2018). These differences also correspond to differences in the management of pests and diseases usually present in the olive groves on these two regions, both in terms of technical knowledge and in terms of application of control measures. In particular, the use of pesticides has a higher intensity in the olive groves of the southern region, partly because of the high tree density and extensive surface areas.

31.2 Portuguese Olive Main Pests and Diseases

The main pests in the Portuguese olive grove are, in descending order of importance, the olive fruit fly *Bactrocera oleae*, the olive moth *Prays oleae* and the black scale *Saissetia oleae*. Other potential pests, like the olive-tree pearl *Palpita vitrealis*, the olive pyralid moth *Euzophera pinguis*, the olive bark beetle *Phloeotribus scarabaeoides*, the olive psyllid *Euphyllura olivina* and the olive thrips *Liothrips oleae*, have usually secondary or residual importance, being localized or occasional pests (Table 31.1) (Pereira et al., 2003; Torres, 2007; Torres & Gonçalves, 2013; EDIA, 2018). Nevertheless, this classification as secondary or residual presence could result from the pesticide applied to control the olive moth and olive fly, which have a broad spectrum of action.

Table 31.1 Pests and diseases of the main Portuguese olive regions

		Presence		
Group	Species	Main	Secondary	Residual
Pests	*Bactrocera oleae*	√		
	Prays oleae	√		
	Saissetia oleae	√		
	Euphyllura olivina		√	
	Palpita vitrealis		√	
	Euzophera pingüis		√	√
	Phloeotribus scarabaeoides			√
	Liothrips oleae			√
Diseases	*Colletotrichum* sp.	√		
	Spilocaea oleaginea	√		
	Pseudomonas savastanoi pv. *Savastanoi*		√	
	Pseudocercospora cladosporioides		√	
	Verticillium dahliae		√	
	Viruses		√	
	Camarosporium dalmaticum		√	√

Regarding diseases, olive anthracnose *Colletotrichum acutatum* and *Colletotrichum gloeosporioides*, olive peacock spot *Spilocaea oleagina*, verticillium wilt of olive *Verticillium dahliae*, olive tuberculosis *Pseudomonas savastanoi* pv *savastanoi*, *Cercospora* leaf spot *Pseudocercospora cladosporioides*, olive fruit rot *Camarosporium dalmaticum* and viruses are the most referred olive diseases in Portugal (Table 31.1). However, the olive anthracnose must be highlighted due to its impact on the quality of the olive oil produced (Torres, 2007).

For most of the described key and secondary pests and diseases, monitoring methodologies, economic threshold levels and control measures are officially indicated and framed within the scope of integrated pest management (IPM) programmes, which must be followed by olive growers (DGADR, 2010).

Although the main pests and diseases are present in the two main Portuguese olive-growing regions, there are differences in the incidence of the olive fruit fly between these two regions. Olive fruit fly usually occurs with a less significant intensity in the northern olive region than in the south region, where this pest is considered the most important and harmful for olive production, especially affecting its quality. This seems to discourage the organic production of olives in the southern region where the usual intense presence of the olive fruit fly hinders this type of production. In fact, in this region and for 2019, only 1.2% of the olive area was on organic production mode and about 29% in integrated production mode (EDIA, 2019). The usual pressure associated to the olive fruit fly in olive groves of the Alentejo region is the main cause for the use of organophosphate insecticides. Still, the use of biopesticide based on *Bacillus thuringiensis* has obtained positive results for olive moth control, especially during the occurrence of the second generation of this pest.

In the northern region, given the less presence of the olive fly, the olive moth usually has greater importance as pest, and the existence of efficient alternative control measures favours olive organic production. Concerning usual secondary enemies, some serious outbreaks have been registered on super-intensive olive orchards, related to *Euzophera pinguis* and olive tuberculosis, possibly related to wounds resulting from mechanical pruning (EDIA, 2018).

Currently, there is great concern regarding the bacterium *Xylella fastidiosa* given the impact caused on the Italian olive grove (Saponari et al., 2013). Dispersing itself over short distances through insect vectors and over long distances by the movement of contaminated plants, the risk of dispersion of *X. fastidiosa* in Portugal is considered high, due to the import of vegetative material, the existence of asymptomatic infections that hinder the early detection of the disease and the presence of insect species capable of dispersing it (Neto, 2018; Popova, 2020).

31.3 Current and Future Perspectives for Olive Protection Against Their Enemies

Currently, the protection of olive groves against their key enemies is essentially supported by pesticides, especially cupric and organic fungicides which are essential for disease control such as the olive anthracnose. Globally, in 2019 the total consumption of these pesticides was 8% of the total pesticides consumed in Portugal (EDIA, 2019).

According to the assessment effected by EDIA in 2019, pesticide substances most applied on intensive/super-intensive olive orchards of Alentejo include the following:

- Herbicides: during the vegetative phase, glyphosate + oxyfluorfen (1.0 + 0.4 kg/ha), glyphosate (1.44), diflufenican + glyphosate (0.16 + 0.64), oxyfluorfen (0.96) and fluroxypyr (0.3) are applied, whilst during flowering and until fruit setting it is diflufenican + glyphosate (0.28 + 1.44).
- Regarding fungicides, during the vegetative rest until flowering, tebuconazole (0.3) is commonly used, whilst for the stage of growing fruits until ripening, treatments are based on copper (3.0).
- Last, and regarding insecticides, during the vegetative rest until flowering, it is deltamethrin (0.045), lambda-cyhalothrin (0.0075), acetamiprid (0.3) and spinosad (0.012), whilst for the stage of growing fruits until ripening, treatments are based on deltamethrin (0.045), cypermethrin (0.01), phosmet (0.21) and spinosad (0.012).

Regarding pest control, until 2020 the use of the organophosphate dimethoate has been predominant in the control of the olive fly and olive moth, due to its systemic properties. The inclusion of dimethoate in the EU Priority List of Hazardous Substances, in 2020, has led to discontinuation of its use from European countries, due to health and environmental risks. Due to its intensive use, when there were no official limits for the number of annual applications, an increasing presence of genes that provide dimethoate resistance on olive fruit fly populations has been identified in several Mediterranean countries, including Portugal (Pereira-Castro et al., 2015; Nobre et al., 2019). Alternatively, dimethoate was replaced mainly by pyrethroids. Spinosad is also an alternative for dimethoate, although its use should be evaluated carefully, mainly in conditions that could increase arthropod biocontrol agents' propensity for feeding on it (Gonçalves et al., 2011).

Nevertheless, differences can be found in the number of pesticide sprays related to the density of plants per orchard, with the ones with less olive trees per hectare (conventional rainfed) registering the lowest number of sprays, in opposition to the ones with the highest numbers of trees (super-intensive), which could be about four times higher to that seen in the conventional olive grove.

According to the characterization report published by EDIA in 2018, the following types and regimes for use of pesticides are prevalent for each of the main olive orchard types found in Alentejo:

- In the case of rainfed conventional olive groves (100–150 trees/ha), fungicides and insecticides are generally applied one or twice along the year, with key substances being copper and pyrethroids (deltamethrin and lambda-cyhalothrin).
- In intensive olive groves (200–600 trees/ha), both herbicides (2–3 applications yearly, with preference for glyphosate and fluroxypyr meptyl) and fungicides and insecticides (5–8 applications yearly), generally including copper and pyrethroids (deltamethrin and lambda-cyhalothrin).
- Lastly, in super-intensive olive groves (1000–2500 trees/ha), the same products and regimes are applied as in intensive olive groves, although the minimum number of yearly applications of fungicides and insecticides is raised from 5 to 6, with the maximum number staying at 8.

The olive culture is and will be facing several and serious challenges, namely, (i) the growing pressure from consumers against the use of pesticides, in accordance to the increasing knowledge about their negative effects for human health and the environment, which has resulted in a continued reduction of available pesticide substances, especially for pest control such as the olive fruit fly and diseases such as the olive anthracnose; (ii) the growing opposition against super-intensive olive groves for their ecological and environmental impact; (iii) the effects associated with climate change, which are currently observed and that in the southern region is reflected either by the increasing reduction in annual rainfall or by higher maximum temperature and heat waves periods; and (iii) by the relatively low olive value paid to the production, a factor that has driven the installation of super-intensive olive groves in the southern region. These challenges have evident repercussions in the management of the protection of the olive grove against its enemies, requiring the use of alternative control methods.

31.3.1 The Olive Fruit Fly Case

Still regarding the olive fruit fly, as the main and most challenging pest of olive groves to manage, several alternative control methods could be already considered. Amongst those is the use and increment of biological conservation, as a high diversity of natural enemies is being present in organic olive groves in both national main growing regions. Measures are needed to increase their action either by reducing the application of synthetic insecticides and by incrementing food availability and shelter in the orchard towards increased populations of beneficial insects. An example on how this can be achieved is through the selection of spontaneous herbaceous plants associated to the olive ecosystem, as recently done in our country for the olive fruit fly parasitoid *Psyttalia concolor* (Furtado et al., 2016). However, although the aforementioned control measures could have application in low-density and smaller olive groves, it does not seem to be viable in high extensive olive groves with high planting densities, since the presence of beneficial insects, such as predators and parasitoids, is not compatible with regular application of pesticides.

Several biotechnological methodologies have been developed in the last decade such as the massive trapping through a high number of traps, as the OLIPE design. However, due to the high number of trees per hectare (which can be more than 2000 plants), massive trapping proves to be difficult to manage and carry out in high density-olive groves, requiring an extremely high number of traps to be successful. To overdue this limitation, several olive fruit fly trap designs are currently under development, in the MED Entomolab (https://www.med.uevora.pt), such a large-scale station trap with autonomous operating and active attractive volatile dissemination (Figs. 31.1 and 31.2), to reduce the number of traps per hectare as well as their operational maintenance, in comparison with the OLIPE reference traps (Rei et al., 2019a, b). These traps, although specifically developed for the olive fly, could be used for the olive moth as well. Another design allows the use of a very flexible and long horizontal tubular trap (up to 6 m), which can be placed along the trees of each line of an intensive olive grove (Fig. 31.3) or around each olive tree in conventional rainfed olive groves (Rei, 2019). Another biotechnological methodology relates to the use of kaolin, as a repellent or/and anti-ovipositional for the olive fly (Caleca et al., 2010).

An also emerging field in the context of sustainable agriculture practices that can respond to modern challenges is the use of symbiosis as a management tool. Symbiosis is ubiquitous and has been a major factor of insect evolution, for instance, by providing them capacities to overcome plant defences or to provide sources of food, as observed on *B. oleae* (e.g. Capuzzo et al., 2005; Estes et al., 2014). Several symbiotic microorganisms were associated through evolution time for sustained

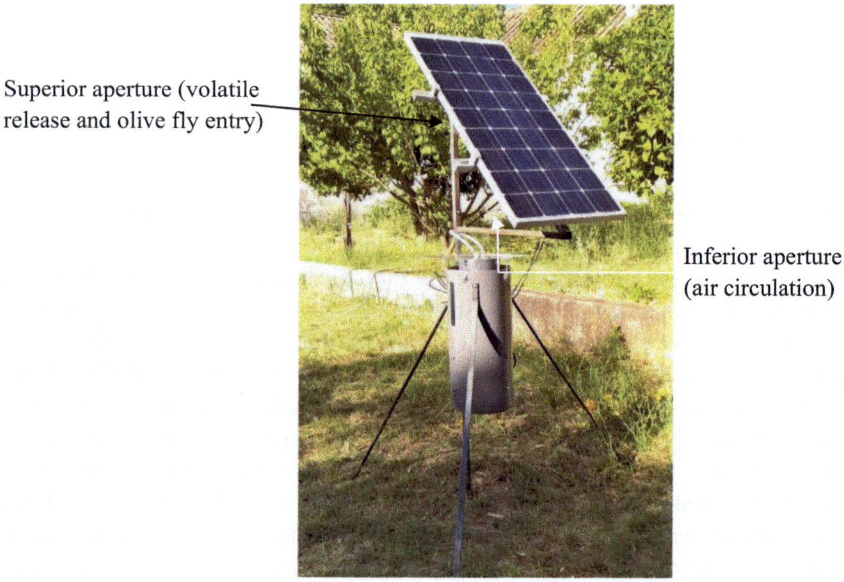

Fig. 31.1 Mass capture station trap for the olive fly, observing its large vertical cylindrical shape and the opening at the top

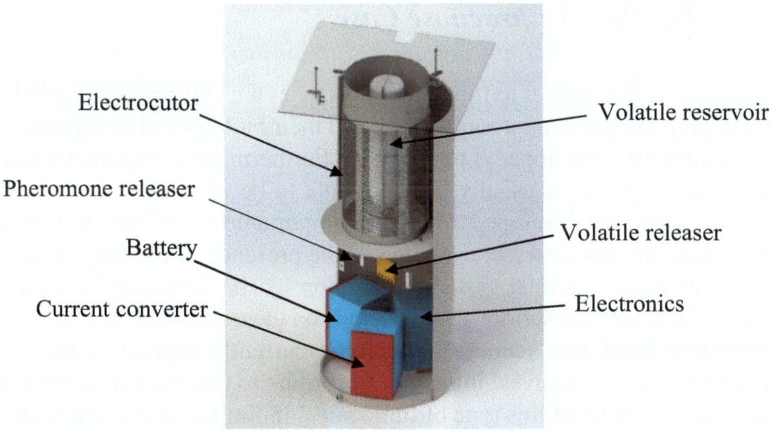

Fig. 31.2 Interior of the 'electrocutor' mass capture station trap

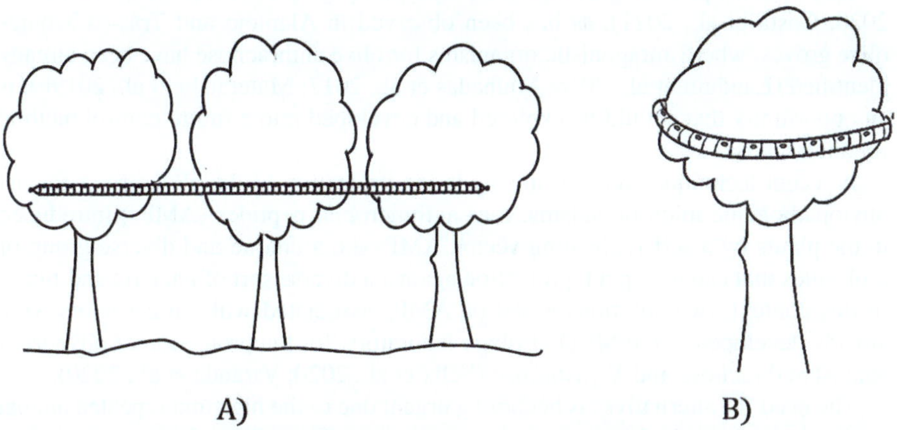

Fig. 31.3 Arrangement of the 'horizontal tubular' mass capture trap, fixed to the canopies of several olive trees on intensive/super-intensive orchards (**a**) or being able to encircle the canopy of a single olive tree (**b**) on rainfed conventional orchards

insect growth, reproduction and survival, so their elimination or manipulation could be a way forward to manage insect population numbers and prevent them to become pests. However, for this objective, there is a notorious need for more knowledge on the symbionts including their function, as it is being pursued in the MED Entomolab, for the olive fruit fly (Nobre, 2019).

Non-chemical future alternative control methods could also be achieved by the management of plant volatiles (Shrivastava et al., 2010), such the ones already identified on Portuguese olive cultivars (Malheiro et al., 2015).

31.3.2 The Olive Anthracnose Case

For the diseases, alternatives to fungicide substances are still scarce, largely because they are phytopathogenic microorganisms, and their biology and ecosystem are also difficult to understand compared to the pests. Furthermore, being microorganisms, their presence in plants is usually only possible to be identified when the disease symptoms are already developed. Thus, it is important to fulfil this important gap. Recent studies on olive anthracnose referred the presence of the fungus in 'closed' flowers, revealing a potential systemic infection by the fungus (Materatski et al., 2018). This fact can have serious repercussions on the traditional protection method with inorganic fungicides (copper), which are currently applied in late summer/early autumn, to prevent olives' infection. According to this, a rigorous observation/evaluation of the effect of this type of fungicides' in limiting anthracnose is recommended, and higher effectiveness should be achieved if it is used in association with organic fungicides, with different modes of action.

However, the profuse presence of endophytic fungi in olive trees (Nicoletti et al., 2020; Costa et al., 2021), as has been observed in Alentejo and Trás-os-Montes olive groves, where antagonistic organisms for olive anthracnose have been already identified (Landum et al., 2016; Malhadas et al., 2017; Materatski et al., 2019b), is an opportunity that should be explored and developed into a future control method for anthracnose.

A recent technique, very promising in the protection of the olive grove against phytopathogenic microorganisms, uses antimicrobial peptides (AMPs) introduced in the plants by a self-replicating vector. AMPs are a unique and diverse group of molecules that can give plant protection against a diverse sort of bacteria and fungi. In this context, two solutions based on AMPs associated with viral vectors were already developed in the MED virology laboratory, for the protection of olive trees against anthracnose and *X. fastidiosa* (Félix et al., 2020; Varanda et al., 2020).

The need for alternatives is becoming urgent due to the high and repeated annual application of cupric fungicides which, being insoluble, have been cumulatively deposited on the olive groves soil, essentially by rain drag, and are toxic in water environments. In addition, the continued use of fungicides has also promoted the presence of more virulent strains of the olive anthracnose in treated orchards, as was observed in Portugal (Materatski et al., 2019a).

31.4 Olive Protection and Climate Change Effects

Concerning the effect of climate change, major constraints are expected for the olive survival, especially for the intensive and super-intensive olive groves, considering the expected reduction in rain and subsequent water availability for irrigation. The overall rising of temperatures, changes in precipitation patterns and water shortages will affect the fierceness and propagation of insect pests in largely unknown ways. How the pests and diseases will respond to the climate change effects is widely unknown and highly species specific, as either organism has its own optimum of temperature and moisture and its own degree of dependence of the olive tree.

Taking the olive fly as a model, high mortality could occur in juvenile instars, whereas adults substantially reduce their activity, and their mortality is high under long periods of summer heat (Wang et al., 2009). In these conditions, only one generation of this pest could be found in the southern region of Portugal, registered successively later, at the end of summer until autumn. However, the continuous period increase associated with high summer temperatures constitutes a second stress factor for the pest, forcing adults to their maximum resilience. This situation is responsible for a reduced quantity of olives available for oviposition, since most of them are already very mature, when the temperatures drop enough to allow the females activity in the olive groves. In fact, it was observed that later cultivars were the most infested by the olive fly, where the presence of olives with more than ten oviposition marks was usually found due to the reduced number of olives. However, a significative presence of predators and parasitoids was observed in the olive orchard ecosystem, after extreme summer weather conditions such the ones that occurred in 2016, the highest since 1930, with several heat waves and severe drought conditions (Neto, 2018; Quadrado, 2020; Ramires, 2020). These conditions could be considered very similar to those predicted by some climate change models for the region of the Iberian Peninsula (Cardoso et al., 2019). In fact, despite these stress conditions, in the beginning of autumn it was possible to observe not only the presence of potential *X. fastidiosa* vectors but also some of them parasitized, both collected from the scarce herbaceous plants present, due to the absence of water.

Regarding the diseases, there is usually less fungal activity during summer months, with its most active periods occurring during spring and autumn, associated to the presence of rain and moisture needed for the germination of the spores. But if olive anthracnose could exist systemically in the olive tree (Materatski et al., 2019b), its survival would then directly relate with the survival of its host. Nevertheless, its spread on olive orchards may be compromised under climate change conditions.

31.5 Conclusion

Undeniable is the need for alternative control measures for pesticides, to be integrated as possible and determined by previously monitoring pests and diseases, in accordance with their economic thresholds, as determined by national rules (DGADR, 2010). The decision to apply control measures should never be based on pest and disease occurrence history, specially under climate change conditions, a situation that still occurs repeatedly in the field. This situation necessarily leads to repeated application of control measures that mismatch the olive enemies' presence, resulting in the ineffectiveness of the control measures and the emergence of

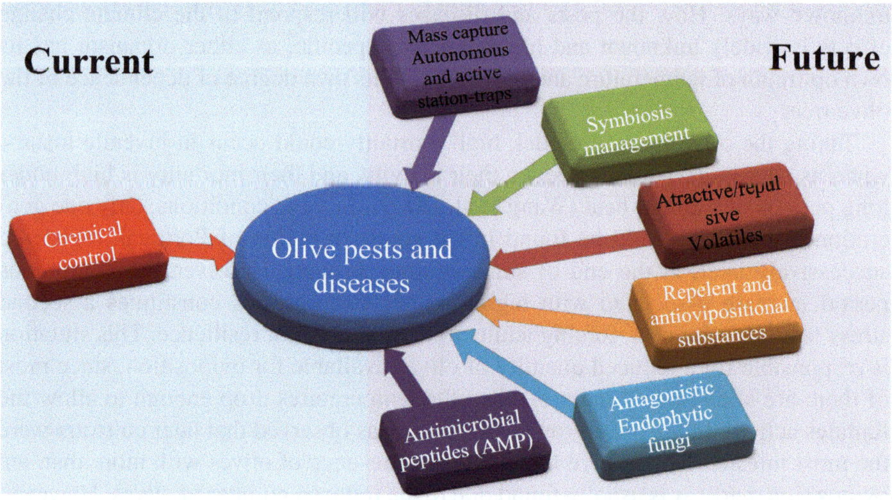

Fig. 31.4 Current and future methodologies for the phytosanitary protection of the Portuguese olive grove

resistance phenomena (Pereira-Castro et al., 2015; Nobre et al., 2019), plus economic losses and environmental impact.

Nevertheless, it should be noted that the olive grove protection must be considered as an integration of various control methods, as recommended by IPM, not being supported by only one method as has been the traditional use of organic pesticides. Thus, to be a viable and effective alternative to the use of chemical pesticides, non-chemical methodologies like those mentioned in this chapter (Fig. 31.4) should preferably be used in an integrated manner, maximizing their individual effect, sometimes partial, on limiting pest and diseases populations.

Acknowledgements This work was funded by FCT-Portugal (UIDB/05183/2020 as member of MED-Universidade de Évora and LA/P/0121/2020 as member of the CHANGE- Associated Lab-Universidade de Évora).

References

Caleca, V., Lo Verde, G., Lo Verde, V., Piccionello, M. P., & Rizzo, R. (2010). Control of *Bactrocera oleae* and *Ceratitis capitata* in organic orchards: Use of clays and copper products. *Acta Horticulturae, 873*, 227–234.

Capuzzo, C., Firrao, G., Mazzon, L., Squartini, A., & Girolami, V. (2005). '*Candidatus* Erwinia dacicola', a coevolved symbDiotic bacterium of the olive *Bactrocera oleae* (Gmelin). *International Journal of Systematic and Evolutionary Microbiology, 55*, 1641–1647.

Cardoso, R. M., Soares, P. M. M., Lima, D. C. A., & Miranda, P. M. A. (2019). Mean and extreme temperatures in a warming climate: EURO CORDEX and WRF regional climate high-resolution projections for Portugal. *Climate Dynamics, 52*, 129–157. https://doi.org/10.1007/s00382-018-4124-4

Costa, D., Fernandes, T., Martins, F., Pereira, J. A., Tavares, R. M., Santos, P. M., Batista, P., & Lino-Neto, T. (2021). Illuminating *Olea europaea* L. endophyte fungal community. *Microbiological Research, 245*, 126693. https://doi.org/10.1016/j.micres.2020.126693

DGADR. (2010). *Produção Integrada do Olival*. 2ª edição. Ministério da Agricultura, do Desenvolvimento rural e das Pescas. Direcção-geral de Agricultura e Desenvolvimento Rural, p. 105. https://www.dgadr.gov.pt/images/docs/prod_sust/normas_pi/Prodi_olival.pdf.

EDIA. (2018). *Olival em Alqueva. Caracterização e perspectivas*. Empresa de Desenvolvimento e Infraestruturas do Alqueva (EDIA) ed., p. 136.

EDIA. (2019). *Anuário agrícola de Alqueva*. Empresa de Desenvolvimento e Infraestruturas do Alqueva ed. Direcção de economia da água e promoção do regadio, p. 197.

Estes, A. M., Segura, D. F., Jessup, A., Wornoayporn, V., & Pierson, E. A. (2014). Effect of the symbiont *Candidatus* Erwinia dacicola on mating success of the olive fly *Bactrocera oleae* (Diptera: Tephritidae). *International Journal of Tropical Insect Science, 34*, 1–9. https://doi.org/10.1017/S1742758414000174

Félix, M. R., Varanda, C. M. R., Materatski, P., Campos, M. D., & Peixe, A. (2020). A viral molecule for protecting plants against *Xylella fastidiosa*. Patent pending n° 20165419.1 European Patent Office.

Furtado, C., Belo, A. F., Nunes, F. M., Ganhão, E., Muller, C. T., Torres, J., & Rei, F. T. (2016). Evaluating potential olive orchard sugar food sources for the olive fly parasitoid *Psyttalia concolor*. *BioControl, 61*, 473–483. https://doi.org/10.1007/s10526-016-9732-5

Gonçalves, M. F., Santos, S. A., & Torres, L. M. (2011). Efficacy of Spinosad bait sprays to control *Bactrocera oleae* and impact on non-target arthropods. *Phytoparasitica, 40*, 17–28.

INE. (2019). *Estatísticas agrícolas 2018*. Instituto Nacional de Estatística ed., p. 155.

Landum, M. C., Félix, M. R., Alho, J., Garcia, R., Cabrita, M. J., Rei, F., & Varanda, C. M. R. (2016). Antagonistic activity of fungi of *Olea europaea* L. against *Colletotrichum acutatum*. *Microbiological Research, 183*, 1–9.

Malhadas, C., Malheiro, R., Pereira, J. A., Pinho, P. G., & Baptista, P. (2017). Antimicrobial activity of endophytic fungi from olive tree leaves. *World Journal of Microbiology and Biotechnology, 33*, 46. https://doi.org/10.1007/s11274-017-2216-7

Malheiro, R., Casal, S., Cunha, S., Baptista, P., & Pereira, J. A. (2015). Olive volatiles from portuguese cultivars cobrançosa, madural and verdeal transmontana: Role in oviposition preference of *Bactrocera oleae* (Rossi) (Diptera: Tephritidae). *PLoS One, 10*, e0125070. https://doi.org/10.1371/journal.pone.0125070

Materatski, P., Varanda, C., Carvalho, T., Dias, A. B., Campos, M. D., Rei, F., & Félix, M. R. (2018). Diversity of *Colletotrichum* species associated with olive anthracnose and new perspectives on controlling the disease in Portugal. *Agronomy, 8*, 301. https://doi.org/10.3390/agronomy8120301

Materatski, P., Varanda, C., Carvalho, T., Dias, A. B., Campos, M. D., Gomes, L., Nobre, T., Rei, F., & Félix, M. R. (2019a). Effect of long-term fungicide applications on virulence and diversity of *Colletotrichum spp.* associated to olive anthracnose. *Plants, 8*, 311. https://doi.org/10.3390/plants8090311

Materatski, P., Varanda, C., Carvalho, T., Dias, A. B., Campos, M. D., Rei, F., & Félix, M. R. (2019b). Spatial and temporal variation of fungal endophytic richness and diversity associated to the phyllosphere of olive cultivars. *Fungal Biology, 123*(1), 66–76. https://doi.org/10.1016/j.funbio.2018.11.004

Neto, A. C. M. (2018). Potential vectors of *Xylella fastidiosa* in Portuguese olive orchards: Survey in Alentejo region and control measures. Mestrado em Ecologia e Gestão Ambiental. Universidade de Lisboa, Faculdade de Ciências, p. 130.

Nicoletti, R., Di Vaio, C., & Cirillo, C. (2020). Endophytic fungi of olive tree. *Microorganisms, 8*, 1321. https://doi.org/10.3390/microorganisms8091321

Nobre, T. (2019). Symbiosis in sustainable agriculture: Can olive fruit Fly bacterial microbiome be useful in Pest management? *Microorganisms, 7*(8), 238. https://doi.org/10.3390/microorganisms7080238

Nobre, T., Gomes, L., & Rei, F. T. (2019). A re-valuation of olive fruit fly organophosphate-resistant ace alleles in Iberia, and field-testing population effects after in-practice dimethoate use. *Insects, 10*, 232. https://doi.org/10.3390/insects10080232

Pereira, J. A., & Bento, A., Lousão, C., Cabanas, E., & Torres, L. (2003). Pragas secundárias do olival Transmontano: Biologia, inimigos naturais e estratégias de protecção. III Simpósio Nacional de Olivicultura. Castelo Branco, 29–31 Outubro, 17.

Pereira-Castro, I., van Asch, B., Rei, F. T., & Costa, L. T. (2015). *Bactrocera oleae* (Diptera: Tephritidae) organophosphate resistance alleles in Iberia: Recent expansion and variable frequencies. *European Journal of Entomology, 112*(1), 20–26. https://doi.org/10.14411/eje.2015.019

Popova, G. (2020). *Identification of potential vectors of Xylella fastidiosa in Portuguese olive orchards and weeds*. Mestrado em Ecologia e Gestão Ambiental. Universidade de Lisboa, Faculdade de Ciências, p. 134.

Quadrado, M. F. (2020). *The diversity of parasitoids in the Alentejo olive grove ecosystem and its potential contribution to the limitation of olive tree pests*. Mestrado em Engenharia Agronómica. Universidade de Évora. Escola de Ciências e tecnologia, p. 88.

Ramires, I. H. (2020). *Olive fly management today: the role of predators*. Mestrado em Ecologia e Gestão Ambiental. Universidade de Lisboa, Faculdade de Ciências, p. 76.

Rei, F. T. (2019). Mass trap model 'Tubular-horizontal'.000 Patent pending n° EP19213409.6. European Patent Office.

Rei, F. T., Gomes, L., Leonardo, A., Tlemcani, M., Peça, A. O., & Dias, A. (2019a). Mass Station-trap model 'Electrocutor'. Patent pending n°. EP19190294.9. European Patent Office.

Rei, F. T., Gomes, L., Leonardo, A., Tlemcani, M., Peça, A. O., & Dias, A. (2019b). Mass station-trap model 'Adesive'. Patent pending n°. EP19190286.5. European Patent Office.

Saponari, M., Boscia, D., Nigro, F., & Martelli, G. P. (2013). Identification of DNA sequences related to *Xylella fastidiosa* in oleander, almond and olive trees exhibiting leaf scorch symptoms in Apulia (Southern Italy). Disease note. *The Plant Pathology Journal, 94*, 688.

Shrivastava, G., Rogers, M., Wszelaki, A., Panthee, D., & Chen, F. (2010). Plant volatiles-based insect pest management in organic farming. *Critical Reviews in Plant Sciences, 29*, 123–133. https://doi.org/10.1080/07352681003617483

Torres, L. (2007). *Manual de Protecção Integrada do Olival*. João Azevedo ed., p. 378.

Torres, L., & Gonçalves, F. (2013). *A protecção contra pragas em olivicultura biológica*. Projecto OlivaTmad, p. 32.

Varanda, C. M. R, Félix, M. R., Materatski, P., Campos, M. D., & Peixe, A. (2020). A viral molecule for protecting plants against *Colletotrichum acutatum* and *C. gloesporioides* fungi. Patent pending n° 20165418.3. European Patent Office.

Wang, X., Johnson, M. W., Daane, K., & Nadel, H. (2009). High summer temperatures affect the survival and reproduction of olive fruit fly (Diptera: Tephritidae). *Physiol Ecology, 38*(5), 1496–1504.

Chapter 32
Agricultural Intensification in the Alentejo Region: Effects on Water Quality and Water Suitability for Irrigation

Martinho Murteira, Maria Morais, and Maria Helena Novais

Abstract Agriculture is an important sector in southern Portugal, intensified in recent years by the construction of the Alqueva Dam on the Guadiana River. The Alqueva Multipurpose Project was designed and built to supply the driest region in Portugal, including the global irrigation system. This global change in the agricultural system (i.e. from dry to irrigated land) led to a significant increase in olive grove areas. Since 2009, olives have been the most representative crop, exceeding 50% in the entire irrigated area, and since 2013 Portugal has become self-sufficient in the olive oil sector. The Alqueva irrigation perimeter is expected to increase in the next years to 170,000 ha, with a predicted olive grove increase, if no restrictive measures are implemented at a national level. In this context, a temporal analysis of water quality for irrigation data is carried out (2010–2019), based on both the Portuguese regulations (DL 236/98) and the FAO guidelines. Therefore, the present work aims (i) to verify the effect of the land use change on the aquatic ecosystem integrity and (ii) to evaluate soil salinity risks and potential yield losses due to salinization of the most representative crops grown.

Our results show a possible degradation of the aquatic ecosystem integrity, given that persistent non-compliances were identified since 2011, for total phosphorus, pH, dissolved oxygen and chlorophyll *a* (biomass indicator), according to the Water Framework Directive guidelines. Furthermore, the breaches for bentazon, total phosphorus and chlorophyll *a* were more persistent in the last years (2018–2019), probably related to the olive grove super intensification in the irrigation blocks. Regarding the suitability of water for irrigation, most values were below the thresh-

---

M. Murteira (✉)
Empresa de Desenvolvimento e Infraestruturas de Alqueva S.A., EDIA, Beja, Portugal
e-mail: mmurteira@edia.pt

M. Morais
Department of Biology | Institute of Earth Science, University of Évora, Évora, Portugal
e-mail: mmorais@uevora.pt

M. H. Novais
Renewable Energies Chair | Institute of Earth Sciences, University of Évora, Évora, Portugal
e-mail: hnovais@uevora.pt

J. Muñoz-Rojas, R. García-Ruiz (eds.), *The Olive Landscapes of the Mediterranean*, Landscape Series 36,
https://doi.org/10.1007/978-3-031-57956-1_32

old of maximums recommended, even though some non-compliances were detected, namely, in the electrical conductivity, pH and bicarbonates.

32.1 Agricultural Changes in the Alentejo Region: The Alqueva Project

In recent decades, Europe has experienced great transitions in the rural landscape driven by multiple factors, perceived at different scales by a major change in the mode of agricultural production and land use (e.g. Carvalho-Ribeiro et al., 2013; Pinto-Correia & Kristensen, 2013). In the Portuguese context, the Alentejo region is experiencing a rapid agricultural intensification (from a predominantly extensive mode to an intensive production), mainly due to a long-standing public investment in the Alqueva irrigation system (Silveira et al., 2018).

The Alqueva project, with a long history (the first Alentejo Irrigation Plan dates to 1957), was designed and built to supply the driest region in Portugal, most prone to water scarcity and physical and human-induced desertification processes (Sanches & Pedro, 2006). The main objective of the initial project was the irrigation of the Alentejo region and to implement a productive agriculture, in order to combat desertification and underdevelopment of the region (Veiga et al., 2008). In fact, the long period that elapsed between the first studies (1957) and the closing of the dam floodgates in 2002 made "Alqueva" a myth for the local population, depressed since the 1930s "Wheat Campaign", through which the region was transformed into the "barn of the country" (Cutileiro, 2004). This campaign led to an intensive agricultural land use (cereal monoculture) with a consequent alteration of the landscape, soil depletion and erosion, simultaneously contributing to enormous social conflicts, repression, unemployment, poverty and migration to the coastal region and abroad. Consequently, the Alqueva project appears as a hope for Alentejo, with 1/3 of the Portuguese territory but with only 4.7% of its population and with high rates of human-induced desertification (PNPOT, 2018).

Currently, the Empresa de Desenvolvimento e Infra-estruturas do Alqueva, S.A. (EDIA) is responsible for the global irrigation system, based on the Alqueva Multipurpose Project (EFMA) that interconnects dams and covers an area of approximately 10,000 km^2, representing, therefore, a national strategic water reserve with the guarantee of water availability for a minimum of a consecutive period of 3 years. EFMA serves a total of 20 municipalities, supplying about 200,000 inhabitants, producing energy for 500,000 inhabitants and irrigating about 120,000 ha.

The Alqueva global irrigation system is divided into three subsystems, according to the different water sources and irrigating areas that belong to the hydrographic basins of the Guadiana and Sado Rivers (Fig. 32.1):

- Alqueva subsystem—origin at the Alqueva reservoir, with the water level raised 90 m at the Álamos pumping station, allowing the irrigation of ~64,000 ha (Guadiana Basin).

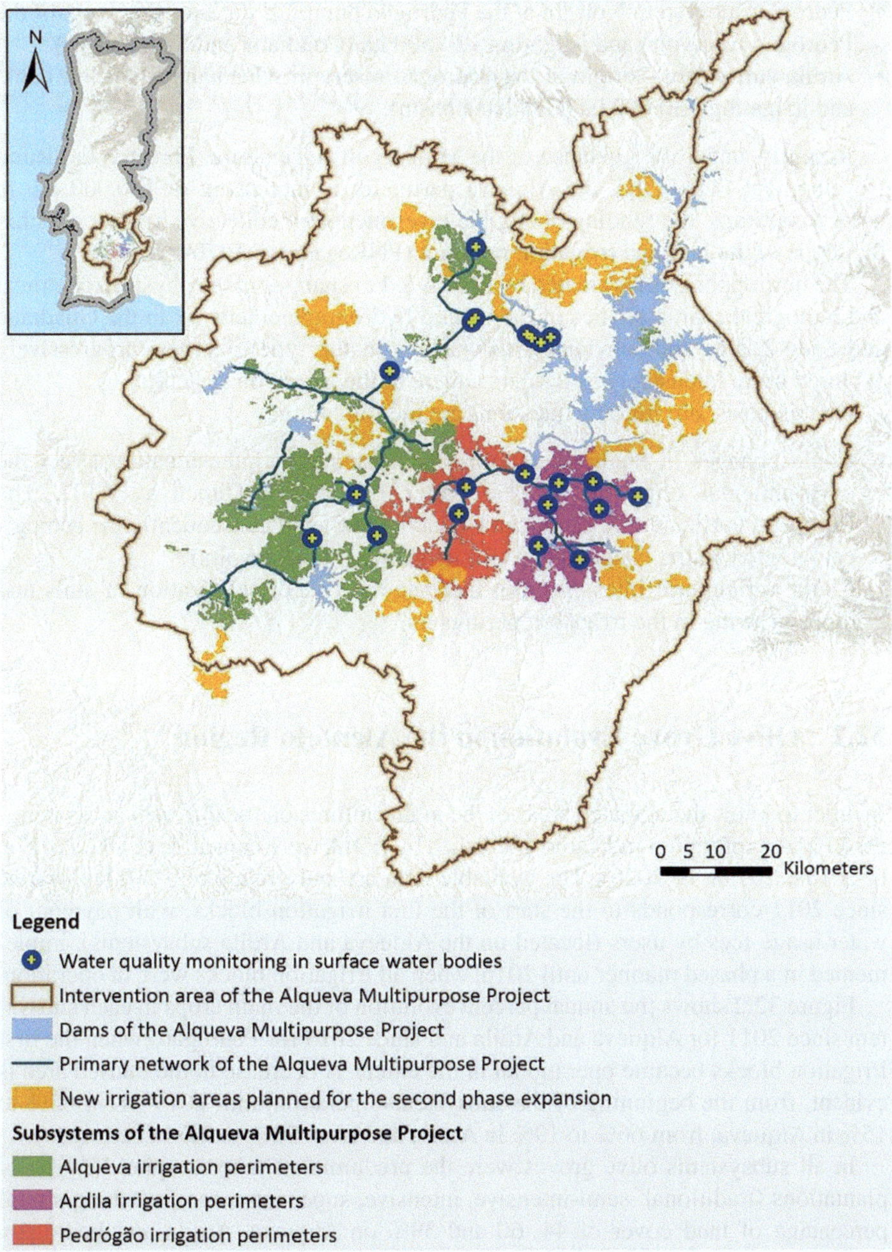

Fig. 32.1 The Alqueva global irrigation system in Portugal, showing the three subsystems: Alqueva, Ardila and Pedrógão. (This figure was produced by Martinho Murteira with original data from EDIA, 2020)

- Pedrógão subsystem—origin at the Pedrógão pumping station (right bank of the Pedrógão reservoir) and irrigating ~24,500 ha (Guadiana and Sado basins).
- Ardila subsystem—origin at the Pedrógão reservoir, Guadiana River left bank, and irrigating ~30,000 ha (Guadiana basin).

Recently, under the guidance of the Ministry of Agriculture, Forestry and Rural Development (MAFDR), the Alqueva perimeter is increasing to 170,000 ha, in order to optimize the funding of eligible investments for collective irrigation within the scope of the national irrigation program (PNRegadios) (EDIA, 2018b).

The new opportunities created by EFMA led to changes in the physical, chemical and biological characteristics of the natural hydrographic network in the Guadiana and Sado basins and, obviously, to changes in the type of crops, progressively explored more intensively as farmers adhere to the irrigation system.

On this basis, our guiding questions are the following:

- Could changes in land use patterns and agricultural intensification affect the environmental objectives of the Water Framework Directive (WFD, The European Parliament & European Council, 2000) and consequently the ecological integrity in the water sources for irrigation (i.e. reservoirs)?
- Could agricultural intensification increase the risk of salinization of soils and crops growing in the irrigation perimeters?

32.2 Olive Grove Evolution in the Alentejo Region

In order to know the irrigated areas of the main cultures on the different subsystems, the EFMA exploration indicators published by EDIA were consulted (EDIA, 2017a, b, 2018a, 2019a, b, 2020). The available data are only relative to 2011 onwards, since 2011 corresponds to the start of the first irrigation blocks, with payment of water usage fees by users (located on the Alqueva and Ardila subsystems), implemented in a phased manner until 2016, when all irrigation blocks were in operation.

Figure 32.2 shows the annual percent evolution of the main crops in each subsystem since 2011 for Alqueva and Ardila and since 2013 for Pedrógão (when the first irrigation blocks became operational in the latter). A decrease in the unused area is evident, from the beginning of the subsystems operation until 2019 (from 65% to 15% in Alqueva, from 66% to 19% in Ardila and from 72% to 9% in Pedrógão).

In all subsystems olive groves were the predominant crop, despite the type of plantations (traditional, semi-intensive, intensive, super-intensive), reaching a total percentage of land cover of 44, 60 and 59% on Alqueva, Ardila and Pedrógão, respectively.

It is worth noticing that the most representative plantation type is the intensive regime with a percentage cover of 21, 39 and 37%, respectively. Another interesting aspect is the almost complete disappearance of the traditional olive grove and its replacement by the super-intensive regime that had a significant presence since 2016, simultaneously with the almond groves that reach 14, 7 and 10%,

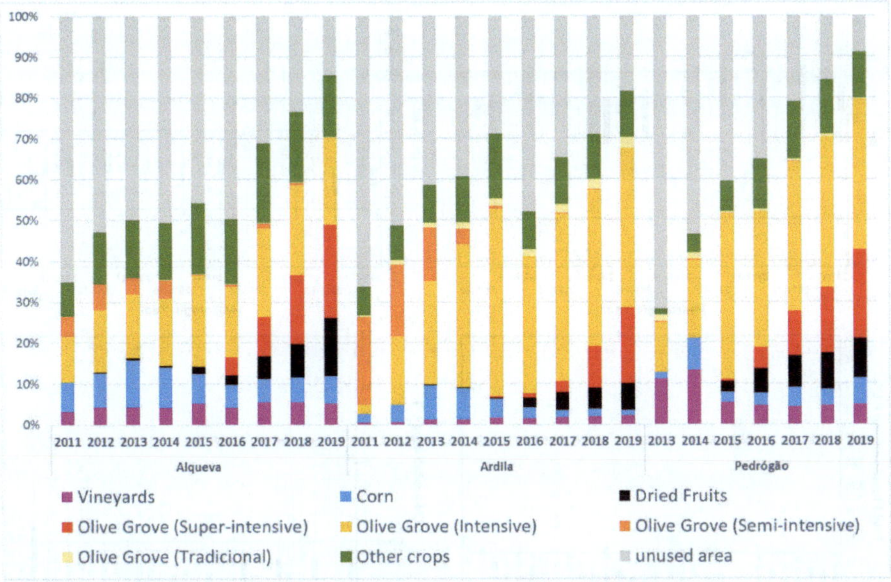

Fig. 32.2 Annual evolution (%) of the main crop change in the Alqueva, Ardila and Pedrógão subsystems, since the first irrigation blocks started operation

respectively, being the second most popular crop. Both groves (almond and super-intensive olive plantations) utilize a similar agricultural mechanization system.

32.3 Olive Grove Intensification Effects on Aquatic Ecosystems

Data from the EFMA "Primary Network Surface Water Resources Monitoring Program" conducted by the EDIA since 2002 in all the subsystem reservoirs were analysed taking into account the compliance with the WFD good ecological potential (The European Parliament & European Council, 2000) and the guidelines for Portugal (APA, 2016).

An analysis of the WFD supporting parameters for the ecological potential assessment of the reservoirs integrating the three subsystems revealed non-compliance for total phosphorus (> 0.07 mg P/L), pH (> 9), dissolved oxygen, DO (> 140%) and chlorophyll *a* (> 9.5 mg/mm^3), systematically from 2011 to the present (Fig. 32.3), whilst nitrate concentrations were always within the limit for the good ecological potential (≤ 25 mg NO_3/L). These breaches were mostly detected in the summer during periods with high temperatures (e.g. 42.0 °C Tmax in July, 2019) and absence of precipitation. To test for differences in these parameters between hydrological years, a Kruskal-Wallis one-way analysis of variance on

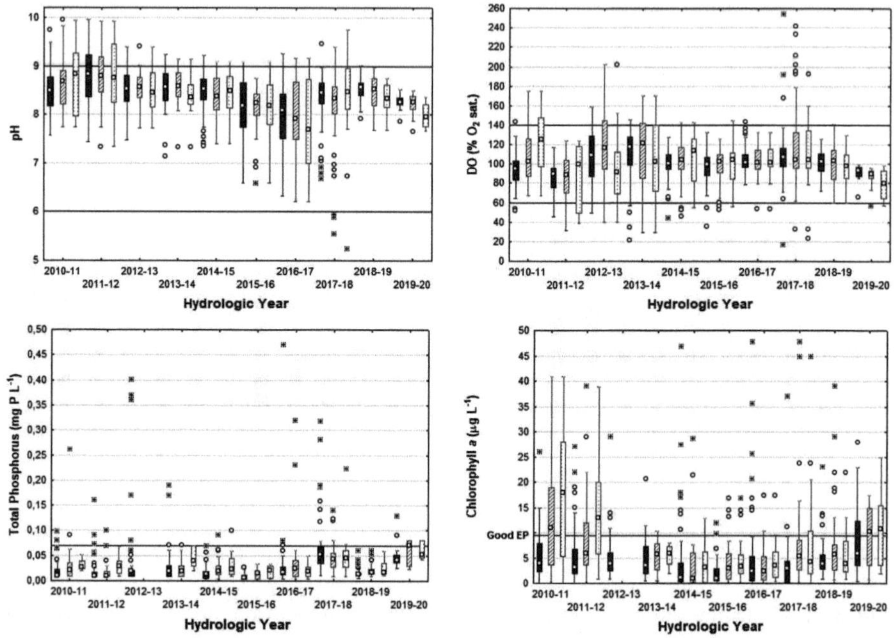

Fig. 32.3 *Box and whiskers* graphs for pH, DO (% O_2 sat.), total phosphorus and chlorophyll *a* (parameters that present values in non-compliance with the WFD requirements) along time (between 2010/2011 and 2019/2020) in the three subsystems. Dark grey, Alqueva; diagonal lines, Ardila; dots, Pedrógão. Median, box, 25–75%; whiskers, non-outlier range; ○, outliers; cross, extremes; horizontal lines, WFD limits for the good potential quality

ranks (Kruskal-Wallis test) was carried out (since the data did not pass the normality and/or the equal variance tests), using SigmaPlot 12.0 (Systat Software Inc., Chicago, IL). The Kruskal-Wallis tests revealed differences for the non-compliance parameters (pH, dissolved oxygen, total phosphorus and chlorophyll *a*) in the three subsystems, when all years were considered, but no clear pattern could be observed in the pairwise multiple comparison analysis (Dunn's method). However, in Fig. 32.3 it is possible to notice an increasing pattern for total phosphorus and chlorophyll *a,* for the last year in all the subsystems, with median values higher than the threshold for the good ecological potential, particularly in the Pedrógão and Ardila subsystems.

In addition, the results for specific pollutants showed non-compliance for bentazon (broad-spectrum herbicide) in the Alqueva and Pedrógão reservoirs, during 2018 and 2019. Palma et al. (2009), working in the Alqueva reservoir, had already detected values of the pesticides atrazine and diuron that were above the annual average proposed by the European Union legislation.

These findings could explain the effect of an agricultural intensification on the water quality of reservoirs (i.e. the origin of water for the irrigation system), which could compromise the compliance of the environmental objectives in the context of the Water Framework Directive. The water quality degradation could also affect the irrigation system by the corrosion and scale formation in pipelines and irrigation equipment, due to an increase in organic matter from primary production and eutrophication processes.

32.4 Olive Grove Intensification Effects on Water Suitability for Irrigation and the Salinization of Soils

Data from the EFMA "Primary Network Surface Water Resources Monitoring Program" conducted by the EDIA since 2002 in all the subsystem reservoirs were also analysed taking into account the compliance with the water quality for irrigation purposes, following the requirements and thresholds established by Annex XVI of the Decree-Law No. 236/98 and the FAO guidelines (salinity of water, EC, sodium risk, sodium adsorption ratio (SAR), residual sodium carbonates (RSC) and ion toxicity) (Ayers & Westcot, 1985).

Regarding the suitability of water for irrigation purposes, most values were below the threshold of maxima recommended. However, there were some non-compliances: the conductivity is over 700 μS/cm in about 5% of the samples, without ever exceeding 3000 μS/cm; pH exceeds the recommended maximum in about 10% of all samples; and bicarbonates are over 90 mg/L in about 18% of the total samples analysed.

In relation to ionic toxicity, the results of boron and chloride did not exceed the established thresholds (0.7 mg/L and 14 mg/L, respectively) with no risk for the cultures. The diagrams proposed by the US Salinity Laboratory (USSL) (Richards, 1954 *in* Zamna et al., 2018) showed that the majority of samples were in class C2-S1 (danger of medium salinity (C2), low sodium levels (S1)), indicating that water can be used in all soils for the crops currently practiced in the blocks with no risk of salinization.

In order to assess the sodium risk, associated with calcium carbonate ($CaCO_3$) and magnesium carbonate ($MgCO_3$) precipitation, the residual sodium carbonates (Na_2SO_3, RSC) were analysed. The obtained results confirmed that all values were below 1.25 me/L, indicating good-quality water for irrigation (Wilcox et al., 1954 method *in* Zamna et al. 2018).

Finally, the Langelier index with values between −0.8 and 0.9 (variations among slightly corrosion but non-scale forming, balanced/neutral and smooth corrosion) revealed the absence of risk of calcium carbonate ($CaCO_3$) film deposition or scales in the pipes.

32.5 Conclusions

The results for water quality for irrigation in all the subsystem reservoirs revealed that, despite the fact that no relevant breaches were detected, there were persistent non-compliances, according to the good ecological potential guidelines (WFD criteria) for total phosphorus, pH, dissolved oxygen and chlorophyll *a*. In the latest years (2018 and 2019), the situation seems to be getting worse since breaches for bentazon, total phosphorus and chlorophyll *a* were more persistent. These non-compliances may be related to the intensification of agriculture in the irrigation blocks, particularly with the super intensification of olive groves, which showed a > 50% increase compared to 2016 in the three subsystems.

In fact, several studies have demonstrated the existence of a direct relationship between olive groves (under-intensive and super-intensive monoculture cultivation) and ecosystem integrity (e.g. Silveira et al., 2018; Matono et al., 2013), compromising the fulfilment of the WFD environmental objectives and accelerating ecosystem degradation.

This situation may worsen with repercussions on soil degradation, given climate change scenarios for the region predicting a decrease in precipitation and an increase in temperature (IPCC, 2018). In this sense, it is necessary to implement a more sustainable agricultural system, with a greater diversity of cultures. Otherwise, we can relive the environmental situation of the beginning of the last century (1930s), now with soil depletion and erosion due to the escalation of an intensely fertilized monoculture of olive trees.

References

APA. (2016). Planos de Gestão de Região Hidrográfica, Parte 2 – Caracterização e Diagnóstico. Agência Portuguesa do Ambiente, I.P. Retrieved from https://apambiente.pt/index.php?ref=16&subref=7&sub2ref=9&sub3ref=848

Ayers, R. S., & Westcot, D. W. (1985). *Water quality for agriculture*. FAO of the UNITED NATIONS, Rome, Italy, 97.

Carvalho-Ribeiro, S. M., Madeira, L., & Pinto-Correia, T. (2013). Developing comprehensive indicators for monitoring rural policy impacts on landscape in Alentejo, southern Portugal. *Geografisk Tidsskrift-Danish Journal of Geography, 113*, 87–96. https://doi.org/10.1080/00167223.2013.84649

Cutileiro, J. (2004). Ricos e Pobres no Alentejo. (Uma sociedade rural portuguesa). Livros. Horizonte: Lisboa, p. 316.

Decreto-Lei n.o 236/98 de 1 de Agosto, Águas para consumo humano, águas para suporte da vida aquícola, águas balneares, água de rega. Diário da República-I série-A N°176. https://dre.pt/application/conteudo/430457

EDIA S.A. (2017a). Anuário Agrícola de Alqueva 2016. BEJA February 2017. Retrieved from https://www.edia.pt/wp-content/uploads/2019/05/anuario_agricola2016.pdf

EDIA S.A. (2017b). Anuário Agrícola de Alqueva 2017. BEJA December 2017. Retrieved from https://www.edia.pt/wp-content/uploads/2019/05/Anuario_Agricola_Alqueva_2017.pdf

EDIA S.A. (2018a). Anuário Agrícola de Alqueva 2018. BEJA December 2018. Retrieved from https://www.edia.pt/wp-content/uploads/2019/05/anuario_agricola-alqueva_2018.pdf

EDIA S.A. (2018b). Report and Accounts Fiscal Year 2018. Retrieved from https://www.edia.pt/wp-content/uploads/2020/05/rc2018_en.pdf

EDIA S.A. (2019a). Anuário Agrícola de Alqueva 2019. BEJA December 2019. Retrieved from https://www.edia.pt/wp-content/uploads/2020/02/AnuarioAgricolaAlqueva2019.pdf

EDIA S.A. (2019b). Relatório da Campanha de Rega dos Perímetros de Alqueva I 2018. BEJA September 2019. Retrieved from https://www.edia.pt/wp-content/uploads/2019/09/Relatorio_campanha_regaAlqueva_2018.pdf

EDIA S.A. (2020). Relatório da Campanha de Rega dos Perímetros de Alqueva I 2019. BEJA May 2020. Retrieved from https://www.edia.pt/wp-content/uploads/2020/06/2019_Relatoriocampanha2019.pdf

IPCC. (2018). Summary for policymakers. In V. Masson-Delmotte, P. Zhi, H.-O. Pörtner, D. Roberts, J. Skea, P. R. Shukla, A. Pirani, W. Moufouma-Okia, C. Péan, R. Pidcock, S. Connors, J. B. R. Matthews, Y. Chen, X. Zhou, M. I. Gomis, E. Lonnoy, T. Maycock, M. Tignor, & T. Waterfield (Eds.), *Global warming of 1.5 °C. an IPCC special report on the impacts of global warming of 1.5 °C above pre-industrial levels and related global greenhouse gas emission pathways, in the context of strengthening the global response to the threat of climate change, sustainable development, and efforts to eradicate poverty* (p. 32). World Meteorological Organization, Geneva, Switzerland.

Matono, P., Sousa, D., & Ilhéu, M. (2013). Effects of land use intensification on fish assemblages in Mediterranean climate streams. *Environmental Management, 52*, 1213–1229.

Palma, P., Kuster, M., Alvarenga, P., Palma, V. L., Fernandes, R. M., Soares, A. M. V. M., López de Alda, M. J., Barceló, D., & Barbosa, R. (2009). Risk assessment of representative and priority pesticides, in surface water of the Alqueva reservoir (South of Portugal) using on-line solid phase extraction-liquid chromatography-tandem mass spectrometry. *Environment International, 35*, 545–551.

Pinto-Correia, T., & Kristensen, L. (2013). Linking research to practice: The landscape as the basis for integrating social and ecological perspectives of the rural. *Landscape and Urban Planning, 120*, 248–256. https://doi.org/10.1016/j.landurbplan.2013.07.005

PNPOT. (2018). Estratégia e Modelo Territorial e Agenda para o Território (Programa de Ação), DIAGNÓSTICO Versão para Discussão Pública 30 abril 2018, p. 203. https://participa.pt/contents/consultationdocument/imported/2210/335538.pdf

Sanches, R., & Pedro, J. O. (2006) Alqueva: empreendimento de fins múltiplos/Rui Sanches, José Oliveira Pedro. Lisboa: EDIA – Empresa de Desenvolvimento e Infra-Estruturas de Alqueva, 2006. 581, [2] p.: il. ; 30 cm.

Silveira, A., Ferrão, J., Muñoz-Rojas, J., Pinto-Correia, T., Guimarães, M. H., & Schmidt, L. (2018). The sustainability of agricultural intensification in the early 21st century: Insights from the olive oil production in Alentejo (Southern Portugal). In A. Delicado, N. Domingos, & L. de Sousa (Eds.), *Changing societies: Legacies and challenges. Vol. III. The diverse worlds of sustainability* (pp. 247–275). Imprensa de Ciências Sociais. https://doi.org/10.31447/ics9789726715054.10

The European Parliament & European Council. (2000). Directive 2000/60/EC of the European Parliament and of the Council establishing a framework for the community action in the field of water policy. Off. J. Eur. Communities L327, 1–73.

Veiga, B., Duarte, L., & Vasconcelos, L. (2008). A Barragem do Alqueva para quem? Por uma contextualização pluridimensional do desenvolvimento no Alentejo -Portugal. In IV Encontro Nacional da ANPPAS (pp. 19). Brasília: Anais do IV Encontro Nacional da ANPPAS. Disponível em www.anppas.org.br/encontro4/cd/ARQUIVOS/GT12-376-513-20080518172237.pdf

Zaman, M., Shahid, S. A., & Heng, L. (2018). Irrigation water quality. In *Guideline for salinity assessment, mitigation and adaptation using nuclear and related techniques* (Vol. 12, pp. 113–131). Springer International Publishing. https://doi.org/10.1007/978-3-319-96190-3_5

Chapter 33
An Ecosystem Service-Based Approach to Adapt Olive Production to Climate Change

José M. Herrera, Bruno Silva, Miguel Ferro, António Cordeiro, Carla Inês, Gerardo Jiménez-Navarro, Inés Guise, José Muñoz-Rojas, and Maria F. Duarte

Abstract Climate change poses major challenges for agriculture. Considering that crop yield and quality will likely decline under less favourable temperature and future precipitation regimes is therefore becoming a major concern for scientists, producers and policymakers alike. Hereby, we describe a conceptual framework aimed to provide a climate-smart pest management strategy for olive (*Olea europaea* subsp. *europaea*) growing. This strategy focuses on taking up on the opportunities arising from the phenotypic diversity related to climate—or climate diversity—currently detected

J. M. Herrera (✉)
Instituto Mediterrâneo para a Agricultura, Ambiente e Desenvolvimento (MED), Universidade de Évora, Évora, Portugal

IVAGRO, Universidad de Cádiz, Puerto real, Spain
e-mail: josemanuel.herrera@uca.es

B. Silva · G. Jiménez-Navarro · I. Guise
Instituto Mediterrâneo para a Agricultura, Ambiente e Desenvolvimento (MED), Universidade de Évora, Évora, Portugal
e-mail: bsilva@uevora.pt; gjnavarro@uevora.pt; inesguise@uevora.pt

M. Ferro · M. F. Duarte
Centro de Biotecnologia Agrícola e Agro-Alimentar do Alentejo (CEBAL)/Instituto Politécnico de Beja (IPBeja), Beja, Portugal

Instituto Mediterrâneo para a Agricultura, Ambiente e Desenvolvimento (MED), Beja, Portugal
e-mail: Miguel.ferro@cebal.pt; fatima.duarte@cebal.pt

A. Cordeiro · C. Inês
Instituto Nacional de Investigação Agrária e Veterinária (INIAV), Elvas, Portugal
e-mail: antonio.cordeiro@iniav.pt; carla.ines@iniav.pt

J. Muñoz-Rojas
MED-Mediterranean Institute for Agriculture, Environment and Development CHANGE-Associated Lab for Global Change & Sustainability Department of Landscape, Environment and Planning, Universidade de Évora, Valverde - Évora, Portugal
e-mail: jmrojas@uevora.pt

J. Muñoz-Rojas, R. García-Ruiz (eds.), *The Olive Landscapes of the Mediterranean*, Landscape Series 36,
https://doi.org/10.1007/978-3-031-57956-1_33

within *O. europaea* and how this diversity may be best used for adapting to climate change. To achieve this challenging goal, a fine-scale ecological and economic assessment of climate change impacts on the abundance and incidence of olive pests will need to be performed by combining climate projections with biocontrol service-based management strategies to identify which varieties may be best positioned to maximize the agronomical goals of olive farmers and managers.

Keywords Biocontrol services · Conservation management · Pests · Vertebrates · Warming

33.1 Background

Climate change is widely recognized as a major threat to olive growing. This being the case, knowing that olive yields and olive oil quality could decline under new temperature and precipitation regimes has therefore become a major concern to olive farmers across the overall Mediterranean. Exploiting intra-specific crop diversity (i.e. varieties) is acknowledged as an effective way to adapt olive farming to shifting climate conditions because of its potential to increase resilience without requiring to shift crops (Wolkovich et al., 2018). However, climate resilience not only depends upon whether crops will physiologically tolerate shifting temperatures and precipitation regimes but also on concomitant alterations caused in the distribution of the multiple species with which crops, directly and indirectly, interact.

For instance, it is well known that climate change will significantly increase olive pest incidence (Fig. 33.1), driving dramatic economic impacts (Ponti et al., 2014). These assessments, however, neglect that pest incidence is widely known to be variety specific and, consequently, that the economic impact of climate change will significantly vary between crop varieties (Meehan et al., 2011). Moreover, these assessments anticipate that agroecosystems will become reliant on higher agrochemical inputs to control pests, neglecting the diversity of alternative management

Fig. 33.1 Damage on olive fruits produced by the olive fly *Bactrocera oleae*, a major olive pest worldwide, which is suggested to significantly increase in the face of climate change and recognized to drive olive yield and olive oil properties

strategies that can be deployed to enhance existing naturally occurring ecosystem services and, in turn, effectively reduce pest populations under environmentally and biodiversity-friendly farming practices (Power, 2010).

In this chapter, we describe an approach intended to provide a climate-smart pest management strategy for sustainable olive growing by supplying effective and alternative pathways to withstand changing climate regimes in olive-producing regions. Its ultimate goal is to indicate those varieties that maximize farmers' aims by performing a fine-scale ecological and economic assessment of climate change impacts on the abundance and incidence of major olive pests and their natural predators, combining climate change projections with environmentally sustainable farm management strategies. This plan comprises two phases and five steps: a resilience assessment phase (including Steps 1–3) and a scenario analysis phase (Steps 4–5) (Fig. 33.2). The resilience assessment phase results in essential baseline information, including spatially explicit distribution maps of olive varieties both currently (Step 1) and projected in response to shifting climate (Step 2) as well as a comprehensive assessment of effects of pest incidence on olive yields and oil quality rates linked to diverse varieties (Step 3). Insights resulting from the resilience assessment phase are then integrated into the scenario analysis phase. Subsequently, Step 4 aims at estimating a service-providing unit (expressed as the density of predators needed to reduce a given number of pests), which—combined with data from Step 3—allows to determine the effect of increasing the population density of predators on olive yields and olive oil quality. Lastly, Step 5 consists on a fine-scale ecological and economic assessment of climate change on olive pest infestation which—combined with Step 4—leads to the identification of olive varieties that maximize farmers' goals.

33.1.1 Step 1: Determining the Current Distribution of Olive Varieties

The first major requirement for the successful application of this approach is to count on spatially explicit distribution maps available of—at least—the most common varieties cultivated in the olive-producing region under study. The importance of these maps is twofold. First, different olive varieties are adapted to specific climatic, edaphic and lithological conditions. Second, each olive-producing region harbours dozens or even hundreds of olive varieties, each one linked to specific life history traits, ultimately influencing olive economic value. Moreover, such a value is closely linked to infestation levels by insect pests such as the olive fruit fly *Bactrocera oleae* and the olive fruit moth *Prays oleae*, both of which are known to be effectively controlled by natural enemies including invertebrate and vertebrate species (de Roince et al., 2010; Dinis et al., 2016; Alberdi et al., 2020).

These maps are already available for many olive-producing regions, including some of major relevance such as Andalusia, Spain. This database is indeed one of

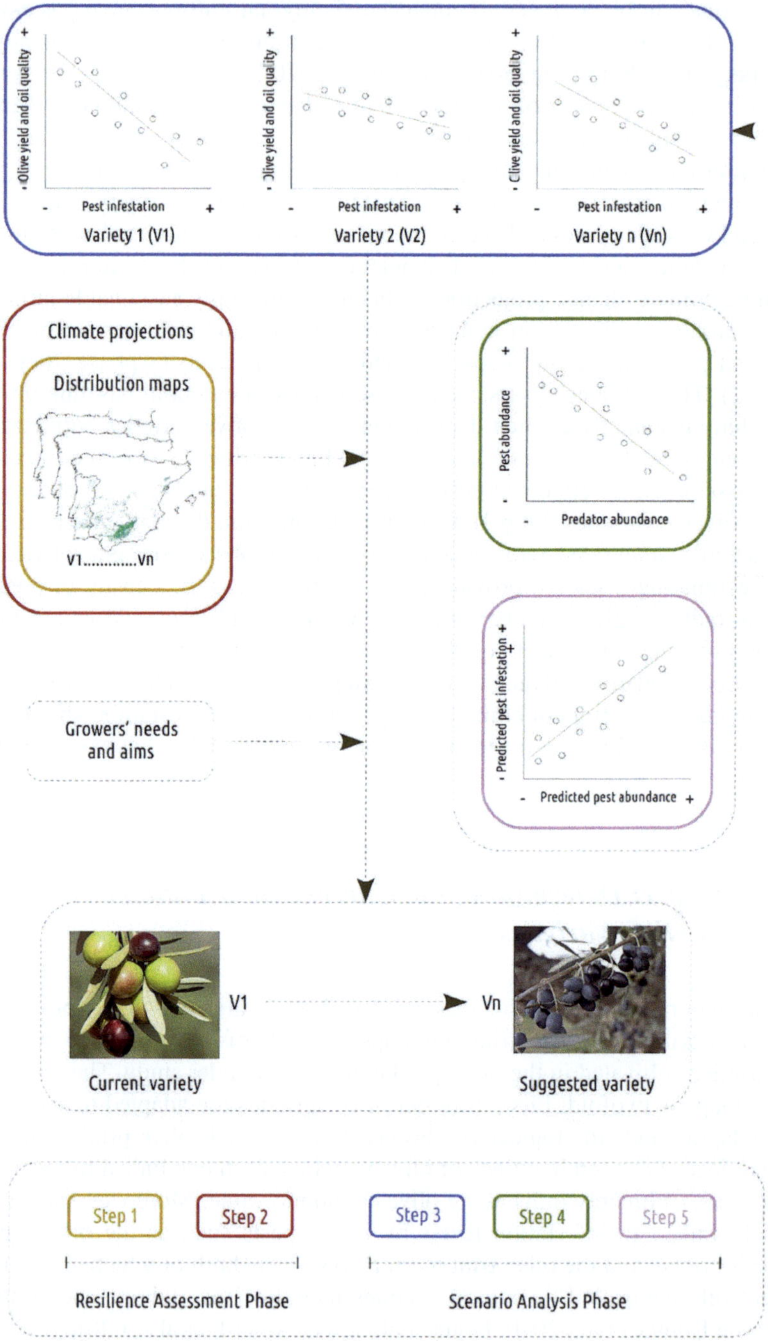

Fig. 33.2 Framework for selecting olive varieties in different regions under future climate regimes

the most complete ones worldwide regarding olive groves. However, most olive-producing regions lack these essential data, and many such regions are climate change hotspots, such as Alentejo in Portugal, the region of highest olive production in Portugal and one of the highest in Europe (Neves & Pires, 2013). In all those olive-producing regions for which variety-specific maps do not exist, they need to be unavoidably constructed. Establishing direct contacts with relevant stakeholders, including both individual landowners and enterprises, could result in great utility to achieve this challenging goal. Stakeholders will need to be asked—at least—to identify their farms and the olive variety they currently grow, in order to collect detailed and spatially explicit information to be used in subsequent modelling procedures (Step 2).

33.1.2 Step 2: Projecting the Distribution of Olive Varieties

A second major requirement in our approach is to gain knowledge on the potential future distribution of the different olive varieties. To achieve this, all those environmental variables characterizing the current distribution of each variety cultivated need to be firstly determined and then followed by the projection of future distributions using climate change scenarios. Species distribution models (SDMs) are particularly useful to perform this task. Research literature suggests that to determine the current and future distribution of olive varieties, at least four groups of explanatory variables should be considered: (i) soil properties, (ii) geomorphology, (ii) water balance and (iv) (bio)climatic conditions. Remarkably, most of these variables are freely available from national and international institutions.

Projections can be accurately modelled using the 1.8 °C warming scenario ($w_{+1.8}$) from the Intergovernmental Panel on Climate Change (IPCC). The reason behind this is twofold: firstly, 1.8 °C warming is a widely applied scenario for assessing risks associated with global warming and as a policy reference; secondly, it is the level of warming that will most likely occur in the Mediterranean Basin between 2030 and 2060 (Giannakopoulos et al., 2009). In any case, other projections applying weaker and greater warming scenarios could be used for comparison purposes. The potential future distribution of the different olive varieties will be determined using the projected variety-specific climatic suitability and the sign and magnitude of any potential variation using the relative change between current (AC) and future (AF) environmental suitability levels as estimated by $100 \times (AF - AC)/AC$. Maps depicting both the future distributions of olive varieties and the relative changes in environmental suitability can, and should, be constructed using 1 × 1 km grid cells as this spatial scale seems to be particularly suitable to relate species' occurrence data and environmental variables over broad spatial extents.

33.1.3 Step 3: Determining the Economic Impact of Pests on Olive Oil Chemical Properties

The retail price of an olive oil may be directly related with its chemical properties. Major pests such as *Bactrocera oleae* and *Prays oleae* in Portuguese olive groves pose considerable threats to the current economic market, affecting both olive oil yields due to fruits' loss and its biochemical and nutritional value. It is recognized that pests affect the final level of olive oil quality by accelerating certain metabolic mechanisms, leading to the degradation of chemical constituents and promoting the oxidation and consequent loss of health-beneficial compounds. According to our approach, in order to access the weight of these pests on the chemical composition and subsequent economic value, olive oil properties may be analysed along an increasing gradient of pest infestation whilst simultaneously considering a control sample (0% infestation). Through this proposed methodology, the influence of a certain infestation percentage may be directly related to a specific chemical alteration in the olive oil quality. Therefore, according to the specific level of pests present in a determined olive grove, and their consequent levels of infestation of olive fruits, this approach will enable to quantify the effects of the infestation over the chemical parameters of the olive oil. Thus, in order to analyse monovarietal olive oils along the proposed gradient of pest infestation, both non-infested and infested olives should be collected within a short period of time, thus considering the same defining harvest conditions such as ripening stage, avoiding the influence of naturally ripening, and the subsequent olive fruit matrix degradation.

33.1.4 Step 4: Linking Natural Enemies and Pests

This fourth step is aimed to determine the ecological impact that natural enemies may bear on pest incidence. Because of the relationships between pest damage on olive yields and olive oil properties established in Step 3, this step would therefore enable to indirectly relate natural enemies to olive's economic values. This economic value can be determined, for example, by assessing expenses related to agrochemical inputs that can be saved because of biocontrol services and/or the value of the crops that remained undamaged (e.g. Cleveland et al., 2006). Moreover, by implementing the concept of service-providing unit, defined as the amount of elements necessary to ensure a particular function that brings about a desired service, the abundance/density of natural predators needed to maintain pest populations below target thresholds and, in turn, olive yield and olive oil properties can be determined. For example, Puig-Montserrat et al. (2015) estimated the number of soprano pipistrelle bats (*Pipistrellus pygmaeus*) per hectare needed to maintain the striped rice borer (*Chilo suppressalis*) populations below two different thresholds that are planted locally by farmers to trigger pest treatments: no treatment needed (<3500 moths/ha) and aerial treatment needed (>8500 moths/ha). Using an ingenious approach based on estimations of the nightly consumption of female moths per bat, these authors estimated that between 9 and 16 bats per hectare are needed to reduce the number of stems affected by the striped rice borer to less than 0.85 affected stems/m^2 (aerial treatment threshold) and between 42 and 67 bats per hectare to further lower the moth density below any treatment-triggering threshold (0.35 affected stems/m^2).

33.1.5 Step 5: Bioeconomic Assessment of Climate Change on Olive

Lastly, and to assess the potential economic impact of climate change on olive yields and olive oil quality, mechanistic physiologically based demographic models (PBDMs) can be applied. We suggest PBDMs (instead of, e.g. SDMs) because they are increasingly recognized as explicitly useful in assessing whether a species should be considered as a current or potential pest in a specific geographic area (e.g. Ponti et al., 2014). PBDMs will use therefore extant (w_0) and climate change ($w_{+1.8}$) weather scenarios to forecast (under a spatial resolution of 1 km^2 square grids) rates of olive yield per tree, cumulative pest infestation per tree as well as the percentage of infested fruits in the absence of pest control. The net simulated change in yields due to climate warming at a grid point will be that forecasted for the w_0 scenario minus $w_{+1.8}$ and, when divided by the maximum difference across all locations, will become directly useful to forecast as well the relative change in yield at each location.

Since most of the olive harvest is used for oil production, crop yields (kg) will be converted into litres of oil by multiplying any given forecasted crop yield by a variety-specific factor that will be determined based on the information obtained from application of Step 3. To compute the net change in profit, both the change in quality and price with changes in % infestation levels and control costs will be incorporated. Variety-specific declines in the price of oil with increased infestation can then be later extracted. Remarkably, insights provided by this Step 5 will be those used to guide decisions about which olive variety should be prioritized on the basis of the expected climate-driven economic losses due to specific pests. By complementing climate projections with biocontrol service-based management strategies (Step 4), conclusions from PBDMs can be reformulated to maximize growers' aims.

33.2 Concluding Remarks

Scientists are increasingly highlighting the benefits of integrating biocontrol services into farming practices. Policymakers likewise are fostering their implementation as this may provide an environmentally friendly pathway towards food security. Yet apart from subsidized management practices such as agri-environment-climate schemes, very few farmers currently integrate effective biocontrol services into daily farming practices, even in organic farming. The lack of adoption by most olive farmers of biodiversity-friendly management schemes may be due to the fact that, so far, mainstream biodiversity science has largely focused on the ecological processes and consequences associated with biodiversity decline, highlighting their policy and social consequences. However, in order to have societal impact and increase the uptake of biodiversity management practices by the agricultural sector, we need a better understanding of both financial benefits and costs of biodiversity management practices by farmers.

To increase the interest of farmers in biodiversity management, this approach provides key incentives that we expect will effectively motivate farmers to integrate biodiversity into daily farm management practices. Incentives need to be both economic (higher crop yield and oil quality) and societal (better reputation by using wildlife-friendly farming practices). This will undoubtedly help bridge the large communication gap that exists between the scientists investigating biodiversity-based farming practices and the farmers that implement them.

Acknowledgements This work is funded by the projects PTDC/BIA-CBI/1365/2020 and UIDB/05183/2020, both of which have received funds from the Portuguese National Funding Agency for Science, Research and Technology (FCT). Authors were supported by the FCT through the contract IF/00001/2015 (JMH) and the predoctoral fellowships SFRH/BD/137803/2018 (BS), SFRH/BD/133017/2017 (GJ-N) and SFRH/BD/140083/2018 (MF).

Acknowledgements and Funding This work was funded by the SUSTAINOLIVE project (Grant n° 1822; PRIMA EU Programme) and by FCT-Portugal (UIDB/05183/2020 as members of MED-Universidade de Évora and LA/P/0121/2020 as members of the CHANGE- Associated Lab-Universidade de Évora).

References

Alberdi, A., et al. (2020). DNA metabarcoding and spatial modelling link diet diversification with distribution homogeneity in European bats. *Nature Communications, 11*(1), 1–8.

Cleveland, C. J., et al. (2006). Economic value of the pest control service provided by Brazilian free-tailed bats in south-Central Texas. *Frontiers in Ecology and the Environment, 4*, 238–243.

de Roince, C. B., et al. (2010). Fonctionnalité des auxiliaires vertébrés et invertébrés dans le contrôle des ravageurs du pommier. *Première étape. Infos CTIFL, 263*, 10–15.

Dinis, A. M., et al. (2016). Suppression of *Bactrocera oleae* (Diptera: Tephritidae) pupae by soil arthropods in the olive grove. *Journal of Applied Entomology, 140*(9), 677–687.

Giannakopoulos, C., et al. (2009). Climatic changes and associated impacts in the Mediterranean resulting from a 2 °C global warming. *Global and Planetary Change, 68*, 209–224.

Meehan, T. D., et al. (2011). Agricultural landscape simplification and insecticide use in the Midwestern United States. *Proceedings National Academy of Sciences. United States of America, 108*, 11500–11505.

Neves, B., & Pires, I. M. (2013). The impact of agricultural policies and growing investment in olive sector in the Alentejo region. In *Polıticas de Base e Recuperaçao Económica*. Universidade do Minho.

Ponti, L., et al. (2014). Fine-scale ecological and economic assessment of climate change on olive in the Mediterranean Basin reveals winners and losers. *Proceedings. National Academy of Sciences. United States of America, 111*, 5598–5603.

Power, A. G. (2010). Ecosystem services and agriculture: Tradeoffs and synergies. *Philosophical Transactions of the Royal Society B, 365*, 2959–2971.

Puig-Montserrat, X., et al. (2015). Pest control service provided by bats in Mediterranean rice paddies: Linking agroecosystems structure to ecological functions. *Mammalian Biology, 80*, 237–245.

Wolkovich, E. M., et al. (2018). From pinot to Xinomavro in the world's future wine-growing regions. *Nature Climate Change, 8*, 29–37.

Chapter 34
Market Challenges and Marketing Strategy of Olive Oil in Alentejo

Raquel Lucas, Rui Fragoso, and Maria de Belém Costa Freitas

Abstract Over the last years, Portuguese olive oil sector grew in area, production and productivity. This growth was accompanied by a change in production systems, technical assistance and commercialization/concentration processes and induced a positive trade balance. The main producing region that concentrates most of the olive groves area and olive oil production is Alentejo. Under the current market conditions, olive oil from Alentejo is competitive for two main reasons: the unitary cost of production and its quality standard. Two business models can be identified—in one of them olive oil is traded as a commodity, and in the other one companies differentiate their product with private brands. The main challenge for olive oil in Alentejo is to be recognized as the biggest contributor for the national strategy to positioning Portugal as the "sustainability hot spot" of the world's olive oil market in the coming years. This chapter suggests the strategy basis to meet this challenge, identifying the main bottlenecks that should be avoided and proposing a management solution within the actual actors in the sector. The core of this strategy is the development of a volunteer olive oil sustainability programme for Alentejo.

R. Lucas (✉)
MED —Department of Management and Mediterranean Institute for Agriculture, Environment and Development, Universidade de Évora, Largo dos Colegiais, Évora, Portugal
e-mail: mrlucas@uevora.pt

R. Fragoso
CEFAGE—Department of Management and Centre for Advanced Studies in Management and Economics, University of Évora, Évora, Portugal
e-mail: rfragoso@uevora.pt

M. d. B. Costa Freitas
MED —Mediterranean Institute for Agriculture, Environment and Development and University of Algarve, Faro, Portugal
e-mail: mbfreitas@ualg.pt

J. Muñoz-Rojas, R. García-Ruiz (eds.), *The Olive Landscapes of the Mediterranean*, Landscape Series 36,
https://doi.org/10.1007/978-3-031-57956-1_34

34.1 Introduction

The Portuguese olive oil sector registered a positive evolution over the last years, in spite of the stagnation of the world demand of olive oil (Parras Rosa & Veja-Zamora, 2020) and its low price level (Mielke, 2019). Several reasons should be mentioned to characterize this evolution. first is the expansion of modern, intensive and/or super-intensive olive groves, much more productive and competitive than traditional ones, the majority located in Alentejo. Second is the spread of assisted production and management and consequent costs reduction and more efficiency use of resources (such as water, fertilizers, soil and workforce). Third is the high quality of the extra virgin olive oils produced in Portugal and the increasing implementation of quality management and certification systems to access specific markets and consumers. Finally is the establishment of different cooperation processes to gain commercial size, such as companies' joint sale and purchase agreements, strategic alliances between privates and cooperatives and also the creation of the inter-professional organization, which are initiatives to reduce costs and/or increase revenue, differentiate supply, get market scale and access profitable market opportunities.

Although, as stated by Muñoz-Rojas and Pinto Correia (2018), this process of change has been widely accepted by Portuguese regional and local society and policy, a controversy remains: are these systems ecologically sustainable? How do they change biodiversity, ecosystems services or resources' use efficiency? The olive oil sector today needs to know exactly where it stands in what concerns sustainability and what should be its position in the future. Different systems are adequate for different conditions. The focus of the Alentejo olive oil sector's strategy should be that each producer and/or group of producers, with its own particularities and business model, is able to identify its position in what concerns sustainability and to imagine the path to reach more ambitious sustainability goals. The analysis and reflections in this chapter are a contribution to address these challenges.

Data has been collected through different methods and information sources. A literature review gathered the extant ideas and theoretical constructs about the topic, and the statistical data grouped the more relevant information about production, trade and markets. The empirical data was obtained through 15 qualitative semi-structured in-depth interviews conducted with experts and representatives behind the main organizations working with olive groves and olive oil in Alentejo and in Portugal. This information allowed to know the production and marketing perceptions and the strategies, sector challenges and future plans. The organizations interviewed were seven: OLIVUM,[1] CEPAAL,[2] AIFO,[3] CASA DO AZEITE,

[1] Associação de Olivicultores do Sul (Southern Olive Growers Association).

[2] Centro de Estudos e Promoção do Azeite do Alentejo (Alentejo Olive Oil Promotion and Study Center).

[3] Associação Interprofissional da Fileira Olivícola (Interprofessional Association of the Olive Branch).

SOVENA,[4] GALLO,[5] Cooperativa de Moura e Barrancos and Fundação Eugénio de Almeida. The experts were seven with different skills.

Apart from this sources, other relevant documents and reports were studied. This material functioned as a benchmark to the interviews, allowing comparisons with experts' opinions and building a holistic view of the Alentejo olive oil strategy. This triangulation of methods and sources ensures a higher level of validity. A thematic coding approach guided the content analysis and interpretation of the collected data, organized by relevance.

34.2 Overview of the Olive Oil Sector

This section encompasses a summarized overview of the olive oil sector at international and national context.

34.2.1 Olive and Olive Oil Production

Between 2016 and 2018, an average of 20.337 million tonnes of olive per year were produced in the world, supported by an area of 10.185 million hectares and account for a yield of 1.997 tonnes/ha. Spain is the major producer, representing 38.4% of the olive production in the world and 25.1% of the area of olive grove. The second place in the ranking is occupied by Italy with 10.7% of the olive production and 11.2% of the area. In this ranking, Portugal is in the ninth place in both dimensions (olive production and olive grove area). Spain has the highest land productivity with a yield of 3063 tonnes/ha, which is 53% above the world average. Portugal shows a good competitive advantage since its yield is the seventh highest in the ranking (FAO, 2020).

In general, Mediterranean countries, namely, Spain, Italy, Greece, Turkey, Morocco and Portugal, have a competitive advantage, both in olive growing and olive oil production.

According to FAO statistics, from 2014 to 2018 world production of olive oil increased almost 0.5 million tonnes. Over the last 10 years Portugal is also among the countries where olive oil production grew much (1.4% a year). Additionally, the International Olive Council stated that in Portugal olive oil sector generates a revenue of 95.5 million euros, which represents 1.36% of the agricultural product (International Olive Council, 2017). According to Avillez (2016) this sector represents 5.1% of food industry and involves 471 manufacturing industries and 1431

[4] Grupo português pertencente à Nutrinveste, uma holding do sector agro-industrial do grupo Jorge de Mello (Portuguese group belonging to Nutrinveste, a holding company in the agro-industrial sector of the Jorge de Mello group) .

[5] Victor Guedes – Indústria e Comércio, S.A.

Table 34.1 Production of olives and olive oil and N°. of mills per Portuguese agrarian region in 2019

	Olive		Olive oil		Productivity	Mills	
	t	%	hl	%	hl/100 Kg	No.	%
Portugal	914,504	100%	1,537,546	100%	0.17	488	100%
Entre Douro e Minho	3060	0.3%	2948	0.2%	0.10	11	2.3%
Trás-Os-Montes	73,735	8.1%	115,980	7.5%	0.16	112	23.0%
Beira Litoral	47,365	5.2%	51,696	3.4%	0.11	81	16.6%
Beira interior	30,410	3.3%	40,208	2.6%	0.13	110	22.5%
Ribatejo e Oeste	59,883	6.5%	84,177	5.5%	0.14	68	13.9%
Alentejo	696,501	76.2%	1,236,656	80.4%	0.18	98	20.1%
Algarve	5300	0.6%	5882	0.4%	0.17	8	1.6%

Produced by authors with data available in INE (2021) (National Agricultural Census 2019)

workers. Olive groves are spread throughout the territory, and olive oil production can be found in all regions.

Table 34.1 presents the production of olives and olive oil, productivity and the number of mills in Portugal per agrarian region. In 2019, Portugal produced 914,000 tonnes of olives to be transformed into olive oil, which represents an increase of 31% on the average value of the period 2016–2018 presented before. This is in line with the path of increase that is a consequence of massive investments in irrigation systems and the olive oil sector, which led to the modernization of production techniques and important quality improvements (Evangelista, 2017).

Although olive groves are spread all over the country, most production of olives and olive oil is concentrated in Alentejo that represents 76.2% of the national olive production and 80.4% of the olive oil production. Among the remaining regions, olive oil production is also important in Trás-os-Monte, Ribatejo-Oeste and Beira Litoral representing 8.1%, 6.5% and 5.2% of the olive production and 7.5%, 5.5% and 3.4% of the olive oil production. The average productivity is 0.17 hl of olive oil per 100 kg of olives, being this value higher in Alentejo (0.18 hl/100 kg). In 2019, the number of mills in the country was 488. They are scattered mainly by five regions, being Trás-os-Montes the region with more mills (112 mills, which present 23% of the total). Alentejo has 20% of the mills, meaning that their mills are larger than in the other regions, and hence this region benefits from important economies of scale.

34.2.2 *Production Systems*

Most olive grove areas in the world are concentrated in the Mediterranean region, where olive harvest is between October and April and cultivation systems are extremely diverse. A study sponsored by IOC (2015) identified and characterized seven different cultivation systems of olive growing in the Mediterranean area based on water use, slope grove density and percentage of area. According to this study, traditional rainfed systems with less than 180 trees per hectare occupy the most area (63.8%), which is divided in almost equal parts between areas of steep slope and moderate slope. Although irrigation systems represent a tier of the area, 22% regards modern irrigation systems, namely, intensive irrigated (17.9%) and super-intensive irrigated (4.2%) systems. In Portugal the recent increase on the olive oil sector is mainly associated with the adoption of these types of systems (Muñoz-Rojas and Pinto Correia, 2018).

In terms of production costs, Italy has a higher cost than the average of the 15 countries of IOC, presenting Turkey with the lowest unitary cost (1.54/kg). In the intensive cultivation system, Portugal and Tunisia are the most competitive countries, presenting both a unitary production cost of 1.75 €/kg. Recently, the super-intensive system has developed well in several Mediterranean countries, including Morocco, Portugal, Spain and Tunisia. Among the main producing countries, Portugal is the country with the fourth lowest weighted production cost, which means a comparative advantage regarding Spain, Italy and Greece.

34.2.3 Trade

Between 2015 and 2017 international trade was almost balanced. Olive oil imports were in average almost 1.9 million tonnes with a value of 7.8 billion dollars, and exports represent 1.8 million tonnes and 7.26 billion dollars (FAO, 2020). Country analysis reveals that Italy is the largest importer: 28% in volume and 25.9% in value. Also, other important producing countries, such as Spain, Portugal and Greece, show high import shares. Portugal imports 5.2% in volume and 4.5% in value. These imports are not only related with domestic consumption but also with these countries' international players' positioning. Particularly Italy concentrates a much higher share of final product trading than its own production would justify and has the highest reputation in the international branding olive oil market. Leading exporter countries are Spain and Italy. Portugal is the sixth exporter country in the world (6.7% in volume and 6.6% in value).

The most important markets for Portugal are Spain, Brazil and Italy with 42.9%, 28.1% and 16.2% of quantity and 35.1%, 36.1% and 15.1% of value, respectively. Among these three markets, the most interesting for Portugal is Brazil where the product is well valued and traded at an average price of 5.32 dollars per kilo; In Spain and Italy the average price is, respectively, 3.37 and 3.85 dollars per kilo (FAO, 2020). Besides various types of costs, Leonidou (2004) states as strong barriers to the export process different customers' habits and access to distribution channels and reliable representation in foreign countries. In fact, according to Muñoz-Rojas and Pinto Correia (2018), who cite the Portuguese Bureau of Planning, Policies and General Administration and the National Institute of Statistics, the official figures for olive oil production worldwide that place Portugal, with around 1.5% of the olive oil production, in a situation that is still comparatively lower to Greece and Africa or Asia and well beyond Spain (40%) and Italy (20% according to the official statistics that are based on olive oil traded under the Italian denomination), may be distorted by the fact that some of these countries, particularly Italy, have a competitive advantage in export markets worldwide related to the market represented by the more than 50,000 Italian restaurants distributed around the globe; most of these restaurants demand olive oil which appears with the denomination of Italian product of EU origin, while much of it is reality of Portuguese origin.

Despite the cyclic price volatility, in the markets of Spain and Italy, the Portuguese olive oil is less valued. In both cases, and as stated before, most imports from Portugal are considered raw material to be incorporated in the domestic production and traded under notoriety brands at high prices, which translates an important discrepancy between quality and notoriety perceived by foreign customers. However, the production standard of Portuguese olive systems could grant them a competitive and technological advantage in international markets. The current prices and quality standards of olive oil, mostly classified as "virgin" or "extra-virgin" and excellent organoleptic characteristics and levels of acidity, are clearly advantageous (Fernandes, 2018).

Nevertheless, a certain oligopoly exists in the market that diminishes producers bargaining power and their individual market access: in the producers' side, two main operators—SOVENA and GALLO—hold almost 50% of the market; in the retailers' side two or three big firms represent an extremely large share of the olive oil retail market, both with private brands and with their own retailers' brands. Both sides are also present in international markets. When olive oil is considered as a hedonic food product, a brand strategy can be enacted associating the country of origin to the olive oil product that could overcome market power problems (Lans et al., 2001) This is effective as the origin and price and olive oils, along with their categorization are the most important determinants of olive oil purchase by retailers (Herculano, 2012). Price and origin are also important factors influencing consumption (Herculano, 2012).

34.3 Exploring the Future: How Does the Olive Oil Sector Match the Challenges?

"The olive oil sector has evolved more in the last 20 years than in the previous 5000 years". The sentence is of one of the big players in the Alentejo olive sector and says a lot about the olive oil sector attributes in 2020. Back in 1998, Fontanazza et al. (1998) stated that the capacity of olives to survive trough centuries has preserved the ancient olive production systems that rarely have been conveniently modernized. This situation changed over the last 20 years. In fact, Avillez (2016) states that the evolution of the olive oil sector in Portugal over the last 20 years has been characterized by a huge increase in production that doubled as a consequence of productivity increase in the new irrigated groves in Alentejo, a burden both in production technologies and in transformation and concentration capacity of industrial mills, interesting policies to support investment and production, a positive evolution of prices and a national and international recognition of Portuguese olive oil qualities. These statements were corroborated by all the players interviewed.

According the interviews, 95% of the olive oil in Alentejo is virgin and extra virgin olive oil and has a quality recognized by markets. This olive oil quality is supported by innovation, in what concerns both the groves and the transformation, by a professional management and a professional technical support, both in the field and oil extraction. The interviewers also corroborate this and underline that the main challenge, in this days, is to lower costs, which will maintain economic sustainability. The price in the market is mostly influenced by Spanish price, and most of the Spanish grove is a traditional grove, with high production costs—the price established by the market under these conditions is favourable for Alentejo. Additionally, Portugal is the first country to harvest olives and therefore sells first than the other countries (Fernandes, 2018). The question is for how much longer, considering the expectable long-term supply growing resulting from the Spain conversion of this traditional extensive olive groves into modern and intensive/super-intensive ones which could lead to a drop in market price. Also, the lack of recognized Portuguese brands and suitable distribution channels is a problem that leads foreign customers to choose products from other countries with much notoriety, such as Italy or Spain (Fernandes, 2018). The vision conveyed by the interviewees is that this implies different kinds of investment. Firstly, an information system is required that would act as a database to support real-time decision-making. This is specially relevant in view of gaining access to new markets and market channels and to improve market knowledge through digitalization. Furthermore, this would be an important step to enhance consumers' training, specially in non traditional consumer markets, and concerning olive oil benefits and their labelling information. Specialists are consensual in that new opportunities exist: in what concerns current markets, an effort should be made on the increase of exports, with intensification of geographical and cultural segments (e.g. Brazil and "saudade" market), and exploring others, as lifestyles and values (e.g. healthy foods, gourmet) and sustainability concerns and responsible consumption (e.g. organic—the organic olive oil niche is increasing in several international markets being very minor nationally—fair trade, ecosystems service). Also consensual is the importance of revitalizing AIFO, which aggregates the representative associations of production, processing and marketing, as a fundamental tool for the sector's organization and development: domestic and international olive oil promotion, value chain information and regulation and I&D activities.

This focus on economic sustainability by the costs side and the professionalization and innovation, both in grove management and olive oil production, contributes to environmental sustainability: the efficiency is one of the main concerns of the sector, and a big effort is being put in water use efficiency and energy efficiency (Cichoń, 2018). In what concerns agrochemicals the control is very efficient as olive oil markets are very demanding in regard to the existence of residues, and producers do not risk; olive production is also a somehow resilient production in what concerns climate change—olive groves support very different climatic conditions, can be produced in warmer conditions (as can be seen by its production in North Africa) and are very resilient to water stress.

The dynamic in what concerns qualified employment contributes for the social sustainability of the sector. In spite of the changes that have ruled out the high need for undifferentiated work, the actual olive oil sector assures employment in management, in the field, in laboratories and in olive mills, contributing to raise the pattern of qualification in agri employment.

The sector is very diversified, and production network is also very differentiated. There are big, small and medium producers, big olive mills that only transform their olives, others that buy olives from producers and eventually buy also the olive oil or just sell services, and there are cooperatives that can act also as service providers or can sell the olive oil with their own brand. This means all the producers have solutions to their specific case—they want to sell in bulk (and have a faster income), they want to sell under a label (in cooperatives), and they want to use their own label for all the olive oil produced or just for a part of it. These attributes must be used to meet the biggest challenges: be sustainable while lowering the costs and choose how to enter the markets and how to instruct the markets towards the quality produced.

34.4 Marketing Strategy for Alentejo Olive Oil Sustainability

Previous olive oil sector diagnostics shows its great relevance in the Portuguese food and beverage sector, both in added value and in employment, with a positive effect on the balance of trade and a real opportunity in the performance of the national economy. As showed before, the main contributor for this performance is Alentejo. Despite investment intensity, the rate of return on investments should be improved through a better valourization of extra quality of virgin olive oil in foreign markets witch pay for it.

Two different production models (extensive and intensive/semi-intensive) and, consequently, two business strategies and distinguishing performances can be identified in Alentejo. They are centralized, respectively, in two organizations, the CEPAAL and the OLIVUM, both mutually valuable and holding dimensions that are important for the future of the sector. Overall, even if their strategies are clearly different, they should work together with AIFO towards market orientation and higher added value. Two main approaches can be identified in the traditional/extensive producers and cooperatives, affiliated under the CEEPAL and related to traditional olive groves varieties and olive oil private brands: raise competitiveness through gradual intensification and differentiation with higher added value (e.g. branding and organic certification) and the multi-functionality combining with tourism, food and wines. Otherwise, in the intensive and super-intensive business model, mostly with unmarked bulk sales, the strategy is to warrant market share, competitiveness and profit from economies of scale and market control, integrating other activities within the chain of services of manufacturing olive oil for others and marketing and export services. Despite the prevalence of different business strategies, the consensus is that the sector will still grow in the future and will stay resilient facing foreseeable changes in the European policies towards lower subsidies and more market-oriented policies. However, a higher internationalization level is indispensable, with new markets entering and exploring new segments, as well as a better and more specialized and professionalized marketing strategies.

The proposal presented in this chapter suggests that the basis for a national strategy should be to position Portugal as the "sustainability hotspot" of the world's olive oil markets in the coming years, supported on an extra quality product (extra virgin axe), in compliance with sustainability standards, close relationship with consumers and the digitalization and balance of power in the value chain. This strategy identifies four levels: (1) country branding under the responsibility of AIFO; (2) regional brands, such as organic olive oil and designations of origin and geographical indications, under the responsibility of private certification bodies; (3) private brands under the responsibility of cooperatives, companies or other economic actors in the sector; and (4) unmarked bulk sales. These different levels and organizations can compete with each other, either because they propose to promote Portuguese oils or because they compete for the same financial envelope and funds from common agricultural policy.

Consumers are regularly seeking products which are careful stewards of environmental issues, business models oriented towards energy and water conservation, composting, canopy management, minimal use of chemicals, social responsibility within the community and so much more. This should be the basis for the future Olive Oil of Alentejo Sustainability Programme (OOSP), focused in making sustainability the core of business model and the distinctive mark of Portuguese olive oils. With similarities with Wines of Alentejo Sustainability Programme (WASP), the OOSP should be a voluntary membership programme with the goal to "reduce costs and increase economic returns through a proactive approach to environmental pressures and social concerns". Using data from AIFO information system and adopting the best practices from WASP, the OOSP should be developed with the scientific support from universities/research centres and other specialized contributions and operationally managed by in a partnership between OLIVUM and CEPAAL.

34.5 Final Remarks

This chapter provides an overview of the Alentejo olive oil sector, supported in statistical data and qualitative in-depth interviews. The findings suggest that the sector has a high potential for further growth, particularly in international markets, deepening the actual ones or exploring new markets and new consumer segments.

The sector is diverse and modern and is increasing its competitiveness and capacity to undertake/explore new market opportunities but also faces a number of internal and external challenges. Thus, to explore the future, marketing strategies should be developed to increase internal and external consumption, through reduced costs/increased income, supply differentiation and diversification within complementary activities and products, greater market orientation and better marketing performance, digital transformation and sustainability standards fulfilments.

The Olive Oil of Alentejo Sustainability Programme can be the way to integrate the entire olive oil sector in a philosophy of social, environmental and economical well-being at a regional level with high emphasis for Portuguese national strategy. All the actors should be called to have a cooperative role, share information and follow eco-efficiency principles.

Acknowledgements and Funding Work by BCF was partly funded by FCT-Portugal (UIDB/05183/2020 as members of MED-Universidade de Évora and LA/P/0121/2020 as members of the CHANGE- Associated Lab-Universidade de Évora).

Funding This research was funded by National Funds through Fundação para a Ciência e Tecnologia, grant numbers UID/04007/2020 and UIDB/05183/2020.

References

Avillez, F. (2016). *A importância económica do sector do azeite. Communication presented at the Seminar O Sector do Azeite em Portugal.* Crescimento e competitividade, Lisbon. Accessed 28 Nov 2016.

Cichoń, E. (2018). *Mechanisms to promote circular economy in agricultural sector – life cycle assessment of olive oil production in Portugal.* Thesis to obtain the master of science degree in energy engineering and management, Instituto Superior Técnico, Lisboa.

Evangelista, F. L. G. F. (2017). *The impact of new trends in supermarket environment: The case of the olive oil sector.* Dissertation submitted in partial fulfillment of requirements for the MSc in marketing, ISCTE – Business School, Instituto Universitário de Lisboa.

FAO. (2020). *FaoStat.* Available at http://www.fao.org/faostat/en/#data/QC

Fernandes, A. M. (2018). *Export promotion agency for the Portuguese olive oil sector: A boost in exports of branded olive oil.* Dissertation submitted in partial fulfillment of requirements for the MSc in management with specialization in strategy & entrepreneurship, Universidade Católica Portuguesa, Lisboa.

Fontanazza, G., Cappelletti, M., Cimato, A., Iacoboni, N., Loussert, R., & Trigui, A. (1998). Aspectos Genéticos e Técnicas de Propagação para Cultivo Intensivo. In *COI – Enciclopédia Mundial da Oliveira, COI* (pp. 113–144).

Herculano, H. (2012). *Fatores de decisão na compra de azeite embalado nos mercados grossista e retalhista (Msc in Oliviculture and Olive Oil).* Universidade de Évora. http://hdl.handle.net/10174/15964

Instituto Nacional de Estatística - INE (2021). Agricultural Census of Portugal 2019. [Available at: https://www.ine.pt/xportal/xmain?xpid=INE&xpgid=ine_publicacoes&PUBLICACOESpub_boui=437178558&PUBLICACOESmodo=2&xlang=en] Last accessed: 30/09/2024.

IOC – International Olive Council. (2015). *International olive oil production costs study: Results, conclusions and recommendations.* International Olive Council. Available at https://www.internationaloliveoil.org/wp-content/uploads/2019/11/INTERNATIONAL-OLIVE-OIL-PRODUCTION-COSTS-STUDY-.pdf

Lans, I. V., Ittersum, K. V., Cicco, A. D., & Loseby, M. (2001). The role of the region of origin and EU certificates of origin in consumer evaluation of food products. *European Review of Agricultural Economics*, 451–477.
Leonidou, L. (2004). An analysis of the barriers hindering small business export development. *Journal of Small Business Management, 42*(3), 279–302.
Mielke, T. (2019). Olive Oil. In G. M. B. H. ISTA Mielke (Ed.), *Oil world annual 2019*. [https://www.oilworld.biz/t/publications/data-base]
Muñoz-Rojas, J., & Pinto Correia, T. (2018). *H2020-SFS-2014-2 SUFISA Portugal National Report*.
Parras Rosa, M., & Veja-Zamora, M. (2020). Análisis de la Demanda. In *Informe Anual de Coyuntura del Sector Oleícola* (1st ed.) March, Caja Rural de Jaén, I.S.B.N.: 978-84-09-18981-6.

Part VI
Olive Landscapes Across the Mediterranean: Regional Heterogeneity and Case Studies–Olive Groves and Landscapes in Tunisia

Chapter 35
A Historical and Social Insight of Olive Groves and Landscapes in Tunisia

Olfa Boussadia

Abstract The olive tree was brought from the Middle East to Tunisia by the Phoenicians. Many Mediterranean civilizations, Phoenicians, Greeks, Carthaginians, Romans, and then Arabs took turns throughout history in a tradition handed down from father to son to transmit the know-how of the culture of the olive tree. Tunisia has thus become the heir to 2000 years of know-how in the cultivation of olive trees.

Tunisia's olive resources are estimated at over 118 million olive trees, grown on 1.9 million ha, of which 251,569 ha is used for certified organic crops. The average production is 176,250 tons of olive oil per year over the past 5 years. The quantity of olive oil exported represents around 70% of national production.

The olive tree contributes to the income of 309,000 farmers and generates 50 million workdays per year, which is equivalent to more than 20% of employability in agriculture.

Mean orchard density in Tunisia ranges between 100 and 150 trees per hectare in irrigated orchards and 40 trees per hectare in rainfed orchards farmed for oil production. As a rule, there are 100 olive trees per hectare in the north, compared to 60 trees per hectare in the center and 20 trees per hectare in the south. In 2000, a "high-density" system (> 1000 olive trees per hectare) was introduced in Tunisia by private entrepreneurs.

Modern agriculture is relatively intensive. It is located in the irrigated areas of the fertile north as well as in some of the most favorable regions of the center. Traditional agriculture employs extensive rainfed farming techniques on poor and marginal land in the south and parts of the center of the country. It depends only on precipitation.

In the sector of table olives, although smaller than the area used for olive oil production, table olives are mainly cultivated in the northern and central part of Tunisia and cover 25,000 ha.

Keywords History · Socioeconomic importance · Olive groves

O. Boussadia (✉)
Olive Institute, Unit of Sousse, Sousse, Tunisia
e-mail: io.olfa.boussadia@gmail.com

J. Muñoz-Rojas, R. García-Ruiz (eds.), *The Olive Landscapes of the Mediterranean*, Landscape Series 36,
https://doi.org/10.1007/978-3-031-57956-1_35

35.1 History: Tunisia, a Millenary Olive Grove

Olive oil has marked all the great civilizations that have spread to the Mediterranean and Tunisia for thousands of years. Rich in its history in the region of the Mediterranean basin, the olive tree was brought from the Middle East to Tunisia by the Phoenicians, founders of Carthage. Many Mediterranean civilizations, Phoenicians, Greeks, Carthaginians, Romans, and Arabs have taken turns throughout history in a tradition handed down from father to son to transmit the know-how of olive tree cultivation (ONH, 2020).

Tunisia has thus become the heir to 2000 years of know-how in the cultivation of olive trees and therefore "an olive grove par excellence," with plantations spread over a large part of the territory, starting with the island of Cyraunis (Kerkennah) passing by Byzacene (Cap Bon) and Hadrumète (Sousse) to finish in the steppes of Kasserine and, in the South, around Zarzis and the island of Djerba (IOC, 2017).

At the time of the Carthaginians, a true culture of the olive tree began to spread thanks to the advantages granted to the peasants who created olive grove Then, the Romans developed it further, by intensifying irrigation on this land where the rain was scarce and by inventing the technique of oil extraction, as evidenced by the excavations of Sbeitla and El Jem, the many mosaics of Roman discoveries in Sousse, and archaeological and ethnographic objects and utensils collected, which certify the importance of olive oil in everyday life and the art of living of peoples through the history of Tunisia (IOC, 2017).

35.2 Socioeconomic Importance

Tunisia is the southern Mediterranean country best known in the field of olive growing and of the export of olive oil after the European Union. Olive growing constitutes for Tunisia the main agricultural activity, and its socioeconomic role has capital importance.

Olive oil is the most important agricultural product in Tunisia, as it represents 50–60% of the country's agri-food exports and more than a million people depend on the sector. Indeed, olive oil exports contribute almost 5.5% to the national economy, reaching 7% in certain years, as was the case in 2005. Despite the importance of the oil sector olive in the national economy of Tunisia, certain difficulties limit the growth of the sector and the exploitation of its true potential (FAO, 2019; Allouche et al., 2021).

The olive oil sector is characterized by a large number of operators: around 309,000 farmers, more than 1700 oil mills, 50 packaging companies, and a large number of exporters who act usually individually. Olive growing is, therefore, the main activity for 60% of farmers nationally, whilst the olive oil sector represents the direct or indirect livelihood for over a million people, and generates between 20 and 50 million working days per year, which is more than 20% of employability in agriculture (IOC, 2017).

Olive growing contributes greatly to regional balance since it is often the only viable crop in the least favored areas. It makes it possible to fix the populations in areas that, in another way, would suffer the negative impact of the rural exodus (Gharbi et al. 2014; Allouche et al., 2021).

In addition to this large number of operators, we are witnessing a multiplicity of public development and supervision structures from the Ministries of Agriculture, the Ministry of the Environment, the Ministry of Industry, and the Ministry of Commerce. However, the sector is characterized by a lack of horizontal and vertical integration of the various actors as well as a lack of coordination of the various development agencies of the state, and this is particularly visible at the level of production where there are absent cooperatives or other structures to bring farmers together (FAO, 2019).

35.3 Olive Groves and Landscapes in Tunisia

Olive resources in Tunisia are estimated to be over 118 million olive trees. Adapted to soil conditions and climate of Tunisia, the olive tree extends over the whole agricultural land and currently occupies 1.9 million hectares. A large part of this area is cultivated in a dry environment, which affects the productivity considered to be the lowest in the Mediterranean. Indeed, olive oil production has varied over the past 5 years between 100,000 tons and 325,000 tons with an average of almost 176,250 tons of olive oil (DGPA, 2020).

Olive populations are present in all regions of Tunisia, from north to south and from east to west. At the north and in certain areas of the center, they are cultivated in association with other annual crops (cereals or trees fruit trees such as citrus, vine, or almonds), while that in the south, they are exclusively cultivated in monoculture (IOC, 2012).

The density varies considerably depending on the rainfed or irrigated crop regime. The number of tree per hectare varies in rainfall regime from north to south according to the precipitation averages ranging from 100 olive trees/ha in the north where the rainfall is greater than 450 mm/year to 70 olive trees/ha in the center with rainfall between 200 and 350 mm/year and at 24–17 olive trees/ha in the far south where the rainfall is less than 200 mm/year (ONH, 2017).

In irrigated cultivation, it is necessary to distinguish several types of farms:

- Extensive plantations having been carried out in rainfed cultivation and which have been converted to irrigation, to undertake other intercropping crops (center and south), such as potatoes, tomatoes, peppers, cucurbits, peas, barley, alfalfa, wheat, and barley. This cultivation method characterizes several areas using water from the groundwater, to maximize profitability per hectare.
- Classic intensive plantations with planting densities varying from 200 to 280 olive trees per hectare, depending on the vigor of the cultivated varieties. This mode is present throughout the national territory from north to south and uses both groundwater and deep water.

- Dynamic intensive plantations with densities between 300 and 625 olive trees per hectare. Most of these plantations are newly created with foreign varieties, mainly Arbequina, Arbosana, and Koroneïki. Most of these plantations use water from the deep aquifer.
- Hyper-intensive plantations at high densities. This system was introduced by a private company in 1999 on an area of 200 ha. Subsequently, it expanded rapidly, reaching an area estimated at 4500 ha in 2010. The densities adopted in this method of cultivation vary between 1250 and 2500 olive trees/ha. This system is considered to be a large consumer of chemical inputs and water; this mode, which should be fully mechanized, has experienced stagnation in areas and currently only has around 5000 ha. The water used in this cultivation method often comes from deep aquifers (FAO, 2019).
- The Tunisian olive grove is made up of 20% of young plantations (1–20 years), 50% of plantations in production (20–70 years), and 30% of old plantations (over 70 years). Of this age category, at least 10% is planted before 1920 and are over 100 years old. These senescent plantations are mainly found in the Sahel and the south of the country (FAO, 2019).

Organic olive growing is a dynamic and evolving component of the Tunisian olive sector. The olive-growing area having obtained organic gone from 127,250 ha in 2015 to 251,569 ha in 2019 (FAO, 2019).

The production of olive oil in organic mode has not stopped increasing according to the years, firstly because of the olive-growing infrastructure adapted for this mode of cultivation, by the fact that the majority of olive growers proceed neither to phytosanitary treatments nor to fertilization, and then because of the added value of the finished products of this culture and finally of the encouragement of the state concerning this mode of culture (FAO, 2019).

Table olive tree growing covers currently more than 25,000 ha. The main variety is Meski (60% of new plantations) covering more than 12,000 ha. This variety, whose fruit is highly appreciated for its organoleptic qualities, has a low production potential due to self-pollen incompatibility (DGPA, 2020).

35.4 Conclusion

In Tunisia, the olive sector constitutes a main component of the agricultural and agri-food sectors. It contributes to the achievement of national objectives of economic growth, food security, job creation, increased export earnings, and the preservation and enhancement of increasingly rare and vulnerable natural resources (Allouche et al., 2021). Particular attention should be paid to olive oil production to have better quality and a tastier savor. Great care is necessary at each stage, from the planting of the olive tree to the final consumer.

References

Allouche, F. K., Abu-hashim, M., & Negm, A. (2021). *Agriculture productivity in Tunisia under stressed environment*. Springer. ISBN: 978-3-030-74659-9.

DGPA (Direction Générale de la Production Agricole) - Ministere De L'Agriculture Et Des Ressources Hydrauliques De La Tunisie. (2020). *Biobase, base de données documentaire en AB* [https://abiodoc.docressources.fr/index.php?lvl=publisher_see&id=2268]. Accessed 29 Sept 2024.

FAO. (2019). *La filière oléicole tunisienne à l'horizon 2030* (p. 222). Projet TCP/TUN/3701 (November 2019).

Gharbi, I., Issaoui, M., & Hammami, M. (2014). La filière huile d'olive en Tunisie. *OCL, 21*(2), D202.

IOC. (2012). *International Olive Oil Council, 2012*. Politique—Tunisie. Indicateurs macro-économiques et agricoles, Unité 2000–2005. Tunisie. 108/Doc. n° 4 Mise à jour n° 32 Réf: 29–36.

IOC. (2017). International olive oil council. Aperçu du secteur oléicole en Tunisie. *OLIVAE Journal Officiel du Conseil Oléicole International*. N° 124 Éd. Française. NOVEMBRE 2017.

ONH. (2017). *Office National de l'huile*. https://www.onh.com.tn. Accessed Dec 2017.

ONH. (2020). *Office National de l'huile*. https://www.onh.com.tn. Accessed 20 Feb 2020.

Chapter 36
Pedoclimatic and Landscape Conditions of Tunisian Olive Groves

Olfa Boussadia and Mortadha Ben Hassine

Abstract Tunisia is simultaneously a Mediterranean and a Saharan country, with soils reflecting such climatic, morphological, and geological diversity. According to the French system of soil classification, the soils of Tunisia are classified as podzols, vertisoils, red Mediterranean soils, calcic-magnesic soils (dominant soils), brown and isohumic soils, and saline and hydromorphic soils, along with poorly evolved soils in the most arid regions. The climate is characterized by a soft winter and a hot summer. Precipitations are irregular and present a gradient between north and south with the highest precipitation in the northwest (1500 mm/year). The center of the country is characterized by a semiarid climate, and mean annual precipitation is moderate (200–400 mm) and occurs between October and May. The southern climate is dry (60 mm/year).

Olive tree crops include deep brown or alluvial soils with a well-balanced texture. These soils are found across various bioclimatic areas. Olive trees extend over 1.8 million ha.

Clay soils are placed mainly in the north where bioclimatic areas range between humid and semiarid. In semiarid regions, soils are poor in organic matter (less than 1%), rich in calcium carbonate, with alkaline pH (>7.8), and moderately calcareous (>15%). Also, the cation exchange capacity is low, and the nutrient circulation and penetration are limited.

Tunisian soils are subjected to severe climatic conditions, including drought (regularly affecting over 3/4 of the surface of the country), short duration torrential rainfall, and increasingly impactful anthropogenic factors along the coast. All these factors turn the soils fragile and increase their sensitivity to their degradation, either mechanical or chemical. Olive tree crop lands were traditionally protected from runoff through hydraulic infrastructures (terraces, tabias, Meskat system, and Jessours), benefiting from additional water along the rainy periods in the central and southern regions. All the above conditions allowed for olive trees to be planted in

O. Boussadia (✉) · M. B. Hassine
Olive Institute, Unit of Sousse, Sousse, Tunisia
e-mail: io.olfa.boussadia@gmail.com; bnhmtd@unife.it

J. Muñoz-Rojas, R. García-Ruiz (eds.), *The Olive Landscapes of the Mediterranean*, Landscape Series 36,
https://doi.org/10.1007/978-3-031-57956-1_36

marginal zones characterized by alternating production. The challenge currently is to adopt novel strategies such as agroforestry based on cropping and green manures, in order to improve soil structure and enhance soil fertility rates as basis for enhancing the sustainability of olive grove landscapes.

Keywords Tunisia · Climate · Soil · Marginal zones · Olive sustainability

36.1 Introduction

In Tunisia, tree crops represent more than two thirds of the cultivated land (Mtimet, 1999). Among this tree crops, olive tree represents the most important crop in terms of cultivated area (1.8 million ha equivalent to 34% of total cultivated land). It has a nutritional value in the Tunisian diet and a high socioeconomic importance. In order to identify the main factors of the olive tree sustainability in Tunisia, the study of the different areas of olive cultivation should be linked to the different levels of the soil-plant-atmosphere system. It should specify the classes of soil and its interaction with the climate and the landscape.

36.2 Pedoclimatic Conditions of Tunisian Olive Groves

Tunisia belongs to the Mediterranean basin and has a large area under Saharan conditions. The French system of soil classification shows that Tunisian soils are classified under seven different classifications such as podzols, vertisoils, red Mediterranean soils, calcic-magnesic soils (dominant soils), brown and isohumic soils, saline and hydromorphic soils, and also poorly evolved soils (Mtimet, 1999). Moreover, the Tunisian climate is characterized by a soft winter and a hot summer and presents a bioclimatic diversity among regions and irregularity of precipitations. The gradient of precipitation is from north to south where the northwest part of the country has the highest amount of precipitation (1500 mm)/year (Ben Ahmed and Ben Rouina, 2019).

The north has three bioclimatic areas such as the humid, subhumid, and semi-arid. In these areas, soils can be rich in organic matter and show cracking during dry season (topomorphic and lithomorphic vertisoils) or have a fine texture, a well-developed polyhedral structure, and calcareous accumulations (Mediterranean red soils in Beja, Boukornine, and Djebel Zaghouan). Moreover, soils of the north exhibit crusted horizons or contain coarse calcareous elements (calcic-magnesic soils). Poorly evolved (alluvial) soils are in the large valleys of extended watersheds of Mellegue, Mejerda, and Tassa. The region undergoes salinization processes downstream in the low valley areas of Mejerda (Mornaguia, Kalaat el-Andalous) or along the Beni Khiar, Kélibia (Cap Bon) plain (Mtimet, 1999). The best soils and highest precipitation are in the north, but this area is more dedicated to cereals rather than olive trees. Furthermore, the smallest area of olive trees is in the north (16%) compared to the south and center.

The center is dominated by the semiarid bioclimatic stage, but some areas are under the arid climate. The mean annual precipitation is moderate (200–400 mm) and occurred between October and May (Ben Ahmed and Ben Rouina, 2019). There is variability according to the altitude that allows identifying four different regions: the high steppes, the low steppes, the Sahel, and the Kairouan region. The soil is variable between an isohumic (brown soils with coarse texture and 0.5–1.5% of organic matter in the low steppe and Sahel) and calcic-magnesic, alluvial, and saline and resulting from erosion in the high steppe of Kasserine and the southwestern part of the ridge. Some areas like Sidi El Hani, Monastir, El Gharra, and Charita are characterized by saline-sodic soils found in depressions and in the Sebkhas. The center has the largest areas of olive cultivation that represents 63% of total land of olive trees in Tunisia.

Ben Ahmed and Ben Rouina (2019) indicated that the south climate is dry (60 mm/year occurring in 30–40 days). Soils in this region are calcareous containing gypsum and limestone (calcic-magnesic), subdesertic (isohumic), saline-sodic, and poorly evolved (alluvial soils). These soils are poor in organic matter (less than 0.5%) and have silty-loam to sandy texture. In the areas of Nefzaoua, Gherib, Jerid, and Gafsa, soil shows spots of water logging and salinity (Mtimet, 1999).

In general, the structure of Tunisian soils is poor due to the low organic matter contents (less than 1%), amount of calcium carbonates, pH alkalinity (>7.8), and moderate calcareous features (>15%). Also, the cation exchange capacity is low, and the nutrient circulation and penetration are limited. The objective is to convert existing sandy soils that are squeletic to well-balanced soils. A soil rich in humus will allow the olive tree an expansion of the roots and better stability compared to a poor soil. Olive trees will be able to fix the soil and reduce soil water and wind erosion. In fact, actually, Tunisia losses 25,000 ha of soil/year due to erosion. Harsh climatic conditions of drought or short duration of torrential rainfall make the soil structure fragile and very sensitive to the degradation at surface and deep layers. For olive tree, factors of non-deep soil depth (<80 cm), high slope, fine structure, and crust layer close to the surface (<40 cm) will affect negatively the growth and development of olive tree productivity and sustainability (Mtimet, 1999). The same author indicated that, in order to keep the olive production stable in particular and sustainable in general, the Tunisian olive production system should adopt news strategies based on the agroforestry.

Jendoubi et al. (2019) indicated, in a study conducted in northwestern Tunisia, that agroforestry system by combining an olive plantation with annual field crops has increased soil organic carbon (SOC) content in the most vulnerable areas. This system started in 1980 by a program developed by the Office Development Sylvo-Pastoral Nord Ouest (ODESYPANO) and financed by the World Bank implemented in order to remedy the degrading effects of monoculture of annual cropping, deforestation, and overgrazing on pastures and forests. Before applying the agroforestry system, there was a predominance of field crops in 1985 as land use (82%). This practice was reduced to 71% in 2016, while the forest plantation increased from 3.9% to 9% of the watershed from 1985 to 2016.

In the center and the south of Tunisia, the soil organic matter is very low, and soil fertility should be enhanced by using green manures to improve the soil structure. In fact, the ground cover is important for preventing erosion by reducing its velocity and increasing infiltration into the soil. Also, compared to row-cropped fields, nutrient uptake can be higher and reduce nutrient loss into streams. Carbon sequestration is an environmental goal that can be reached by applying an agroforestry strategy. It increases carbon stocks and can maintain the wildlife habitat.

36.3 Landscape Conditions of Tunisian Olive Groves

Mediterranean agricultural landscapes provide ecosystem services and disservices that are driven by land use pattern dynamics, the latter of which results from the crop spatiotemporal distribution (Mekki et al., 2018). In Tunisia, olive tree is appreciated for its great importance in the preservation of green area landscape in arid region, its limitation of desertification and the prevention of soil erosion and land degradation, and so its involvement in the maintenance of land durability (Ben Ahmed et al., 2009).

Since the twentieth century, in southern Tunisia, strong anthropic and environmental pressures have resulted in several asynchronous phases of occupation and enhancement of space, visible in the landscape through water and soil conservation works, called Jessours. These small earth dams, sometimes consolidated by a dry stone wall, close the thalwegs of the ravines in which they are located. Then it forms a flat cultivable surface, concentrating flows and preserving water and soil. In general, it includes a weir, central or lateral, sometimes several, whose function is to evacuate the overflow toward the plots downstream and to protect the system from destruction (Blond et al., 2019).

In Tunisian olive groves, we can identify different landscapes that are in close relation to precipitations that follow a gradient of precipitation from north to south. In the north, olive groves are characterized by high density (100 tree/ha). As the gradient of precipitation decreases from north to south and the soil fertility decreases, the olive groves density decreases. It is 50–60 olive tree in the center and 17–20 olive tree in the south (FAO, 2015).

The low density of olive groves in the center and south is explained by the fact that 85% of olive groves are located in the drier regions of the center and the south belonging to the semiarid and arid climate (FAO, 2015). In addition, the traditional system of olive groves is rainfed and extensive and does not rely on inputs. Only 3% of olive groves area is irrigated. The transition from extensive to intensive agriculture started slowly. It will require the irrigation of olive groves and planting olive cultivars that are adapted to the super-intensive agriculture using more inputs and therefore changing the structure of olive groves and increasing the plantation densities. The surface area of olive groves in the north slightly increased from 2008 to

2012 suggesting that the north is attractive for farmers able to control the production cost efficiently and ready to invest more, especially to irrigate their olive groves (FAO, 2015).

36.4 Conclusion

The sustainability of olive tree in Tunisia depends not only on pedoclimatic conditions but also on landscape conditions through water and soil conservation works. Furthermore, the agroforestry land use system should be adopted. Also, conserving the traditional and ancestral hydraulic works such as the terraces, cordons, tabias, Meskat system, and Jessours is essential to protect soils from runoff. It has a beneficial effect as it collects water in the rainy periods.

References

Ben Ahmed, C., & Ben Rouina, B. (2019). Tunisia: Water resources' management and sustainability of agriculture. In *Encyclopedia of Environmental Health* (Vol. 6, 2nd ed.). https://doi.org/10.1016/B978-0-12-409548-9.11854-8

Ben Ahmed, C., Ben Rouina, B., & Boukhris, M. (2009). Chapter 10: Olive tree growth in Tunisia; types, limitations and influences. In *Forest management*. ISBN 978-1-60692-504-1.

Blond, N., Jacob-Rousseau, N., Ouerchefani, D., & Callot, Y. (2019). Étude de l'évolution duravinement dans les jessour du Sud Tunisien grâce aux images aériennes. *Cybergeo: European Journal of Geography, Environment, Nature, Paysage, 906*. https://doi.org/10.4000/cybergeo.32495

FAO. (2015). *Tunisie Analyse de la filière oléicole* (p. 186). https://www.fao.org/3/i4104f/i4104f.pdf

Jendoubi, D., Liniger, H., & Ifejika Speranza, C. (2019). Impacts of land use and topography on soil organic carbon in a Mediterranean landscape (North-Western Tunisia). *The Soil, 5*, 239–251. https://doi.org/10.5194/soil-5-239-2019

Mekki, I., Bailly, J. S., Jacob, F., Chebbi, H., Ajmi, T., Blanca, Y., Zairi, A., & Biarnès, A. (2018). Impact of farmland fragmentation on rainfed crop allocation in Mediterranean landscapes: A case study of the Lebna watershed in Cap Bon, Tunisia. *Land Use Policy, 75*, 772–783. https://doi.org/10.1016/j.landusepol.2018.04.004

Mtimet, A. (1999). In Ministère de l'Agriculture (Ed.), *Atlas des sols tunisiens* (p. 165) ISBN: 9973-914-05-8.

Chapter 37
Varieties and Olive Grove Production Systems in Tunisia

Sofiane Abdelhamid, Amal Omri, Amira Araouaki, and Khaled Hibar

Abstract The olive tree (*Olea europaea* L.) is a species characteristic of the Mediterranean landscape, and Mediterranean countries clearly dominate world olive oil and table olive production and consumption. Tunisia is the largest southern Mediterranean country producing olive and olive oil and is considered the fourth producer of olive oil in the world. At present, olive trees are planted in all regions from the north to the south, and a rich assortment of varieties exists in the country due its geographical location and the crossroad trade between the East, West, and Europe. However, the two main varieties grown in the country are Chétoui in the north and Chemlali in the south. Others, called secondary varieties, are more specific to small regions, especially Oueslati, Chemchali, Zalmati, Zarrazi, and Gerboui varieties. Olive culture is a complex agrosystem with variability in production systems, cultivation techniques, and genetic resources. Most of the orchard density ranges between 100 and 150 trees/ha in the north compared with 60 trees/ha in the center and 20 trees/ha in the south generally in rainfed orchards farmed for oil production. Since the 2000s, traditional planting systems have been transformed into more intensive groves, and some traditional orchards are reconverted into more intensive systems. At present, there are more than 2000 ha of super-intensive orchards. Nevertheless, traditional olive production systems are the most expended in the country as these systems contribute to rural development, landscape conservation, and protection of the environment. Tunisia is deploying restructuring efforts to modernize and improve the quality of olive oil, accompanied by a significant expansion of intensive and high-density groves. The challenge of Tunisia is to develop its competitive strategies of olive sector allowing it to strengthen its competitiveness compared to European countries.

Keywords Tunisia olive sector · Olive growing · Production system · Cultivars · Chemlali

S. Abdelhamid (✉) · A. Omri · A. Araouaki · K. Hibar
Olive Institute, Unit of Sousse, Sousse, Tunisia
e-mail: amal.omrii@yahoo.com; khaled_htn@yahoo.fr

J. Muñoz-Rojas, R. García-Ruiz (eds.), *The Olive Landscapes of the Mediterranean*, Landscape Series 36,
https://doi.org/10.1007/978-3-031-57956-1_37

37.1 Introduction

Tunisia is the most important olive-growing country of the southern Mediterranean region, and over 30% of its cultivated land is dedicated to olive growing (1.88 million ha, DGPA, 2019). The olive oil production is marketed to local demand for around 20%, while the rest 80% is exported to 50 international destinations, in which the main market is the European Union, with about 70% of the Tunisia's olive oil exports. Tunisia's olive resources are estimated at over 95 million olive trees, grown on 1,880,000 ha, of which 1,790,000 ha is produced in rainfall (95.2%) and 95,800 ha in irrigated orchards (4.8%). Approximately 254,411 ha is for certified organic crops, producing around 40% of world production of organic olive oil and has managed to glean more than 100 medals in quality competitions, and 6500–7000 ha is in hyper-intensive system. The number of trees cultivated in the rainfall is 77,130,000 trees (80%) and 1,885,0000 (20%) in irrigated systems. In rainfall area, the annual productivity of olive oil is considered as among the lowest in the Mediterranean. In fact, the production of olive oil varied during the last decade between 120,000 tons and 350,000 tons with an average of almost 180,000 tons of olive oil. To improve the variability of production from one year to another, the authority launched a strategic plan to increase the area of the irrigated olive grove.

The geographic distribution of olive growing in the country shows that the number of olive trees is mainly concentrated in the center regions (Sousse, Kairouan, Monastir, Mahdia, Sfax, Kasserine, and Sidi Bouzid) which represent 39.7% of the total number of trees in the country and participate with 48% of the total of olive oil produced. In the north (Tunis, Ariana, Ben Arous, Nabeul, Bizerte, Béja, Jendouba, Kef, Siliana, and Zaghouan), there is 34.5% of olive trees, participating with 21% of olive oil production. In the south (Gafsa, Gabès, Medenine, Tozeur, Kébili, and Tataouine), the number of tree represents 25.8% with a production of olive oil by 20% (Table 37.1).

Olive tree covered all the country from the north to the south, and olive orchards can be found under very different bioclimatic stages, from humid, subhumid, to semiarid and even arid and at different rainfall conditions, from the desert in the south "Ras Jedir" which is a very marginal area and to humid or subhumid climates. Semiarid and arid areas are characterized by a low precipitation and high evapotranspiration. These climatic factors cause long periods of water deficit. Olive has an exceptional capacity to resistance to the most difficult climatic conditions (Ben Rouina et al., 1995). The morphological and physiological behavior of the olive tree

Table 37.1 Repartition of olive tree and olive oil production in Tunisia

	Number of olive tree (%)	Production of olive oil (%)
North	34.5	21
Center	39.7	48
South	25.8	20
Total	100	89

DGPA (2019)

shows a great capacity adaptation due to the morphological particularities of its leaves and as well as its root system (Lavee & Wodner, 1995). In Tunisia, olive culture is a complex agrosystem with variability in production systems, cultivation techniques (irrigation), and genetic resources.

37.2 Traditional Olive Plantations

In general, olive cultivation occurs in two different and distinct systems of production: (1) traditional, extensive, rainfed orchards and (2) modern, high-density, intensive, irrigated orchards (Zipori et al., 2020). Traditional olive-based production systems are the most widespread in the country, located in ancient olive-growing areas and using local varieties (Fig. 37.1). Most of the Tunisia's olive orchards are rainfed, and traditional orchard density ranges between 100 and 150 trees/ha in the north compared by low planting densities with 60 trees/ha in the center and 20 trees/ha in the south generally in rainfed orchards farmed for oil production. When intended for table olives, orchard densities vary from 200 trees/ha in irrigated conditions to 100 trees/ha under dry farming. In rainfed orchards, growth rates and yield levels are much lower than in irrigated systems. Many plantations in production are between 20 and 70 years old, and sometimes they are centennial olive trees. In this system, olive harvesting operation is considered as a highly consuming hand labor task.

Most of the traditional olive orchards are produced under rainfed condition and are characterized by low planting densities (low water availability), and cultivated varieties were obtained by local growers through local selection across centuries or through the exchange of superior genetic material among different growers (Trigui, 1996). This system is characterized by low productivity, low inputs, and a very high biennial bearing but with a high cultivar biodiversity (Fernández Escobar et al., 2013).

Fig. 37.1 Olive orchards growing in traditional system

In this system, organic farming is proposed as another option to increase the benefit of olive growers since traditional olive growing is largely chemical-free orchards. Tunisia is the largest exporter of organic olive oil in the world and has around *250,000 ha of organically certified land.* The center of Tunisia is considered the most important Tunisian organic olive-growing regions in surface, in production, and in farmers' number. The total production of organic olive oil in Tunisia in 2017 exceeded 30,000 tons, accounting for approximately 10% of the total production.

In arid and semiarid areas, olive growers applied traditional and extensive olive plantation. These regions are characterized by the scarcity and the irregularity of the precipitation with high temperature and evapotranspiration. In order to overcome the deficit of water, olive growers applied some traditional soil and water conservation practices. Since antiquity, on the region of semiarid area (Sousse), growers applied the "Meskat and Mankâa" systems, and in the arid region (Medenine), they applied the "Jessours" system (Majdoub et al., 2015). Farmers applied these methods to cope with water shortage and especially to manage the runoff water.

37.3 Intensive Olive-Growing System

In Tunisia, since the 2000s, olive cultivation has converted from traditional, extensive, widely spaced, low-density, and rainfed orchards to intensive, moderate to high-density, closely spaced, and irrigated ones, leading to an increase in olive production and olive oil production (Zipori et al. 2020). The evolution from traditional to more intensive production systems is to reduce unproductive periods and costs. Planting densities varied from 200 to 625 olive trees per hectare depending on the availability of water from irrigation. Converted intensive plantations in which densities varied from 300 to more than 600 olive trees per hectare are newly created with introduced foreigner varieties from Spain and Greece such as Arbequina, Arbosana, and Koroneïki, respectively (Fig. 37.2). This intensive olive-growing system based on the intensified traditional plantations uses new technologies, artificial fertilizers, and pesticides and intensive management for weed control and soil management (Fernández Escobar et al., 2013). These changes in intensive growing system were accompanied with a higher mechanization for orchard management and also for harvesting. This intensification led to increased tree density whose number of trees per hectare depends on the edaphoclimatic conditions and mainly water availability and irrigation infrastructures. This has allowed an intensification of olive growing with an increase on tree density, up to 400 trees/ha, with final spacing determined by water availability, edaphoclimatic conditions, harvest system, and cultivar.

Fig. 37.2 Converted olive orchards from traditional system to intensive plantation

37.4 Hyper-Intensive Plantations

Intensification of olive orchard continues by the establishment of a new process based on high-density planting. This system was introduced in Tunisia in the 2000s by private Tunisian companies using foreign varieties, mainly Arbequina, its clone "Arbequina I-18" and Arbosana (from Spain), and Koroneiki (from Greece), because of their low vigor and erected habit which are the main criteria required for this type of planting system (Allalout et al., 2009).

Subsequently, high-density olive plantations have rapidly increased, covering approximately 5000 ha with densities of 1250–2500 trees/ha (Fig. 37.3). The main advantage of this system is the ease and the full mechanization of harvesting by the use of over-the-row mechanical harvesters, which reduce harvest costs in comparison to manual and even trunk shaking harvest and the rapid return on investment.

This system is considered advantageous, could reduce the labor needed for harvesting, and could obtain an olive oil with high quality for the rapid and earlier harvesting, but the process of intensification needs large quantity of chemical inputs and water use (Rallo, 2013). Additionally, with this system many problems could be observed especially related to deficient pruning and to performing monocone shape of trees from the beginning of plantation (Fernández Escobar et al., 2013).

Nowadays, breeding programs and research are undertaken in order for Tunisian olive varieties to adapt to this system and to new cropping technologies.

Fig. 37.3 Olive orchards growing on hyper-intensive system

37.5 Tunisian Olive Genetic Resources

The Tunisian olive tree belongs to the Mediterranean form *Olea europaea*, subspecies *europaea*, which includes wild (*Olea europaea* subsp. *europaea* var. *sylvestris*) and cultivated olives (*Olea europaea* subsp. *europaea* var. *europaea*), is a diploid species (2n = 2x = 46), and predominantly allogamous.

Tunisia olive varieties exhibit a great morphological differentiation and phenotypic plasticity to environmental factors which are marked by the existence of a considerable number of different olive cultivars. More than 150 cultivars are characterized by morpho-pomological traits related to the pit, fruit, leaf, and tree leading to the publication of catalogue entitled "olives of Tunisia: autochthonous varieties and local types" (Trigui & Msallem, 2002). Two major varieties are cultivated in the country: "Chemlali" in the south and the center of the country and "Chétoui" in the north (Trigui, 1996). These varieties account for 95% of the total olive tree orchards and contribute more than 90% of the national production of olive oil. As regards olive table varieties, we can find Meski, Besbessi, Bidh Hmam, Limi, and Limouni.

However, other secondary varieties were found which are cultivated in specific regions like Oueslati in Ouislatia, Chemchali and Zarrazi in Gafsa, Zalmati in Medenine, and Gerboui and Sayali in the north (Kef, Siliana, Béja…).

The "Chétoui" variety is the main variety cultivated in the north. It produces fruity and not very sticky oil, very appreciated for its taste and its organoleptic characteristics. Its cultivar is very rich on phenolic compounds and antioxidants.

The "Chemlali" variety is the main oil variety of the Tunisian olive grove, it is cultivated in the arid and semiarid areas (center and south), and it shows a great ability to adapt to soil and severe climatic conditions. It occupies almost 70% of the olive-growing areas and contributes more than 80% to the national production of olive oil. It presents a great phenotypic variability especially with regard to the shape and the color of the fruit. The oil produced is appreciated by consumers which is not very bitter and very spicy.

The identification of olive germplasm has been traditionally carried out by morphological descriptors related to the pit, fruit, leaf, and tree. Due to the great number of varieties and the existence of the problem of synonymy and homonymy (Abdelhamid et al., 2017), a powerful method of genetic analysis based on molecular markers has been developed and applied for studying the variability of olive cultivars and their origin, providing a more powerful tool to guarantee varietal identification. Several studies on Tunisian olive tree varietal identification and characterization based on genetic molecular markers such as microsatellite or Simple Sequence Repeat (SSR) (Abdelhamid et al., 2013), Amplified Fragment Length Polymorphism (AFLP) (Grati-Kamoun et al., 2006), Random Amplified Polymorphic DNA (RAPD) (Zitoun et al., 2008), and Single Nucleotide Polymorphism (SNP) (Rekik-hakim et al., 2010) are performed.

Several breeding programs were developed in Tunisia. The first breeding program was initiated by breeders and researchers in the olive institute, Sfax, since 1989 to improve the quality of olive oil of Chemlali and olive table quality of Meski varieties (Ben Amar et al., 2018). The National Olive Genetic Resources Conservation located in Boughrara-Sfax performed and maintained diverse germplasm resources of olive worldwide.

In conclusion, the olive oil sector constitutes a head component of the agricultural and agri-food sectors. Particular attention should be paid to olive agriculture and olive oil producers in order to save the profitability of the sector. The major challenges for olive production are to improve olive oil quality, develop complementary industries, and improve the productivity and yields in order to maintain their competitiveness on the international oil market.

References

Abdelhamid, S., Grati-kamoun, N., Marra, F. P., & Caruso, T. (2013). Genetic similarity among Tunisian cultivated olive estimated through SSR markers. *Science in Agriculture, 70*, 33–38.

Abdelhamid, S., Omri, A., Grati-kamoun, N., Paolo Marra, F., & Caruso, T. (2017). Molecular characterization and genetic relationships of cultivated Tunisian olive varieties (*Olea europaea* L.) using SSR markers. *Journal of New Sciences Agriculture Biotech, 40*, 2175–2185.

Allalout, A., Krichène, D., Methenni, K., Taamalli, A., Oueslati, I., Daoud, D., & Zarrouk, M. (2009). Characterization of virgin olive oil from super intensive Spanish and Greek varieties grown in northern Tunisia. *Scientia Horticulturae, 120*, 77–83.

Ben Amar, F., Arfa, H., & Yengui, A. (2018). Agronomic characterization of selected clones of Tunisian table olive (*Olea europaea.* L) Variety 'MESKI'. *Journal of Fundamental and Applied Sciences, 11*, 82–91.

Ben Rouina, B., Yousfi, M., M'Laouah, M., & Louizi, A. (1995). L'entretien des sols à olivier en milieu aride: conséquences agronomiques et maîtrise par l'agriculteur. In S. Zekri & A. Laajimi (Eds.), *Agriculture, durabilité et environnement* (Cahiers Options Méditerranéennes; n. 9) (pp. 125–134). CIHEAM.

Direction générale de la production végétale (DGPA). (2019). *Ministère de l'agriculture et de la pêche (Tunisie).*

Fernández Escobar, R., de la Rosa, R., Leon, L., Gomez, J. A., Testi, F., Orgaz, M., Gil-Ribes, J. A., Quesada-Moraga, E., & Trapero, A. (2013). Evolution and sustainability of the olive production systems. In N. Arcas, F. N. Arroyo López, J. Caballero, R. D'Andria, M. Fernández, R. Fernandez Escobar, A. Garrido, J. López-Miranda, M. Msallem, M. Parras, L. Rallo, & R. Zanoli (Eds.), *Present and future of the Mediterranean olive sector* (Options Méditerranéennes: Série A. Séminaires Méditerranéens; n. 106) (pp. 11–42). CIHEAM / IOC. International Seminar: Present and Future of the Mediterranean Olive Sector, 2012/11/26–28, Zaragoza (Spain). http://om.ciheam.org/om/pdf/a106/00006803.pdf

Grati-kamoun, N., Lamy Mahmoud, F., Rebaï, A., Gargouri, A., Panaud, O., & Saar, A. (2006). Genetic diversity of Tunisian olive tree (*Olea europaea* L.) cultivars assessed by AFLP markers. *Genetic Resources and Crop Evolution, 53*, 265–275.

Lavee, S., & Wodner, M. (1995). The effect of growing region, maturation and fruit handling on oil quality of cv. «Nabali» olives in West Bank Mountains. *Agricoltura Mediterranea, 125*, 395–403.

Majdoub, R., Ben Salem, A., M'sadak, Y., Khlifi, S., Boujnah, D., & Gouiaa, M. (2015). Etude du comportement végétatif d'une oliveraie en Meskat. Revue « Nature & Technologie ». *B-Sciences Agronomiques et Biologiques, 12*, 03–10.

Rallo, L. (2013). *The olive growing in Spain and its genetic improvement. la rivista di scienza dell'alimentazione, numero 1.*

Rekik-hakim, I., Grati-kammoun, N., Makhloufi, E., & Rebaï, A. (2010). Discovery and potential of SNP markers in characterization of Tunisian olive germplasm. *Diversity, 2*, 17–27. https://doi.org/10.3390/d2010017

Trigui, A. (1996). L'amélioration quantitative et qualitative de la production oléicole en Tunisie: l'incournable nécessité et les perspectives de l'identification et de l'amélioration génétique de l'olivier. *Olivae, 61*, 34–40.

Trigui, A., & Msallem, M. (2002). Oliviers de Tunisie. In: *Catalogue des variétés autochtones & types locaux, identification variétale & caractérisation morpho-pomologique des ressources génétiques oléicoles de Tunisie* (vol. 1). (Fr) Ministère de l'Agriculture, IRESA, Institut de l'Olivier, Tunisie.

Zipori, I., Erel, R., Yermiyahu, U., Ben-Gal, A., & Dag, A. (2020). Sustainable Management of Olive Orchard Nutrition: A review. *Agriculture, 10*, 11.

Zitoun, B., Bronzini de Caraffa, V., Giannettini, J., Breton, C. M. M., Trigui, A., Maury, J., Gambotti, C., Marzouk, B., & Berti, L. (2008). Genetic diversity in Tunisian olive accessions and their relatedness with other Mediterranean olive genotypes. *Scientia Horticulturae, 115*, 416–419.

Chapter 38
Soil Management Practices, Fertilization, Pruning, and Harvesting of Olive Groves in Tunisia

Hechmi Chehab, Olfa Boussadia, Khaled Hibar, and Sofiane Abdelhamid

Abstract In Tunisia, oleiculture represents one of the principal economical and agricultural strategic sectors. In fact, about 80 million olive trees are spread over 1.8 million hectares. More than 97% of this area is conducted under dry farming system. Dry-farmed olive orchards of arid and semiarid areas are traditionally widely spaced to take full advantage of the stored water from winter rains for spring and summer growth. Additionally, soils of Tunisian olive orchards have low water holding capacity and low soil fertility are and are exposed during the dry season to long periods of soil water deficit, coinciding with the active fruit growth and oil synthesis. In the Tunisian low-density olive orchards, mechanical tillage still represents the most common technique of soil management. In the north, soil tillage is limited to two interventions per year using disc plow, while in the center and south, the soil was mechanically tilled 3–4 times per year at a depth of 0.15–0.25 m and perpendicular to the slope using a cultivator with dovetail.

In rainfed olive grove farm, manure is the most organic amendment used by farmers to improve olive grove nutrition (Haifa Group, 2013). But during the last decade, the use of olive mill wastewater as an organic amendment is widely recommended and applied every 2/3 years at the rate 50m^3.ha^{-1}. A few farmers used green manure, especially leguminous mainly in the northern areas in which rains exceeds 400 mm/year.

Pruning is necessary to adjust the trees to the climatic conditions of the area and increase the productivity of a plantation. The aims of pruning are (1) to balance vegetation with fruit yield, (2) to minimize the nonproductive period, (3) to prolong the productivity of the trees, (4) to delay senescence, and (5) to save soil water, a critical factor in nonirrigated orchards.

The most applied pruning types:

- Regulated pruning. It aims to develop the tree's frame and is of great importance in the first years of the tree's life.

H. Chehab · O. Boussadia · K. Hibar · S. Abdelhamid (✉)
Olive Institute, Unit of Sousse, Sousse, Tunisia
e-mail: cheheb2003@yahoo.fr; io.olfa.boussadia@gmail.com; khaled_htn@yahoo.fr; sofiane.abdelhamid@icloud.com

J. Muñoz-Rojas, R. García-Ruiz (eds.), *The Olive Landscapes of the Mediterranean*, Landscape Series 36,
https://doi.org/10.1007/978-3-031-57956-1_38

- Pruning for fruiting. The aim of this pruning is to induce productive branches to form fruits, leaving the structural branches unaffected. Additionally, it maintains uniform production in terms of yield and quality, a feature that is particularly important in table olive varieties.
- Renovating pruning. This aims to stimulate sprouting in order to rejuvenate senescent trees.

In the irrigated olive orchards, short pruning shape used is, namely, the "short cup." In the former shape, branching takes place very close to the ground, at a height of 40–60 cm, while in the latter no pruning is done in the first 3 years. Afterward, only weak shoots and top branches are removed.

For table olives, Meski and Picholine are the main table olive cultivarś used. Fruits that are to be processed as green olives are picked while they are still green but have reached full size, generally early in October. Ripe olives bruise easily and are harvested by handling with care. For oil olives, harvesting start at the second week of November, and the most methods used include manual harvesting, such as small hand rakes, picking bags with a harness, and pneumatic olive harvesters.

Keywords Tillage · Olive mill wastewater · Vase shape

38.1 Soil Management

In Tunisia, oleiculture represents one of the principal economical and agricultural strategic sectors. In fact, about 80 million olive trees are spread over 1.8 million hectares. More than 97% of this area is conducted under dry farming system. Dry-farmed olive orchards of arid and semiarid areas are traditionally widely spaced to take full advantage of the stored water from winter rains for spring and summer growth. Also, sparse tree canopies maintained by pruning and weed control by tillage are also the main agronomic practices that have been used to ensure the survival of rainfed olive orchards. For the dry-farmed olive orchards, soil management practices depend on the bioclimatic and edaphic conditions as well as the production system in each olive-growing area. Mechanical soil tillage is the most soil management applied in olive orchards (Ben Rouina et al., 1995). In the north the soil is generally loam and/or clay, and the mean annual rainfall exceeds 400 mm, and soil management is limited to 2–3 tillages per year to destroy spontaneous herbs, mainly dicotyledonous and grass species. Plowing is carried out with shallow disc tools, but it negatively affects the quality of the soil by forming the plowing sole in the soil superficial layer at 20–25-cm soil depth. After a long-term period, the soil in these olive orchards has become impermeable and limits water infiltration and root growth and caused soil erosion. Thus, the majority of olive trees in the north have small or medium vigor and consequently low olive yields. In the olive orchards in the center and Sahel, the soil is sandy loam, the mean annual rainfall varied between 250 and 350 mm, and soil tillage is applied four times per year, after the autumn rain.

Farmers tilled their soil by disc tools, mainly the offset type one time at 20–25-cm soil layer depth to destroy spontaneous weed. Three soil tillages are also applied in autumn and spring to save water and limit the soil evaporation by a superficial tool as cultivator with dovetail at a 10-cm soil layer. In the south, olive trees grow on sandy soil, the mean annual rainfall is less than 200 mm, and the soil of these olive orchards is mainly occupied by the spontaneous herbs that negatively affected the olive growth and water availability and reduced the olive production by about 30% (Ben Rouina et al., 1995). To destroy this spontaneous grass species, soil is frequently tilled, about six times per year. Plowing is carried out by a tool developed by the Sfax olive growers called el "mhacha" specified to keep the soil clean and limit soil evaporation (Abichou, 2016). Trigui (1987) found that the frequent soil tillage in the region of Sfax induced the soil erosion after high rain amounts. In the local production system of the Sahel "Meskat" (Majdoub et al., 2014) and the mountains of the south in "Jessours" in slope soil, tillage is applied 2–4 times per year using a cultivator with dovetail. In the irrigated olive orchards including high-density system, soil is tilled 3–4 times per year using a disc tool and a cultivator with dovetail. More recently, Tekaya et al. (2016) reported the importance of the soil tillage under rainfed condition to destroy weed, enhance the leaf photosynthetic functioning, avoid summer water stress, and increase the olive and oil yields of olive trees cv. Chemlali as compared to zero-tilled soil.

38.2 Fertilization

Traditionally, the olive tree was planted in marginal with poor soils to valorize the adaptive capacity of this species. Indeed, the high frequent soil management practices not adapted to the soil structure have caused soil degradation, decreasing the soil organic matter that reached actually a very low amount, not exceeding 0.6% in rainfed olive orchards (Gargouri & Mhiri, 2003; Boussadia, 2004). Small olive farms represent more than 70% of the olive-growing area and have less than 5 ha. In a small rate of them, fertilization mainly occurrs by the biannual application of about 20 t/ha of sheep manure, especially in the Sahel and the center. This type of management promotes a good development of olive trees and guarantees an acceptable production and oil with an appreciable quality. But for the majority of other olive growers, the fertilization of olive trees is marginalized and has little importance given the unavailability and the expensive prices of manure. Taking into account these considerations, along the most commonly accepted recommendations for olive nutrition and fertilization (Haifa Group, 2013), and given the importance of the olive sector in the economy of the country, the Ministry of Agriculture subsidized the nitrogen fertilization of olive trees. A large number of olive growers benefited from these interventions by the contribution of 3 kg of ammonium nitrate (33% N)/tree split into two intakes (1 kg/tree in autumn and 2 kg/tree in spring). Nitrogen application contributed to the improvement of olive yields and the production of the sector, but they were not generalized because they are linked to the rains and the soil quality of the olive orchards (Trigui, 1987). After a long period of studies and tests on the development of the olive mill wastewater as fertilizer, the agricultural ministry has authorized since 2011 the spreading of the olive mill wastewater (OMWW) in olive groves as fertilizer at the rate of 50 m^3/ha. The effects of OMWW on soil characteristics, trees nutrient status, vegetative activity, and on yield as well as on oil quality have been well documented by several Tunisian authors (Mechri et al., 2009; Chehab et al., 2019). But the use of OMWW represents a serious problem by increasing soil salinization (Mekki et al., 2013; Chehab et al., 2020). The increase of sodium toxicity in olive leaves as well as the accumulation of polyphenolic compounds in the soil was reported by Abdennbi et al. (2023). Furthermore, Mechri et al. (2011) found that the application of OMWW decreased the abundance and arbuscular mycorrhizal fungi biomass and consequently the nutrient uptake by the olive trees.

The use of compost as an organic fertilizer in olive orchards has been tested under field condition in different olive-growing areas of Tunisia. The main compost used in olive orchards is made of manure (40%), olive pomace (40%), and shredded olive branches (20%).

Results are very promising in relation to the soil quality (increased soil water holding capacity and plant nutrient uptake and tree performance (increased the shoot and tree growth)) as well as yields and oil production without affecting the oil quality (Chehab et al., 2019, 2020). But farmers refuse the introduction of this material as soil organic amendment in their olive orchards, based on the high financial charges of compost production and land spread.

A few farmers in the north applied the cover crop based on the use of some leguminous species (Chehab et al., 2019). Since 2002, the application of some fertilizers (N, P, K) by spraying on olive orchards under rainfed conditions has been well studied in Tunisia (Tekaya et al., 2013). Indeed, for the intensive and super high-density orchards, the basic norms in relation to the olive yield and the soil quality as well as leaf diagnostics reported by Gargouri (2012) are strongly recommended.

38.3 Pruning

The pruning practices of olive trees in Tunisia depend on the production system (rainfed or irrigated) and tree age. Taken in to account the tree age, three types of pruning are applied. During the first period of 3 years for irrigated and 5 years for rainfed orchards, pruning aimed for the formation of olive tree trained at 50–60 cm on the ground with three branches and to maximize the tree volume. The second period aimed for the pruning of fruiting (5–70 years), and it was applied biannually after the *on* year (year of production) by renewing the branches that have previously produced and favoring the ventilation and penetration of light into the canopy. In the third period, after 70-year-old trees or older, when the production declined, pruning is aimed for the rejuvenation of trees by the removal branches of high size (Ayachi et al., 2012). Generally, pruning of the olive trees is done manually, while mechanization is recently introduced. It is limited to the use of pneumatic scissors in some large olive farms, from which the pruning time decreased by about 60% (Ayachi et al., 2012).

38.4 Harvesting

In Tunisia, harvesting of olive trees is often done manually using plastic combs and nets from November to January and can reach March or April in the high-yielding years. About 120,000 workers are mobilized during the period of harvesting. For table olives, Meski and Picholine are the main table olive cultivars used. Fruits that are to be processed as green olives are picked while they are still green but have reached full size, generally at early October. Ripe olives bruise easily and are harvested by handling with care. For olive oil, mechanization of harvesting still until now is very limited and in their first steps. The mechanization of harvesting significantly decreased the time of harvest and the financial costs by about 65% and 50%, respectively (Haggi, 2016). Taking into account the high financial input of the manual harvest as well as their negative effects on the oil quality (later harvest induced high free acidity and low phenol contents), the Ministry of Agriculture is advised to promote and encourage the introduction of the mechanization in harvesting olive trees at large scale.

References

Abdennbi, S., Chaieb, M., & Mekki, A. (2023). Long-term effects of olive mill waste waters spreading on the soil rhizospheric properties of olive trees grown under Mediterranean arid climate. *Soil Research, 62, SR23102*. https://doi.org/10.1071/SR23102

Abichou, M. (2016). *L'oléiculture Pluviale Quel Avenir? Dans un contexte de Changements Climatiques, Edaphiques, Biologiques et socioéconomiques* (p. 45). Power point presentation.

Ayachi, M., Ben Rouina, B., Laabidi, F., Meddeb, K., & Attia, L. (2012). La taille de l'olivier en Tunisie: Bases Agronomiques et types pratiques. *Doccument technique*, 26.

Ben Rouina, B., Yousfi, M., M'Laouah, M., & Louizi, A. (1995). L'entretien des sols à olivier en milieu aride: conséquencesagronomiques et maîtris e par l'agriculteur. In S. Zekri & A. Laajimi (Eds.), *Agriculture, durabilité et environnement Zaragoza: CIHEAMCahiers Options Méditerranéennes; n. 9* (pp. 125–134).

Boussadia, O. (2004). *De la fertilité du sol oléicole dans le semi-aride* (p. 120). Mémoire de Master, INAT.

Chehab, H., Tekaya, M., Ouhibi, M., Gouiaa, M., Zakhama, H., Mahjoub, Z., Laamari, S., Sfina, H., Chihaoui, B., Boujnah, D., & Mechri, B. (2019). Effects of compost, olive mill wastewater and legume cover crops on soil characteristics, tree performance and oil quality of olive trees cv. Chemlali grown under organic farming system. *Scientia Horticulturae, 253*, 163–171.

Chehab, H., Tekayab, M., Hajlaouic, H., Abdelhamid, S., Gouiaaa, M., Sfinaa, H., Chihaouia, B., Boujnaha, D., & Mechri, B. (2020). Complementary irrigation with saline water and soil organic amendments modified soil salinity, leaf Na+, productivity and oil phenols of olive trees (*cv. Chemlali*) grown under semiarid conditions. *Agricultural Water Management, 237*, 106183.

Gargouri, K. (2012). Programme de ferti-irrigation d'une parcelle d'olivier selon l'age. Présenté à l' occasion de la journée de formation adressée aux ingénieurs des CRDA. *Ben Arous* 2010.
Gargouri, K., & Mhiri, A. (2003). Relationship between soil fertility, phosphorus and potassium nutrition of the olive in Tunisia. *Options Méditerranéennes, 50*, 199–204.
Haggi, M. (2016). *Rentabilité économique de la mécanisation cueillette olives & taille arbres fruitiers – lisam* (p. 30). Power point présentation.
Haifa Group. (2013). *Crop Guide: Nutritional recommendations for Olives* (p. 83). https://www.haifa-group.com/files/Guides/Olive_Booklet.pdf
Majdoub, R., Ben Salem, A., M'sadak, Y., Khlifi, S., Boujnah D., & Gouiaa, M. (2014). Etude du comportement végétatif d'une oliveraie en Meskat. *Revue « Nature & Technologie ». BSciences Agronomiques et Biologiques, 12*(Janvier 2015), 03–103.
Mechri, B., Issaoui, M., Echbili, A., Chehab, H., Ben, M. F., Braham, M., & Hammami, M. (2009). Olive orchard amended with olive mill wastewater: Effects on olive fruit and olive oil quality. *Journal of Hazardous Materials, 172*, 1544–1550.
Mechri, B., Chehab, H., Boussadia, O., Attia, F., Ben, M. F., Braham, M., & Hammami, M. (2011). Effects of agronomic application of olive mill wastewater in a field of olive trees on carbohydrate profiles, chlorophyll a fluorescence and mineral nutrient content. *Environmental and Experimental Botany, 71*, 184–194.
Mekki, A., Dhouib, A., & Sayadi, S. (2013). Review: Effects of olive mill wastewater application on soil properties and plants growth. *International Journal of Recycling of Organic Waste in Agriculture*, 2–15.
Tekaya, M., Mechri, B., Bchir, A., Attia, F., Chehab, H., Daassa, M., & Hammami, H. (2013). Effect of nutrient-based fertilisers of olive trees on olive oil quality. *Journal of the Science of Food and Agriculture, 93*, 2045–2052.
Tekaya, M., Mechri, B., Bchir, A., Attia, F., Chehab, H., Daassa, H., Boujnah, D., Attia, F., & Hammami, M. (2016). Improving performance of olive trees by the enhancement of key physiological parameters of olive leaves in response to foliar fertilization. *Acta Physiologiae Plantarum, 38*, 101–109.
Trigui, A. (1987). *Relations entre le climat, le sol et la production de la variété d'olivier Chemlali dans la région de Sfax (Tunisie)*. Thèse de Docteur es sciences. Académie de Montpellier.

Chapter 39
Challenges for Sustainable Water Resource Management: Irrigation and Salinity

Nada Mzid and Olfa Boussadia

Abstract Tunisia is widely considered as deficitary country in water resources, and water scarcity is evident in certain regions of the country. In the long term, this situation could become more exacerbated, considering the increased risk of aridification. However, Tunisia counts the total amount of water resources around 4860 million m^3/year. Surface water resources are stressed by challenges of water quantity and quality. The distribution of these resources among the regions of the country showed the superiority of the extreme north which provides an amount of 930 million m^3, despite its low area (3% of the total area). In contrast, the center and the south representing more than 75% of total national area provide only 18% of mobilizable resources. Regarding water quality, the groundwater resources are generally of low water quality. The Tunisian government has recognized the need to evolve toward a demand-oriented approach (improving overall efficiency and favoring the most productive uses of water). Moreover, great emphasis is placed on forecasting rainfall, more efficient irrigation management, and water quality (salt concentration), with the aim of increasing water use efficiency. A broad range of solutions could be considered to tackle the problems of salinization and irrigation-water scarcity. To avoid the multiple risks arising, it is important to control soil salinity and keep it below plant salinity tolerance thresholds. Nowadays large quantities of marginal water, including saline water, are still used for tolerant crops' irrigation. Olive groves are now being fostered, as they have been found to develop more efficient physiological adaptations to cope with water and salt stress compared with alternative crops. Gaining resilience within communities was traditionally sought through water saving infrastructure. These hydrological systems have been considered as ancient techniques, for use only in semiarid Tunisian regions. The conservation of these traditional practices is however of a great importance for a sustainable development and allows considering water issues as part of life in sustainable landscapes. Investments in irrigation infrastructure in Tunisia have reached 35% of all agricultural investments along the past decades. Two-thirds

N. Mzid (✉) · O. Boussadia
Olive Institute, Unit of Sousse, Sousse, Tunisia
e-mail: nada.mzid@hotmail.com; io.olfa.boussadia@gmail.com

J. Muñoz-Rojas, R. García-Ruiz (eds.), *The Olive Landscapes of the Mediterranean*, Landscape Series 36,
https://doi.org/10.1007/978-3-031-57956-1_39

of Tunisia's irrigated area have drip or sprinkler technology for on-farm effective irrigation. In addition, the country is now expanding its program of innovative practices, including (i) artificial aquifer recharge, (ii) deficit irrigation under field conditions, and (iii) developing alternative, nonconventional resources, including treated wastewater reuse and brackish and seawater desalination.

Keywords Water use efficiency · Desalination · Drip irrigation · Wastewater reuse

39.1 Introduction

The Mediterranean region, including Tunisia, is one of the most prominent "hot spots" in future climate change projections (Giorgi & Lionello, 2008). This is due to an expected future warming stronger than the global average and to a pronounced increase in precipitation interannual variability. Tunisia is afflicted by water scarcity and shows a growing water deficit (Saadi et al., 2018), due to the combined effect of the growth in water demand (soaring demography and irrigated areas extension) and the reduction of resources (temporary droughts exacerbated by climate change).

Tunisia's climate, including both rainfall and temperatures, varies considerably from north to south. It is strongly influenced by the Mediterranean Sea to the north and east and by the Sahara Desert to the south and southwest. Rainfall is irregular and varies considerably from the relatively humid coastal area to the desert (Trabelsi et al., 2019; Verner et al., 2018). The past decade has seen a variety of noteworthy climate events in Tunisia, including numerous droughts, heat waves, and the occasional heavy rainfall and flooding. These recurrent climate events have been strongly influenced by El Niño Southern Oscillation (ENSO), the North Atlantic Oscillation (NAO), and the Mediterranean Oscillation (MO) (Verner et al., 2018). Drought and salinity are among the most significant outcomes of these climate dynamics and are gradually becoming more frequent. In forthcoming years, changes in policy and water management will become mandatory to secure a more efficient and sustainable use of water resources (WWAP, 2012; Amri et al., 2014; Pereira et al., 2002). The optimization of water resources management needs reliable information and knowledge on water resources and water requirements. Remote sensing and geographic information system techniques are used to estimate the irrigation water requirements (IWR), for the summer season, when the peak of the water consumption is reached (Guermazi et al., 2016). The Tunisian government is expected to rationalize water management by fostering the appropriate irrigation techniques and improving field irrigation management (MARE-DGGRRE, 2007). In fact, subsidies of 40–60% of equipment costs are already provided in order to promote trickle irrigation in the country (Frija et al., 2015; Saidi et al., 2016). Thus, 70% of the irrigated areas are nowadays equipped with water-saving systems (Saidi et al., 2016).

This chapter synthesizes current scientific evidence on climate variability and water scarcity and examines how this will likely affect the occurrence and severity of drought and salinity across Tunisia. Within the agricultural sector, the chapter explores climate-related impacts. In addition, it examines the Tunisian government's options devised to better protect its water resources and the agricultural sector from drought, salinity, and climate variability.

39.2 Climate Variability and Water Resources Risk in Tunisia

39.2.1 Future Drought and Salinity Prospects

Climate projections indicate that Tunisia is facing a high likelihood of continued warming above the global mean rate (Tanasijevic et al., 2014). The Intergovernmental Panel on Climate Change (IPCC, 2007) projections indicate a 0.50 °C decadal rate of increase since the 1970s, greatly exceeding the global average of 0.15 °C. An increase between 3 °C and 7 °C increases is projected by the end of the twenty-first century, with the largest increases (4 °C–7 °C) occurring during summer months. Under the same scenario, precipitation is projected to decline 10–40% annually, including 10–30% in the wet season and 10–40% in the dry season (Verner et al., 2018). This directly imposes an increase of the evaporation (Ghrab et al., 2013) and, consequently, surface soil moisture will tend to decrease leading to a parallel decrease of soil water quality and thus an increase of salinity levels (Saidi et al., 2016). A combination of the arid climate and the poor quality of its water and soils is leading Tunisia to face high risks of soil salinization (Hachicha, 1998; Verner et al., 2018). Salt-affected soil covers already occupy about 1.5 million hectares (10% of the land) and are located mainly in the center and south of Tunisia (Hachicha, 2007).

Climate change is affecting Tunisia at both the national and farm levels, according to stakeholder observations gathered by the International Center for Biosaline Agriculture (ICBA) and the University of Nebraska Lincoln (UNL) in 2016[1] Insufficient or highly variable rainfall likewise tends to result in increased pressure on often already declining groundwater resources. Approximately 60% of Tunisia's water resources are concentrated in the northern regions, which represent 17% of the national territory (Verner et al., 2018). Meanwhile, 18% of water resources are in the central regions, which occupies 32% of the national territory, and 22% of water resources are in the south, which is 51% of the country's land area (Verner et al., 2018).

[1] Check on International Center for Biosaline Agriculture ICBA Annual Report 2016. Available at: https://www.biosaline.org/sites/default/files/Annualreportpdf/annual_report_2016_en_web.pdf.

Table 39.1 Impacts of climate change on olive crop evapotranspiration and irrigation requirements in Tunisia, according to Tanasijevic et al. (2014)

Suitability for olives cultivation 2050	Suitable
Olives flowering date 2050 vs. 2000 (days)	−11–10
Olives crop evapotranspiration 2050 vs. 2000 (mm)	41–50
Olives near irrigation requirements 2050 vs. 2000 (mm)	50–100

39.2.2 *Implications on the Olive Sector*

Climate irregularity drives droughts and increased levels of salinity, which negatively affect key rainfed crops, including olives. Tunisian agriculture uses 80% of renewable water resources available (Verner et al., 2018). In the future, surface water may not meet farmers' irrigation needs, so they increasingly will rely on groundwater. This will likely intensify pressure on already-stressed aquifers.

Agricultural practices in Tunisia need to better respond to both the short-term trends related to climate variability and the long-term ones associated with climate change (Frija et al., 2015; Verner et al., 2018). The design of tools providing regional estimates about evapotranspiration (ETp) may help the sustainable management of water resources (Saadi et al., 2018). As a characteristic crop of the Mediterranean basin, olive (*Olea europaea* L.) is often grown in water-limited agroecosystems (Abboud et al., 2019). Olive growing in Tunisia is a considered as a strategic socio-economic sector. Currently, Tunisia's olive-growing potential is estimated at nearly 90 million trees, covering an area of 1.8 million hectares or 79% of the total arboreal area (GDAP, 2015). About 90% of this area can only be cultivated under rainfed conditions and receives annually no more than 350 mm of rainfall. The results of Tanasijevic et al. (2014) (Table 39.1) indicate Tunisia as a suitable area for olive cultivation in 2050. According to these authors, this scenario would imply for the flowering date to advance by only a few days, compared with other Mediterranean countries where a great anticipation of flowering could be expected. Within the same study, Tunisian olives showed also the smallest increase of close irrigation requirements and crop evapotranspiration by the same date of 2050.

39.3 Challenges

39.3.1 *General Approaches to Water Risk Management*

The Ministry of Agriculture, Water Resources, and Fisheries (MoAWRF) is the public institution in charge of officially declaring drought in Tunisia. The Ministry assesses drought at the start of each agricultural season based on three main indicator groups: (i) the lack of summer breaking rainfalls (August–October) and a prior rainy season precipitation deficit higher than 40%; (ii) a significant water storage depletion of dams; and (iii) agricultural indicators, such as affected pastures, dehydration of olives, inflation in feed prices, and delayed farmer practices related to soil tillage. The government of Tunisia is currently undertaking measures to address the rising frequency and severity of climate variability and associated drought events (Verner et al., 2018).

It is nowadays considered in the public interest to develop scientifically sound climatological drought onset indicators that reflect climate conditions, both nationally and regionally. To facilitate this, a data-sharing platform ought to be established, connecting datasets from different agencies, and integrating agricultural, hydrological, and meteorological information atomatedly. To monitor weather and water-related conditions, the National Institute of Meteorology (INM) and the MoAWRF's General Directorate of Water Resources (DGRE) maintain monitoring stations throughout the country to measure river flows, temperatures, precipitation, and groundwater levels. The rainfall measurement network consists of 850 stations and 100 pluviographs (collected data available since 1969). Data are published on a daily and monthly basis. The flow monitoring network consists of 60 stations and 74 measurement points (historical data available since 1974–1975) (Nouiri et al., 2014; Verner et al., 2018).This is especially useful for monitoring the water conditions for olive grove cultivation.

39.3.2 Rainfed-Scale Management Approaches

The Tunisian government, through the Ministry of Agriculture, Hydraulic Resources and Fisheries, annually reserves a budget for the creation and maintenance of water and soil conservation actions and to program how such actions will be set within the framework of territorial development programs (Verner et al., 2018). The goals are, on the one hand, the conservation of water and soil and, on the other hand, the sustainable management of land use. The remaining ancestral techniques, which are considered in public programs, constitute a capital to be managed well according to the edaphoclimatic and socioeconomic conditions and more particularly according to the region considered.

Hereby, we provide an overview of the different techniques of good water and soil conservation practices that have proven their effectiveness and have given spectacular results in terms of sustainable land management in Tunisia:

- The Meskat symbolizes the ancestral system of the Sahel, characterized by an annual rainfall ranging from 200 to 350 mm. It recharges groundwater levels and decreases stream flow, facilitating mitigation of struggle against erosion by water and soil retention, and also collects rainwater for (rainfed) olive cultivation in semiarid regions (Ben Salem et al., 2013).
- The Jessours is a method particularly practiced at the massif level of the Matmatas. The yields obtained for crops grown in these Jessours are at times

spectacular. In addition, they have proven their effectiveness in soil conservation and the fight against erosion.

- The Mgouds are techniques aiming at orienting runoff water, in particular, at the level of a catchment area, and their collection by benches of spills in an adjacent Oued. Their ultimate goal consists in diverting them to previously prepared spreading fields.

Driven by increasingly drier trends and the reliability of quality olive supply required by globalized markets, supplemental irrigation is likely to increase in the future, even in these areas that have been traditionally rainfed. In some Tunisian regions with a scarcity of water, supplemental irrigation is practiced to improve olive yields.

39.3.3 Irrigated-Scale Management Approaches

Tunisia is lately adapting its irrigation systems along modern standards (Verner et al., 2018). Investments in irrigation infrastructure have reached 35% of all agricultural investments in the past decades. This has equipped the country with 410,000 ha for irrigation, which is 95% of the estimated potential. Two-thirds of Tunisia's irrigated area has now drip or sprinkler technology for on-farm irrigation (Verner et al., 2018). However, innovative approaches for irrigation management are required to reduce water consumption and make more efficient use of water by maximizing water saving while minimizing impacts on crop productivity. To meet this goal, deficit irrigation is proving to be one of the most promising techniques (Ghrab et al., 2006, 2007).

Late major infrastructure investments have allowed Tunisia to capture most of its scarce water, mobilizing 92% of renewable resources, and deliver it toward where it is most needed. However, the Tunisian government has recognized the need to evolve from a supply side response (increasing water mobilization) toward a demand management approach (improving overall efficiency and favoring the most productive uses of water). The government's 2030 water strategy is in line with the 2030 Sustainable Development Goals and includes the following main lines of action: (i) further increase storage capacity to harvest surface water, including the replacement of outdated and inefficient dams; (ii) increase water transfer capacity and safety; (iii) renew old boreholes; (iv) foster and facilitate artificial aquifer recharge and recycling of water; and (v) develop alternative, nonconventional resources, including treated wastewater reuse and brackish and seawater desalination (Verner et al., 2018).

39.4 Conclusion

As described in the chapter, Tunisia has been lately undertaking a number of steps to address the rising frequency and severity of climate negative events, including their variability and change. Improved information sharing and establishment of triggers to determine what types of actions need to be taken, in what precise locations, and in accordance with the evolution of local drought conditions, including their extension over time and severity, should also become an essential component of such a monitoring system. The development and implementation of expert networks to provide feedback and strengthen current water scarcity monitoring efforts and information products could greatly help improve mapping accuracy as well as support ownership and trust in the reliability and capacity of maps to improve water scarcity-informed management and decision-making. The urgent need is also agreed to reach beyond exclusive dependence on information from satellite images and model outputs and more closely engage the experts who are able to identify on the ground the signs of water challenges at local levels.

References

Abboud, S., Dbara, S., Abidi, W., & Braham, M. (2019). Differential agro-physiological responses induced by partial root-zone drying irrigation in olive cultivars grown in semi-arid conditions. *Environmental and Experimental Botany, 167*, 103863.

Amri, R., Zribi, M., Lili-Chabaane, Z., Szczypta, C., Calvet, J. C., & Boulet, G. (2014). FAO-56 dual model combined with multi-sensor remote sensing for regional evapotranspiration estimations. *Remote Sensing, 6*, 5387–5406.

Ben Salem, A., Majdoub, R., M'Sadak, Y., & Khlifi, S. (2013). Importance of the Meskat system and its landscape insertion through the olive groves of Sousse region (Tunisian Sahel). *International Journal of Innovation and Applied Studies, 4*(2), 382–386.

Frija, A., Dhehibi, B., Chebil, A., & Villholth, K. G. (2015). Performance evaluation of groundwater management instruments: The case of irrigation sector in Tunisia. *Groundwater for Sustainable Development, 1*, 23–32.

General Directorate of Agricultural Production GDAP Tunisia. (2015). http://www.agridata.tn/organization/direction-generale-de-la-production-agricole?organization=direction-generale-de-la-production-agricole&page=2. Accessed 30 Mar 2020.

Ghrab, M., Ayadi, M., Gargouri, K., Bentaher, H., Ben Mimoun, M., & Ben Mechlia, N. (2006). Partial rootzone drying (PRD) effects on yield quality of olive tree (cv. Chemlali) irrigated with saline water. *Proceeding of Olive Bioteq, 2*, 83–89.

Ghrab, M., Gargouri, K., Ayadi, M., & Bentaher, H. (2007). Water saving in orchards: Partial root-zone drying strategy on olive tree. In: Bari A (eds) *Assessment of plant genetic resources for Water Use Efficiency (WUE): Managing Water Scarcity* (pp. 33–40). Proceedings of the Biodiversity International/INRA/IDRC/AARINENA Workshop for North Africa and West Asia, Marrakech, Morocco, 10–12 October 2005. Bio-diversity International, Rome, Italy.

Ghrab, M., Gargouri, K., Bentaher, H., Chartzoulakis, K., Ayadi, M., Ben Mimoun, M., Masmoudi, M. M., Ben Mechlia, N., & Psarras, G. (2013). Water relations and yield of olive tree (cv. Chemlali) in response to partial root-zone drying (PRD) irrigation technique and salinity under arid climate. *Agricultural Water Management, 123*, 1–11.

Giorgi, F., & Lionello, P. (2008). Climate change projections for the Mediterranean region, global planet. *Change, 63*, 90–104.

Guermazi, E., Bouaziz, M., & Zairi, M. (2016). Water irrigation management using remote sensing techniques: A case study in Central Tunisia. *Environmental Earth Sciences, 75*(3), 1–14.

Hachicha, M. (1998). *Mise en valeur des sols salés: Organisation, fonctionnement et Evolution des sols salés de la Tunisie*. Thèse Doctorat, INAT.

Hachicha, M. (2007). Les sols salés et leurs mises en valeur en Tunisie. *Sécheresse, 18*(1), 45–50.

Intergovernmental Panel on Climate Change. (2007). The physical science basis. In S. Solomon, D. Qin, M. Manning, Z. Chen, M. Marquis, K. B. Averyt, M. Tignor, & H. L. Miller (Eds.), *Contribution of working group I to the fourth assessment report of the intergovernmental panel on climate change* (p. 996). Cambridge University Press.

Ministère de l'Agriculture et des Ressources en Eau, Direction Générale du Génie Rural et des Ressources en Eau MARE-DGGRRE. (2007). *Rapport de préparation du projet d'investissement dans le secteur de l'eau*. https://projectsportal.afdb.org/dataportal/VProject/show/P-TN-AAC-009?lang=fr. Accessed 15 Apr 2020.

Nouiri, I., Nasri, S., & Tahrani, M. (2014). *Drought Management in Tunisia. National Capacity Development Training of trainers*. Paper presented at the workshop on developing and implementing mitigation and preparedness Water Scarcity and Drought (WS&D) management plans, Zaragoza, Spain, May 2014.

Pereira, L. S., Oweis, T., & Zairi, A. (2002). Irrigation management under water scarcity. *Agricultural Water Management, 57*, 175–206.

Saadi, S., Boulet, G., Bahir, M., Brut, A., Delogu, É., Fanise, P., Mougenot, B., Simonneaux, V., & Chabaane, Z. L. (2018). Assessment of actual evapotranspiration over a semiarid heterogeneous land surface by means of coupled low-resolution remote sensing data with an energy balance model: Comparison to extra-large aperture scintillometer measurements. *Hydrology and Earth System Sciences, 22*(4), 2187–2209.

Saidi, A., Hammami, M., Daghari, H., Boughdiri, A., & Ali, H. B. (2016). Irrigation uniformity impacts on water use efficiency and soil salinity: Case study of tomato crop under trickle irrigation system in the north east of Tunisia. *Journal of Hydrogeology & Hydrologic Engineering, 5*(1).

Tanasijevic, L., Todorovic, M., Pereira, L. S., Pizzigalli, C., & Lionello, P. (2014). Impacts of climate change on olive crop evapotranspiration and irrigation requirements in the Mediterranean region. *Agricultural Water Management, 144*, 54–68.

Trabelsi, L., Gargouri, K., Ben Hassena, A., Mbadra, C., Ghrab, M., Ncube, B., Van Staden, J., & Gargouri, R. (2019). Impact of drought and salinity on olive water status and physiological performance in an arid climate. *Agricultural Water Management, 213*, 749–759.

Verner, D., Treguer, D., Redwood, J., Christensen, J., McDonnell, R., Elbert, C., & Konishi, Y. (2018). *Climate variability, drought, and drought management in Tunisia's agricultural sector*. Available via World Bank Group. https://elibrary.worldbank.org/doi/pdf/10.1596/30604.

World Water Assessment Program (WWAP). (2012). *UNESCO*. https://en.unesco.org/wwap. Accessed 20 Apr 2020.

Part VII
Olive Landscapes Across the Mediterranean: Regional Heterogeneity and Case Studies: Olive Groves and Landscapes in Morocco

Chapter 40
A Historical and Social Insight of Olive Groves and Landscapes in Morocco

Abdelaziz El Mahmoudi, Ouiame Tabet, Julio Antonio Calero González, and Ali Yahya

40.1 Introduction and Historical Development

The first traces of olive pollen appeared in the paleographic data 3.2 MA ago in the northwest part of the Mediterranean zone at the base of the beds of the Miocene.[1]

The presence of oleaster which represents a wild variety of Moroccan olive dates for 10,000 years before present, which is still exploited today to produce oil reputed of good quality for culinary and medicinal uses.

In the era of the Phoenicians was olive growing first introduced and extended in Africa, but the significant development of olive cultivation began with the arrival of Romaine, by creating olive-growing fields and olive presses for the extraction of oil which was used for lighting, bathing, and cooking.[2] Archaeological evidence of these activities exists only in the north of Morocco: Mauretania Tingitana and especially in the regions of Volubilis, Tingis, Lixus, Banasa, and Sala (Lenoir, 1994). An emblematic example is an archaeological site of Volubilis, a Roman ruin site located near Meknes (northern part of Morocco), where some of the best-preserved traditional oil presses (maâsra) in the world have been located; Volubilis alone had more

[1] Origin and evolution of the Mediterranean vegetation and climate in Europe.

[2] Histoire de l'introduction et de la diffusion de l'oléiculture au Maroc.

A. El Mahmoudi · O. Tabet (✉) · A. Yahya
Department of Biology, Abdelmalek Essaâdi University, Khenifra, Tétouan, Morocco
e-mail: abdelaziz.elmahmoudi@etu.uae.ac.ma;
ouiametabet@gmail.com; alii.yahya1995@gmail.com

J. A. C. González
Departamento de Geología, Universidad de Jaén, Jaén, Spain
e-mail: jcalero@ujaen.es

J. Muñoz-Rojas, R. García-Ruiz (eds.), *The Olive Landscapes of the Mediterranean*, Landscape Series 36,
https://doi.org/10.1007/978-3-031-57956-1_40

than 55 oil mills. The olive press of Roman origin is a system that is currently used with some differences.[3]

Olive cultivation in Morocco has known three phases of development: the phase of Roman diffusion from the first to the third century (Roman era), a phase of extensive development from the tenth century (medieval), and a phase of intensive diffusion from the nineteenth century and during the twentieth century (modern).[2]

Botanists recognize two varieties of olive trees in Morocco: *Olea europaea* var. *sativa* which includes all cultivars and *Olea europaea* var. *sylvestris* Mill., commonly known as the oleaster, which groups wild and silted forms. The cultivation of the olive tree dates at least to the Neolithic on the shores of the Mediterranean; on the other hand, the domestication of the tree is only effective since the Bronze Age.

The history of Morocco shows that olive growing in central Morocco (Marrakech, Sijilmassa, and the Souss) developed much later (between the twelfth and seventeenth centuries). This situation is probably due to the place occupied by the argan tree, *Argania spinosa*.[3]

The spread of the olive tree has been intensified under the Protectorate and since Independence benefiting not only from the creation of new irrigated perimeters, from the adoption of intensive propagation techniques in the tree nursery (the technique of sowing grafting and the technique of semiwoody cutting), and the increase of the capacity of crushing by the adoption of mechanical presses but also economic and political factors. The number of trees increased from 5.3 million in 1930 to 13.7 million in 1960. The geographical distribution shows three regions where the olive tree dominates: Fez, Marrakech, and Rabat Nord.[4]

40.2 Socioeconomical Insight

The olive tree represents the main fruit species in Morocco where it occupies more and more important areas. Olive growing is currently undergoing a major expansion with a continuous increase in olive-growing areas. The objective of the Green Morocco Plan (GMP) is to reach 1,220,000 ha by 2020, which is double the area of 2010. The corollary of this dynamic is the evolution toward the modernization of orchards and olive growing practices, which will accentuate the marginalization of orchards and traditional practices.[5]

Olive growing in Morocco is a central socioeconomic sector in agricultural development and in particular in the framework of the Green Morocco Plan where

[3] L'oléiculture au Maroc de la préhistoire à nos jours: pratiques, diversité, adaptation, usages, commerce et politiques M. Ater, L. Essalouh, H. Ilbert, A. Moukhli, B. Khadari.

[4] Code des Investissements Agricoles, Discours Royal de 1986, Plan National Oléicole, Plan Maroc Vert).

[5] Olive Cultivation Programme Contract, the Ministry of Agriculture (MAPM, 2008).

the extension of olive orchard areas to an area of 1.07 million hectares and two million tons of olive production in 2020.[6]

The olive grown in Morocco, from traditional agroecosystems to the most modern orchards including those driven in high density, is probably the result of a succession of public policies based on massive plantations of the Moroccan Picholine variety from the Roman Empire to the Green Morocco Plan.

Thus, we can consider that Moroccan olive cultivation can be characterized, among other things, by two important traits: (i) the dominance of "Moroccan Picholine" variety and (ii) the extension of areas and the modernization of olive growing practices with the intensive use of olive plants produced in nursery.

The northern region of Morocco, and particularly the Tingitana Peninsula, is characterized by the dominance of the agro-silvopastoral production system and the practice of traditional subsistence agriculture. Indeed, the traditional agroecosystems of this region are characterized by the importance of agro-diversity and the persistence of local varieties and traditional practices.[7]

The olive tree is an essential crop for the domestic economy and food security of the farmers of these regions. It is a structural species of the local arboreal landscape where olive groves occupy the first place with important areas. Traditional orchards in these regions differ in the persistence of local traditional varieties (Haouane 2012).

The olive tree, due to its products and centuries-old uses as well as its multiple functions of erosion control, development of agricultural land, and settlement of populations in mountain areas, is the main fruit speculation cultivated in Morocco.[8]

40.3 The National Strategy to Develop the Moroccan Olive Sector

In 2008, the Department of Agriculture launched a new agricultural development strategy entitled Green Morocco Plan (GMP), based on a global and contractual approach that integrates all stakeholders. The aim of this strategy is to establish the agricultural sector as a real lever for socioeconomic development in Morocco, in harmony with the country's major projects, which are job creation, the fight against poverty, and the protection of the environment and natural resources.

As part of this partnership, the Moroccan State and Inter-Professional Olive Industry (interpretive) signed in April 2009 a program contract for the development and upgrading of the olive sector, during the period 2009–2020, whose main objectives are as follows:

---

[6] Description of the state of olive groves in Morocco. Msc Thesis. Universidad de Jaén. Tabet, Ouiame, 2020 https://crea.ujaen.es/bitstream/10953.1/13172/1/OUIAME-TABET-TFM.pdf

[7] Agrodiversité des agroécosystèmes traditionnels du pays Jbala (Rif, Maroc) et produits de terroirs Hmimsa et Ater, 2008; Ater et Hmimsa, 2008 et 2013).

[8] http://www.agropoleolivier.com/partenaires/#

- Encouraging private investment in the various links of the chain.
- Improving production conditions.
- Increasing the competitiveness of the sectors through improved productivity and quality.
- Improving marketing conditions and export development.
- Improving the framework conditions of the sectors, through support for the upgrading of interprofessional organizations and the strengthening of research and development and training.

40.4 Main Achievements

After 10 years of implementation of this program contract, among the main achievements can be summarized as follows:

- Extension of olive plantations on 272,408 ha, reaching 1,045,000 ha in 2017/2018.
- Equipping 111,200 ha with localized irrigation or 82% of the target set by 2020 (136,000 ha).
- Significant improvement in production from 549,000 tons in 2003–2007 to 1,414,000 tons in 2015–2018, an increase of 158%, with production for the 2018–2019 crop year expected to reach two million tons.
- Increased production of olive oil from 66,000 tons over the period 2003–2007 to 125,600 tons over the period 2015–2018, an increase of 98%. For table olive production, it rose from 88,000 tons to 113,000 tons over the same period, accounting to a 28% improvement.
- Creation of 211 crushing units.[9]

40.5 Current Situation of Moroccan Olive Cultivation: Socioeconomic Importance of the Olive Tree

The olive tree, due to its products and centuries-old uses, as well as its multiple functions of erosion control, the development of agricultural land, and the settlement of populations in mountain areas, provides more than 51 million working days per year or the equivalent of 380,000 permanent jobs, held by 20% women. It is the main fruit sector grown in Morocco and accounts for 65% of the national tree area.

The average annual production of olives is about 1.1 million tons, 65% of which is reserved for crushing and 25% for table olives, and the remaining 10% represents self-consumption and losses. Morocco is the second largest exporter of table olive

[9] Ministère de l'Agriculture, de la Pêche maritime, du Développement rural et des Eaux et Forêts.

by a production estimated at 90,000 tons per year in the world. The main producing regions are Marrakech, Fez, and Meknes.[10]

The products include green olives, black olives, rotary olives, pitted olives, sliced olives, stuffed olives, candied olives, etc. A large part, 64,000 tons per year, is exported to European countries and the United States.[11]

At the national level, the irrigated olive-growing area covers 220,000 ha or nearly 40% of the total area planted with olive trees. Perennial irrigation is located at the level of the areas of action of the Regional Offices of Agricultural Development and extends over an area of about 40,000 ha. The remainder, that is, 180,000 ha, is conducted in supplementary irrigation and mainly covers the regions of Marrakesh, Béni Mellal, Azilal, Taza, El Kela, Boulemane, Oujda, and Nador.[12]

With two million tons of olives harvested this season, Morocco has become the world's fourth largest olive oil producer. Morocco has harvested an estimated two million tons of olives in 2018, an increase by 28% compared to the crop last year.[13,14,15,16,17,18,19]

[10] https://www.fellah-trade.com/fr/filiere-vegetale/chiffres-cles-oleiculture

[11] https://www.fellah-trade.com/fr/filiere-vegetale/chiffres-cles-oleiculture

[12] Séminaire International sur l'olivier, Acquis de recherche et contraintes du secteur oleicole: FILIERE OLEICOLE: SITUATION ET PERSPECTIVES D'AVENIR Mohammed Berrichi Direction de la Production Végétale.

[13] 5th National Fair of the Olive Tree on Saturday, Minister of Agriculture. https://morocco-gold.com/morocco-is-worlds-4th-largest-olive-oil-producer/#:~:text=Morocco's%20principale%20regions%20for%20olive,Fez%2C%20Meknes%2C%20and%20Casablanca

[14] http://ormvah.com/filiere-olivier/

[15] Ministère de l'agriculture de la pêche maritime du développement rural et des eaux et forêtsfilièreoléicole. http://www.agriculture.gov.ma/pages/actualites/le-ministre-inaugure-la-foire-nationale-de-lolivier-ouazzane-et-visite-une-unite-trituration-olives https://leseco.ma/maroc/oleiculture-la-region-de-fes-meknes-cartonne.html

[16] Monographie Régionale de Tanger-Tétouan-Al Hoceima, Direction Régionale de Tanger-Tétouan-Al Hoceima.

[17] https://aujourdhui.ma/economie/production-dolives-a-tanger-tetouan-al-hoceima-nouveau-record-en-vue

[18] https://fr.le360.ma/economie/un-nouveau-record-pour-la-production-dolives-dans-le-nord-180331

[19] Agrodiversité des agroécosystèmes traditionnels du pays Jbala (Rif, Maroc) et produits de terroirs Hmimsa et Ater, 2008; Ater et Hmimsa, 2008 et 2013.

Chapter 41
Pedoclimate and Landscape Conditions of the Moroccan Olive Groves

Ali Yahya, Nard Bennas, Abdelaziz El Mahmoudi, and Kamal Targuisti

Abstract Morocco, a subtropical country, experiences a temperate and warm Mediterranean-type climate, with variations across different regions: oceanic in the west, continental in the center and east, and arid in the south. Annual rainfall ranges from 500 to 1200 mm in the wettest northwestern areas influenced by the Atlantic, 200 to 1000 mm in the west and center, 100 to 200 mm in the east, and less than 100 mm in the south. This climatic and pedoclimatic diversity has positioned Morocco as a key player in olive cultivation, with the olive tree thriving across a wide range of bioclimatic zones, from mountainous regions to arid and Saharan areas. The country's olive production is concentrated in three major areas: mountainous zones, irrigated regions, and favorable bour lands, distributed across the north, center, and south. Additionally, Morocco's varied soils support olive cultivation, particularly vertisols, calcimagnesic soils, isohumic soils, regosols, and fluvisols, each with distinct characteristics that enhance the suitability for olive growth.

Keywords Morocco · Pedoclimate · Climate · Mediterranean climate · Rainfall · Geographical distribution · Olive groves · Temperature · Soil

---

A. Yahya (✉) · A. El Mahmoudi
Département de Biologie, Faculté des Sciences de Tétouan, Université Abdelmalek Essaâdi, Tétouan, Morocco
e-mail: alii.yahya1995@gmail.com; abdelaziz.elmahmoudi@etu.uae.ac.ma

N. Bennas · K. Targuisti
Département de Géologie, Université Abdelmalek Essaâdi, Faculté des Sciences de Tétouan, Tétouan, Morocco
e-mail: nbennas@uae.ac.ma; ktarguisti@gmail.com

J. Muñoz-Rojas, R. García-Ruiz (eds.), *The Olive Landscapes of the Mediterranean*, Landscape Series 36,
https://doi.org/10.1007/978-3-031-57956-1_41

41.1 Olive Growing: Distribution Patterns and Climate Influence

Olive cultivation holds significant importance in all Moroccan agro-systems due to its adaptability across various bioclimatic zones. Olive growing extends throughout Morocco, with the exception of coastal and desert regions. The country is characterized by five major biogeographic regions: mountainous areas, plains, and hills with rainfed cultivation models, semiarid and arid plains and plateaus, large irrigated areas, and pre-Saharan and Saharan zones (CGDA, 2008). By 2018–2019, the area planted with olive trees at the national level exceeded 1.07 million hectares, accounting for more than 55% of national orchards.

Morocco has a temperate, warm Mediterranean-type climate, with oceanic influences in the west (temperatures ranging from 10 °C to 26 °C), a more continental climate in the center and east, and arid or desert conditions in the south (with contrasting temperatures between 0 °C and 40 °C) (Fig. 41.1). The country's Mediterranean climate is shaped by a combination of oceanic, mountainous, and Saharan influences. Morocco experiences two distinct seasons: a hot, dry summer from May to September, and a short winter characterized by sudden, concentrated rainfall.

Morocco's climate is characterized by significant annual and interannual variability. Precipitation generally decreases from north to south and is more concentrated in mountainous regions. The average annual rainfall varies as follows: 500–1200 mm in the wettest areas of the northwest, influenced by the Atlantic; 200–1000 mm in the west and center, also under Atlantic influence; 100–200 mm in the eastern regions; and less than 100 mm in the southern parts of the country.

The geographical distribution of olive orchards highlights three distinct olive-growing areas: mountain regions (270,000 ha or 27% of the production), irrigated areas (440,000 ha, accounting for 44% of the production), and rainfed areas (90,000 ha, or 9% of the production). Olive production is concentrated in three key areas: (1) the Rif region in northern Morocco (Tanger, Tetouan, Chefchaouen, and Ouazzane), which produces around 10% of national olive output, with a dominant mountainous olive groves; (2) the central plains and hills mainly under rainfed conditions (Fes, Meknes, Taounate, and Taza), contributing over 40% of the national production; and (3) the southern large irrigation perimeters (Haouz-Marrakesh, Tadla-Azilal, Safi, and Essaouira), responsible for over 22% of olive production.

The pedoclimatic conditions of olive groves in Morocco vary based on the geographical location of key olive-growing regions. In the wetter, humid mountainous areas, olive groves receive over 400 mm of rainfall annually, while those in arid mountainous regions receive less than 400 mm. Olive groves in these mountain agricultural systems are often found on moderate to steep slopes.

In rainfed areas, olive groves thrive on plains and hills with rainfall exceeding 400 mm. These regions have generally fertile soils, and the risk of soil degradation is minimal. This vast crescent region, largely exposed to oceanic influences, benefits from relatively high rainfall and moderate summer temperatures. Alongside

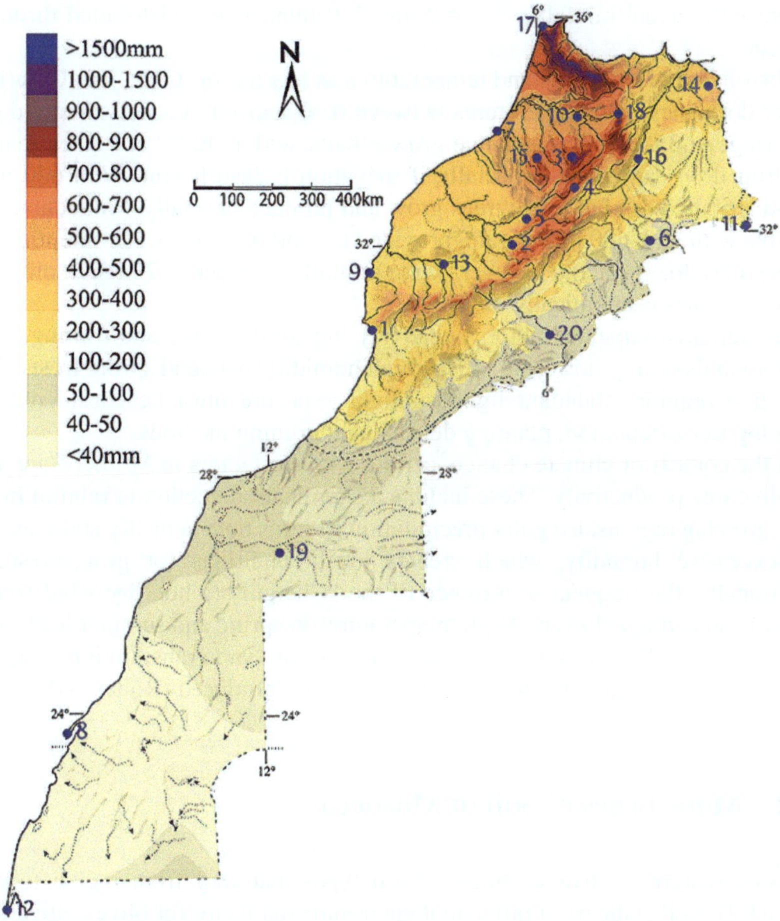

Fig. 41.1 Annual rainfall map of Morocco

favorable climatic conditions, the soils are well suited for cultivation, including Mediterranean red soils, clay soils of the Meknes plateau and pre-Rif, and clayey black vertisols of the Rharb and Chaouïa regions.

In the coastal plains, rainfall averages around 300 mm and decreases toward the south, placing these regions in zones of irregular and extensive rainfed cultivation. Favorable climatic factors are limited but include high air moisture, occasional occult precipitation, and moderate temperatures in both summer and winter.

In the plains along the middle courses of the Atlas rivers (Tadla, Srarhna, and Haouz), aridity is pronounced, with less than 300 mm of rainfall annually. This is exacerbated by factors such as strong seasonal variability, extreme inter-annual irregularities, and high summer evaporation. Conversely, the Atlantic influence provides the Souss plains with mild winters and soils rich in clayey alluvium, which offer significant agricultural potential. However, aridity remains a major limiting factor, with annual rainfall not exceeding 250 mm, poorly distributed throughout the year.

The olive tree can withstand temperatures as low as −8 °C to −10 °C during its winter dormancy, but temperatures between 0 °C and −1 °C can severely damage flowering. At 35–38 °C, vegetative growth halts, and at 40 °C or higher, leaf burn and fruit drop can occur, especially if irrigation is insufficient. With 600 mm of well-distributed rainfall, olive trees grow and produce normally. Production is still possible with 450–600 mm of rain, provided the soil has good water retention (such as deep clay-loam soils). However, when rainfall drops below 200 mm, olive cultivation becomes economically unviable.

Several environmental factors negatively impact flowering and fruiting, including hot winds during flowering, mists, high humidity, hail, and spring frosts. As the olive tree requires abundant light, sunlight exposure must be considered when choosing tree orientation, planting density, and pruning methods.

In the context of climate change, certain climatic factors in Morocco are affecting olive tree productivity. These include a significant reduction in rainfall in some olive-growing regions, irregular precipitation patterns both annually and seasonally, and excessive humidity, which creates ideal conditions for pest infestations. Additionally, the frequent occurrence of the "Chergui" (a hot, dry wind from the Sahara) can damage flowers. High temperatures in spring and summer lead to early fruit drop and slow fruit growth due to excessive evapotranspiration, negatively impacting both the quality and quantity of olive oil produced (Rotten, 2007).

41.2 Main Types of Soil in Morocco

Morocco features a diverse range of soil types that vary from region to region (Fig. 41.2). Soil is the most influential environmental factor for olive cultivation, as it determines the nutritional status of the crop and mediates the effects of precipitation and temperature on the tree. The major soil types in Morocco include:

1. Slightly evolved erosional soils (Entisols) are found in the inland mountain areas. These underdeveloped soils are poorly weathered, shallow, and prone to erosion, retaining the characteristics of the parent rock.
2. Poorly developed soils (entisols and aridisols), containing alluvial deposits along rivers, share characteristics with erosional soils but are better weathered.
3. Vertisols are present north of Fez and south of Casablanca and El Jadida. They have a high clay content (35–40%) and excellent water retention, with a high cation exchange capacity of 35 to 40 me/100 g.
4. Calcimagnesic soils (inceptisols and mollisols), located mainly in the northeastern provinces of Nador and Taza. These soils are rich in organic matter and over 90% saturated with calcium and magnesium.

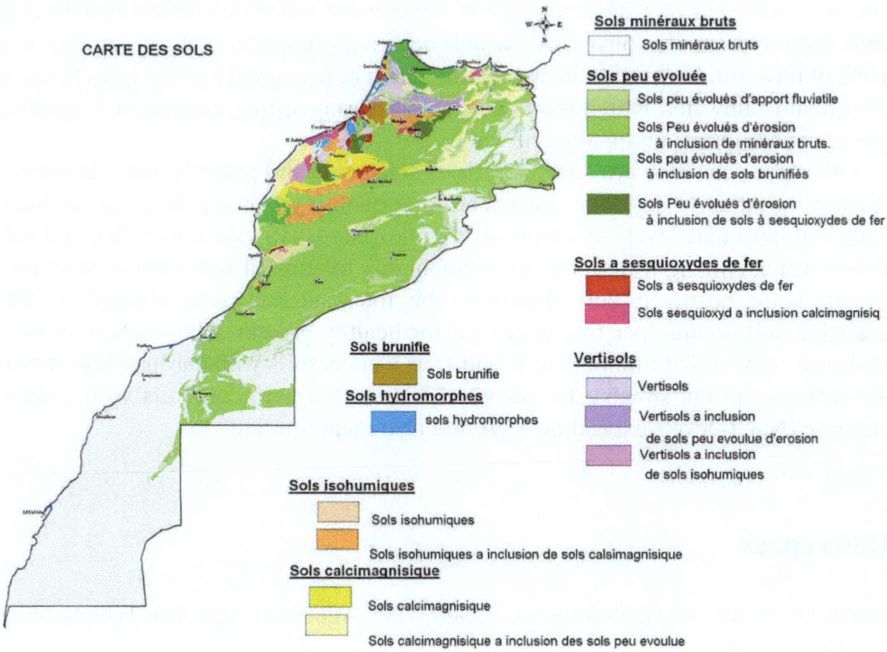

Fig. 41.2 Map of main soils in Morocco

5. Iron sesquioxide soils (alfisols) stretch along the coast from Tangier to El Jadida. They are gray or brown due to minerals, with high base saturation (over 50%) and low organic matter content.
6. Isohumic soils are found in Fez, Beni Mellal, Marrakech, and Taroudant. Developed on limestone, they feature an A[B]C or ABC profile with more than 15% organic matter in the A horizon, and high levels of exchangeable calcium and magnesium.
7. Brown soils, located between Khemiset and Kenitra, are characterized by intense biological activity in the humus layer, with less pronounced influence from bedrock due to the humus and clay presence.
8. Hydromorphic soils are found in Benslimane and south of Casablanca. These soils form under anaerobic conditions due to hydromorphic processes, leading to high organic matter accumulation.

The major soil types in Morocco that support olive cultivation include vertisols, known for their high water retention capacity and significant clay content (35–40%). Additionally, calcimagnesic soils (inceptisols and mollisols) feature a light texture and are rich in organic matter bound to minerals. These soils are more than 90% saturated with calcium and magnesium. Olive farming is also common in isohumic soils, which contain high levels of organic matter, as well as in regosols and fluvisols.

However, olive trees do not thrive in heavy clay soils (over 40% clay), as these soils cause root asphyxiation during rainy seasons. In summer, such soils develop cracks, leading to root desiccation and subsequent water deficiency, resulting in fruit drop and smaller olive size, which negatively impacts both the quality and yield of olive oil. In contrast, deep soils are much better suited to olive trees because they retain rainwater, which the trees can use during spring to support vegetative growth and improve oil quality and yield.

Olive trees can also grow in poor, dry, calcareous, and gravelly soils. However, the most productive soils for consistent, high-quality yields are deep, sandy-loam soils with adequate levels of nitrogen (N), phosphorus (P), potassium (K), and sufficient water. Among fruit trees, olives are highly tolerant of water stress, with root density being positively correlated with this tolerance and water absorption. The available soil volume per tree is crucial for healthy growth. Clay soils with high moisture content that immobilize K and P, as well as soils with hardpan layers near the surface, are not suitable for olive cultivation. Furthermore, soils with sodium chloride (NaCl) levels exceeding 1 g/L are also inappropriate.

References

CGDA. (2008). *Atlas de l'agriculture marocaine* (p. 135). Ministère l'Agriculture la Pêche Marit. Published online.

Rotten, B. (2007). *Guide du producteur de l'huile d'olive*. ONUDI, Published online.

Chapter 42
Varieties and Olive Grove Production Systems

Ali Yahya, Fatima Raji Allah Bodinier, Etabek Mounim, and Kamal Targuisti

Abstract In Morocco, more than 90% of olive orchards are dominated by the 'Moroccan Picholine' variety, but according to the National Bureau of Food Product Health Safety Office (ONSAA) and the National Agronomic Research Institute (INRA), many other varieties are approved. These include native varieties, such as 'Haouzia', 'Menara', 'Dahbia', and 'Meslala', but also some European-introduced cultivars such as 'Arbequina', 'Arbosana', 'Koroneiki', 'Picholine de Languedoc', 'Frantoio', 'Manzanilla', and 'Hojiblanca', among others. The agronomic performance of these varieties differs in this context. In this sense, several studies have been conducted, including one by INRA which proves that 'Ascolana Dura', 'Lecino', 'Haouzia', and 'Manzanilla' are the most productive varieties in the two farming systems dominating in Morocco (rainfed and irrigated). The most alternating varieties are 'Arbequina', 'Frantoio', 'Picholine de Languedoc', and 'Menara'. The production performance achieved depended on the varieties and the cropping system. In an irrigated system, the production was generally higher, and all the varieties responded positively to the irrigation inputs. Overall, in Morocco, traditional olive groves are still dominant, with an average farm size of 1.5 ha, representing 52% of the olive groves, spreading across mountain and rainfed agricultural regions. Intensive olive groves, in flatter areas and with irrigation, would account for up to 44%, while super-intensive orchards would still account for less than 5% of the total (about 40,000 ha) surface currently occupied by olive groves nationally.

A. Yahya (✉) · E. Mounim
Département de Biologie, Faculté des Sciences de Tétouan, Université Abdelmalek Essaâdi, Tétouan, BP, Morocco
e-mail: alii.yahya1995@gmail.com

F. R. A. Bodinier · K. Targuisti
Département de Géologie, Université Abdelmalek Essaâdi, Faculté des Sciences de Tétouan, Tétouan, BP, Morocco
e-mail: tismonine.tiout@gmail.com; ktarguisti@gmail.com

J. Muñoz-Rojas, R. García-Ruiz (eds.), *The Olive Landscapes of the Mediterranean*, Landscape Series 36,
https://doi.org/10.1007/978-3-031-57956-1_42

42.1 Olive Varieties Cultivated in Morocco

Moroccan olive growing is still dominated by the 'Moroccan Picholine' variety which until a decade ago represented more than 90% of the plantations (MAPMDREF, 2014). This is a variety with a double purpose, table olive and olive oil. It is characterized by an average richness in oil, and a productivity potential of 50–60 kg per tree. Alas, it is also very sensitive to diseases, presenting a strong index of alternation. This local variety is particularly well adapted to the diversity of climatic contexts across Morocco and therefore enjoys great agronomic plasticity. It is a polyclonal variety, with very heterogeneous characters, due to some clonal mutations linked to intensive multiplication over a long period. The remaining olive production, 10%, is made up of several autochthonous and foreign varieties (Hadiddou et al., 2015; Kartas et al., 2015). National varieties include 'Haouzia' and 'Menara', which are clonal varieties of the 'Moroccan Picholine', and also 'Dahbia'. Other varieties of Spanish origin are also found, namely 'Arbequina', 'Hojiblanca', 'Manzanilla de Sevilla', 'Picual', and 'Gordal'. The varieties of Italian origin include 'Carolea', 'Frantoio', and 'Ascolana Dura'. From Greece, 'Koroneiki', and 'Mastoidis' are also planted in Morocco, and finally also 'Picholine de Languedoc' of French origin.

The first selection work carried out by INRA made it possible to conserve two clones ('Haouzia' and 'Menara') from the 'Moroccan Picholine', which were mainly distributed to professionals as well as also planted in irrigated farms, such as those encountered in Bour systems (Hadiddou et al., 2015; Kartas et al., 2015). The newest plantations carried out within the framework of the Green Morocco Plan have been performed with certified olive trees, of which the 'Haouzia' and 'Menara' varieties constitute the majority. The share of these varieties increased especially after the year 2018. In the Haouz region, the 'Haouzia' and 'Menara' varieties represented nearly 25% of the region's olive trees in 2016. Generally, efforts to diversify the varietal profile are ongoing. Indeed, the National Institute of Agronomic Research (INRA) has been carrying on research on genetic cross-breeding of the 'Haouzia' and 'Menara' varieties on one hand, and 'Arbequina', 'Leccini', 'Manzanilla', and 'Picholine de Languedoc' on the other, which has made it possible to obtain five new high-performance and more tightly adapted varieties: 'Tassaoute', 'Agdal', 'Dalia', 'Mechkate', and 'Baraka'. These are being distributed to plant nurseries for their propagation (MAPMDREF, 2018).

The strategic indications of the Ministry of Agriculture, Maritime Fisheries, Rural Development and Water and Forests (MAPMDREF) for the development of the olive sector include the propagation of plantules adapted to the following three olive grove management systems:

I. Rainfed olive groves

- National varieties: 'Moroccan Picholine', 'Haouzia', and 'Menara'
- Foreign varieties (only with annual rainfall>400 mm): 'Picholine de Languedoc', 'Leccino', 'Hojiblanca', 'Picual', and 'Frantoio'

II. Intensive irrigated olive groves

- Table olives: 'Haouzia', 'Menara', and 'Picholine de Languedoc'
- Olives for crushing: 'Haouzia', 'Menara', and new INRA-developed varieties

III. Super intensive olive groves

- Foreign varieties: 'Arbequina', 'Arbosana', and 'Koroneiki'
- National varieties: Hybrid variety developed by INRA obtained from 'Dalia' and 'Agdal' through multiplication[5].

In terms of yield ratios, 'Menara' is the most effective variety with 24%, followed by another local variety 'Haouzia', whose yield ratio is 23%, then 'Picual' with 22%, followed by 'Moroccan Picholine' and 'Manzanilla' both with a yield ratio of around 20%, 'Hojiblanca' with 17.1%, and lastly, we find 'Picholine de Languedoc' with a yield ratio of between 15% and 18%. Furthermore, all cultivars showed low resistance to cold, except for 'Picual' and 'Mastoidis'. In contrast, 'Houazia' and 'Menara' olive trees are the most resistant ones to diseases that afflict olives, along with the Greek varieties, 'Mastoidis' and 'Koroneiki'. In contrast, the remaining varieties suffer from great sensitivity to pest diseases.

Many studies have been conducted on evaluating the performance of various olive varieties in Morocco, while perhaps the most important one was the one jointly conducted by INRA and the Meknes Regional Center for Agronomic Research, to evaluate 14 olive varieties cultivated in Morocco under two different agronomic management systems (rainfed and irrigated).[4] According to this study, 'Leccino', 'Menara', 'Manzanilla', 'Ascolana Dura', and 'Haouzia' were the most productive varieties (40–60 kg/tree) for rainfed olive groves. Irrigation has, however, significantly improved the yields across all varieties, which, nonetheless, are not characterized by the same water use efficiency rates. The most productive varieties across all management systems include 'Ascolana Dura', 'Leccino', 'Haouzia', and 'Manzanilla'. The heterogeneity of productivity rates characterized all the varieties but with a higher average of heterogeneity in rainfed (0.28) compared to irrigated (0.22) olive groves. The most heterogeneous varieties include 'Arbequina', 'Frantoio', 'Picholine de Languedoc', and 'Menara'. The oil content of the table olive varieties ('Gordal Sevillana' and 'Ascolana Dura') did not exceed 6%, while for the exclusively oil-based ones ('Arbequina', 'Leccino', 'Picual', and 'Frantoio') these values ranged between 15.7% for 'Arbequina' and 25.6% for 'Frantoio'. For multipurpose varieties, the oil contents ranged between 20% and 23.5%. Irrigation has helped increase the oil content across all varieties. The average weight of the olives differed depending on the variety, ranging between 1.2 g ('Arbequina') and 8.3 g for 'Gordal' under irrigation, and 1.2 and 6.2 g, respectively, for the same varieties in rainfed olive groves. The average gain in weight generated by irrigation for olives of all varieties was about 3.5%. The average stone of olives had a weight which varied between 1.15 g for 'Gordal' and 0.28 g for 'Arbequina'. The average increase in stone weight, caused by irrigation, considering all varieties, was 6.3%. This increase was especially marked for the 'Gordal' variety (29.2%).

The production performance achieved depended on the varieties and the agronomic management systems. Under irrigation, the production rates were generally higher, and all the varieties responded positively to the inputs of irrigation. However, 'Leccino', 'Menara', 'Manzanilla', and 'Haouzia' olive trees produced the highest rainfed yields, demonstrating to be more suited to rainfed olive growing (Hadiddou et al., 2015).

42.2 Olive Grove Production Systems

In Moroccan oliviculture, a large number of small farms are found to be scattered across the whole country with mainly rainfed management practices (generally known as Bour). These are traditional agronomic systems. In contrast, we also find farms of several hundred hectares with irrigated olive trees under intensive or super-intensive tree densities (with high tree densities, drip irrigation, and mechanized harvesting) (MAPMDREF, 2014). According to MAPMDREF (2018), traditional olive groves, with an average farm size of 1.5 ha, represent 52% of the olive groves in Morocco, mainly in both mountains and agricultural hilly regions. Intensive olive groves, generally located in flatter areas and with irrigation, would account for up to 44% of them, while super-intensive groves would account for less than 5% of the total (about 40,000 ha) (MAPMDREF, 2018).

In the mountainous regions of Morocco, olive trees are generally tendered without the application of training and/or maintenance pruning, and in varying densities ranging from 3 × 4 m to 12 × 12 m. This variability in planting densities is explained by the lack of information received by the farmers, especially regarding the optimal planting densities allowing for improved physiological development of the tree, thus potentially conducting the olive plantations towards an optimality in yield production. In addition, they may also be influenced by the presence of one or several intercrops. Since the Moroccan Picholine variety, which constitutes the majority of the olive groves, is a vigorous variety, many trees reach an excessive height and occupy an increasingly important space resulting in a reduction of the interception of the light and consequently a decrease in their agronomic productivity. The densities frequently practiced in extensive Moroccan olive groves in mountainous areas are between 80 and 120 trees/ha, with an average density of 100 trees/ha. These olive groves are frequently fragmented and widely dispersed plots, occurring in sloping areas with poor and shallow soils. These trees, for which phytosanitary protection is difficult to achieve, are rarely pruned and generally receive little maintenance, including measures to combat weeds. These traditional mountain groves are characterized by their heterogeneity and low productivity rates, which are linked to their aging. Many older plantations result from the restructuring of trees with wide spacing that thereby do not develop a large volume of foliage, and hence a very low fruiting area per unit of land area is achieved. Occasionally, the planting pattern is carried out according to topographic contour lines, in order to limit soil erosion. Irrigation is hereby only possible in the slightly sloping olive groves. In the case of irrigation, the tree densities implemented are higher (200 trees/ha or more depending on the farmer's mastery of planting and management techniques). The cultivation techniques generally applied include soil maintenance using light plowing, gravity irrigation, seasoned fertilization, and pruning, as well as phytosanitary protection. The expected yields in these irrigated olive groves are, therefore, not too fluctuating and generally equal to or greater than 3 tonnes of olives/ha. There are also many olive groves planted along with intercropping, mainly beans, common wheat, barley, and lentils. In these areas, which are defined by a highly contrasting average annual rainfall rate (330–1000 mm/year), the yields obtained vary from 1 to 5 tonnes depending on the density of the olive grove. The olive production levels obtained increase with tree densities, and the optimum results obtained for a tree density between 200 and 250 trees/ha is a yield of 4 to 5 t/ha under a rainfall amount greater than 500 mm/year. However, the increase in tree densities to more than 150 trees/ha, especially in areas of low rainfall (<500 mm), affects agronomic productivity, the optimization of which requires the provision of supplemental irrigation.

This leads us to state that irrigated and properly managed olive groves undergoing full production can reach a yield estimated at three times the yields generally observed in rainfed and traditional Moroccan groves when rainy conditions are average (Kaji et al., 2017).

Alongside traditional agronomic practices, intensive or modern olive growing has lately developed in Morocco exponentially, as has been the case all around the Mediterranean. The intensification of crops rests in the planting densities which vary between 200 and 600 trees per hectare, generally equipped with an irrigation system providing water quantities between 1500 and 2500 m^3 ha^{-1}. This process of intensification has received great attention in recent years in Morocco thanks to its positive effects in improving the economic profitability of farms. This new cultivation system is marked by very high yields of between 8000 and 12,000 kg ha^{-1} of olives, which also reached the higher levels of mechanization associated. Furthermore, this has also contributed to improving the quality of Moroccan olive oil, with a significant income for the farmers through the reduction of production costs, the harvest of which is mechanized using trunk vibrators (INRA, 2023). Super-intensive irrigated cropping systems were originally developed in Spain and adapted to the Moroccan context (INRA, 2023).

This type of production system makes it possible to achieve very high production rates in a limited time interval thanks to the considerable number of olive plants. The size of the trees does not exceed 2.5 m in height. Planting densities in this system can reach over 1500 trees ha^{-1}, resulting in very high yields (between the third and seventh year of plantation, average yields of 8000–13,000 kg ha^{-1} can be achieved), with harvesting efficiency rates significantly increased by the introduction of thresher or harvesting machine, and also resulting in the reduction of harvesting periods and related financial and energy costs. Furthermore, some studies show how the adoption of these agronomic models ends up improving the quality of the olive oils obtained (Melgar et al., 2008). Despite all of this, super-intensive cultivation systems in Morocco are currently subject to controversy from the point of view of economic and social sustainability (low local workforce required for the harvest) along with their higher environmental impacts (application of systematic chemical treatments against the fly, irrigation effects, a large area in olive tree monoculture, and consequent loss of biodiversity), and reduced varietal alternatives (local varieties vs imported varieties) (INRA, 2023).

Currently, the degree of implementation of this latest management system by Moroccan olive farmers is still weak, with a rate that does not exceed 5%, most of which are located in central Morocco, in the plains along the Fez-Meknes region (INRA, 2023).

In view of all of the data presented in this chapter, and subsequent arguments regarding the distribution and preference for certain varieties and production systems across Morocco, the main challenge facing the Moroccan olive sector is how to adapt to a highly competitive and globalized market, through the modernization of olive farms (thus adopting intensive and super-intensive systems) while maintaining the ability to generate rural employment and secure sustainable use of resources (MAPMDREF, 2019).

References

Hadiddou, A., Oukabli, A., Moudaffar, C., et al. (2015). Performances of 14 national and Mediterranean olive (*Olea europaea* L.) varieties conducted under two contrasts systems (rainfed and irrigated) in Morocco. *Al Awamia. 2013*(127), 23–43 https://www.inra.org.ma/sites/default/files/12702.pdf. Last retrieved: 04/09/2024.

INRA. (2023). *Filière L. 1. La Filière Olivier au Maroc.* https://www.inra.org.ma/fr/content/filiere-olivier. Last retrieved: 04/09/2024.

Kaji, Mamouni, A., Razouk, R., Alghoum, M., & Mekkaoui, A. (2017). Oléiculture : Densité plantation de l'olivier et production. Agriculture-Maroc. http://www.agri-mag.com/2017/06/oleiculture-densite-plantation-olivier/. Last retrieved: 04/09/2024.

Kartas, A., Chliyeh, M., Touati, J., et al. (2015). Performances and potentialities of introduced varieties and local types of olive trees (*Olea europaea* L.) grown in the Ouazzane areas (North of Morocco). *International Journal of Recent Scientific Research, 2*(February 2016), 2571–2586. ISSN: 0976-3031.

MAPMDREF. (2014). *Guide de l'investisseur filière oléicole*. Published online 2014:18. https://www.fellah-trade.com/fr/filiere-vegetale/fiches-techniques/olivier. Last retrieved: 04/09/2024.

MAPMDREF. (2018). *Rapport annuel de la Cour des comptes au titre de l'année 2018. 2020*(1), 257–286. https://www.courdescomptes.ma/publication/rapport-annuel-au-titre-de-l-annee-2018/. Last retrieved: 04/09/2024.

MAPMDREF. (2019). *National Sustainable Ddevelopment Strategy for the Kingdom of Morocco (NSDS) The national strategy to develop the Moroccan olive sector*, pp. 15–16. https://www.greenpolicyplatform.org/sites/default/files/downloads/policy-database/ENG-SNDD_RESUME%20EXECUTIF-V24-D%20(1).pdf. Last retrieved: 04/09/2024.

Melgar, J. C., Mohamed, Y., Navarro, C., Parra, M. A., Benlloch, M., & Fernández-Escobar, R. (2008). Long-term growth and yield responses of olive trees to different irrigation regimes. *Agricultural Water Management, 95–8*, 968–972. https://doi.org/10.1016/j.agwat.2008.03.001

Chapter 43
Soil Management Practices, Pruning, and Harvest of Olive Groves

Salaheddine El Azzouzi, Ali Yahya, and Kamal Targuisti

Abstract The cultivation of olive trees in Morocco is characterized by a great diversity of practices, techniques, and traditions variable and transitional between the different regions of the country. In this text, we retain that the olive trees, apart from the intensive crops that are beginning to invade the plain, remain a source of life for farmers, including small farmers in the mountains and plains. Indeed, in the latter context we present techniques and practices of olive tree cultivation as well as the process of valorization of the olives "from the tree to the table."

43.1 Techniques and Practices of Olive Tree Cultivation in Morocco

Farmers and producers carry out several practices and tillage techniques related to the cultivation of olive trees in different regions of Morocco depending on the nature and size of the plots, farms, and olive trees.

S. El Azzouzi
Département de Biologie, Faculté des Sciences, Université Ibn Toufail, Kenitra, Ouezzane, Morocco
e-mail: salaheddine.azz@gmail.com

A. Yahya
Département de Biologie, Faculté des Sciences de Tétouan, Université Abdelmalek Essaâdi, Tétouan, BP, Morocco
e-mail: alii.yahya1995@gmail.com

K. Targuisti (✉)
Département de Géologie, Faculté des Sciences de Tétouan, Université Abdelmalek Essaâdi, Tétouan, Morocco
e-mail: ktarguisti@gmail.com

J. Muñoz-Rojas, R. García-Ruiz (eds.), *The Olive Landscapes of the Mediterranean*, Landscape Series 36,
https://doi.org/10.1007/978-3-031-57956-1_43

43.1.1 Multiplication of Olive Trees

There are several techniques for the multiplication of olive trees which are characterized by both traditional and modern methods.

At the traditional level, the olive tree is multiplied by woody cuttings, division of stumps, and grafting on the oleaster.

43.1.1.1 Cutting

It is a production technique of olive plants which is characterized by traditional practices that respect the law of the olive tree nature: carefully choosing good twigs from selected olive trees (quality, variety, period, nursery locations, and know-how). Sow the selected shoots in small holes in the tray, pot, or in the previously prepared multiplication pit and then follow up with watering and natural fertilization.

43.1.1.2 Oleaster Grafting

The oleaster is a wild olive tree that grows in the wild in forest ecosystems. This beautiful oleaster tree is called in local dialect "EL Berrri البري".[1] It is a forest tree that develops from seeds that grow in forest trees "neurces" such as the pistachio lentisk tree.

The young olive tree blooms when it is walled at the level of the trunk through different techniques, among which is the grafting by the griffin. In this technique, we take a small part of the branch of a healthy olive tree, which is called "EL MEROUED" المرود[2] (Fig. 43.1). This process is carried out in spring, following the preparation of a 10–15 cm claw, resulting from a quality olive tree and a cutting of the trunk of oleaster (grafting base, Fig. 43.2). Sticking the grafting and the rootstock at the level of the trunk by a wire of palm tree of doum is called CHRIT (الشريط)[3] (Fig. 43.3), which is biodegradable according to the growth of olive tree grafted on the oleaster (Fig. 43.4).

Sectional grafting, called "CHAK الشق",[4] preferably for core trees, is another grafting technique, preferably used for core trees, though it is less commonly applied.

43.1.1.3 Modern Greenhouse Nursery Production

It is an intensive method that is characterized by the sowing of nucleus followed by grafting semi-linear cuttings with hormonal treatment. This olive tree breeding is done in equipped and adapted greenhouses.

---

[1] A common word, "wild oleaster".

[2] A common word, "olive tree griffon to do the grafting".

[3] A common word, it is the yarn prepared from doum palm and is naturally biodegradable.

[4] A common word, it is grafting technique.

Fig. 43.1 Graft preparation

Fig. 43.2 Bonding of grafting onto the rootstock at trunk level

Fig. 43.3 Operation finishing by straw soil

Fig. 43.4 Olive grafted on the oleaster

43.1.1.4 Plantation

The geographical area of Morocco is characterized by a Mediterranean climate, which is well adapted to the cultivation of all the original and introduced varieties of olive trees. The planting is done according to state programs or by the need of agriculture. The olive tree is to be planted in the open ground, from winter to spring, after preparatory work for planting either in modern or peasant cultivation.

43.2 Tillage Techniques

43.2.1 Labor

In the plowing season at the beginning of November, farmers generally intercrop mainly cereals (wheat, barley, and oats) and vegetables (beans, chickpeas, and lentils). Thus, olive trees are associated also with other trees such as figs, vines, almonds, and carobs.

In the mountains, the majority of olive farmers work the soil in a traditional way, using the plow "EL Mehret[5] in wood," and in the large farms of olive trees, there is monoculture oriented towards semi-intensive cultivation with mechanized plowing by tractor and the use of chemical products (treatment and fertilizers).

Hoeing the soil is done around trees by practicing impluvia and small stone terraces to ensure the accumulation of rainwater or irrigation especially around young olive trees. It is also a way of curbing soil erosion.

43.2.2 The Association or Intercropping

Usually, the olive plots in the mountains are characterized by a diversity of crops (Fig. 43.5). The farmers associate crops such as cereals (wheat, barley, and oats) and vegetables (beans, chickpeas) depending on the season as well as the integration of livestock to the olive fields especially in summer (Fig. 43.6).

Fig. 43.5 Association of crops to olive fields

[5] A common word, it is the tool for plowing the fields.

Fig. 43.6 Landscape of diversity and associations in arboriculture in Ouezzane

43.2.3 Olive Hoeing

It is a technique to turn the ground around the olive thrones with tools like a pickaxe and sapping on a limited surface of 3 m by 3 m in max, depending on the size of the tree (Fig. 43.7).

43.2.4 Impluvium

It is a dyke around a tree in the shape of a semicircle to stop soil erosion and water accumulation around olive trees (Fig. 43.8).

43.2.5 Stone Terraces

It is a technique to break the slope and stop soil erosion in order to facilitate the accumulation of rainwater or irrigation especially around young olive trees (Fig. 43.9).

Fig. 43.7 Action of hoeing the olive tree and adding manure in winter

Fig. 43.8 Half impluvium of water accumulation around the olive trees

43.2.6 Fertilization of the Olive Tree Soil

The majority of farmers add chemical fertilizers (NPK) when plowing olive fields, especially in relation to intercropping (cereal, leguminous, and fodder). In this operation, the olive trees benefit from the presence of either chemical or natural fertilizers.

The period of concentration of chemical fertilizers is from November to March. In large olive farms, the producers also use urea and chemical fertilizers directly to the olive trees. On the other hand, the peasant agriculture of olive trees is based on polyculture, which benefits from the falling leaves of other trees such as fig trees,

Fig. 43.9 Growing olive trees on terraces around the olive trees. Region of Khemisset

vines, pomegranate, carob, almond, etc. This diversity of crops reproduces a balanced ecosystem in the production plots especially in the mountains (Fig. 43.10).

Another level of natural agriculture is characterized by wild olive trees that are grafted (no plowing, no irrigation, no phytosanitary treatment).

Measures of either chemical or organic fertilization include the following:

Young trees: 20–40 kg of manure, 80–100 g of N per tree and per year of age, 60–80 g of P205 per tree and per year of age, and 80–120 g of K20 per tree and per year of age.

Mature trees in production: 60–80 kg of manure, 600–1500 g N per tree (5–7 kg ammonia sulfate), 800–1000 g P205 per tree (1.8–2.2 kg super triple 45%), and 1000–1500 g K20 per tree (2–3 kg of potassium sulfate).[6]

[6] https://www.fellah-trade.com/fr/filiere-vegetale/fiches-techniques/olivier

Fig. 43.10 Addition of manure at the olive tree

43.3 Olive Tree Pruning Techniques

43.3.1 Type of Pruning

43.3.1.1 Growth Pruning

This type of pruning is done in the early years of tree growth, forming the tree branches of olive; it grants an important step in tree life (Fig. 43.11).

43.3.1.2 Fruiting Pruning

This type of pruning is practiced after the olive harvest to clean the dead branches, rejects, etc. (Fig. 43.12).

43.3.1.3 Rejuvenation Pruning

This is done for old trees that produce little or none at all. This operation is important to regenerate the olive tree life (Fig. 43.13).

Fig. 43.11 Olive growth pruning

43.3.1.4 Tree Care

Farmers usually maintain their olive trees during the schooling period (cut the bad branches, damaged branches); they use an ax to cut the branches because of the lack of pruning tools.

43.3.2 Pruning Period

The best period of olive tree pruning is after the olive harvest between the end of November and the end of January.

43.3.3 Wood Energy

Wood energy can be generated from olive trees in the form of branches or wood, with small pieces coming from productive trees and larger pieces from old trees).

Fig. 43.12 Olive tree fruiting pruned

43.4 Traditional Techniques and Practices of Olive Preparation

43.4.1 Olive Harvesting Technique and Transferring Olives from the Field to Home

Over a period of 3 months, from the beginning of November to January, women and men climb tall olive trees to drop the olives with special tools such as a wooden stick of 2–3 m. It is well prepared for this operation which is called "Sakkat"[10] (Fig. 43.14); it is the common name used by Jbala (mountain inhabitants of the Western Rif) in the country.

Also, the farmers use other tools to drop the olives like reeds from 3 to 5 m, which are prepared and intended for this harvesting operation. It is called "Kasba[7] of olives"; the usefulness of this tool either from the ground or from the trees depends on the size of the olive tree.

[7] A common word of Arabic origin (سقط ، السقوط), which means a tool for dropping olives.

Fig. 43.13 Olive tree rejuvenation pruning

The highest Kasba is called "Allaya[8]," a common name meaning the tallest trunks to drop the olives that are very high in the olive tree. The person who drops the olives by the wooden stick is called "EL Sakkat."

Hand-harvesting is a very important technique (Fig. 43.15). It is done to pick up the olives from young olive trees, either from the ground or climbing the young trees with care so as not to break the branches. The tree is climbed with a bucket and a tool called "Jennaya,"[9] meaning the tool to pick olives.

In olive fields, especially in the mountains, farmers use mules to move their olives to home, to the olive crushing unit (Al Maassara), or to the "Rha" stone mill of the olives (Fig. 43.16).

---

[8] A common word of Arabic origin (العلو /), which means a tool from a higher reed to drop the olives.

[9] A common word of Arabic origin (جني، الجني) which means a tool to pick up olives from the tree.

Fig. 43.14 Dropping down of olives by "El Sakkat"

43.4.2 Sorting, Processing, and Conservation of Olives

Farmers often bring their olives home to sort them on several levels (Fig. 43.17). Black olives are essential for table olives. They are well sorted and cleaned to be preserved with salt in bags or jerry cans.

The green and/or violet olives are a type of olives well prepared and added with salt and lemon, if necessary. The producer takes care of breaking the olives by a stone or cutting the olives on three ("cut olives") with a knife to speed up the removal of bitterness. These olives are intended for the table, either for home consumers or in restaurants. However, a large amount of olives is destined for crushing to make table olive oil as needed (Hammadi, 2002).

The national production of olives is destined to 65% for crushing and 25% for canning. The remaining 10% are the losses caused by the various manipulations and for self-consumption. The processing of olives is done by a modern sector composed by industrial or semi-industrial units and a traditional sector by stone mill (AL Maâsras[10]).

[10] The olive crushing unit.

Fig. 43.15 Hand-harvesting and sorting of olives

Fig. 43.16 A traditional olive press Al Maassara or Rha

Fig. 43.17 Sorting of olives

43.5 Conclusions

Farmers do soil work and activities throughout the year. After the harvest, the farmers process the olives either in the modern units ("Al Maassara") or in the traditional stone mills called "*Rha*," which is less used recently. Also, the producers realize the different types of olive pruning, growing, fruiting, and rejuvenation of old trees.

One agricultural practice involves grafting onto wild oleaster and also taking cuttings in a nursery to grow young olive trees.

Traditionally in Morocco, families prepare various types of olives using artisanal methods, including green olives preserved with lemon and salt, broken green olives preserved in salt, and black olives conserved with salt. However, a large amount of harvested olives is intended for the production of table olive oil.

The consumption of prepared olives and olive oil reflects a unique food culture both locally and in the Mediterranean Basin, thanks to the heritage of traditions and know-how related to gastronomy in the countries with olive trees.

The economic valorization of the olive tree shows several interesting products: prepared olives, olive oil, wood energy, waste from the pomace, marginal, and compost.

References

Hammadi, C., 2002. *Amélioration de la qualité de l'huile d'olive, restructuration et modernization des unites de trituration des olives*. Rapport du Séminaire internationale sur l'olivier, Acquis de la recherche et contraintes du secteur d'olivier, INRA, pp. 335–351.

Webography

http://www.agri-mag.com/2017/06/oleiculture-olivier-et-changements-climatiques/
https://www.fellah-trade.com/fr/filiere-vegetale/fiches-techniques/olivier

Chapter 44
Pests and Diseases and the Main Control Strategies

Ali Yayah, Bennas Nard, Salaheddine El Azzouzi, and Kamal Targuisti

Abstract Olive tree and its products can be damaged by the presence of numerous pest insects (phytophagous invertebrates). A large group of those phytophagous invertebrates' species that afflict olive trees is composed of polyphagous species, and the other group made up of monophagous or oligophagous species. This phytosanitary problem of the olive tree constitutes the main obstacle to the productivity of olive growing. As is the case in all countries of the Mediterranean basin due to the big similarity in climatic conditions, the most important and harmful pests that afflict olive trees in Morocco are olive moth (*Prays oleae*), the olive fly (*Bactrocera oleae*), the olive psyllid (*Euphyllura olivina*), and the black cochineal (*Saissetia oleae*). In addition to these pests, microorganisms are also involved in attacking Moroccan olive trees. The main fungi identified are *Fusicladium oleagineum*, responsible for peacock eye; *Verticillium dahliae*, the causative agent of verticilliosis; and *Fumago salicina* and *Capnodium oleaginum*, involved in the appearance sooty mold. Among the bacteria involved, *Pseudomonas savastanoi* pv. *savastanoi* has been identified as responsible for tuberculosis in olive trees, characterized by the presence of tubers on the branches and stems, as well as *Xylella fastidiosa*, provoking olive quick decline syndrome. Mostly, pesticides are used to protect the olive harvest from pests and diseases. However, the worrying issue is the negative effects of pesticides on the environment, human health, and product quality. So among the tools available for the management

A. Yayah
Département de Biologie, Faculté des Sciences de Tétouan, Université Abdelmalek Essaâdi, BP, Tétouan, Morocco
e-mail: alii.yahya1995@gmail.com

B. Nard · K. Targuisti (✉)
Département de Géologie, Faculté des Sciences de Tétouan, Université Abdelmalek Essaâdi, BP, Tétouan, Morocco
e-mail: nbennas@uae.ac.ma; ktarguisti@gmail.com

S. El Azzouzi
Département de Biologie, Faculté des Sciences, Université Ibn Toufail, Kenitra, Ouezzane, Morocco
e-mail: salaheddine.azz@gmail.com

J. Muñoz-Rojas, R. García-Ruiz (eds.), *The Olive Landscapes of the Mediterranean*, Landscape Series 36,
https://doi.org/10.1007/978-3-031-57956-1_44

of olive pests and diseases, certain biological practices make it possible to limit the spread of a specific type of disease in crops and new management systems such as integrated production and organic farming.

44.1 The Main Pests and Diseases Affecting Olive Trees in Morocco

The national olive grove is made up of more than 96% of the variety Moroccan Picholine population which, despite its adaptability and its dual-purpose (production of oil and olive preserves), has certain drawbacks, in particular, a large susceptibility to certain diseases (Rotten, 2007). Indeed, the various studies carried out nationally have revealed that the olive tree and its products can be damaged by the presence of numerous pest insects (phytophagous invertebrates) (Rhimini et al., 2014). A large group of those phytophagous invertebrates' species that afflict olive trees is composed of polyphagous species, each of them has many host plants in addition to the olive tree, those species are usually occasional pests, the second group is made up of monophagous or oligophagous species, which are particularly adapted to the olive tree, and they may cause economic losses and pose a serious risk for the annual olive yield (EIP-AGRI Focus Group Pests and diseases of the olive tree, 2020).

This phytosanitary problem of the olive tree constitutes the main obstacle to the productivity of olive growing. As is the case in all countries of the Mediterranean basin due to the big similarity in climatic conditions, the most important and harmful pests that afflict olive trees in Morocco are olive moth (*Prays oleae*), the olive fly (*Bactrocera oleae*), the olive psyllid (*Euphyllura olivina*), and the black cochineal (*Saissetia oleae*).

The symptoms of these pests are diverse and can affect all organs of the olive tree (twigs, leaves, flowers, and fruits), the damage of which can range from premature leaf drop to a reduction in the quantity and quality of the oil. Thus, they can cause economically significant damage (Barguigua et al., 2020). A majority of these pests spread rapidly, and they move from field to field by air or with plowing operations; therefore, it is not possible to eliminate a pest in a field without cleaning up the whole neighboring fields, which explains the difficulty of eliminating these pests completely (https://www.ts3a.com/bi2a/افات-الزيتون-الامراض-شجرة-الزيتون/).

The olive fruit fly (*Bactrocera oleae*) is the first pest on the olive crop in Morocco, and it causes heavy economic losses, especially when controlling it is neglected. It is a specialized pest, as it affects only olive fruits. It affects most varieties of olives in all areas of its cultivation.

The female lays her eggs individually in slanted cavities that dig them directly under the skin of the fruit, the eggs hatch, and the larvae feed on the pulp of the fruit, and when it is fully grown, they come out of the fruit where they fall to the ground to turn into a barrel-shaped virgin of light-brown color. In the absence of olive fruits on the trees, the stages of the insect (larva) remain in a dormant state until the next season. Likewise, the adult insect remains and extends its life to the next olive season to be able to lay eggs to restore its life cycle. Its symptoms are characterized by the presence of black spots on the fruits, as a result of being stabbed with an egg-laying machine, as the fruits lose their marketing value. Fruit rot is a result of contamination of the tunnels dug by the larvae in the pulp of the fruits during feeding, and the areas of infection on the fruits become soft spongy, with the fruits often falling out as a result of infection. The percentage of oil in infected fruits decreases with the high percentage of acidity of the oil, as the acidity of the oil resulting from the infected fruits may reach four times the acidity of the oil produced from healthy fruits. The infection dates differ according to the different olive varieties and the region in which the late varieties are grown in which the infection is later than the early varieties. Additionally, this pest can infect olive crops every year if the appropriate environmental conditions are provided (Technologie TDE, 2009; Fellah-Trade, n.d.). In addition to these pests, microorganisms are also involved in attacking Moroccan olive trees. The main fungi identified are *Fusicladium oleagineum*, responsible for peacock eye; *Verticillium dahliae*, the causative agent of verticilliosis; and *Fumago salicina* and *Capnodium oleaginum*, involved in the appearance sooty mold. Among the bacteria involved, *Pseudomonas savastanoi* pv. *savastanoi* has been identified as responsible for tuberculosis in olive trees characterized by the presence of tubers on the branches and stems, as well as *Xylella fastidiosa*, provoking olive quick decline syndrome (Barguigua et al., 2020).

Among all these diseases, tuberculosis is considered the main bacterial disease of the olive tree in Morocco. It develops wherever the olive tree is cultivated and mainly causes damage in the humid regions of northern Morocco, in Meknes, Fes, and Zerhoun. The intensity of the damage caused is strongly related to the number of tumors per tree, as indicated in a study carried out by the National Institute of Agronomic Research "INRA" on the degree of attack of tuberculosis to the olive tree in olive orchards in Morocco (Benjama, 2003). This study was carried out on 20 trees with an average age of 35 years, of the same phenotype, and of equivalent foliage. They were chosen at random from a stand of olive trees of the variety "Meslalla" in northern Morocco, infested with tuberculosis in the natural state. As a result, the number of tumors systematically counted on all of these 20 trees observed varied between 205 and 1500 tumors per tree. The average degree of tree attack was 741 tumors per tree with a standard deviation of 370 tumors per tree. The olive trees studied, made up on average of 25 structural units per tree, presented an average of 44.5 tumors per structural unit (Benjama, 2003).

Thus, a close correlation was demonstrated between the number of tumors per linear meter of shoots and the growth of these shoots. As the number of tumors increases, the growth of the twig is stunted, and the number of fruit clusters is affected (Benjama et al., 1993). The disease also causes leaf desiccation. For the Meslalla variety, desiccation and defoliation begin as soon as the number of tumors per tree reaches or exceeds a threshold of 600 tumors (Benjama, 2003).

Likewise, the study of the behavior of four varieties towards this disease had shown that Moroccan Picholine was tolerant, while Gordale was immune; Meslalla was very sensitive, and Picholine du Languedoc was sensitive (Benjama et al.,

1993). These two varieties (Moroccan Picholine and Gordale) could therefore be used as a reference in a breeding program for olive tree varieties resistant to tuberculosis.

In 2012, surveys conducted in the Ouazzane region revealed another disease affecting olive orchards of the Moroccan Picholine variety. This disease manifested in both ripe and immature fruits, exhibiting brown to black rot spots resembling chocolate truffles of various sizes. These spots affected either a small portion of the fruit or a significant part of it, leading to the partial or complete mummification of the olives and, ultimately, their premature fall.

This anthracnose disease is one of the most important leaf fungi of the olive tree and the main disease of the fruit in Morocco, which is qualified as very dangerous because of the enormous damage it causes on olive production. The most characteristic symptom of the disease is the rotting and mummification of olives. These attacked organs constitute an important reservoir of inoculum from which the conidia are dispersed to other healthy organs (secondary infection) via, among other things, splashing rain and wind.

Subsequently, the impact of the spread of this disease increased, especially in 2015, when the disease spread to several sites in the Taounate region (Benimlid, Ain Aicha), Moulay Driss Zerhoun-Meknès (Moussaoua), and Khénifra, with incidences oscillating between 4% recorded in the Khénifra site and 30% in the Taounate region. In Moulay Driss Zerhoun, the incidence is around 10%. The disease may have spread after 3 years to other distant sites.

These mummifications and rots on olives have a negative impact on the quality of production and especially on the quality of the oils extracted. In fact, they lead to very acidic oil and a deplorable organoleptic quality, and the oils extracted from the fruits infected with the fungus contain toxins (Agriculture-Maroc, 2017). The incidence and severity of mycosis vary considerably depending on the favorable environmental conditions (abundant rains, high humidity, and heavy dew, temperatures below 25 ° C), which explained the strong presence of this disease in northern Morocco more than in the South. As for olive peacock disease, which is considered one of the most significant diseases affecting olive trees in Morocco, its symptoms primarily manifest on the leaves. In cases of severe infection, the leaves may drop prematurely, leading to a general weakening of the tree and a reduction in yield (Lopez-Villalta, 1999). A study (Rhimini et al., 2014) that was conducted on the spread of this disease in Morocco proved that the severity of the infection and the percentage of infected leaves are very high in the north of Morocco compared to the south, where the humidity is a contributing factor in the growth and spread of the disease. In addition, the greater the distance from the sea, the lesser the spread and severity of the disease. In "Kenitra" near the sea, the percentage of infection is about 12.1%, while the percentage of infection in "Jamaa Lalla Mimouna" far from the sea is 6.5%, which confirms that continentality and humidity are a key factor in the spread of this disease in Morocco (Rhimini et al., 2014). In addition to the altitude, it has been observed that the prevalence of the disease in high-altitude areas is higher than in other areas with relatively low altitudes. The higher the altitude, the more severe the disease spread, as the altitude compensates for the loss of humidity due to the distance from the sea. The injury rate in Ouezzane, which is around 300 meters high, is 1.16%, while the injury rate exceeds 4.24% in the Al Masmouda center, which is 400 meters above (Rhimini et al., 2014).

44.2 The Main Control Strategies

Current olive pest and disease management strategies in Morocco are still based on the chemical pest and disease control (pesticides). But sustainable management strategies are also involved.

The consumption of phytopharmaceutical products varies from year to year, depending mainly on the vicissitudes of the climate, the pressure of diseases and insect pests, but also according to regions, farming methods, and crops.

According to statistics from the exchange office, imports of phytopharmaceutical products have continued to increase since 2005. In 2005, Morocco imported about 14,679 tonnes of phytopharmaceutical products, equivalent to 548 million dirhams, and this increase in imports continued. In 2009, Morocco imported 17,519 tonnes of phytopharmaceutical products worth 955 million dirhams. Through these figures, the growth of imports in value has been sustained since 2005, on average 15% per year, while in volume it is only 4.5% per year. This reflects Morocco's orientation towards less and less organophosphates and more and more towards so-called soft plant protection products that are effective and more respectful of the environment, which are used at increasingly low doses

In Morocco, the consumption of pesticides for agricultural crops is 1.46 kg per hectare, according to datafrom the Food and Agriculture Organization (FAO) for 2017 (Agrimaroc.maroc-pesticides.pdf). However, insecticides take up the great part, with percentages ranging from 40% to 55%, followed by fungicides from 35% to 45%, then come the herbicides which can oscillate between 10% and 15%, depending on the year and weather conditions.

While the olive sector ranked 21st among the different crops in Morocco in terms of pesticide consumption, in 2018 according to Croplife Maroc, the estimated olive consumption of agricultural pesticides is equivalent to 0.63% of agricultural pesticides used at the national level (Agriculture-Maroc, 2020).

The repeated use of pesticides based on the same active substance would promote the development of a resistant population of pests, so the pesticides would become less and less effective which will inadvertently push farmers to put in an additional amount. This explains the admission of noncompliance with the recommended dose (Naamane et al., 2020). And the problem posed with the frequent use of pesticides is that the chemicals used cannot be completely harmless, which hinders the quality of the oil extracted. Indeed, any pesticide residue present in the olive can end up in the oil as is the case with fat-soluble pesticides. The problem of pesticide residues arises much more when consuming these raw oils.

Integrated pest and disease control is also involved depending on the use of various control techniques (biological, genetic, physical, mechanical, etc.) such as predators, parasites, cultural control strategies, insect pathogens, and nonpathogenic antagonistic or competitive microorganisms, the reasonable use of pesticides, crop maintenance, appropriate management of crop waste and residues, the adequacy of structures, efforts to raise awareness, and training producers.

The major components of IPM systems are identification of pests, diseases, and natural enemies; monitoring of pests, diseases, damages, and natural enemies; selection of one or more management options on the basis of monitoring results and action thresholds, from a wide range of pesticide and non-pesticide options; and use of selective pesticides targeted at the pest or disease, for instance, pesticides that will interfere the least with natural enemies, targeted only at infested trees or parts of trees.

Although it is not at the level of pesticide use, recent projects and programs have been launched by the concerned ministry to encourage the application of the integrated management (PLAN DE PROTECTION ET DE PRODUCTION INTÉGRÉES DES CULTURES (PPPIC), 2010) (Mimina, 2014).

References

Agriculture-Maroc. (2017). http://www.agri-mag.com/2017/06/oleiculture-lantrachnose/

Agriculture-Maroc. (2020). http://www.agri-mag.com/2020/07/le-marche-des-pesticides-au-maroc-realites-et-defis/

Barguigua, A., Zahir, I., Youss, S., Fikri, N., & Youss, B. (2020). Prospection des maladies microbiennes de l' olivier dans la région Tadla-Azilal. *Revue Marocaine des Sciences Agronomiques et Vétérinaires, 8*(3), 331–338.

Benjama, A. (2003). Méthode d'évaluation rapide du degré d'attaque de l'olivier par la tuberculose causée par Pseudomonas savastanoi pv. Savastanoi, en verger au Maroc. *Fruits., 58*(4), 213–219. https://doi.org/10.1051/fruits:2003009

Benjama, A., Atbi, M., Tahiri, J., & Aznadi, M. (1993). Situation actuelle de la tuberculose de l'olivier dans la région de Zerhoun au Maroc. *ElAwamia, 83*, 153–160.

EIP-AGRI Focus Group Pests and diseases of the olive tree. 2020. Final Report https://ec.europa.eu/eip/agriculture/sites/default/files/eipagri_fg_pests_diseases_olive_tree_final_report_2020_en.pdf

Fellah-Trade. (n.d.). https://www.fellah-trade.com/fr/filiere-vegetale/fiches-techniques/olivier

Lopez-Villalta, C. (1999). *Contrôle des parasites et des maladies de l'olivier* (pp. 151–162). Conseil Oléicole international, Principe de Vergara.

Mimina, H. M. (2014). La Lutte Intégrée : historique, complexité, principes et mise en œuvre Integrated Control : history, complexity, principles and implementation. *Revue Marocaine de Protection des Plantes, 6*, 1–20. Published online.

Naamane, A., Sadiq, A., Belhouari, A., Iounes, N., & El Amrani, S. (2020). *Enquête sur l'utilisation des engrais et pesticides chez les agriculteurs de la région de Casablanca-Settat*. IAV Hassan II.

PLAN DE PROTECTION ET DE PRODUCTION INTÉGRÉES DES CULTURES (PPPIC). Projet, 2010.

Rhimini, Y., Chliyeh, M., Touati, J., Touhami, A. O., & Benkirane, R. (2014). Influence of certain cultural practices and variable climatic factors on the manifestation of Spilocaea oleagina, olive peacock spot agent in the northwestern region of Morocco. *International Journal of Pure & Applied Bioscience, 2*(5), 1–9.

Rotten, B. (2007). *Guide du producteur de l ' huile d ' olive Guide du producteur de l' huile d' olive*. ORGANISATION DES NATIONS UNIES POUR LE DÉVELOPPEMENT INDUSTRIEL https://ec.europa.eu/eip/agriculture/sites/default/files/eipagri_fg_pests_diseases_olive_tree_final_report_2020_en.pdf

Technologie TDE. (2009). *Les principaux ravageurs de l' olivier* (0537) (pp. 77–80). MAPM/DEFR.

Chapter 45
Challenges for Sustainable Water Resource Management: Irrigation, Salinity and Contamination

Dorsaf Bouharat, Farida El Youssefi, Ali Yahya, and Kamal Targuisti El Khalifi

Abstract In the countries around the Mediterranean where water needs are increasing and resources tend to decrease, lack of water is one of the most serious threats to agriculture and the environment in general and that of the olive tree in particular. In Morocco, the situation is particularly difficult in the southern regions, which suffer from a chronic insufficiency of water resources (reduced precipitation, overexploitation of groundwater, salinization and contamination of aquifers). The olive tree is relatively resistant to water stress and moderately tolerant to the salinity of water irrigation. The olive industry in Morocco is responsible for huge environmental problems caused by pollution and by liquid effluents or vegetable water commonly known as "margine". It is estimated that the annual output of margine in this sector is around 250,000 m^3.

45.1 Description of Olive Systems in Morocco in Terms of Water Resources

The culture of olive cultivation is ancestral in Morocco. With one million hectares spread across 400,000 farms, the olive sector is vast and significant, with 60% of the olive area located in arid and semi-arid zones (Fig. 45.1). However, the olive tree is

D. Bouharat · F. El Youssefi
Département de Chimie, Faculté des Sciences, Université Abdelmalek Essaâdi, Tétouan, Morocco
e-mail: dorsaf.bouharat@gmail.com

A. Yahya
Département de Biologie, Faculté des Sciences, Université Abdelmalek Essaâdi, Tétouan, Morocco
e-mail: alii.yahya1995@gmail.com

K. T. El Khalifi (✉)
Département de Géologie, Faculté des Sciences, Université Abdelmalek Essaâdi, Tétouan, Morocco
e-mail: ktarguisti@gmail.com

J. Muñoz-Rojas, R. García-Ruiz (eds.), *The Olive Landscapes of the Mediterranean*, Landscape Series 36,
https://doi.org/10.1007/978-3-031-57956-1_45

Fig. 45.1 Olive grove in the region of Amizmiz (Province of Marrakech)

relatively resistant to water stress and tolerates water deficit well, although from a critical limit, vegetative growth and yield decrease significantly. The response to water stress affects the quality of oil by increasing the concentration of phenols while reducing that of chlorophyll and certain fatty acid. Nevertheless, the free acidity of oil, which is the most important criterion in defining its quality, is not affected by the water stress.

In fact, for a normal, favourable and regular fruiting, the olive trees require a sufficient water supply for a good accomplishment of its cycle. Thus, the rainfall is one of the factors allowing a favourable production. However, the following facts should be taken into consideration:

- With 600 mm of well-distributed rain, the olive tree vegetates and produces normally.
- Between 450 and 600 mm, production is possible provided that the soil's water retention capacities are sufficient (deep clay-luminous soil).
- With a rainfall of less than 200 mm, olive growing is economically unprofitable.

Although the olive tree is found throughout the whole national territory, with the exception of the Atlantic coast, the geographical distribution of this heritage brings out three large areas that are very distinct depending on the level of precipitation:

- The North: Chefchauen, Taounate and Ouezzane, which present poor soils, rugged lands and a rainfall level of 1000 mm/year.
- The Centre: Taza, Fez and Meknès which have rich and deep soils, flat lands and a rainfall between 450 and 500 mm/year.
- The South: Haouz, Tadla, Safi and Essaouira which present very rich soils, a flat land and rainfall of less than 400 mm/year. The plantation receives perennial or supplemental irrigation.

Consequently, the geographical distribution of this heritage highlights three large very distinct olive-growing areas:

- The mountainous areas: 250,000 Ha with an area of 37% of olive trees in Morocco, occupying the provinces of Taounate, Taza, Ouezzane, Chefchaouen, Azilal, Khénifra, Al hoceima, Amizmiz, etc.
- The irrigated areas: 250,000 Ha, with an area of 37% of olive trees in Morocco, occupying the regions of Al Haouz, Tadla, Oasis, etc.
- Favourable Bour areas: 180,000 Ha, with an area of 26% of olive trees in Morocco, encompassing the provinces and regions of Meknès, Fès, Sefrou, El Hajeb, Gharb, Loukkos, etc.

Irrigated cultivated areas, at the national level, are 250,000 Ha, nearly 37% of the total area planted with olive trees.

45.2 Irrigation Strategies for Olive Groves in Morocco

In 2008, Morocco has implemented a new agricultural development strategy, the Green Morocco Plan. Precisely, the olive sector has benefited from a strong dynamic in the framework of the GMP, which aims at the long-term transformation of this sector with the ambition of making agriculture a real driving force of economic growth and a vector of social cohesion at the national and territorial level.

45.2.1 Measures Taken to Mobilize the Water Resources

The measures taken to mobilize the water resources will allow the development of irrigation systems and the evaluation and protection of hydraulic installations, stressing the importance of controlling the water flow to meet the water needs during the drought years, thanks to the construction of the necessary hydraulic installations (Fig. 45.2).

Nowadays, several regions of Morocco benefit from a development programme which is part of the development of the olive sector among small farmers, registered in the pillar II of Green Morocco Plan. This project benefits the setting up of a network of canals for regular and efficient transport of water which will improve the irrigation of olive trees, ensuring a regular and efficient water supply without losses.

Indeed, to manage water in a rational and independent way at all levels and to design ways for farmers to increase their production capacity, the project has a dual objective: to convert the non-cereal area into olive tree plantations and to improve the yield in terms of quantity and quality of oil that meets standards. The average target yield is 4 tons per hectare, double the current average yield of 2 tons.

In Morocco, the challenges related to the availability of water resources impose a certain strategy when it comes to irrigation. There are two irrigation systems:

Fig. 45.2 Terraced bund to break the slope, slow down soil erosion and facilitate the accumulation of rainwater or irrigation

perennial irrigation practiced in the areas of action of the Regional Agricultural Development Offices, such as Haouz, Tadla, Errachidia, Souss-Massa, etc., and supplemental irrigation, mainly present in the regions of Marrakech, Beni-Mellal, Azilal, Taza, El Kelâa, Boulemane, Oujda and Nador.

Among these methods is "localized irrigation", whose initial investments are subsidized by the State. In drip irrigation, water is supplied at a fixed frequency to maintain a wet bulb under the drippers at all times. The doses are variable according to the water needs of the olive tree (Fig. 45.3).

Given that the olive tree is more sensitive to hydromorphic conditions than to drought, it is therefore necessary to think about irrigation, in conditions of abundant water, in order to limit water stagnation.

However, in times of water shortage, another less widespread yet very effective technique is the "controlled water deficit". This method consists of reducing irrigation water inputs to a minimum at the least critical phenological stages in terms of production. Then, during the most critical stages of growth and development of the olive tree, in spring (April), when the tree needs irrigation, it is flowering, and also at the beginning of summer (late May to mid-July), which is the period from fruit set to the hardening of the stones.

In arid and semi-arid areas where 60% of the national olive-growing area is located, and where the water supply is provided by low water reserves (wadis or wells of low capacity), supplemental irrigation is possible and consists of providing a small amount of water during the critical phases of olive tree growth (Fig. 45.4).

Fig. 45.3 Drip irrigation

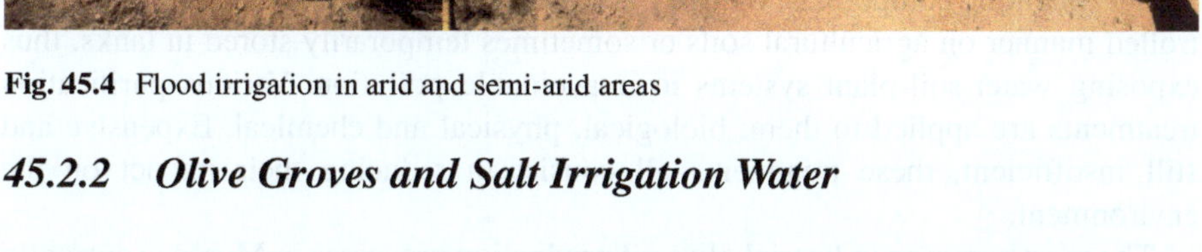

Fig. 45.4 Flood irrigation in arid and semi-arid areas

45.2.2 Olive Groves and Salt Irrigation Water

The olive tree is a plant moderately tolerant to the salinity of irrigation water. Symptoms and damages on the tree vary according to the cultivars. The most sensitive plants will begin to show their intolerance from an electrical conductivity (EC) of 2.5 and 4 dS/m^{-1}.

The productivity of the olive grove irrigated with saline water is lower than that irrigated with normal water, due to a lower fermentation of the fruit, which is then returned to the oil. Olives irrigated with salt water, however, show a significant increase in the content of phenolic compounds (+58%) compared to the control. In order to avoid problems related to the salinity of the water, it is important to have good drainage. This allows the salts transported in depth by the leaching to be removed. It also favours leaching during moments of low evaporation. Another solution to control salinity is to play with the acidity of the water. For example, it is possible to add sulphuric acid to solubilize the calcium salts and facilitate the leaching of the sodium salts.

45.3 The Environment in the Face of the Olive Industry in Morocco

45.3.1 Environmental Problems Caused by Vegetable Water Commonly Known as "Margine"

The national olive production reached almost two million tons in 2018–2019, 65% of which is destined for crushing and 25% for canning. The transformation into olive oil is done through modern, semi-modern and traditional sectors: Morocco has 948 crushing units and 11,000 maâsras. The preservation of olives, for its part, is ensured by 75 modern units and other crafts. The olive industry in Morocco is responsible for huge environmental problems caused by pollution from these liquid effluents commonly known as "margines". It is estimated that about 250,000 m^3 are produced annually in this sector.

These volumes of margines are produced in very short time intervals from November to April. The discharge of these effluents without prior treatment is detrimental to the environment (Fig. 45.5), particularly to the wadis (Fig. 45.6) and the water in the dam which become highly loaded with organic matter and pollutants and lose their capacity to self-purify. The pollutant load can reach 200 g of COD at the exit of the olive industry.

Marginal waters, or vegetation water, are liquid discharges very rich in organic matter (phenolic compounds, lipids), often spread as they are in nature in an uncontrolled manner on agricultural soils or sometimes temporarily stored in tanks, thus exposing water-soil-plant systems to unavoidable pollution. Various purification treatments are applied to them: biological, physical and chemical. Expensive and still insufficient, these treatments all consist in reducing their impact on the environment.

The most common industrial olive oil production processes in Morocco currently include a continuous extraction system design with a vertical centrifugation and a horizontal centrifugation that separate the paste into three phases: the oil, the olive rind and the olive pomace. However, the traditional process is based on a discontinuous system (Figs. 45.7 and 45.8), using hydraulic presses or stone mills that exert pressure on the olives and allow the production of three phases: oil, vegetation water and pomace.

The disadvantage of these two systems is that they generate 60–70 L/100 kg of olives (liquid waste), in addition to the pomace (solid waste). Some manufacturers have therefore converted the three-phase continuous system into a two-phase continuous system with decanters allowing the production of two phases, which are the olive oil and a plastic paste (mixture between the olive oil and the olive pomace).

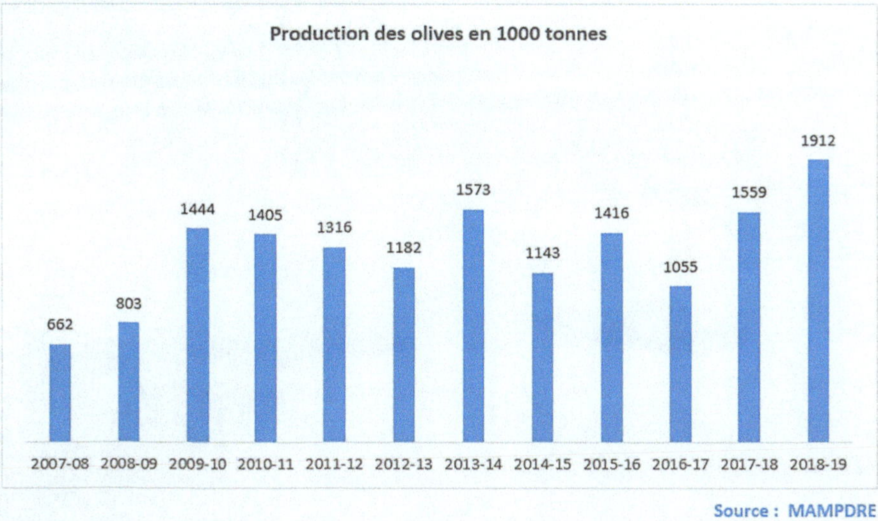

Fig. 45.5 Production and wild deposit of vegetable waters (margines) in an olive trituration plant

Fig. 45.6 Vegetable waters (margines) dumped in wadis

The physicochemical characterization of the margines is generally dependent on the techniques and systems of extraction of olive oil; it differs from one country to another. In general, the vegetable oils contain a variety of organic and mineral compounds of very different nature and concentration.

Fig. 45.7 Hydraulic press process

Fig. 45.8 Manual press process

Concerning the mineral fraction of the margines, they contain significant quantities of mineral salts, 80% of which are soluble (phosphates, sulphates and chlorides) and 20% insoluble (carbonates and silicates). The most representative elements are potassium (47%), carbonates (21%), phosphates (14%) and sodium (7%). However, the organic part of the curtain walls has two fractions: an insoluble part consisting essentially of olive pulp, which represents the suspended and colloidal matter, and a fraction soluble in the aqueous phase, which contains sugars, lipids, nitrogen compounds, vitamins, organic acids and phenolic compounds.

45.3.2 How to Value the Proposed Vegetable Water (Margine)

The valorization of vegetable water (margines) has been the subject of several studies; some of these studies are still at the experimental stage. Among the methods used is fertigation, which involves spreading olive pomace (OP) and compost on agricultural land. This simple, effective, and inexpensive technique helps eliminate by-products from olive crushing and provides appropriate fertilization for olive trees, vines, and certain annual crops without posing risks to the environment or the crops. In the same way, the valorization of the vegetable gardens can be used as cattle feed, in the production of proteins of unicellular organisms, in the production of enzymes, in the recovery of natural antioxidants (natural polyphenols) and in the production of compost.

Following the physicochemical characterization carried out for the margines, it turns out that they are rich in organic and mineral matter (nitrogen, phosphorus, sugars, etc.) which can be used as a culture medium for microalgae. However, the high concentration of phenolic compounds that they contain can be detrimental to the growth of algae. Therefore, it is necessary to treat these effluents beforehand to eliminate the phenolic compounds, which are the main contributors to pollution. We therefore propose the adsorption process for the treatment of the margines, which shows great efficiency for the elimination of organic pollutants that are difficult to biodegrade as well as the discoloration of the water. After having cultivated the microalgae in the treated (decolourized) margines, we propose to harvest the produced biomass and to extract biodiesel from it, which is part of the third-generation biofuels.

45.4 Conclusions

The olive tree is planted all over the country, with the exception of the Atlantic coast. The geographical distribution of this heritage highlights three large, distinct olive-growing areas, which are the mountainous, irrigated and the Bour favourable areas. In Morocco, water stress is particularly acute in the southern regions although the olive tree is a relatively resistant tree to water stress and tolerates water deficit to a certain extent. For this reason, several regions of Morocco benefit from a project to develop the olive sector especially among small farmers by benefiting from the establishment of a network of canals for regular and efficient transport to improve the irrigation of olive trees in times of water stress.

National olive production reached almost two million tonnes in 2018–2019; however, the processing of these olives is responsible for huge environmental problems caused by pollution and by these liquid effluents commonly known as "margine". The valuation of marginals is the subject of several studies. So, we propose the system of adsorption, for the treatment of marginals, which shows great efficiency for the elimination of organic pollutants difficult to biodegrade as well as the discoloration of water. After cultivating microalgae in treated (discoloured) marginals, we propose harvesting the biomass produced and extracting biodiesel from the latter which is part of the third-generation biofuels.

Chapter 46
Management Practices Boosting Ecosystem Services

Ahmed Taheri and Nard Bennas

46.1 Introduction

While natural ecosystems occur as a result of spontaneous natural reaction, agriculture is a manmade, production-driven system where men intervene to alter biological and physicochemical parameters through various practices, like irrigation, fertilization, and the control of bioagressors such as weeds, pests, and diseases. The goal of these practices is to ensure agricultural production and improve its quantity and quality while maintaining the natural ecological functions of the system.

In Morocco, olive trees represent the main fruit crop and make up 65% of the total orchards (Berrichi, 2005; www.agriculture.gov.ma). The lands for olive tree cultivation cover almost the entire country with the exception of the Atlantic coastal strip, where weather conditions are less favorable. The olive agroecosystem is managed for the purpose of producing olives with little focus on its by-products.

In Morocco, there has not been an official evaluation of these good and service yet, as the implementation of this foreign concept in the case of agricultural ecosystems is not clearly planned out or is still in its early stages. However, there exist some olive cultivation practices that directly benefit the farmers, such as soil structuring, supply of nitrogen and other nutrients to cultivated plants, water storage and restitution, erosion control, pollination of cultivated species, and management of

A. Taheri (✉)
LB2VE, Department of Biology, Faculty of Sciences, Chouaib Doukkali University, El Jadida, Morocco
e-mail: taheri.ahmed@ucd.ac.ma

N. Bennas
LESCB URL-CNRST N° 18, Department of Biology, Faculty of Sciences, Abdelmalek Essaadi University, Tetouan, Morocco
e-mail: nbennas@uae.ac.ma

J. Muñoz-Rojas, R. García-Ruiz (eds.), *The Olive Landscapes of the Mediterranean*, Landscape Series 36,
https://doi.org/10.1007/978-3-031-57956-1_46

weeds, insects, and pests, as well as benefits for society (natural attenuation of pesticides by soils, regulation of water quality, storage and release of blue water, regulation of the global climate) (Tibi & Therond, 2018).

Olive-growing practices, such as soil and biomass management, have a huge impact on soil and biodiversity. They actively shape the ecosystem services over the years which can lead to either positive or negative outcomes depending on the kind of practices that were performed by farmers. And while the outcomes are generally instantaneous, there can also be long-term side effects. For example, adding organic compounds can be beneficial to ecosystem services that provide nutrients, but can also have a negative impact on the ecosystem services responsible for structuring soil.

In Morocco, modern orchard maintenance techniques are not widely applied as farmers practice more outdated methods, which is the same for phytosanitary protection and fertilization (Evaluation de la filière oléicole, 2018). Mechanical harvesting of olives is rarely used as galling remains the most widely practiced harvesting technique, which causes considerable harm to the environment as it damages trees, decreases productivity, and degrades the quality of the olives and oil produced (Evaluation de la filière oléicole, 2018). And in the case where heavy mechanicals are used, the passage of the machines can lead to soil compaction either on the surface or in depth depending on mechanical pressure, which can last in the medium and long term if not treated immediately by tillage practices that recreate porosity.

46.2 A Strategy Dedicated to Modern and Solidarity Agriculture

The Green Morocco Plan, launched in April 2008, is a strategy that was established with the aim to make the Moroccan agricultural sector a real lever of socio-economic development in the country, by accelerating growth, reducing poverty, and enforcing the integration of agriculture into national and international markets. The strategy is built around a comprehensive approach that includes all stakeholders while taking into account their own objectives. It is based on two major pillars: modern and solidarity agriculture.

For modern agriculture, the objective of the Green Morocco Plan is to develop efficient agriculture, adapted to market rules, through a new wave of private investments, organized around new models of equitable aggregation.

As for solidarity agriculture, the goal is to build an approach aimed towards fighting against poverty, by significantly increasing the agricultural income of vulnerable farmers, especially the ones living in peripheral areas.

The strategy is built around several components, such as developing sustainable Moroccan agriculture, by integrating the "climate change" dimension in the design of projects, conversion of nearly one million hectares of cereals crops to fruit plantations (including olive trees) as they are more likely to protect agricultural lands, endorsing the use of water-saving irrigation systems, and supporting the use of renewable energies that could be used in agriculture practices.

The Green Morocco Plan introduced nine main levers through which it integrated the elements of climate change and sustainable development and, thus, the promotion of an agriculture resilient to the impacts of climate change and low carbon. The levers are as follows:

- Adoption of a major proactive program to save irrigation water.
- Reuse of unconventional water such as through seawater desalination.
- Enforcing good agricultural practices that are resilient to climate change.
- Adoption of a program that aims to convert nearly one million hectares of cereals crops to fruit plantations.
- The Fund for Agricultural Development and the incentive for the adoption of good agricultural practices resilient to climate change and sustainable development.
- The development of an agricultural insurance (multi-risk and multiline).
- Carrying out of two strategic studies: the first is on strategic environmental assessment of the Green Morocco Plan, and the second is a study on the assessment of mitigation by the Green Morocco GHG Plan, GEF (Global Environment Facility), and FAO (Food and Agriculture Organization) support to Pillar II of the Green Morocco Plan in the area of climate change and sustainable development.

46.3 Olive Growing in Bour Zones and Mountains Requires Special Attention

In 2016/2017, Bour zones (favorable Bour zones or the mountains) made up 64% of the total lands where olive growing is practiced in Morocco (Evaluation de la filière oléicole, 2018). This high percentage of Bour zones is due, for the most part, to the increase of the number of projects aimed at expanding this type of crop in the "Green Morocco Program." The massive cultivation of olive trees, especially in semiarid zones, done in accordance with the Green Morocco Plan would undoubtedly have a positive impact on the protection of mountain soils especially in the case of the Rif Mountains, known for their fragile and vulnerable environment and where water erosion causes enormous land losses and excessive sediment production (Tribak et al., 2012; Evaluation de la filière oléicole, 2018). Some orchards already play an important role in erosion control because they occupy slopes that would otherwise be devoted to valueless and soil-destroying cereals (Evaluation de la filière oléicole, 2018). Other traditional olive cultivation practices in the Moroccan mountains strengthen the regulatory ecosystem function (erosion control and soil structuring) of the olive tree, for example, plowing by animals (Fig. 46.1), the installation of small rainwater retention basins near the feet of olive trees (Fig. 46.2), and surrounding sloping trees with large stones to prevent loss of the ground (Fig. 46.3). These last two ancestral practices also favor the incorporation of a biodiversity of auxiliary fauna and soil microorganisms, thus increasing other ecosystem services

Fig. 46.1 Plowing soils in olive groves using animals

Fig. 46.2 Small rainwater retention basins near the feet of olive trees

Fig. 46.3 Stone surrounding sloping trees to prevent loss of the ground

such as the supply of nutrients and the regulation of pests and the pollination of other species of plants cultivated along the olive trees.

In the mountains, olive trees do not benefit from any special care, except for certain pruning and trimming. Tillage in these areas is mainly intended for intercropping (cereals and legumes). As for favorable Bour zones, the most practiced techniques are related to tillage, pruning (annual or biennial), and fertilization intended, in part, for the intercrops represented in these areas by market gardening, legumes, and cereals (Fig. 46.4), (Berrichi, 2005). The nonuse or the limited use of herbicides, insecticides, fertilizers, and boreholes makes olive cultivation in Bour zones beneficial to the environment and also requires minimum use of groundwater (Fig. 46.5).

For irrigated zones, almost all olive plantations benefit from soil work for weed control and the making of basins for irrigation, nearly all the plantations are pruned every year or every 2 years, and half benefit from an application of cover fertilizer and sometimes basic fertilizer (Berrichi, 2005).

46.4 Irrigation of Olive Plantations

Under the current technological and economic conditions, the available hydraulic potential is estimated at around 20 billion m^3, with 16 billion m^3 being from surface water and the rest 4 billion m^3 being from groundwater (agriculture in figures, 2018). In the case of olive plantations, only 36% of the land dedicated to olive

Fig. 46.4 Cultivation of legumes in an olive tree field (Rif, Morocco)

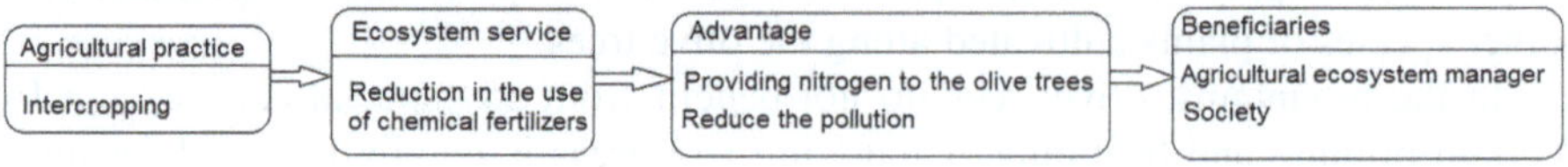

Fig. 46.5 "Chain" of specification of ecosystem services applied to the case of intercropping. (Adapted based on original in Tibi & Therond, 2018)

cultivation is irrigated with an estimated 9% land expansion from 2008 to 2016 (Evaluation de la filière oléicole, 2018). Despite this low percentage, olive growing remains as one of the main causes of high water wastage, due to wasteful water use practices and the lack of use of the most recent water-saving technologies.

Analysis of the data provided by the Ministry (Evaluation de la filière oléicole, 2018) shows that over the period from 2009 to 2016, the olive plantations watered by drip irrigation that profited from the subsidy represent only 23% of the total subsidized planted lands, which are estimated at a total of 34,079 ha, compared with 41% of plantations with low density that are irrigated with other than the localized system and the remaining 36% of plantations located in Bour zones.

The latest assessment of the olive sector in Morocco advised assigning more interest to the matter of irrigation and maintaining the government program for equipping olive orchards with local water-saving systems.

46.5 Agricultural Practices and Biodiversity

It is quite evident that intensive agriculture has a negative effect on the aboveground and underground biodiversity and, therefore, on the ecosystem services provided by the soil (Bender et al., 2016; Emmerson et al., 2016; Sánchez-bayo & Wyckhuys, 2019; El Mujtar et al., 2019). In Morocco, it is assumed that reduced working practices and additions of mineral fertilizers and synthetic phytosanitary products, especially in areas Bour and mountains, could favorably alter the soil structure of the olive grove and promote the diversification of species of bacteria, fungi, and animals in the ground. These facts have been widely studied around the world but not in Morocco (i.e., Emmerson et al., 2016; Sánchez-bayo & Wyckhuys, 2019). It should also be noted that the new pest control practices by using exogenous biosolutions are less frequently used in Morocco.

In addition, in Morocco, the traditional integrated crop-livestock system seems to be a relevant lever for optimizing the functioning of the olive grove ecosystem in terms of biodiversity and the provision of ecosystem services. This type of cropping system makes it possible to achieve yield levels equivalent to those of conventional systems if these conditions are met (Garbach et al., 2016). Several studies have shown that these integrated crop-livestock systems can be developed at the scale of the crop-livestock production system or at the scale of farmer collectives or the territory (Ryschawy et al., 2017; Moraine et al., 2016, 2017). This can allow the development of ecosystem services for biological regulations, nutrient supply, carbon storage, and maintenance of soil structure (Martin et al., 2016; Moraine et al., 2016, 2017).

46.6 Few Data on the Effects of Climate Change on Olive-Growing in Morocco

Morocco is among the countries that would experience the most negative impacts on agronomic performances due to climate change. Droughts are very costly: the drought of 1994/1995 caused a decline in the PIBA by 45% and the national GDP by 8%.

In 2006, the country witnessed a loss of 60% productivity in intensive due to a lack of cold (climate change) in Sefrou (Moroccan Middle Atlas), and the loss exceeded 80% for Arbequina and 90% in the region of Tadla for Arbosana and on the super intensive exceeds 80% (Olea, 2016).

In common with climate change, the olive tree, whose cultivation area is limited by the −10 °C isotherm, would progress a little more in altitude (towards the foot of the mountain) (Ait Houssa et al., 2017).

46.7 The Utilization of Olive Oil By-Products: A Prevalent Challenge!

Vegetable waters and olive pomace that remain following the process of olive crushing are a matter of huge concern in Morocco, as the disposal and the processing of this biomass cause severe environmental damages. These residues are either released directly into watercourses or are amassed and then disposed of in random discharge sites (Évaluation de la filière oléicole, 2018). Not only that, the most recent two-phase trituration method and the use of certain substances in cosmetics as well as the food industry have failed to solve this problem—a failure that is mainly due to the heavy investments required and the additional costs involved in the use of these new technologies (Ouazzani, 2018).

The national legal and regulatory arsenal aimed at environmental protection and sustainable development compiles all sources of pollution, including olive waste. It has been set up by the departments responsible for the environment and water since the 1980s in consultation with the other departments concerned, including the Department of Agriculture.

Given the complexity of this problem and the multiplicity of stakeholders, the reduction of the negative effects of olive waste on the environment requires a long period of time.

The Ministry chose to encourage manufacturers to emphasize oil by-products through the conclusion in 2015 of a specific agreement with the environment department for the implementation of a program for the recovery of waste from the olive sector. This agreement plans to finance projects that involve agriculture spreading of vegetable water on agricultural land and the valorization of wet olive pomace (Evaluation de la filière oléicole, 2018).

46.8 A Special Cultural Service of the Olive Tree

Throughout the Mediterranean basin, olive oil has been considered, since antiquity, as an essential food ingredient; a real medicine capable of treating the stomach, liver, intestines, and respiratory tract; and a key ingredient in cosmetic treatments that target the skin and hair.

In Islam, olive holds a sacred place as the Quran devotes verses to this fruit, making it a product of "blessing."

46.9 Conclusion

One of the major challenges in the development of cropping systems heavily based on ecosystem services is to design a set of coherent practices, that means a cropping system, which optimizes the functioning of the ecosystem in terms of "goods and services" with maximum benefit for the farmer and society.

To this day, the olive sector in Morocco is heavily reliant on the practices adopted by farmers and the conditions of olives during the stages of harvesting, transporting, and storing. The production is also heavily affected by other factors as it experiences significant inter-annual fluctuations due to crop rotation, the irregularity of rainfall, and the lack of maintenance of the orchards located in Bour zones which affect productivity (Evaluation de la filière oléicole, 2018).

References

Ait Houssa, A., Drissi, A., Asehraou, S., Asfers, A., Oubaki, A., & Chraibi, H. (2017). Changements climatiques au Maroc: Quels systèmes de culture pour s'y adapter. *Revue Marocaine des Sciences Agronomiques et Vétérinaires, 5*(3), 210–221.

Bender, S. F., Wagg, C., & van der Heijden, M. G. (2016). An underground revolution: Biodiversity and soil ecological engineering for agricultural sustainability. *Trends in Ecology & Evolution, 31*(6), 440–452.

Berrichi, M. (2005). *Filière oléicole: situation et perspectives d'avenir. Séminaire international sur l'olivier. Acquis de recherche et contrainrtes du secteur oléicole* (p. 446). Ministère de l'agriculture du développement durable et des Pêches Maritimes.

Cour des Comptes. Royaume du Maroc. Evaluation de la filière oléicole. (2018). *Rapport annuel de la Cour des comptes au titre de l'année 2018, Maroc* (pp. 257–286). https://www.courdescomptes.ma/wp-content/uploads/2023/01/Evaluation-de-la-filiere-oleicole.pdf

El Mujtar, V., Muñoz, N., Mc Cormick, B. P., Pulleman, M., & Tittonell, P. (2019). Role and management of soil biodiversity for food security and nutrition; where do we stand? *Global Food Security, 20*, 132–144.

Emmerson, M., Aavik, T., Batáry, P., et al. (2016). How agricultural intensification affects biodiversity and ecosystem services. *Advances in Ecological Research, 55*, 43–97.

Garbach, K., Milder, J. C., DeClerck, F. A. J., Montenegro de Wit, M., Driscoll, L., & Gemmill-Herren, B. (2016). Examining multi-functionality for crop yield and ecosystem services in five systems of agroecological intensification. *International Journal of Agricultural Sustainability, 5903*, 1–22.

Martin, G., Moraine, M., Ryschawy, J., Magne, M. A., Asai, M., Sarthou, J. P., Duru, M., & Therond, O. (2016). Crop-livestock integration beyond the farm level: A review. *Agronomy for Sustainable Development, 36*, 53.

Moraine, M., Grimaldi, J., Murgue, C., Duru, M., & Therond, O. (2016). Co-design and assessment of cropping systems for developing crop-livestock integration at the territory level. *Agricultural Systems, 147*, 87–97.

Moraine, M., Melac, P., Ryschawy, J., Duru, M., & Therond, O. (2017). Participatory design and integrated assessment of collective crop-livestock organic systems. *Ecological Indicators, 72*, 340–351.

Olea. (2016). *CR de la campagne sur la production de l'Olivier en Super Intensif.* Unpublished Internal Report.

Ouazzani, N. (2018). Available via https://www.agrimaroc.net/wp-content/uploads/2018/06/valorisation_sous_produit_olivier.pdf. Accessed 15 Aug 2020.

Ryschawy, J., Martin, G., Moraine, M., Duru, M., & Therond, O. (2017). Designing crop–livestock integration at different levels: Toward new agroecological models? *Nutrient Cycling in Agroecosystems*. https://doi.org/10.1007/s10705-016-9815-9

Sánchez-bayo, F., & Wyckhuys, K. A. G. (2019). Worldwide decline of the entomofauna: A review of its drivers. *Biological Conservation, 232*, 8–27.

Tibi, A., & Therond, O. (2018). *Services écosystémiques fournis par les espaces agricoles: Caractériser et évaluer* (p. 186). Editions Quae.

Tribak, A., El Garouani, A., & Abahrour, M. (2012). L'érosion hydrique dans les séries marneuses tertiaires du prérif oriental: agents, processus et évaluation quantitative. *Revue Marocaine des Sciences Agronomiques et Vétérinaires, 1*, 47–52.

Chapter 47
Market Challenges

Laila Ajouaou and Ahmed Taheri

Abstract The Mediterranean region is home to some of the world's most economically significant olive producing areas, with Morocco standing out as one of the largest producers. Despite the Moroccan government's efforts, including initiatives like the "Green Morocco Plan", the country's olive oil Market continues to face several challenges. This chapter aims to assess the Moroccan olive oil Market by highlighting key obstacles and global trends in olive oil production and consumption. Moreover, to ensure sustained growth, the chapter proposes some policy recommendations for Morocco's olive oil sector to overcoming market fragmentation, increasing competitiveness and addressing global and domestic demand shifts.

47.1 Introduction

In the Mediterranean area, olive trees are one of the most economically significant oil-producing crops (Ben Ayed et al., 2016). In Morocco, the olive tree is the most cultivated fruit representing 65% of the total tree-crop area of the country. Morocco became the fifth-largest producer of olive oil in the 2020/2021 crop year, its production level reaching over 140,000 tons of olive oil.

Despite that Morocco continues to grow with support from the government and related institutions' efforts and support, various challenges still face the Moroccan olive market.

L. Ajouaou (✉)
Department of Private Law, FSJES-T, Abdelmalek Essaâdi University, Tangier, Morocco
e-mail: laila.ajouaou@etu.uae.ac.ma

A. Taheri
LB2VE, Department of Biology, Faculty of Sciences, Chouaïb Doukkali University, El Jadida, Morocco
e-mail: taheri.ahmed@ucd.ac.ma

J. Muñoz-Rojas, R. García-Ruiz (eds.), *The Olive Landscapes of the Mediterranean*, Landscape Series 36,
https://doi.org/10.1007/978-3-031-57956-1_47

This chapter aims to assess the Moroccan olive oil market in order to appoint the challenges and to suggest a range of policies that the Moroccan olive oil sector should carry out to face the challenges. The chapter begins with an overview of macro-trends in worldwide olive oil supply and demand.

47.2 Macro-Trends in Worldwide Olive Oil Supply and Demand

47.2.1 Worldwide Supply Side

From the supply side, there is a tendency toward progressive growth in production, especially in large producer countries, such as Spain, Greece, and Italy (Anania & Pupo D'Andrea, 2011). In fact, the world's olive oil production has tripled in the last 60 years, reaching 3,262,000 tons in the campaign of 2018/2019, according to the International Olive Council (2019). This growth is mainly due to new orchards and reconversion of existing plantations into more productive, as well as restructuring of the first (milling) and second (refining and packing) industrial transformation stages (Mili & Rodríguez Zúñiga, 2003).

Moreover, a notable rise in productivity has been achieved, mainly through advanced cultural practices, the development of drip irrigation, and the appreciable removal of marginal plantations. Additionally, advances have been made to convert to extraction continuous systems instead of traditional extraction systems (Mili & Rodríguez Zúñiga, 2003).

Technological advancements have led to a gradual adaptation of supply to the growing demands for quality, resulting inter alia in a progressive increase in the share of virgin oil suitable for consumption.

Although, this restructuring dynamic was about the adoption of technical innovations in product and process, it has scarcely contributed to convert the business management and organization practices.

47.2.2 Worldwide Demand Side

From the demand side and according to the International Olive Council (2019), in the major consumer countries, which are also the principal producers, global olive oil consumption remained practically stable in the last decade.

In the long term (ILO, 2014), the olive oil market is expected to show a stable growth of 3–5% over the next 5 years. This growth is mostly due to changes in the consumption patterns of consumers, like the rising interest in healthier cooking using olive oil and the Mediterranean diet in various countries worldwide, especially new users. The best opportunities for suppliers can be reached in France, the UK, Germany, the Netherlands, Switzerland, and Belgium (Mili, 2006).

However, the consumption rate observed in aggregate data for countries or groups of countries, apart from not reflecting seasonal fluctuations, hides some relevant facts in a trend analysis of consumption (IOC, 2019); a notable dynamism with respect to internal composition of distinct product categories can be noticed. For example, there is evidence that in these markets, the demand is being moved toward higher-quality segments, especially the extra virgin category, which represents almost 82% of total olive oil consumption in Greece, 70% in Italy, 37% in Portugal, and 20% in Spain. Secondly, aggregate figures do not reflect adequately the evolution in countries where consumption is not significant in terms of quantity and quality. In fact, consumption in nontraditional markets has increased almost exponentially. Trends in purchases in these markets over the last decade point to an appreciable, sustained growth in imports, although the situation varies across countries (IOC, 2019).

To sum up, the analysis of worldwide supply and demand tendency leads to expect an increase in both magnitudes in the near future, though the increase in supply will probably exceed the potential demand (Bernini Carri & Sassi, 2007). This could generate surpluses, which could generate a reduction in prices received by producers and decrease marketing margins in the absence of strategies to increase demand.

However, there are various reasons to predict future growth in worldwide consumption, which could successfully offset the predicted increase in supply, as long as the industry works on policies to boost the demand. This implies that quality policy should emerge as central axis in any strategy of demand expansion, but without diminishing the efforts to pressure prices downward through a reduction in production and organizational costs. We refer not only to the intrinsic quality of the product and production processes, in which significant progress has been made as discussed earlier, but also, and especially, to the commercial logistics quality (Mili & Rodríguez Zúñiga, 2003).

47.3 The Moroccan Olive Oil Market and Its Challenges

Morocco is one of the largest producers of olive oil in the world. According to the International Olive Council (2019), the country ranks as the sixth-largest oil producer in the international market. Thanks to the Green Morocco Plan's strategy, which the Moroccan government launched in order to deal with several constraints facing the agricultural sector in general, Morocco has doubled its production of olive oil in the past 6 years (Directorate of Financial Studies and Forecasts. Kingdon of Morocco (2014).

Despite the efforts made by different actors in the olive oil sector and the improved performance achieved, Morocco continues to face various market challenges, which run in parallel to those faced by other Mediterranean producing countries (Turkekul et al., 2010).

47.3.1 *Rarity of Water Resources and Unpredictability of Rainfall Levels*

Scarcity of water resources remains the biggest challenge facing the Moroccan olive oil sector, especially in the presence of the changing climatic conditions, which has caused increased unpredictability of rainfall levels in recent years. Climatic changes make it difficult to estimate seasonal production at a fair degree of accuracy. The problem is serious, especially due to the lack of adequate rainwater harvesting and storage techniques and methods to use when rainfall levels are high.

47.3.2 *High Input Costs*

Field research findings show that the majority of olive farmers and producers in Morocco complain about the high input costs and how this tightens considerably their profit margins. High costs can be particularly seen in the cost of labor, freight charges, and organic fertilizers and pesticides. However, despite the notion that growers usually complain from high input costs, this does not actually appear to be a key challenge.

However, Covid-19 containment measures bring overall operating costs up since they enforce how many workers should be present at one time and control their movement across regions. Moreover, workers would normally commute to the groves in crowded transport trucks, but as a result of the virus, transportation costs would likely double or triple. Also, increasing costs for producers are due to the low level of mechanical innovation in the fields and mills.

47.3.3 *Price Issues*

The prices in the olive oil market fluctuate from 1 year to another, depending on production rates, which are related to climate variations; huge production tends to be followed by a lower production in the following year depending on precipitation rates. This fluctuation in production combined with market heterogeneity in terms of production systems (traditional and modern) leads to sharp price fluctuations in the olive oil market (Rahmani, 2017).

In addition to price fluctuations due to the dependence on rainfalls, there is no clear or unified pricing strategy that is adopted by Moroccan farmers and producers, which are considered to be a weak link within the olive value chain due to their divergence and lack of proper representation. The way the farmer's olive prices are set depends just on the bargaining power of wholesale traders.

47.3.4 Low Domestic Consumption

According to the Ministry of Agriculture and Fisheries, the average consumption of olive oil for a Moroccan habitant is estimated to be around 3,5 L per year.

Compared to Greece, Spain, and other producers, Moroccan olive oil domestic consumption rate remains very low. This weakness could be related to income issues. In fact, income is a better determinant of olive oil demand than the price of olive oil, which shows that consumers are more sensitive to income changes than price variation (Elbeydi, 2016). That is to say that a weakness of economic situation could have a negative impact on olive oil demand; in reality, consumers might substitute olive oil by cheaper products, such as sunflower oil and other substitute oils and fats.

In addition to income issues, the new food habits adopted by the new generations, which are characterized by the abundance of fast food, make olive oil consumption rate lower and substitute oils and fats higher.

47.3.5 Consumer Trends and Preferences

In the national market, table olive and olive oil are the main ingredients of the Moroccan cuisine. Therefore, the product is not considered price elastic, and consumers will continue to buy it even with increasing inflation and declining purchasing power (Ilak Peršurić, 2020). However, local buyers are mostly used to "traditional" table olive and olive oil, which is a single product, without any specification of quality. In fact, the average consumer has no real knowledge of the product.

Another major issue is the lack of trust by consumers in domestic industrial producers. This has resulted in many domestic consumers purchasing olive oil directly from known or trusted mills and producers instead of industrial producers.

47.3.6 Lack of Promotion

In many Moroccan regions, the olive oil companies and some public institutions do not spend enough financial resources for brand and generic promotion in order to raise awareness and favorable perception among consumers in the expectation of further demand expansion. For example, in Italy, olive oil producers communicate to consumers the aspects that they prefer and link them to local history and culture to boost the domestic demand and conquer new market segments.

In addition to raising awareness about its health benefits, there is also a lack of promotion for olive derivative product.

47.3.7 Low Export Level

The share held by Morocco in the world market of olive oil oscillated between 0.5% and 4% between 2002 and 2012. In fact, the Moroccan exports of olive oil are characterized by an important fluctuation. In addition, the low technological level and the depreciation of the raw material quality make the Moroccan olive oil insufficiently adapted to international standards. Olive oil and olive-pomace oil are only exported from Morocco once they meet the requirements of trade standards and the required phytosanitary and sanitary traceability measures in force in the destination markets. The standards of the International Olive Council, Codex, and the USA are the main references that Morocco Foodex uses to assess the conformity of olive oils with the requirements of the export markets.

47.4 Policies to Address the Challenges Faced by the Moroccan Olive Oil Market

47.4.1 Improve the Quality of the Product

The quality of the Moroccan olive oil should be enhanced to meet the international standards with lower costs. By improving the quality to meet the international standards, Morocco could improve its exportation rate.

To ameliorate the quality of Moroccan olive oil, technical and technological collaborations in the olive growing sector between the different Mediterranean partners should be strengthened. This will lead to better competitiveness on the international market and a better modern, efficient, and environmentally friendly olive oil that could appeal to new consumers and raise the domestic consumption.

The growth in the quantity and variety of oils consumed, in terms of their qualities, will doubtless depend also on the dynamics of demography and per capita incomes of individual countries, as well as on the effectiveness of marketing campaigns promoting consumption.

47.4.2 Conquer New Market Shares

Companies and public institutions involved in the sector should try to increase the demand in nontraditional consumer countries both inside and outside Morocco, taking into consideration that the price elasticity is usually higher in new markets. They also have to enhance communication about the meaning of the different types of olive oil.

Additionally, olive oil producers should take into consideration that olive oil consumption can be increased by analyzing consumer's preferences, due to its external attributes and hedonic and health dimensions, which prove to be highly important in its rejection or acceptance.

Also, Moroccan olive oil companies should offer a range of other food products, taking profit from scope economies and synergies in logistics, trade, and promotion of different product lines, such as margarine, vinegar, oilseed oils, or table olives.

47.4.3 Make More Promotional Efforts

In many Moroccan regions, the olive oil companies and some public institutions do not spend enough financial resources to promote the Moroccan olive oil. To improve this situation, adequate marketing measures should be taken to meet the domestic and the international demand.

In addition to marketing measures, considerable financial resources should be allocated to spread a good image of Moroccan olive oil—an image based on good quality product, an essential component of the Mediterranean diet, and an environmentally friendly product.

The promotion should also target the new generations, who are the future consumers, by raising their awareness about the nutritional qualities of olive oil in their diet to be in a good shape. Consumer attitudes and behaviors vary significantly between countries and among consumer segments within each country. Olive oil producers should try to adapt the product (type of oil, color, taste, packaging, labeling) to different segments, promotion, and other marketing variables to the specific circumstances in each target market.

47.4.4 Change Price Policies

The prices in the Moroccan olive oil market remain fluctuated and depend on the production rates and climatic variations. The Ministry of Agriculture and Fisheries and all the actors should set up rainwater harvesting and storage techniques and methods to use when rainfall levels are high, to reduce the dependence on rainfalls.

Also, a clear and unified pricing strategy adopted by Moroccan farmers and producers will certainly contribute to gaining consumer's trust in domestic industrial producers and being more competitive in international markets.

Moreover, the production cost of Moroccan olive oil should be reduced. To achieve this, efforts on technical and technological studies should be conducted. This is considered as a main role, which has to be played by the research and development departments in companies and universities.

Fig. 47.1 Full-time job in Al Houda agricultural cooperative located in Ouezzane (Morocco)

47.4.5 Enhance the Role of Farmer Cooperative in Traditional Systems Areas

Enhancing the role of farmer cooperative in traditional systems areas to reach economies of scale through collective procurement and enforce coordination between member farms and the olive mills. This work organization can also introduce a number of value-added services to member farms, such as rendering outsourced postharvest handling services and hosting micro- and small businesses that produce olive by-products, in addition to introducing mobile mills, through the introduction of a business idea to start a company that renders mobile milling to Moroccan olive orchards. Such business can be operated by women and small-scale producers (Fig. 47.1).

47.5 Conclusion

To sum up, the analysis has pointed out interesting aspects. First of all, the most important variables defining the medium-term perspectives of Moroccan olive oil market are the developments on the demand side. In fact, the growth in demand will continue in the years ahead, especially in countries that are not traditional consumers but also in those with a tradition of consumption. The growing demand will be accompanied by a rise in the demand for olive oils differentiated based on quality characteristics.

In parallel, various challenges are still facing the Moroccan market such as fragmentation of the sector, labor issues, and marketing constraints related to branding and packaging, as well as limited financial and management planning know-how. There are also threats that are continuously impacting negatively on the sector, such as changing climatic conditions that are causing unpredictable rainfall; competition in gaining export markets, as well as domestic market challenges related to consumer trust; and declining purchasing power due to the present economic crisis.

Some actions need to be taken in order to improve the global situation of the Moroccan market and to enhance the sector's chances of achieving and sustaining growth.

References

Anania, G., & Pupo D'Andrea, M. R. (2011, January). Olive oil in the Mediterranean area: Production, consumption and trade. *The CIHEAM Watch Letter*.

Ben Ayed, R., Ben Hassen, H., Ennouri, K., Ben Marzoug, R., & Rebai, A. (2016, January 30). OGDD (Olive Genetic Diversity Database): A microsatellite markers' genotypes database of worldwide olive trees for cultivar identification and virgin olive oil traceability. *Database: The Journal of Biological Databases and Curation, 2016*, bav090.

Bernini Carri, C., & Sassi, M. (2007). *Trade and competitiveness of the Mediterranean countries on the olive oil market, 2007.* I Mediterranean conference of agro-food social scientists. 103rd EAAE seminar adding value to the agro-food supply chain in the future euromediterranean space'. Barcelona, Spain, April 23–25.

Directorate of Financial Studies and Forecasts. Kingdon of Morocco. (2014, July). *Performmances and competitiveness of the Moroccan agrifood sector exports, directorate of financial studies and forcasts*, 40pp. https://www.finances.gov.ma/Publication/depf/2014/EN-Note%20 Performance.pdf

Elbeydi, K. R. (2016). Estimating price and income elasticity of olive oil demand in Libya during 1980-2010. *Afaqeqtisadiah Journal, 2*(4), 5–16.

Ilak Peršurić, A. S. (2020). Segmenting olive oil consumers based on consumption and preferences toward extrinsic, intrinsic and sensorial attributes of olive oil. *Sustainability, 12*, 6379.

International Labour Organization. (2014). *Market study and marketing strategy of the olive sector in irbid.* https://www.ilo.org/wcmsp5/groups/public/%2D%2D-arabstates/%2D%2Drobeirut/documents/genericdocument/wcms_319825.pdf

International Olive Council (COI). (2019). The olive grove in Morocco. *OLIVAE: Official Journal of the International Olive Council, 125.* 89 pp.

Mili, S. (2006). *Market dynamics and policy reforms in the EU olive oil industry: An exploratory assessment.* 98th seminar, June 29–July 2, 2006, Chania, Crete, Greece 10099. European Association of Agricultural Economists.

Mili, S., & Rodríguez Zúñiga, M. (2003). The olive oil sector facing new international market challenges. *Journal of International Food & Agribusiness Marketing, 14*(3), 35–55.

Rahmani, M. (2017). *Commercialisation et Marketing de l'huile d'olive au Maroc, Mars 2017.* https://unctad.org/system/files/officialdocument/Commercialisation%20et%20marketing%20 huile%20olive%20Maroc_Rahmani.pdf

Turkekul, B., Günden, C., Abay, C., & Mirant, B. (2010, January). Competitiveness of Mediterranean Countries in the olive oil market. *New Medit.* https://newmedit.iamb.it/share/img_new_medit_articoli/282_41-turkekul.pdf

References

Part VIII
Olive Landscapes Across the Mediterranean: Regional Heterogeneity and Case Studies: Local Innovative Approaches to the Management of Olive Groves and Landscapes Across the Mediterranean

Chapter 48
The Denomination of Origin of Estepa

Moisés Caballero Páez

The Protected Designation of Estepa (https://www.doestepa.es) was born officially in 2004. Estepa is a municipality in Southern Spain (Fig. 48.1) of about 190 km^2 and with 41,500 inhabitants, where the cultivation of olive, both for olive oil and table olives, supports the socio-economic life and culture of its inhabitants, being the olive groves the predominant landscape (Fig. 48.2). Olive oil cultivation has shaped this municipality, and others nearby, for about 2000 years, as witnesses by numerous archaeological remains and sites related to the olive oil sector found in the area.

The DOP Estepa covers another 12 municipalities nearby which share a common vision and objectives and the same way of cultivating the olive groves and producing olive oil and table olives. The embryo of the DOP Estepa has been in the making for the last 40 years with the incorporation of Spain into the European Union (EU). The arrival of EU funds boosted the modernization of the productive structures (Fig. 48.3) of these municipalities. The DOP Estepa arose from the need, and the common vision, to produce a high-quality extra virgin olive oil with high standards of environment-friendly management, to defend and valorized the own personality of the olive oil of the area and to enter into the olive oil value chain of a territory with long history of olive oil cultivation.

The objectives of the DOP Estepa are (i) to guarantee the origin and the quality of the olive oil and table olive which are commercialized under the umbrella of the DOP, (ii) to promote and disseminate both the olive oil and the table olive and the culture linked to the olive cultivation in the area and (iii) to guarantee high standards of environmentally friendly management practices.

M. C. Páez (✉)
Olesoetepa—Consejo Regulador de la Denominación de Origen Protegida de Estepa, Sevilla, Spain
e-mail: secretaria@doestepa.es

J. Muñoz-Rojas, R. García-Ruiz (eds.), *The Olive Landscapes of the Mediterranean*, Landscape Series 36,
https://doi.org/10.1007/978-3-031-57956-1_48

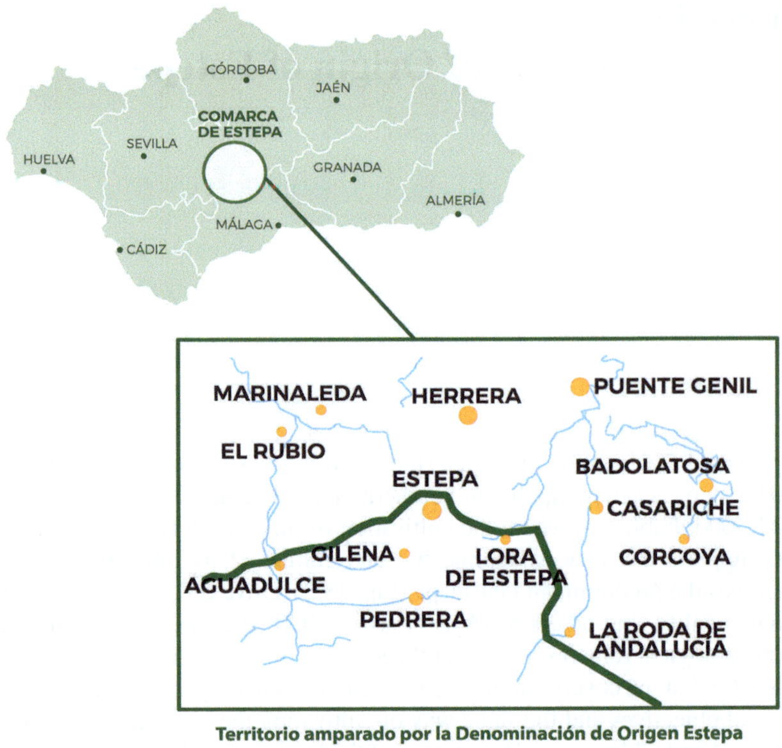

Fig. 48.1 Location of Estepa and other municipalities nearby which are under the umbrella of the DOP Estepa

Fig. 48.2 An aerial view of an olive farm in Estepa

Fig. 48.3 An example of the mechanization level during harvesting

Since the creation of the DOP Estepa, a successful stage has begun in the area which can be specified as follows: (i) an improvement in the olive groves managements, with a more rational and sustainable methods of cultivation. This has been possible thanks to many agricultural technicians specialized in olive cultivation; (ii) the production has steadily increased; (iii) an increase in the olive farmers' income; (iv) an increase in the rustic value of the olive groves; (v) an increase in the quality and the added value of the virgin extra olive oil (Fig. 48.4); (vi) an increase in the sales of the packaged product, with an important expansion of the distribution channels and formats; and (vii) a boost of export with a great increase in the international market.

About 4300 olive farmers cultivate about 40,000 hectares with olives trees, and the olive fruits are processed in 18 olive mills. All of these are under the umbrella of the DOP Estepa, and olive farms include traditional rainfed, traditional irrigated, semi-intensive, intensive and superintensive, all under integrated (90%) or organic (10%) production systems. The five main olive varieties (*hojiblanca*, *picual*, *arbequina*, *manzanilla* and *lechín*) are grown, although the most abundant by far is the *hojiblanca*.

Among the environmentally friendly management encouraged in the DOP Estepa include (i) a rational use of fertilisers and plant protection products throughout the assessment of the agricultural technical staff, (ii) the implementation of temporarily spontaneous cover crops (Fig. 48.5) which are mainly controlled physically and (iii) the shredded tree pruning residues which are applied on top of the soil.

The 18 olive mills of the DOP Estepa produce about 30000 tons of olive oil annually, 90% of which is extra virgin olive oil and is one of the DOP with the highest amount of production of high-quality extra virgin olive oil with homogeneous

Fig. 48.4 An example of packeted extra virgin olive oil of the DOP Estepa

Fig. 48.5 Example of an olive grove with spontaneous cover crop of the DOP Estepa

features. Extra virgin olive oil of the DOP Estepa reaches 25 countries including Korea, Taiwan and Singapore. Not only the monovarietal *hojiblanca* is under the DOP Estepa but also the coupage of *hojiblanca* and *arbequina* and the coupage multivarietal of *hojiblanca*, *arbequina*, *manzanilla*, *picual* and *lechín*.

The commitment of the DOP Estepa for the innovation, dissemination and promotion in the protection of the environment while producing a high-quality olive oil in an optimal way is specified in: (i) the organization of between 2 and 3 capacitation events annually in the topics related to spontaneous cover crops, composted olive mill pomace and strategies to minimize soil erosion and soil rehabilitation, (ii) the hosting of the highly endowed research award Eduardo Pérez (http://www.premiodeinvestigacioneduardoperez.com) and (iii) the participation of local, regional, national and international projects related to sustainability, as it is SUSTAINOLIVE (PRIMA H2020).

Finally, the DOP Estepa has initiated a strategy aimed to increase the level of digitalization of the main management of its olive groves and the development of the oleoturism, as part of the culture of the territory.

Chapter 49
The Agricultural Cooperative of Kalamata

Michael Antonopoulos

The Agricultural Cooperative of Kalamata (ACK) was founded in 1963 and is comprised of over 300 olive oil growers in the municipality of Kalamata which is situated in the region of Messenia (Peloponnese, Greece). Since then, the members of the Agricultural Cooperative of Kalamata have implemented an environmental management system while ensuring the production of premium quality extra virgin olive oil. Today ACK members cultivate an area of 300 ha and produce more than 260 tons of olive oil. Most of this excellent quality olive oil and specifically 90% of the production conform to P.D.O. KALAMATA standards, and about 10% is certified organic olive oil.

The ACK (http://www.kalamata-olive.gr/ and www.kalamata-olivefestival.gr/) is a typical Greek cooperative which conforms to the mandatory national legislation for cooperatives. It is managed by a five-member board and chaired by Mr. Michael Antonopoulos (MSc. Environmental Geologist from Drexel University, USA). The ACK has limited personnel, specifically three permanent employees and seasonal staff.

The priority and the main goal of the Agricultural Cooperative of Kalamata is to provide high quality, excellent standardization, and innovative products such as olive oil rich in oleic acid and polyphenols that have a singular health characterization. Qualified staff provide technical support to the members of the cooperative with on-site visits to the olive groves, soil and leaf diagnoses, and professional directives. Also, the ACK provides its members with high-quality agricultural supplies with an emphasis on organic farming. All producers among others use the certified vegetable oil of the cooperative from olives for the lubrication of their machines in the harvest but also inside the machines of the olive mill. The result is

M. Antonopoulos (✉)
Kalamata Olive Oil Cooperative, Kalamata, Greece
e-mail: asin_kal@otenet.gr

J. Muñoz-Rojas, R. García-Ruiz (eds.), *The Olive Landscapes of the Mediterranean*, Landscape Series 36,
https://doi.org/10.1007/978-3-031-57956-1_49

that we have olive oil without poly-aromatic hydrocarbons (PAHs). In addition, the Agricultural Cooperative of Kalamata is active in the diversification in the cultivation and production of horticultural crops, peanut, and peanut butter as well as table olive P.D.O. Kalamata.

In the region of Kalamata, agricultural land is very small and is subdivided into many plots whilst the average density of its trees is about 350 olive trees per hectare, and for most olive producers, farming is not their main occupation. Furthermore, most olive farms of ACK members are traditional. The olive harvest begins at the end of October and ends at the end of January, depending on the weather conditions and the ripening of the fruit in semi-mechanical ways. The fertilization of the olive tree is based on the soil analyses and the fights against the diseases according to the instructions of the state agronomists. We sell olive oil at least to traders in Messinia. We standardize a large part of the production in bottles of 250, 500, and 750 mL (Fig. 49.1) and in tin cans of 3 and 5 L, which we channel in the market of Central and Northern Greece but also in European countries abroad such as Germany, Malta, Finland, etc.

The ACK has a technologically advanced and efficient olive oil mill (Fig. 49.2), which implements a two-phase cold extraction process. Currently, the production of

Fig. 49.1 Bottle (750 mL) with extra virgin olive oil "P.D.O. Kalamata" by the Agricultural Cooperative of Kalamata

Fig. 49.2 Olive mill of the Agricultural Cooperative of Kalamata

olive oil of the ACK member includes the following basic steps: After being harvested, from late October to early December, the olives are delivered to the processing units for processing as soon as possible, typically in the same day. Olive fruits are transported in large plastic crates with ventilation holes. Transportation with plastic crates is preferred, because the olives can "breathe." The transport of the olives and the cold-press process in the olive mill take place within 12 hours and under the best possible conditions for the preservation of the fruit (storage in the shade and on pallets for the circulation of the air and the avoidance of direct contact with the ground). The olive grove fruit is processed immediately. The olives are first unloaded on the conveyer belt and then transported to the defoliator, where the leaves and other foreign materials are removed. This is followed by a rinsing process to remove foreign matter (dust, dirt, etc.). After being rinsed the fruit is milled in a crusher. Milling is the first part unit operation performed in order to prepare the olive paste for olive oil extraction. After grinding the oil paste is mixed into a homogeneous paste by the malaxer machine (stainless steel double-walled tanks heated by hot water circulation whose temperature should not exceed 25 °C). Malaxation is a key part of the treatment process, usually lasting 30 min, and helps to combine small oil droplets into larger drops of oil. Afterwards, up to 100% of the amount of olive paste is added to the softener (depending on the ripeness of the olive fruit), before the extraction of the olive oil into two-phase centrifugal separators. The temperature of water should not exceed 30 °C. The resulting olive oil is stored in stainless steel tanks at controlled temperatures, away from sunlight.

An onsite laboratory takes a representative sample of each olive batch, in order to check attributes, such as the acidity of the olive oil and the percentage of peroxides. A device is used to measure the oil percentage in olive oil trees before the extraction procedure. Also, the ACK olive oil mill is equipped with a technologically advanced bottling system. The ACK members (farmers) have the appropriate equipment, the knowledge, and the needed experience for olive cultivation. They use equipment such as tractors, brush cutters, off-road vehicles, etc.

The ACK has been certified for (a) food safety management systems ISO 22000:2018, (b) protected designation of origin (Agrocert), and (c) Total System Management Agricultural Production according to AGRO 2.1 and 2.2 (under evaluation).

Recently, the ACK has been funded by 2 national and European projects with more than 900.000 euro. Specifically, the olive oil mill acquired new machinery 8 years ago and recently implemented the requirements of the 611/2014 and 615/2014 directives of the EU.

The ACK participates as a partner in the research project SUSTAINOLIVE, which is funded by the Euro-Mediterranean PRIMA Foundation.

Every year for the past 10 years, the Cooperative organizes the Panhellenic Olive Oil and Table Olive Festival which is supported by The Chamber of Commerce, The Geotechnical Chamber of Commerce, The Greek Ministry of Agriculture, The Agricultural University of Athens, the local government as well, and many more. The festival takes place in the beginning of April and lasts for a few days. It is carried out at the Elite City Resort, which is situated by the sea. Along with the festival, the Kalamata Olive Oil Awards Competition is carried out.

Many visitors attend the festival of the Agricultural Cooperative of Kalamata, since the information and the activities it provides are very helpful for many who are involved with the cultivation of olives and the production of olive oil. Interested parties can also participate either as exhibitors or sponsors [(+30)27210–24,590, fax (+30)27210–24,590, or via email asin_kal@otenet.gr or via website www.kalamata-olive.gr & www.kalamata-olivefestival.gr].

Chapter 50
NILEAS Producers Group Company A.C.

Georgios Kokkinos and Vasiliki Inglezou

Table olives and olive oil have always been the principal products of the Messinian land. Significant archaeological remains from the Palace of Nestor in Pylos, such as the tablets of Linear B script, which is located 3 km west from Chora (seat of NILEAS) is a proof of this activity in Messinia during the ancient period.

Nowadays, olive cultivation in Messinia is almost monoculture, with 14 million olive trees and an olive oil production that is reaching the amount of 55,000 tonnes. The varieties that are cultivated are mainly (a) 'Koroneiki', (b) 'Tsounati', (c) 'Kalamon' for table use and (d) 'Mavrolia', which is the second most cultivated variety in the area where NILEAS operates.

Chora, where NILEAS Producers Group Company A.C. is settled, is a rural community of about 3000 inhabitants which is exclusively engaged in the production of olive oil. The community cultivates about 8100 hectares and 360,000 olive trees, 60% of which is devoted to the 'Koroneiki' variety, and about the other 40% is the 'Mavrolia' variety, which is considered the local variety. The high number of trees over 300 years of age and the high size (more than 500 cm of diameters of the main trunk) of many olive trees testify that this variety has its roots in ancient times.

NILEAS Producers Group Company A.C. (https://www.nileasoliveoil.gr) was established in 2001 by 55 founding members, who are olive oil producers in the wider region of west Messinia (South West Peloponnese Region, Greece). The olive farms of NILEAS Producers Group Company A.C. are distributed in a relatively abrupt and sloppy complex landscape of compound succession of Mediterranean forest areas and olive tree groves. NILEAS' goal was to implement an Integrated Environmental Management System in the cultivation of olive trees and was at the time the first organized group in Greece established for this purpose. In the summer of 2003, the

G. Kokkinos (✉) · V. Inglezou
Nileas Producers' Group, Chora, Greece
e-mail: gkokkinos3@gmail.com

J. Muñoz-Rojas, R. García-Ruiz (eds.), *The Olive Landscapes of the Mediterranean*, Landscape Series 36,
https://doi.org/10.1007/978-3-031-57956-1_50

abovementioned system was certified to ISO 14001 and was in fact the first certification awarded to a European olive grove. Nowadays NILEAS consists of 25 cooperative members and 11 collaborating farmers, cultivating a total of 1630 km^2 or 163.5 hectares of olive groves. The olive farms are relatively small, between 2 and 4 hectares. NILEAS' annual olive oil production is approximately 280 tonnes.

Tree spacing should allow the trees, when fully developed, to intercept the largest possible amount of radiant energy without shading adjacent trees. This is the reason why planting density in NILEAS' member's olive farms is calculated to 170–180 olive trees per hectare. About 50% of the cultivators are using traditional methods of cultivation, while the other 50% is using semi-intensive methods.

Greek cooperatives are characterized by small participation of producers; only 10% of the Greek olive oil producers participate in organized structures. Most olive oil cooperatives have little or no involvement in the production organization. They are mainly engaged in the production of olive oil in a privately owned olive mill.

NILEAS is a typical Greek cooperative that complies with the mandatory Greek legislation. It is governed by a five-member board of directors, with Georgios Kokkinos being the President. It has a limited number of staff, and it owns a big branch crusher machine, a small wheel loader, a platform and olive oil storage facilities. The members have the appropriate equipment for olive tree cultivation, such as tractors, off-road vehicles, brush cutters, etc.

In 2005 NILEAS started trading the olive oil produced by its members, as a branded packaged product that meets the high standards of olive oil quality and, at the same time, of environmental-friendly managements (Fig. 50.1). The olive oil of NILEAS, mostly extra virgin, is distributed mainly in Greece, but a small proportion is also sold abroad. Indicatively, every year 30 tonnes are exported packaged to Austria, 40 tonnes are exported in bulk to Canada, while the rest is sold to Greek export companies.

Ever since, NILEAS continued to implement environmental-friendly managements, in conjunction with conventional and top-quality organic olive oil production. In 2009, NILEAS was registered with EMAS (Eco-Management and Control System). Shortly thereafter followed the environmental statement of the NIELOELED (Environmental Product Declarations), and NILEAS published its carbon footprint (330gr of carbon footprint per litre of olive oil). In 2009 it also participated in the first Environmental Product Declaration (EPD) worldwide. The EPD for extra virgin olive oil was published on 3 May 2011 in ENVIRONDEC online database (www.environdec.com).

The primary goal of NILEAS is to increase the financial results per square meter of olive tree cultivation, as well as to assure farmers' motivation to cultivate following environmental protection policies. Therefore, NILEAS' members are progressive and motivated olive producers who apply the current quality requirements in the production of olive oil as well as environmental protection practices. The organization is also in the final stage of completing its packaging plant. Furthermore, during the arrival of olive oil in its facilities, it implements a traceability system, through a quality control which is based on the acidity and organoleptic characteristics of each batch. The community operates five small-capacity two-phase olive

Fig. 50.1 NILEAS olive oil bottle with the local landscape as background

mills and one small-capacity three-phase olive mill. All of them have modern equipment. Their approximate capacity is calculated between 500 and 1000 tonnes.

The distinctive cultivation practices that NILEAS' members follow include:

- Fertilization: This is based on soil analysis and foliar diagnostics which are performed each 2–3 years' period. In addition, several members are applying manure and home-made compost as main sources of nutrients.
- Plant protection: Producers are trained to recognize and control enemies and diseases and to apply authorized plant protection phyto-chemical products when necessary. Additionally, they promote strategies for biological control such as the presence of temporary spontaneous cover crops.
- Soil management: No-till soil (or zero tillage) is applied. Furthermore, temporary spontaneous plant cover is allowed in the olive groves. Mowing lawn with tractor or manual machines is basically used. Herbicides are used only when the slope of the ground is over 25%. Herbicides are used in 20–25% of the total area. The above two practices have proven to be the most effective strategy to reduce to the minimum the soil lost by erosion, especially in the sloppy landscapes where NILEAS's olive farms are well integrated.
- Pruning: Rejuvenation pruning is carried out in most cases during the harvest period. Moreover, from the beginning of the summer, NILEAS's producers remove the small branches located in the centre of the tree. Pruning remains are

typically burned in the olive farms. However, NILEAS has started a motivational campaign to promote among producers to crush the tree pruning remains and apply them on the soil as a soil improver and soil nutrient retention strategy. For that, NILEAS owns a professional tree branch crusher machine (Fig. 50.2).

- Irrigation: Most of the olive groves of NILEAS are rainfed. Irrigation is carried out in about 25% of the area of NILEAS' members, mainly by drilling. The irrigation network is aerial (based on trees) with the use of storm sprinklers or drippers. The annual amount of rainfall in the area is between 700 and 800 mm, and thus irrigation is usually implemented every year from the middle of July. The sources of the irrigation water are boreholes, and the quality of irrigation water, which is high, is monitored frequently (analysis done in the framework of the LIFE SAGE 10 program that NILEAS participated).
- Harvest: The harvest of the olive fruits from the 'Mavrolia' variety is relatively early in the olive campaign, which typically starts during the first ten days of October. As soon as it is finalized (early to mid-November), the harvest of the olive fruit of 'Koroneiki' starts (Fig. 50.3). In that way, the harvest of both varieties does not overlap. Harvest is carried out by traditional methods. This is the main reason for the high cost (between 1.8 and 2.1 euros per kg olive oil) of the harvesting.

Fig. 50.2 Tree branch crusher machine

Fig. 50.3 Olive harvesting in an olive farm in Chora

NILEAS trains its members in all the above practices, as well as any others, to acquire the necessary specialized knowledge and skills that will help them to become more competitive and to meet the needs of the market and the demands of consumers for quality products in the constantly changing international environment.

The organization's aim is the continuous transmission of scientific and technical information and data for the implementation of modern production processes and competitive agricultural products. Additionally, NILEAS is encouraging its members to gradually switch to organic olive tree farming, where it is achievable.

Furthermore, to achieve its goals, NILEAS participates actively in research funding projects dealing with environmental policies, new technologies and olive tree cultivation, some of which include the SAGE10 program (LIFE09 ENV/GR/000302 SAGE 10 in 2010) and oLIVE-CLIMA program (LIFE11 ENV/GR/942 oLIVE-CLIMA in 2012), which in 2018 won the first prize in the category 'Action for Climate' at a special ceremony held during the EU Green Week (Europe's largest environmental event). Since 2017, NILEAS has been participating in the IoF2020 (Internet of Food and Farm) program that has enabled NILEAS to ascertain the benefits of implementing internet technologies in order to achieve environmental and agricultural sustainability goals (increase efficiency, improve quality and reduce costs).

Last but not least, in summer 2019 NILEAS launched the SUSTAINOLIVE project in which it participates as a partner.

Chapter 51
AIAB

Alessandro Triantafyllidis and Sara Petrucci

The Italian Association for Organic Farming (AIAB) was officially established in 1988; it gathers about 15,000 members (farmers, processors, experts, researchers and consumers). It is organized by regional chapters (15) coordinated by a federal office located in Rome. AIAB promotes organic farming as a model of rural development. During the 1990s AIAB increased its collaboration with governmental agencies and research institutions. AIAB has been involved in many national and E.U. projects on research (FP6, FP7, Horizon 2020), training, promotion and information in the organic farming sector. AIAB staff participates in ministerial committees on organic farming and in IFOAM regional groups and committees where they have the task to represent Italian organic farmers and consumers. AIAB publishes a bi-monthly magazine (*Bioagricultura*, 16,000 subscribers) concerning vulgarization and innovation transfer activity, a weekly newsletter (*B@N*, 8000 subscribers), several technical leaflets and books and has an updated website (www.aiab.it, 4000 single visits per day); it organizes field days, workshops and seminars.

Approximately 3000 operators of AIAB members practice olive growing, for a total of 30,000 hectares. In the field of organic management of olives, AIAB has undertaken several activities, such as seminars, courses, field demonstration events and extension services on the following topics:

- Soil fertility and management: tillage, cover crops, grassing, mulching, fertilization, grazing of the olive grove
- Intercropping

A. Triantafyllidis (✉) · S. Petrucci
Asoociazione Italiana per l'Agricoltura Biológica, Reggio Calabria, Genova, Italy
e-mail: a.trianta@aiab.it; info@sarapetrucci.it

J. Muñoz-Rojas, R. García-Ruiz (eds.), *The Olive Landscapes of the Mediterranean*, Landscape Series 36,
https://doi.org/10.1007/978-3-031-57956-1_51

- Pruning
- Pest and disease control: agronomic prevention and treatments
- Trapping of insects
- Certification process and legislation

Chapter 52
Consorzio Olivicolo Italiano (UNAPROL)

Annunzziato Scaramozzino and Giorgio Pace

UNAPROL—Consorzio Olivicolo Italiano—is an association of the Italian and EU olive oil sector recognized in 2015 by the Italian Ministry of Agricultural, Food and Forestry Policies as the 'Association of Producers Organization' in the olive oil and table olive sector pursuant to Regulation (EU) No. 1308/2013 and Ministerial Decree 86,843 of November 28, 2013.

Active for about 40 years, UNAPROL today represents the interests of over 140,000 associated Italian companies in over 35 territorial economic organizations, with 258 thousand hectares in quality production and 5 thousand companies included in traceability programmes. The Consortium has always been committed to supporting olive production, guaranteed by carrying out multiple activities relating to supply chain traceability, product quality improvement, environmental sustainability of olive growing, promotion, information and education, the analysis of market and marketing support.

With reference to these operational areas, UNAPROL maintains institutional relations at national, community and intergovernmental level, participates in Brussels in the meetings of the Consultative Committee on Fatty Materials at the EU, at Copa Cogeca, and is an observer member of the International Olive Council of Madrid (IOC).

In more detail, the main activities and objectives that UNAPROL carries out and pursues, directly and through the associated olive producer organizations, are:

- Application of rational and environmentally friendly production techniques.
- The qualitative and quantitative improvement of the products of the olive oil production chain.
- Concentration and storage of the offer.

A. Scaramozzino (✉) · G. Pace
UNAPROL Consorzio Olivicolo Italiano, Rome, Italy
e-mail: a.scaramozzino@unaprol.it; g.pace@unaprol.it

J. Muñoz-Rojas, R. García-Ruiz (eds.), *The Olive Landscapes of the Mediterranean*, Landscape Series 36,
https://doi.org/10.1007/978-3-031-57956-1_52

- National and international marketing.
- Optimization of production costs and profitability of the investment in response to applicable environmental standards and stabilization of production prices.
- Assistance and coordination of the activities of the member organizations.
- The enhancement and promotion of quality and consumption through the adoption of traceability and product certification systems.
- The realization of projects and operational programmes at national and international level and also in collaboration with external partners.
- The training of operators in the olive sector in particular on the adoption of environmentally friendly cultivation practices and production techniques that can guarantee product quality improvement and production cost containment.
- Training and consumer information on the characteristics, peculiarities and benefits of extra virgin olive oil and table olives in order to promote and disseminate a true oil culture.
- Carrying out research and developing initiatives on sustainable production methods, innovative practices, economic competitiveness and market trends.
- The correct management of by-products and waste from the olive production chain to protect the quality of water, soils and the landscape and to preserve or promote biodiversity.
- The application of crop-production criteria and methods that contribute to the sustainable use of natural resources and the mitigation of climate change.

UNAPROL's commitment for a more sustainable overall approach to the cultivation of the olive is based on a capillary and multiyear strategy. By the training of operators and through raising awareness among farmers, for example, the organization managed to obtain that more than 60% of its associates today apply pest and disease control through supervised pest management systems based on DSS, in contrast with the old approach based on scheduled pest management (Fig. 52.1).

Fig. 52.1 Setting up of monitoring traps. Pest traps are used to gather data to be used in the preparation of warning bulletins within supervised pest management systems to be sent to UNAPROL-associated farmers

Likewise, every year, thousands of olive growers are trained on rational pruning techniques aimed at the health and productivity of olive trees, thus maximizing their resistance against pests and diseases and reducing the need for external inputs like fertilizers and water (Fig. 52.2).

Finally, UNAPROL promotes the adherence of olive growers to the production methods envisaged by sustainable certification schemes such as organic and integrated agriculture.

Currently about 20% of the farms associated with UNAPROL are certified for organic production, while the share of those adhering to the National Integrated Production Quality System (SQNPI) is no more than 2% of the total, but this percentage is steadily growing year by year (Fig. 52.3).

Fig. 52.2 Ongoing theoretical-practical training session in the field carried out by experts from UNAPROL to demonstrate the application of rational pruning techniques

Fig. 52.3 Application of kaolin powder film as a repellent and anti-ovipositional products in the control of *Bactrocera oleae* in organic farming

Chapter 53
Omniya and Houda Women's Cooperative in Bellouta (COB), Morocco

Salaheddine El Azzouzi

The Omniya and Houda Women's Cooperative in Bellouta (COB) was created in 2005 by a group of women from the villages of Douar Bellouta and Brikcha in the province of Ouazzane in the north of Morocco. The cooperative has been supported by the local development association of Bellouta and has benefited from development projects that are funded by the Provincial Direction of Agriculture affiliated with the National Ministry of Agriculture.

The cooperative brings together more than 30 members, all of whom are women, 5 of whom constitute the permanent board, administrative office. The roles of the cooperative include assuring the harvesting of the olives of the farming members of the cooperative, including those located in Douar Bellouta and in other neighboring villages, as well as the valorization of the quality of olive oil, which is intended for commercialization at the local and national markets (Fig. 53.1).

The joint surface area of the fields belonging to the members of the cooperative is estimated at 10 hectares of olive trees. The cooperative does not own the olive groves; these belong to the families of the various women producers that are partners, while the cooperative is involved in coordinating their management and commercial tasks.

Since the owners of the olive plots have several scattered narrow fields, the number of olive trees planted varies between 50 and 100 per hectare. These producers use intercropping systems and fertilize exclusively using cattle manure (goats, cows, etc.). No chemical or biological treatments are used against the diseases of the olive trees; however, a management scheme for the trees and forest species is provided by the cooperative. In spite of this, no formal organic certification is yet attributed to their production (Fig. 53.2).

S. El Azzouzi (✉)
Département de Biologie, Faculté des Sciences, Université Ibn Toufail,
Kenitra, Ouezzane, Morocco
e-mail: salaheddine.azz@gmail.com

J. Muñoz-Rojas, R. García-Ruiz (eds.), *The Olive Landscapes of the Mediterranean*, Landscape Series 36,
https://doi.org/10.1007/978-3-031-57956-1_53

Fig. 53.1 Administrative office of the cooperative Al Houda

Fig. 53.2 Sheep in the olive groves in Houda's farm

The olive harvesting starts at the beginning of November, normally following the first autumn rains. The harvest is performed by a traditional method: picking by hand from the young olive tree or dropping the olives using wooden sticks from the big, older, olive trees.

Once the harvest is finalized, the olives are either transported on the backs of mules in sacks or, if possible, via motorized means of transport that remains somewhat rare in mountainous areas. In order to avoid the increase of the temperature of the olives during transport from the plots to the molting unit, the bags used during the displacement are aerated. Once in the crushing unit, the olives go through the cold washing section and are then moved to the crusher station; then comes the automatic separation of olive oil and olive pomace, with olive oil filling the aluminum tanks (Fig. 53.3).

The cooperative delivers the service of oil extraction to both the farmer members of the cooperative and for other olive producers of the zone. To secure that the extra virgin olive oil extracted is considered valid for the local or national market, less than 24 h must be spent between the harvest and the crushing. However, in the majority of the cases, the bags of olive can remain waiting for several days before their crushing. To overcome this problem, the cooperative uses a smaller machine that can process up to 100 kg per hour through two phases and cold extraction.

Once olive oil is extracted, a large proportion of it is destined for the self-consumption by the members of the cooperative; however, the cooperative buys olives from other local producers to store a sufficient quantity of olive oil that would be destined for sale on the local or national markets. For this purpose, the cooperative holds bottles and labels of 250 ml, 500 ml, and 750 ml as well as large bottles

Fig. 53.3 Traditional harvesting of olives using manual labor

Fig. 53.4 Example of 500 ml labeled bottles and large 5 liter bottles

of 5 L. The cooperative buys the packaging of these bottles from suppliers who in most cases are subsidized by the state as part of development projects (Fig. 53.4).

The cooperative is engaged in several national and international fairs and has recently participated in the international fair of Meknes organized by the Moroccan Ministry of Agriculture, as well as in the International fair of Berlin, and in several national fairs of social and solidarity economy in Tangier, Chefchaouen, and Tetouan.

Chapter 54
The Center for the Study and Promotion of Olive Oil in Alentejo (CEPAAL)

Gonçalo Morais Tristão, Filipa Velez, and Mariana Teles Branco

Abstract As it has been thoroughly demonstrated along this book (see Chaps. 27 and 33), the olive sector is a crucial and strategic sector in the Portuguese agricultural policy and economy. Currently, Portugal is the seventh-largest producer of olive oil in the world and the fourth-largest exporter, with 96.4% of the olives produced used for olive oil and the remaining 3.6% for table olives. In the 2019/2020 campaign, Portugal produced around 140.5 thousand tons of olive oil (corresponding to about 930 thousand tons of olives), reaching a historic high. Within Portugal, Alentejo, in the southern Mediterranean half of the country, has consolidated itself as the region hosting the largest area of olive groves and, especially, as the most effective producer of olive oil, CEPAAL, created in 1999, strongly contributes to this goal, whilst also innovatively facilitating associativism and cooperation within the sector, a key factor for future success and sustainability.

As it has been thoroughly demonstrated along this book (see Chaps. 27 and 33), the olive sector is a crucial and strategic sector in the Portuguese agricultural policy and economy. Currently, Portugal is the seventh-largest producer of olive oil in the world and the fourth-largest exporter, with 96.4% of the olives produced used for olive oil and the remaining 3.6% for table olives. In the 2019/2020 campaign, Portugal produced around 140.5 thousand tons of olive oil (corresponding to about 930 thousand tons of olives), reaching a historic high.

Although olive trees are cultivated across several different regions of Portugal, the bulk of the production, both agronomically and industry-wise, is in Alentejo, a

G. M. Tristão (✉) · F. Velez
Centro de Estudos para a Promoção do Azeite do Alentejo-CEPAAL, Moura, Portugal
e-mail: cepaal@azeitedoalentejo.pt

M. T. Branco
Dazeite, Montemor o Novo, Portugal
e-mail: office@dazeite.pt

J. Muñoz-Rojas, R. García-Ruiz (eds.), *The Olive Landscapes of the Mediterranean*, Landscape Series 36,
https://doi.org/10.1007/978-3-031-57956-1_54

region with a long tradition in the production and consumption of olive oil. According to the Olive Oil and Table Olives Information System, in the 2019/2020 campaign, Alentejo was responsible for 88% of olive oil production in Portugal (GPP, 2023).

Concerning the area occupied by olive groves, out of the 361,483 hectares of olive groves that exist in Portugal, about 352 thousand are for olive oil (INE, 2022), with Alentejo occupying 52.2% of the total national area.

It is not known for certain when olive oil began to be produced in Alentejo, but we know that it was the Phoenicians and mainly the Romans who introduced planting, grafting, and olive oil extraction improvements. Writings by the Roman historian, Strabo, mention olive oil from Alentejo as a product of excellent quality imported by Rome 2000 years ago.

The olive sector has developed in a particularly powerful way in recent years in Alentejo—the numerous investments that are occurring in the region related to the modernization of the sector are well known and have been synthesized more recently in a report edited by international consultants (Vilar et al., 2019). According to such recent studies, among traditional and modern olive groves, Alentejo is now internationally recognized as a successful model, on one hand influenced by the Alqueva Dam (which significantly increased the availability of water in the region) and, on the other hand, because of the knowledge and experience of the peoples of the region—a region with great tradition in the olive sector from the recognition of the productive potential of the region.

In response to the relevance of the sector and resulting from the multiplicity of olive groves and olive oil producers in Alentejo, the Center for the Study and Promotion of Alentejo Olive Oil (CEPAAL—www.azeitedoalentejo.pt/en) was created in 1999. CEPAAL is an association that has as main goal the enhancement, the study, and the promotion of olive oil and olive grove from Alentejo in particular and from Portugal in general. *Ultimately, CEPAAL is focused on preserving the identity of the sector and region.* CEPAAL is registered as a nonprofit association and is based in Moura (Alentejo Central).

CEPAAL has now 28 Alentejo olive oil producers as members, who trade dozens of olive oil brands (Table 54.1).

In addition to the 28 member-producers, CEPAAL also has the direct institutional support of 12 institutional members, such as municipalities, universities, and public entities.

CEPAAL's work and activities focus on the dissemination, promotion, and enhancement of the entire regional olive oil sector. To enact these goals, CEPAAL fosters the development of promotion and marketing actions, both within internal and external markets, and the preservation of all related historical, cultural, and gastronomic heritage related with the olive sector. Lastly, they are behind other activities, where the valorization of regional and national olive oils assumes a preponderant role in their strategy, seeking for the defense of an extremely rich olive heritage and with undeniable market potential.

CEPAAL carries out innumerous initiatives, such as actions in agricultural fairs, showrooms, and tastings that present Alentejo olive oil and show the health,

Table 54.1 Alentejo Olive Oil producers members of CEPAAL

Private family and individual producers	Agro-industrial producers and corporations	Local and regional cooperatives	Foundations and nonprofits	Others
António Godinho da Cruz Barradas João Luís Martins de Sousa Jordão—Monte de Portugal João Miguel Rego Rosado	AZAL—Azeites do Alentejo, SA. Amareleza Vinhos, Lda Casa Agrícola HMR, SA Cortes de Cima, SA Courela do Zambujeiro, LDA Esporão Azeites, LDA Diterra—Comércio agro-industrial, LDA Empresa Fabril de Moura, LDA Marcolino Sebo wines and oils, SA Monte das Louzeiras, Prod. Agrícola LDA Nesgas da Terra, Lda SICA—Sociedade Industrial e Comercial Sociedade Agrícola de Pias SA—Família Margaça Sociedade Agrícola do Vale de Umbria, SA Sociedade Agro-Pecuária do Mouchão e Cavaca Dourada, SA VIDCAVEA—Azeites da Vidigueira de Azeites SA	Caminhos do Futuro—Cooperativa de Comercialização e Transformação de Produtos Agropecuários de Montemor-o-novo, CRL Cooperativa Agrícola da Granja, CRL Cooperativa Agrícola de Moura e Barrancos, CRL Cooperativa Agrícola de Portel, CRL Cooperativa Agrícola de Reguengos de Monsaraz, CRL Olidal—Olivicultores do Alentejo CRL	Fundação Eugénio de Almeida	Portugal rural, Soc. Agr. Hotelaria e Turismo, LDA

economic, and environmental values of regional olive oils. With a strong focus on the consumer, CEPAAL's actions of promotion intend to enhance value and raise awareness.

Besides promoting Alentejo olive oil products and producers, CEPAAL has a wider scope and intent to enhance the national sector, holding numerous initiatives, of which we highlight:

- The organization of the *Portugal National Olive Oil Competition*, which, since 2007, seeks to promote not only the Portuguese olive sector but also, and mainly, the valorization of national olive oils and producers.
- The organization of the *Virgin Olive Oil Contest at the National Olive Oil Fair* and collaboration in the organization of the Fair in partnership with the municipalities of Moura and Valpaços (Portugal).
- The organization of the *National Olive Oil Congress* that brings together the top professionals in the sector to discuss the foremost challenges, solutions, and trends.

CEPAAL acknowledges how the likely future trend is for the olive grove areas in Alentejo to continue growing and that environmental sustainability poses a number of challenges for the sector that demand more effective and innovative strategies. In response, they have lately become responsible for supporting and advising Alentejo olive farmers and producers across their diversity, by sharing and highlighting good agricultural practices in the sector. That implies that CEPAAL remains at the forefront of information sharing on the sector, especially as liaison between producers and consumers.

Because it is now essential to consider the sustainability of olive groves and the management of by-products, CEPAAL decided to get engaged in the SUSTAINOLIVE project, with the intent to contribute to the promotion of the sustainability of the sector by encouraging the implementation of a set of sustainable solutions based on agroecological concepts and in the exchange of knowledge between the main actors of the sector, not only at a national level but also, and mostly, internationally.

Acknowledgments This work was funded by the SUSTAINOLIVE project (Grant n° 1822; PRIMA EU Programme).

References

Gabinete de Políticas e Projetos – GPP. (2023). *SIAZ—Sistema de Informação do Azeite e Azeitona de Mesa*. https://www.gpp.pt/index.php/sistemas-de-informacao/sistema-de-informacao-do-azeite-e-azeitona-de-mesa. Last Retrieved: 08/2023.

Instituto Nacional de Estatística (INE). (2022). *Estatísticas Agrícola. Olival e Azeite*. https://www.ine.pt/xportal/xmain?xpid=INE&xpgid=ine_destaques&DESTAQUESdest_boui=594695354&DESTAQUEStema=55505&DESTAQUESmodo=2. Last Retrieved: 08/2023.

Vilar, J., Santos, P., Caño, S., Caldeira, B., Dias, P., & Rosário, L. (2019). *Alentejo: A liderar a olivicultura moderna internacional. relatório Final*. https://www.olivumsul.com/_files/ugd/a303d9_5993f29b65054e46a54accff8c90cf7f.pdf. Last Retrieved: 08/2023.

Chapter 55
Implementing Sustainability Strategies and Standards in a Farming Enterprise: Esporão (Alentejo, Portugal)

Ana Gaspar and Ana Carrilho

Abstract Founded in 1973 by José Roquette and Joaquim Bandeira, Esporão is one of the most important wine companies in Portugal and one of the largest producers of organic wine in the world Furthermore, Esporão has also embraced olive oil production, focusing in their southern premisses of Alentejo, and thereby pioneering the production of organic olive oil in Portugal, with a innovative plan for corporate-led sustainability. In this chapter, the key actions and strategies undertaken to achieve higher standards of sustainability in the cultivation, production and comercialization of olive oil at the firm level are discussed.

55.1 Esporão: The Enterprise

55.1.1 An Introduction: Vision and Mission

Founded in 1973 by José Roquette and Joaquim Bandeira, Esporão is one of the most important wine companies in Portugal and one of the largest producers of organic wine in the world.

With a leading role in the national and international consolidation of the agricultural sector of Alentejo, Esporão is also present in the Douro and Vinho Verde regions of Northern Portugal. It develops its activity within the limits of sustainability and building close relationships with customers and consumers around the world.

Along with the main wine business branch, Esporão also produces organic olive oil in its own farm "Herdade de Esporão," located in the municipality of Reguengos de Monsaraz (Alentejo, Portugal).

A. Gaspar (✉) · A. Carrilho
Esporão S.A, Reguengos de Monsaraz, Portugal
e-mail: ana.gaspar@esporao.com; ana.carrilho@esporao.com

J. Muñoz-Rojas, R. García-Ruiz (eds.), *The Olive Landscapes of the Mediterranean*, Landscape Series 36,
https://doi.org/10.1007/978-3-031-57956-1_55

55.1.2 Timeline

- 1973—José Roquette and Joaquim Bandeira founded the company Finagra and bought Herdade do Esporão in Alentejo to focus on wine production.
- 1997—Launching of the first Esporão olive oil.
- 2007—Beginning of the transition to organic farming at Herdade do Esporão.
- 2007—Esporão achieves market leadership in Portugal.
- 2010—Certification of the organic production method at Herdade do Esporão.
- 2015—Launch of the first olive oil with organic certification—Esporão Olival dos Arrifes Organic Olive Oil.
- 2021—Herdade do Esporão's restaurant is recognized with a Green Star from the Michelin Guide.

55.1.3 The Olive Groves

At Herdade do Esporão (Alentejo), there are 100 ha of certified organic olive groves and at Quinta dos Murças (Douro) a further 15 ha under the same production mode. In addition to the olive groves, the firm joins up with more than 60 family-based olive producers in the Alentejo, most of whom belong to families that have been working their farms for several generations.

55.1.4 Our Organic path

Esporão was born in Alentejo with the desire to produce the best wines, then also expanding the business towards the production of olive oil. Over the years, they acknowledged that organic farming contributes to this mission—by protecting the soil and increasing the resilience and quality of the vines and olive groves, which allows to obtain healthier and more balanced grapes and olives, which require less intervention in the cellar and mill and allow to produce wines and olive oils with greater identity and sense of place and without products harmful to health and the environment.

In Alentejo (Herdade do Esporão, Herdade dos Perdigões, and Vinha dos Lavradores), all vineyards and olive groves are certified as organic, as are all also those at Quinta dos Murças in the Douro.

55.2 Herdade do Esporão: The Farm

Herdade do Esporão is located in Alentejo, 170 km SE from Lisbon, in the municipality of Reguengos de Monsaraz. With a total area of 1840 ha and with geographic boundaries unchanged since 1267, it offers adequate agricultural conditions—wide temperature variations, seven different soil types, and an extraordinary biodiversity. The 441 ha of vineyards, the 100 ha of olive groves, and other crops are all certified in organic production. Open since 1997, the tourism premisses at Herdade do Esporão host a wide range of wine and olive oil tastings, as well as a restaurant with a Michelin Green Star.

55.2.1 Agricultural Practices at Herdade do Esporão

Agriculture is responsible for some existing imbalances in ecosystems, altering the relationships between living beings and the environment. The expansion of viticulture, and lately the intensive olive cultivation, in Alentejo has led to a considerable change in ecosystems and landscapes, due to the installation and maintenance of agricultural techniques that are not always respectful of the soil and the wider environment.

The will to continue to improve the quality of farming products and the concerns with the environmental, economic, and social sustainability of the entire production process at Esporão led them to look at management in a more integrated way, which provides for more effective interactions between agriculture and the underpinning ecosystems.

The main objectives of the agronomic techniques applied at Herdade do Esporão include the following: recover, maintain, or increase soil fertility, reduce erosion, promote a balance between pests and their biological enablers, reduce the consumption of water use in irrigation, take advantage of by-products generated in the our activity to produce fertilizers for use in crops, maintain and promote animal and plant variability in ecosystems, and preserve plant material at risk of disappearing from national plants.

The main agricultural management practices implemented at Esporão to facilitate achievement of these goals include the following:

(a) *Topographic survey*. The first operation carried out before the installation of a new vineyard or olive grove is the topographical survey of the plot. The topographic survey is fundamental to carry out land adjustments and to assess the need to build drainage systems (Fig. 55.1).
(b) *Ripping*. In the summer months, following the topographical survey, an operation is carried out to prepare the soil for the installation of a vineyard or olive grove: ripping. Ripping consists of the vertical mobilization of the soil without overturning its layers, avoiding bringing less fertile horizons to the surface, which would lead to its impoverishment and to an increase in the need for corrections. Its purpose is to uncompact and aerate the soil, creating favorable conditions for the development of the future roots of the vine or olive trees.
(c) *Drainage*. Drainage systems, either underground or open-air, are essential to ensure the disposal of excess water from the plot, promote soil aeration so as not to limit plant growth, and facilitate the movement of machines and implements. Drainage is performed following the natural water lines that cross plots and fields. Therefore, when designing the land for planting a new crop, it is

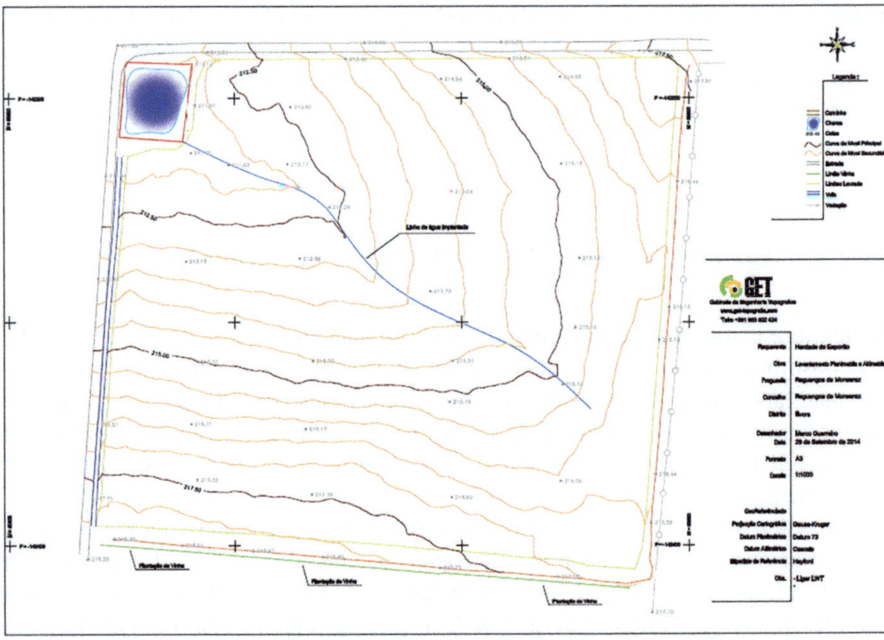

Fig. 55.1 Altimetry map of a vineyard to be installed in 2015 (T082)

essential to preserve the existing water line, if this is the natural one, or to replace this water line through a ditch and its subsequent consolidation with vegetation (Fig. 55.2).

(d) *Afforestation of drainage ditches*. Afforestation of open ditches is essential to prevent the collapse of barriers and soil erosion. The species used for afforestation are chosen according to their ability to fix auxiliaries that promote the natural control of pests in the vineyard or olive grove (Fig. 55.3).

(e) *Protection hedges*. Protective hedges are essential to create ecological corridors within the vineyard or olive orchard, promoting the establishment of auxiliaries and creating excellent nesting sites for birds. Thus, there is an increase in populations and diversity of species. They also intend to reduce crop dehydration by reducing the incidence of prevailing winds which, in the Alentejo summers, are very dry and aggressive (Figs. 55.4, 55.5, and 55.6).

(f) *Hedges for the fixation of auxiliaries*. In 2009, an annual survey and monitoring of farm-level biodiversity was conducted at Herdade do Esporão. Some shortcomings were identified, and a plan was drawn up for the installation of vegetation (hedges, overgrowth, density, bat shelters) that promotes the installation of natural crop aids, helping to control pests.

In 2014, a new strategy for securing and protecting auxiliaries within the wooden crops was implemented. A survey was carried out of the existing pests of Herdade

Fig. 55.2 Drainage ditch executed according to the topographic survey (T082)

Fig. 55.3 Drainage ditch with protective hedge (T061 and T062)

Fig. 55.4 Protection hedge in a vineyard (T015)

Fig. 55.5 Blackbird nest

do Esporão (including species such as yellow spider, green leafhopper, grape moth, cotton mealybug) and their main predators (auxiliaries). After this survey, the most suitable plant species for the fixation of these auxiliaries were studied. Following this, rows were planted with the following species: wild rose, honeysuckle, blackberry, blackthorn, leafy, pomegranate, wildflower, laurel, and elderberry. The

Fig. 55.6 Study of the influence of hedges over crop protection from wind

Fig. 55.7 Hedge within a vineyard, installed in 2014

purpose of these hedges is to encourage the increase of auxiliary species, so that there is greater control over pests, promoting balance in the future (Fig. 55.7).

(g) *Irrigation water control.* Water is a scarce commodity in the context of Southern Portugal, and as such, all measures undertaken to reduce its consumption and preserve its quality are essential to conserve this resource. At Herdade do Esporão, there is enormous control over the water used for irrigation. Before starting a campaign, strict supervision of the entire system is essential to avoid losses due to breakages and poor adjustments. Thus, all pipes are checked (over 2000 km of dripline), and all pressures are adjusted. The installation of flowmeters at the exit of the pumping station and at the exit of the filtering stations was fundamental for the control of ruptures.

(h) *Soil protection.* Healthy, balanced, and "alive" soils are essential for us to achieve quality.

The control of spontaneous vegetation in the planting lines is done mechanically to the detriment of chemical control. Since 2011, no herbicides have been applied at Herdade do Esporão.

The sowing of vegetation cover between the rows (grass) of planted crops is one of the most important activities in the sustainability plan of Herdade do Esporão and aims to better conserve the soil, reduce erosion, increase the biodiversity of ecosystem, and increase organic matter contents. Excluding chemical fertilizers and increasing auxiliary fauna will play a fundamental role in the natural limitation of crop pests.

Herbaceous cover can cover part or all of the area, be permanent or temporary, and be sown with one or more plant species. It can also be natural by allowing the autochthonous flora to develop spontaneously.

For the preparation of the sowing/mulching plan, the following factors are considered: the age of the crop, the fertility of the soil (soil analysis), the quality of the fruits obtained, the quality of the intended production, the vigor of the plant, the incidence of pests and diseases in the previous year, and the fertilization needs.

After assessing these parameters, a sowing plan is drawn up based on the strategy defined for each of the farmed plots. Thus, six types of intervention were outlined (sowing/management of cover crops):

1. 50% mixture of annual clovers +50% peas
2. 50% mixture of annual clovers +50% spontaneous vegetation
3. 100% peas
4. 50% fescue +50% peas
5. 100% spontaneous vegetation
6. 50% red clover +50% peas.

All these strategies are essential for increasing the fertility of soils and reducing the incidence of pests, which will allow us to naturally increase the quality of fruits for the future, consequently raising the quality of wines and olive oil (Figs. 55.8, 55.9, 55.10, and 55.11).

(i) *Mulching*. Mulching aims to prevent the development of spontaneous flora in the planting line, increase fertility, and conserve soil moisture. For the execution of mulching, by-products resulting from within the farming activities and cereal straw resulting from sowing are used (Fig. 55.12).

(j) *Composting*. Composting is a process that promotes the transformation of organic matter present in biodegradable organic waste (Fig. 55.13).

At Herdade do Esporão, all the organic by-products generated by farming activities are reused (stems, vines, olive leaves, olive pomace, and pruning residues) to build a compost pile which in the end will give rise to compost, produced in a natural way, rich in essential nutrients for the crops.

(k) *Irrigation control*. The irrigation strategy applied in the farm foresees a moderate consumption of water, in order to reduce water stress in the plants, which will result in a higher quality of the fruits produced. To this end, the acquisition of equipment to monitor the climate (meteorological stations), soil humidity (Enviroscans and Diviners), and water availability in the plants (pressure chamber) is fundamental for decision-making, to increase the rigor in the applications and to avoid water loss through runoff and drainage (Figs. 55.14 and 55.15).

(l) *Application of kaolin*. The application of kaolin showed a significant reduction in leaf temperature and an improvement in water potential and photosynthetic activity in the leaves, enhancing the quality of the grapes and reducing the

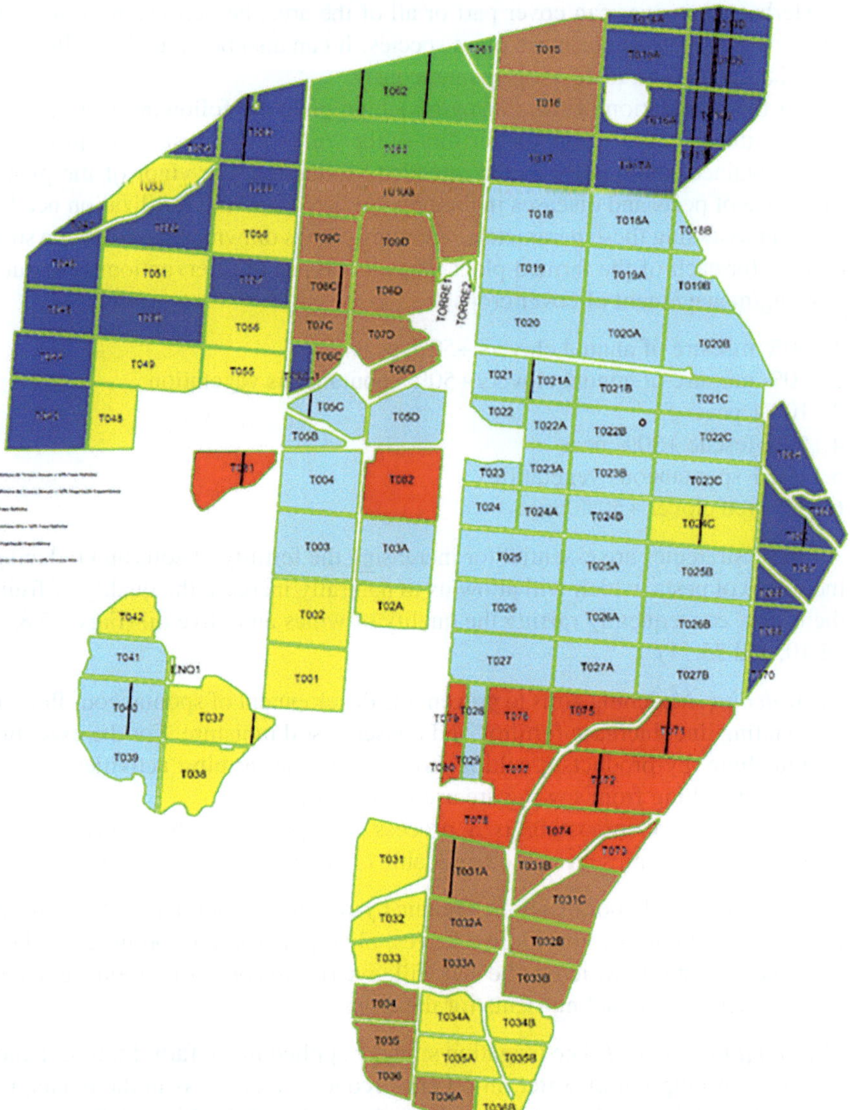

Fig. 55.8 Map showing the plots for sowing cover crops for agronomic year 2016 at Herdade de Esporão

Fig. 55.9 Sowing of lupine for sideration (green manure) (T057)

Fig. 55.10 Covering with spontaneous species (T017)

Fig. 55.11 Sown top dressing (T044)

Fig. 55.12 *Mulching* (with vineyard stalks) in the olive grove at Herdade do Esporão, 2009

Fig. 55.13 Composting pile built in 2013

number of irrigation cycles required, leading to considerable savings in water and energy (Fig. 55.16).

(m) *Pest control*. All aids in pest control are used and enhanced at Herdade do Esporão.

Using the support of recordings and ultrasound analysis, five species of bats were identified. As a result of this, it was observed that the bats systematically fed inside the homestead but then took shelter and reproduced outside of it.

Part of the diet of bats is some species of moths responsible for causing damage to the vineyards and olive groves. As such, shelter boxes for bats were installed, so that they can shelter and reproduce next to the crops. This provides the farmer with another tool for controlling pests in the vineyard. The objective of Esporão is to increase the number of shelters installed, as the result has been a success. At this moment, practically all the boxes are occupied, which contributes to the increase in the population of bats in the farm and the increase and maintenance of the local biodiversity (Fig. 55.17).

(n) *Sexual confusion*. Sexual confusion is a preventive method for combating pest moths, which consists of distributing 500 sexual pheromone diffusers per hectare homogeneously throughout the fields, which will form a pheromone cloud, preventing males from finding the females and mating, turning the eggs unfeasible. This method is aimed solely at a certain species and replaces chemical struggle (Fig. 55.18).

(o) *Release of auxiliaries*. The release of auxiliary pest predators is essential to increase their populations, limiting the impact caused by certain pests of

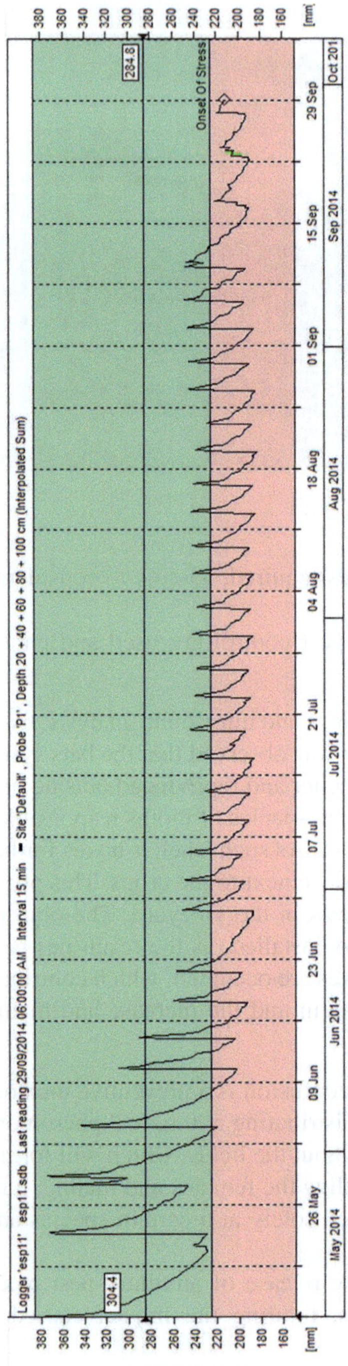

Fig. 55.14 IRRIMAX 2014. Graphical representation of water movements in the soil throughout the vegetative cycle of a plot (T027A) with stress induction. The time period represented expands betwen May 2014 and October 2014, and is exemplary of the regular fluctuations of water availability in an irrigated olive plot along a charactristic Mediterranean seasonal transition between wetter (spring) and more arid (summer) months

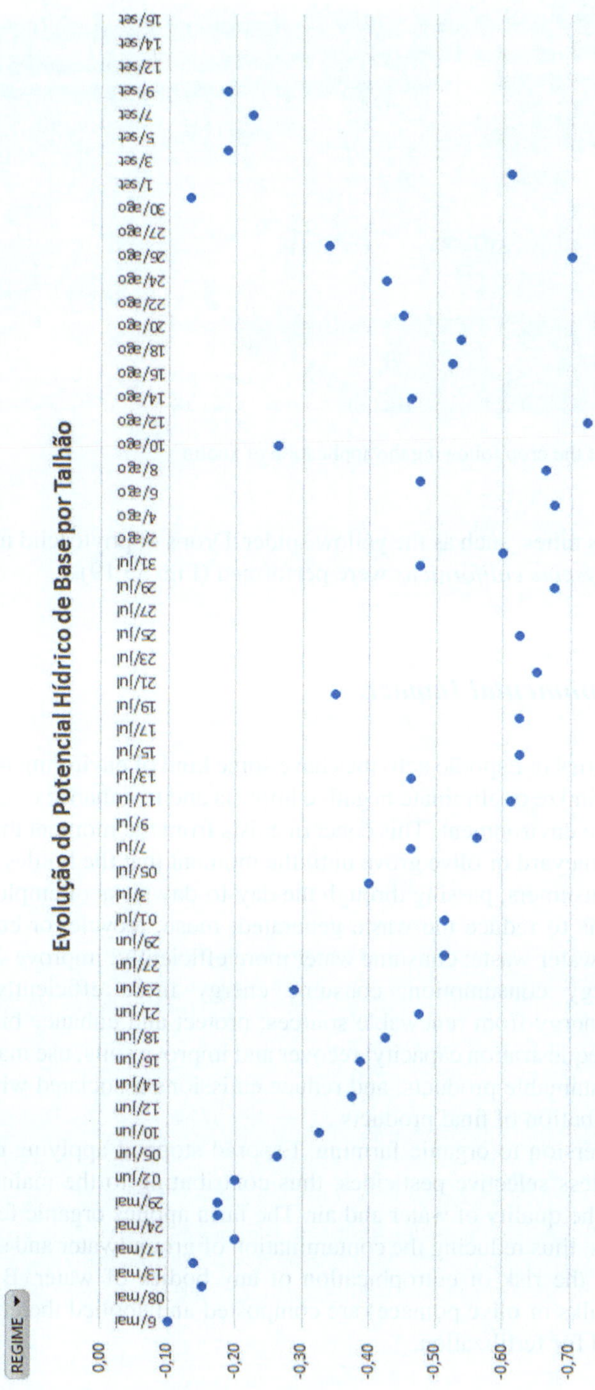

Fig. 55.15 Evolution of base water potentials throughout the vegetative cycle for the period between may and september (thus, thoughout the arid summer sason, where water balance is naturally negative in a Mediterranean climate) (T027A)

Fig. 55.16 Aspect of the crop following the application of kaolin

phytophagous mites, such as the yellow spider. Drops of phytoseiid mites of the species *Amblyseius californicus* were performed (Fig. 55.19).

55.2.2 Environmental Impact

All farming activities in Esporão activities have some kind of environmental impact. The aim is to minimize or eliminate negative impacts and to enhance or create positive impacts on the environment. This concern exists from the moment that a plan is made to plant a vineyard or olive grove until the moment that the bottles of wine or olive oil reach consumers, passing through the day-to-day tasks of employees.

The key goal is to reduce the waste generated; reuse, recycle, or enhance by-products; reduce water waste; consume water more efficiently; improve water quality; reduce energy consumption; consume energy more efficiently; achieve self-sufficiency energy from renewable sources; protect and enhance biodiversity; increase carbon sequestration capacity; recover and improve soils; use materials and produce more sustainable products; and reduce emissions associated with the production and distribution of final products .

With the conversion to organic farming, Esporão stopped applying chemically synthesized and less selective pesticides, thus contributing to the maintenance of biodiversity and the quality of water and air. The farm applies organic fertilizers in adequate amounts, thus reducing the contamination of groundwater and soil, consequently reducing the risk of eutrophication of any bodies of water. By-products (such as grape stalks or olive pomace) are composted and applied the crops, partly reducing the need for fertilization.

Fig. 55.17 Foster box for bats (T032B)

An efficient irrigation system is in place that permits to detect and repair ruptures automatedly, preventing water loss. Water is applied only when strictly necessary. A WWTP (Waste Water Treatment Plant) is in place, which allows the water used in the industrial complex to be treated, introduced back into the farm dam, and reused again.

An aim also exists to reduce waste when products are bought, giving preference to products with less packaging materials or made with recycled materials. The firm is committed to eco-design, trying to use more sustainable materials when bottling and packaging farm products and reducing the weight of materials. With this commitment, in addition to reducing the waste of products, the use of space for transport

Fig. 55.18 Diffuser used to control *Lobesia botrana* through sexual confusion systems

Fig. 55.19 Sachet placed in the vineyard with phytoseiid mites

can be optimized, reducing the number of transports and the associated carbon dioxide emissions.

There are some areas of the farm that are set aside for the conservation of biodiversity, including restoration of water lines and riparian galleries that play an important role in water filtration and quality. Hedges have also been planted in productive areas that promote biodiversity and the establishment of auxiliary fauna, functioning as ecological corridors.

Soil movements are minimized as much as possible, performing them as close to the surface as possible, only when planting crops or, in exceptional cases, when there is high soil compaction. A further aim is to increase the water retention and carbon sequestration capacity of the soil through practices that promote an increase in organic matter in the soil, such as ironing or the maintenance of vegetation cover, which can be either natural or sown. These practices also favor biodiversity and reduce soil erosion and fertilizer needs.

55.2.3 Sustainability at the Restaurant in Herdade do Esporão

Open since 1997 and integrated in the cork oak forest which is located at the heart of Herdade do Esporão, the farm restaurant hosts "terroir" cuisine, of modern confection and interpretation, where the center is the product: The Alentejo, Portuguese, seasonal product—the product of the land. The product is a starting point, where the shape, texture, color, and all its best features are preserved, extracting the maximum flavor.

In 2021, the Herdade do Esporão restaurant was recognized with one Michelin Green Star, with emphasis on the quality of ingredients, the harmony of flavors, mastery of techniques, identity, and consistency.

55.3 Mais.Devagar ("Slow Farming")

In 2019, and with the aim of promoting debate in society about the need to slow down the hectic pace of today, Esporão challenged the Catholic University of Portugal to develop a study on a more relaxed lifestyle in Portugal (Mais.Devagar). The main conclusion indicates that those who slow down tend to be happier.

The Mais.Devagar manifesto cemented the path that Esporão has undertaken to build a better and more sustainable company. The Manifesto can be fully accessed and consulted in the following website, which is entirely dedicated to foster and disseminate our main message: https://maisdevagar.pt/manifesto/.

Acknowledgments This work was funded by the SUSTAINOLIVE project (Grant No. 1822; PRIMA EU Programme).

Part IX
Olive Oil: Diversity, Heterogeneity and Quality Challenges

Chapter 56
Origin and Traceability of Olives and Olive Oils in the Mediterranean Area: Genomic, Metabolomic, and Control Insights

Caterina Agrimonti, Alessandra Bendini, Elena Maestri, Tullia Gallina Toschi, and Nelson Marmiroli

Abstract Olive oil is one of the high-quality products for which Europe/the Mediterranean is renowned. As all prized products, olive oil is the subject of fraudulent practices, which compromise the market positions of European and Mediterranean countries. A series of EU-funded projects has addressed the integrity of the olive food chain, improving the methods for traceability and authenticity protection, based on chemical, genetic analysis, and block chain approaches. The new challenges and the new technologies and the OLEUM databank will be discussed.

Keywords Olive oil fraud · Traceability · PDO and PGI oils · Authenticity · Metabolomic/fingerprinting approach · Genomic tool · Markers · Analytical methods

---

Caterina Agrimonti and Alessandra Bendini are first co-authors; Tullia Gallina Toschi and Nelson Marmiroli are co-last authors.

C. Agrimonti
University of Parma, Interdepartmental Center for Energy and Environment (CIDEA), Parma, Italy

University of Parma, Department of Chemistry, Life Sciences and Environmental Sustainability, Parma, Italy
e-mail: caterina.agrimonti@unipr.it

E. Maestri · N. Marmiroli (✉)
University of Parma, Interdepartmental Center for Energy and Environment (CIDEA), Parma, Italy
e-mail: elena.maestri@unipr.it; nelson.marmiroli@unipr.it

A. Bendini · T. G. Toschi
Department of Agricultural and Food Sciences, Alma Mater Studiorum – University of Bologna, Bologna, Italy
e-mail: alessandra.bendini@unibo.it; tullia.gallinatoschi@unibo.it

J. Muñoz-Rojas, R. García-Ruiz (eds.), *The Olive Landscapes of the Mediterranean*, Landscape Series 36,
https://doi.org/10.1007/978-3-031-57956-1_56

56.1 Overview on Frauds in Olive Oil Sector

The Rapid Alert System for Food and Feed (RASFF) database and the EU Food Fraud Network (FFN) together with the Administrative Assistance and Cooperation System (AAC)—these last two available for Member States from 2015—are the most important tools for exchanging information on food safety and food adulteration issues at the EU level (Casadei et al., 2020). Every year, a report describing the activities carried out by the EU FFN and the AAC is published. Even if it is obviously not exhaustive of the EU noncompliances and suspicions of food fraud, it records and presents trends and informs about the EU initiatives to combat fraud. Moreover, it identifies the products most affected, for which the best fraud prevention and detection actions are needed. From 2016 to 2019, the number of requests for assistance and cooperation reported by the Member States in the FFN Activity Report raised from da 156 to 234 cases, and the category of "fats and oils" was the third-most cited group in 2018 after "fish" (45 cases) and "meat products" (41), becoming the first in 2019 (44) with "olive oil" (OO) as the most notified product in the system. The reason of that is linked with the high economic value of olive oil, in particular when extra virgin olive oil (EVOO) is compared to other vegetable oils, due to its unique sensory, compositional, and nutritional qualities. What is essential to underline is the level of attention and the high request in terms of conformity checks (EU Reg. 29/2019 and 1308/2013) that are currently improving the quality of the OO on the market (Conte et al., 2019), hindering the activities of counterfeiters (European Commission Areté Research, 2020).

As regards OO, one of the most prominent technical issues (H2020-SFS-14a-2014) concerns the need to develop, validate, and standardize methods able to support the organoleptic assessment of virgin olive oils (VOOs). In fact, to gain the "extra" prefix of the olive oil top quality (EVOO), the OO has to be sensory impeccable (0 defects and fruity >0) at the panel test, in addition of passing many "chemical" conformity checks. To give an idea of the problem from data presented by the Italian ICQRF (Bologna, September 2019), 56% of irregularities found for EVOO and VOO in the years 2015–2018 was linked to a noncompliant organoleptic classification. For the purpose of strengthening this control and reducing legal disputes due to misalignments in the organoleptic classifications, a research line called "Quantitative Panel Test" (Barbieri et al., 2020), developed in the framework of the European Horizon 2020 OLEUM project (Grant Agreement No. 635690), was proposed and developed as an innovative approach. The strategy was to enhance and enforce the sensory quality classification, by adding robust and reliable instrumental qualitative and quantitative information on volatile molecules responsible of defects and reference materials (RMs), able, these last, to improve also labs proficiency and global harmonization.

In addition to other fraud of dilution or substitution concerning the quality aspects, i.e., adding soft-deodorized olive oil or lampante (Tibola et al., 2018) to EVOO, the irregularities can deal also with authenticity, the so-called purity of olive oils. In this regard, the frauds are the realization of illegal blends with other vegetable oils such as sunflower, corn, palm, rapeseed, etc. (European Commission Areté Research, 2020). That fraudulent mixture of OOs with other vegetable oils does not usually lead to health-related problems for the consumers, even if the document falsification of a product puts the consumer potentially at risk of safety because it breaks the virtuous chain of traceability. Other discovered frauds are case of EVOO misbranding or mislabeling (Gallina Toschi et al., 2013; Tsimidou et al., 2016) and false declaration of geographical origin. The demonstration of botanical or geographical origin is fundamental, given that at now there is no validated method to determine them, apart from documental traceability. Many methods, as we will see in this chapter, have been proposed or are under study currently, but none has the characteristics to be used for the official quality control. The most promising way that EU is taking to certify quality and authenticity is to combine analytical determinations, data compliance, and reference and calibration materials, stored in a repository of validated data (JRC OLEUM databank), and to put this information in relation with volumes of olive oil produced and its geolocation. The intersection between quality control and traceability ad a "tip" to counteract fraud (Casadei et al., 2020), that seems nowadays futuristic, will be achieved, perhaps even sooner than we can expect, through the collection and certification of physical and temporal data, typical of a block chain scenario.

Fig. 56.1 depicts the overall approach to trace the origin and quality of high-quality oils.

56.2 Metabolomic and Genomic Approaches for the Traceability of Olive Oils

56.2.1 Metabolomic Approach

Metabolomic analyses are classified as targeted or untargeted. Targeted analyses are directed to the identification and quantification of a certain group of metabolites to assess their behavior under determined conditions. In contrast, untargeted metabolomics focuses on the detection of as many groups of metabolites as possible to obtain a fingerprint of a sample without necessarily identifying nor quantifying a specific compound (Cevallos-Cevallos et al., 2009). Based on the specific objectives, metabolomic studies can be classified as (a) discriminative, aiming to find differences among samples that can be used as markers for their classification, and (b) predictive, with the goal of creating a model to predict class memberships (Cevallos-Cevallos et al., 2009).

Discrimination is usually achieved by the use of multivariate data analysis: the simplest tool is principal component analysis (PCA), able of classifying the samples on the basis of the most variable compounds (Kemsley et al., 2007). Predictive studies create statistical models based on metabolite profile to predict the conformity of an unknown sample. One example is the morphotypes that allow the visual representation of the characteristics of a certain olive (Pinatel et al., 2014) (Fig. 56.1).

Fig. 56.1 Overview of the overall approach to assess the conformity of an EVOO or VOO with its label

56.2.1.1 Metabolomic Markers to Assess the Varietal and Geographical Origin of EVOOs

EVOOs have been extensively investigated with regard to fatty acids (FAs), triacylglycerols (TAGs), and phenol composition, which are greatly influenced by cultivar, environment, climate, agronomic techniques, degree of fruit ripening, harvest time, extraction technology, and storage (Baiano et al., 2013; García-González et al., 2009).

Metabolomic studies have been largely employed to find EVOO adulteration with low-quality oils (Bajoub et al., 2018). However, since 2000 they were used also to assess varietal composition and geographical origin of PDO (Protection Denomination Origin) and PGI (Protected Geographical Indication) oils. The projects OLIV-TRACK (http://www.dsa.unipr.it/foodhealth/oliv-track/index.html) and Foodintegrity (http://foodintegrity.eu) improved the portfolio of analytical determinations (Morin & Lees, 2018).

In general, these and other studies reported that metabolomics analyses are able of discriminating olive oils on the basis of cultivars e/o geographical origin, but there is a great difference, among the authors, about the metabolites with the highest power of discrimination. Moreover, the analytical platforms and statistical approach can greatly influence the results.

The variability in many compounds is related to two factors: genetics and environmental conditions of tree growth.

The composition of TAGs and FAs seems to be related to the cultivars. In fact, several authors found very little differences in TAGs and FAs among EVOOs obtained from the same cultivar grown in different geographical areas (Aranda et al., 2004; Bronzini de Caraffa et al., 2008; Galeano Diaz et al., 2005; Kritioti et al., 2018; Portarena et al., 2019; Skiada et al., 2020). However, the influence of orchard location on oil FA composition is not negligible: Bronzini de Caraffa et al. (2008) found significant differences in the content of linoleic acid and the corresponding TAGs (POL, OOL, OLL) in oils made with the same cultivar grown in different locations. Other authors found differences in the oleic and palmitic acid in EVOOs produced in different areas of Tunisia (Issaoui et al., 2010); in palmitic, palmitoleic, and linoleic acids in oils produced in different regions of Australia (Mailer et al., 2010); and in nine FAs in oils produced in different Turkish regions (Gourdeniz et al., 2008). In the last case, the content of FAs depends also on the year of harvest.

The content of secondary metabolites (phenols, sterols, tocopherols, etc.) is more influenced by environmental conditions than cultivar, because their production is modulated also in response to biotic and abiotic stresses.

Phenolic compounds play a key role in providing the healthy nutritional properties of olive oils, such as antioxidant activity and protection against cardiovascular disease (Granato et al., 2016). Six phenols (tyrosol, hydroxytyrosol, 3,4-DHPEA-EDA, 3,4-DHPEA-EA, *p*-HPEA-EDA, pinoresinol) and total phenolic compounds allowed a satisfactory separation of nine Spanish PDO EVOOs (Becerra-Herrera et al., 2018). With a high-throughput nontargeted approach, Ghisoni et al. (2019) found a great variability in phenolic compounds among 50 olive oils produced in Italy and in other areas of Mediterranean basin. Only 9.6% of phenols was common in the oils: 47% was related to the origin and 43.4% to the cultivar. Phenols were useful for the classification of Italian Sabina PDO oils (Nescatelli et al., 2014) and Tunisian and Italian oils (Mohamed et al., 2018).

Several authors found that the phenol content varies with the altitude of location in which fruits are collected (Issaoui et al., 2010), but this correlation was not confirmed by other authors. Some cultivars show a higher variation in phenol content among locations than others (Baiano et al., 2013). Also the ripening stage of fruits and malaxation during oil production greatly influence the phenol content (Diamantakos et al., 2020).

Phytosterols are nutritionally important lipids known for their anticancer preventive action. In olive oil, the content and composition of sterols are influenced by cultivar and agronomic and climatic conditions, fruit ripening, and processing. Sterols were useful in the discrimination of oils produced with cultivars grown in different areas: for example, they determined the separation of EVOOs from six locations in the western part of Turkey (Gumus et al., 2018), from six regions of Italy and other Mediterranean areas (Ghisoni et al., 2019), and from four Spanish municipalities (Sayago et al., 2019). Even though the sterol profiles are considered more influenced by environmental factors than TGAs and FAs (Montealegre et al., 2010), some studies have demonstrated that they are also related to the cultivars (Lukic et al., 2013; Matos et al., 2007; Skiada et al., 2020).

Tocopherols, of which vitamin E is the best-known component, are another class of lipids with remarking healthy properties. Their profile in olive oils is mainly influenced by olive cultivar (García-González & Aparicio, 2010; Rodrigues et al., 2020), but also the environmental factors, in particular the year of production, are relevant (Beltrán et al., 2010).

Beside the organic compounds, inorganic composition may be useful for olive oil classification. In particular, rare earth elements (vanadium, chrome, germanium, arsenic, selenium, thulium, lutetium, hafnium, and tantalum) are robust tracers of geographical origin of olive oil because they depend on the soil, and their concentrations do not change over time (Farmaki et al., 2012). By tagging 55 elements, Sayago et al. (2018) classified 125 oils of 3 Spanish olive oils according to their geographical origin.

56.2.1.2 Metabolomics for Assessing Sustainability of Olive Oil

The sustainability of agriculture is a question becoming more and more urgent, and at this purpose, metabolomics may be useful also to define the nutritional sustainability index (SI) of a certain olive oil. SI derives from two macro-indicators: the first, called "business distinctiveness," considers the conformity to regulations and standards in terms of quality, traceability, and origin of raw materials; the second, called "nutritional quality," considers the contents of compounds like saturated, monounsaturated, and polyunsaturated FAs, vitamin E, polyphenols, oleic acid, etc. (Azzini et al., 2017). For each indicator a scale from 0 (unsustainable) to 10 (very sustainable), with 5 as a sustainability benchmark value, was set: for example, the high content of saturated FAs means a low SI, while a high content of oleic and linoleic acid means a higher SI.

As will be discussed in detail below, the production of organic EVOOs has been favorably accepted by consumers; therefore, instruments for distinguishing organic and conventional oils are necessary. A study conducted over 3 years showed that organic oils contain a lower concentration of pesticide fenthion and dimethoate than conventional ones (Tsatsakis et al., 2003).

Regarding the nutritional value, significant differences in tocopherol and phenol contents, as well fatty acid composition, were found between oils from organically and conventionally cultivated Spanish olives, but the authors remark that the influence of cultivar and the ripening stage of olives is as stronger as the way of cultivation (Jimenez et al., 2014). Significantly higher levels of phenols (Anastasopoulos et al., 2011; López-Yerena et al., 2019) and volatile compounds were found in organic EVOOs (Jiménez et al., 2017). In contrast, the determination of 13 phenolic compounds in 55 monovarietal EVOOs indicated the flavonoid luteolin as the only discriminative marker for organic EVOOs (Kalogiouri et al., 2017). Other studies did not find a significant or reproducible difference in the content of phenols and tocopherols between organic and conventional oils (Dolgun et al., 2010; Ninfali et al., 2008).

56.2.2 Genomic Approach

The genetics of olive (*Olea europaea* L.) has been largely studied. A draft of its genome was published in 2016 (Cruz et al., 2016), and an improved version was recently presented (Julca et al., 2020). The genome that has managed to anchor the sequences to chromosomes is that of the wild variety (Unver et al., 2017), but it is still incomplete and lacks genes; in addition it has been seen that it was not a true wild one but a feral olive tree (Julca et al., 2020). The most complete genome so far is that of Picual (Ramirez-Tejero et al., 2020).

These studies made available a large number of molecular markers which can identify olive cultivars: random amplified polymorphic DNA (RAPDs), amplified fragment length polymorphisms (AFLPs), sequence-characterized amplified regions (SCARs), microsatellites (or SSRs), inter-simple sequence repeats (ISSRs), and single nucleotide polymorphisms (SNPs) (Sebastiani & Busconi, 2017). The development of pyrosequencing (next-generation sequencing) allowed the discovery of a large number of SNPs, associated with useful agronomic traits (Kaya et al., 2019) or tissue transcriptome (Carmona et al., 2015). Some data collected in the studies were transferred in these databases: Istrian Olive Database (http://old.iptpo.hr/iod/index.php?option=com_content&task=view&id=15&Itemid=30); Olea SSRs database (http://www.oleadb.it/ssr/ssr_dca_search.php); and ReprOlive, containing the transcriptome of pollen and stigma (http://reprolive.eez.csic.es/olivodb/) (Carmona et al., 2015).

This dramatic increase of genomic and transcriptomic data constitutes a mine of tools for the application of "food genomics," the comparison of the sequence of DNA traits recovered from food with that of raw material, to olive oil. Food genomics was developed after the PCR development, which allows the amplification of a very small quantity of degraded DNA.

56.2.2.1 DNA Extraction from Olive and Non-olive Oils

At the beginning of the twenty-first century, the extraction of DNA from olive oil was considered impossible, but the results of the project OLIV-TRACK and numerous publications demonstrated its feasibility (Agrimonti et al., 2011; Agrimonti & Marmiroli, 2019; Pasqualone et al., 2016).

The methods for extracting DNA from vegetable oils may be classified in two principal classes: those using cetyl trimethylammonium bromide (CTAB) (Doyle and Doyle, 1987) and those employing commercial kits (Agrimonti et al., 2011; Agrimonti & Marmiroli, 2019; Costa et al., 2012). In general, CTAB-based methods are more economic, but are time-consuming and require a large volume of sample (about 50–100 ml). Despite their higher costs, kits are the most employed methods because they utilize a smaller volume of oils, are of rapid execution, and do not require toxic reagents like phenol and chloroform. The kits most employed are QIAamp Stool (Qiagen), NucleoSpin Food (Macherey-Nagel), and Promega Wizard Food (Promega).

Other methods are based on hexane extraction (Consolandi et al., 2008) or Tris EDTA buffer (de La Torre et al., 2004). Two papers described the direct PCR amplification of oil after its filtration through Hybond membrane (Chiappetta et al., 2017; Muzzalupo et al., 2015). Carvalho et al. (2018) experimented, with satisfactory results, with a washable and reusable miniaturized device in which olive oil DNA is extracted and purified through a microglass fiber filter.

So far, a universal method for extracting DNA from vegetable oils has not been adopted, and there is a great variability among the authors concerning the methods, yields, and initial volume of sample. This last parameter is important because working with low volumes ($\leq$ 2 ml) allows a significant saving of times (avoiding long centrifugations); moreover, the utilization of small reagent volumes reduces the environmental impact of the analysis. For these reasons the researches evolved in this direction, and recent papers showed that it is possible to extract DNA with CTAB-hexane-chloroform method using 0.5–0.6 ml of oil (Kalogianni et al., 2015; Raieta et al., 2015).

In general, the amount of DNA recovered is very low, hardly measurable with UV spectrophotometry, and not visible on agarose gel electrophoresis. The only exception is reported by Raieta et al. (2015) who were able to visualize DNA extracted with a CTAB method on agarose gel and purify it with electro-elution.

56.2.2.2 Molecular Markers for Detecting Frauds in Olive Oil

The first studies on olive oil DNA, started in 2002–2003, were based on end point PCR, carried on until the maximum number of molecules is obtained and amplified DNA (amplicons) are visualized on agarose or acrylamide gel electrophoresis. Differently, real-time PCR (qPCR) monitors the amplification during its occurrence by means of continuous fluorescence detection. qPCR offers the possibility of quantifying template DNA, exploiting the inverse relationship between threshold cycle (CT), when the fluorescence is significantly above the background, and the initial amount of DNA. qPCR may be conducted through hydrolyzable fluorogenic probes (Holland et al., 1991) or through double-strand intercalating dye, like SYBR Green (Ponchel et al., 2003).

56.2.2.3 Adulteration with Non-olive Oils

A frequent fraud occurring during the olive oil production is related to the addition of cheaper oils obtained from other species such as hazelnut (*Corylus avellana*), soya (*Glycine max*), almond (*Prunus dulcis*), maize or corn (*Zea mays*), sunflower

(*Helianthus annuus*), and sesame (*Sesamum indicum*). The identification of adulterant oils is mainly based on barcoding DNA, which exploits the variability within a target sequence for species differentiation (Hebert et al., 2003). Plastidial sequences are mostly utilized: in particular, the high polymorphic chloroplast region trnL (UAA) is able of discriminating a large number of oil-producing species (Spaniolas et al., 2008; Uncu et al., 2017). The amplification of this region, associated to capillary electrophoresis, was able of detecting adulterant species in 11 mixes of oils with a limit of detection of 5% (Uncu et al., 2017).

High-resolution melting PCR (HRM-PCR), able of discriminating until one single nucleotide variation, is often utilized for the detection of adulterant oils. Targeting by HRM the plastidial gene rbcL, Ganopoulos et al. (2013) distinguished olive, sunflower, sesame, soya, canola, and corn oils. Vietina et al. (2013) detected adulterant oils of maize, sunflower, and hazelnut mixed to olive oil in percentages as low as 10% targeting plastidial and nuclear sequences. SNP variations in noncoding spacer region between psbA-trnH and partial coding region of matK of plastid genome were exploited to detect canola and sunflower oils added to olive oils in percentages of 5% (Kumar et al., 2011).

56.2.2.4 Identification of Olive Cultivars Through Molecular Markers

Single-locus SSRs are the most employed markers for traceability of food matrices, because, being rather short, they can be amplified also in degraded DNA. They are repetitions of 2–6 nucleotides (Tautz, 1989) with a length related to the number of repetitions. SSRs isolated in olive were largely used to compare their profile in leaves and oil DNA (Agrimonti et al., 2011; Pasqualone et al., 2016). Some authors (Corrado et al., 2011; Martins-Lopes et al., 2008; Muzzalupo et al., 2015; Pasqualone et al., 2004) reported a good correspondence between monovarietal oil and leaves and drupe SSR profile, while others found several differences (Alba et al., 2009; Ayed et al., 2009; Vietina et al., 2011). After a reviewing of literature, Agrimonti et al. (2011) concluded that correspondence between monovarietal oils and leaves amounted to 51%. A common difference consisted in additive alleles in oil, due to a possible contamination during the milling process and/or to seed embryo that also contains alleles of the pollinator (Doveri et al., 2006). Another difference is a loss of one allele in oil, probably due to the low amount of DNA, which makes it difficult to catch two alleles simultaneously. Differences in allele size can be due also to intravarietal difference among trees (Lopes et al., 2004; Muzzalupo et al., 2009; Pasqualone et al., 2007).

Chloroplast DNA (cpDNA) is more abundant and resistant to degradation than the nuclear DNA, and the pollinator does not interfere with the analysis, but its level of polymorphism is low and can be increased by combining different alleles (Pérez-Jiménez et al., 2013). Recently, Gomes et al. (2018) targeted SSRs to find the botanical origin in commercial PDO oils with success.

Multilocus markers, RAPD, ISSRs, and AFLPs, are largely informative because they monitor several loci with a single PCR reaction. RAPDs have low cost and difficulty of execution, but their profiles, consisting of no more than 10–12 bands, are poorly informative and reproducible. In fact, the fingerprinting of 23 oils with 11 RAPD primers gave a polymorphism of 78%, and only 2 primers gave reproducible patterns. Better results were obtained with ISSRs, both in terms of reproducibility and polymorphism which was 83% (Martins-Lopes et al., 2008).

AFLP markers are more laborious and expensive than RAPDs but more reproducible and informative because their profiles consist of 50–60 bands in a range of 50–400 bp differing up to a single nucleotide, allowing the detection of a large number of polymorphisms (Vos et al., 1995).

AFLPs were used to assess varietal composition of olive oils with different results. Busconi et al. (2003) found a good correspondence between monovarietal oil and leaf AFLP profiles, while Pafundo et al. (2005) found that oil profiles share no more than 70% of fragments with leaves. Moreover, oil profiles have less band and less intensity, and fragments were no longer than 250 bp. On the contrary Montemurro et al. (2008) observed that all bands of oil profiles correspond to those of leaves, and fragments were long up to 350 bp.

Specific fragments derived from AFLP or RAPD profiles can be converted into SCAR markers that have the same characteristics of a single locus marker. Busconi et al. (2006) developed SCAR markers, able of unequivocally discriminating up to 26 varieties of olive, but only few applications of this kind of markers are reported in oil (De la Torre et al., 2004). Pafundo et al. (2007) isolated a SCAR marker directly from a monovarietal oil AFLP profile. Even though the discrimination power is rather low, this work proves that it is possible to isolate SCAR markers directly from oil.

SNP markers are single mutations densely dispersed in the genome of living organisms with a frequency of around 1%, which permits to distinguish even very similar individuals. Moreover, they can be detected within short amplicons and therefore are useful to trace degraded DNA.

In olive SNPs have been isolated in genes involved in metabolic pathways (Consolandi et al., 2007; Reale et al., 2006) or in restriction sites (Bazakos et al., 2012; Uncu et al., 2015). Moreover, the transcriptome analysis allowed the identification of 3384 SNPs across 123 individuals of F1 progeny of Turkey cultivars Gemlik and Edincik Su (Ipek et al., 2017) and 2987 SNPs in 96 genotypes of Turkey trees (Kaya et al., 2013). All SNPs constitute a rich toolbox for varietal differentiation and identification in olive oils. Consolandi et al. (2007) exploited 17 SNPs to genotype a panel of 49 olive varieties with 17 SNPs. Out of the 13 amplicons containing these SNPs, 12 were successfully amplified from oil DNA with full concordance with leaves. The potentiality of SNPs for traceability of olive oil was demonstrated also through PCR-RFLP that detected the olive varieties in a mix of two monovarietal oils with a limit of detection of 10% (Bazakos et al., 2016). A similar approach allowed finding single varieties in olive oil mixes with a limit of detection of 20% (Uncu et al., 2018).

An innovative platform is based on microsphere tagging a high number of SNPs with specific fluorescence analyzed with a flow cytometer, allowing the identification of olive cultivar in olive oil (Kalogianni et al., 2015).

The third generation of PCR (ddPCR) enumerates positive amplifications in a droplet emulsion sprayed with DNA. ddPCR allows the absolute quantification of the DNA target, using end point PCR and Poisson statistics, without the need for external/internal controls or standard curves (Hindson et al., 2011). A work demonstrated that ddPCR has a lower limit of detection than qPCR and a similar accuracy in detecting olive oil DNA (Scollo et al., 2016), but, so far, it requires special expensive apparatus and software.

56.3 New and Improved Analytical Methods Addressed to the Protection of Olive Oil Authenticity: Possible Solutions Proposed to Verify the Conformity to Information Reported on the Label

The European Regulation 29/2012 (published as consolidated text on 6 February 2019) (Commission Implementing Regulation (EU), 2012, N. 29/2012) represents the relevant legislation concerning the information that must or can be reported on the label of olive oil (Fig. 55.1), and, as cited in *Article 1*, "*...lays down specific standards for retail-stage marketing of the olive oils and olive-pomace oils retail stage" where "retail means the sale to the final consumer of oil....*"

In the specific of this Regulation are reported indications for:

(i) Type and capacity of packaging (*Article 2*).
(ii) Description on the oil category, specifically extra virgin olive oil, virgin olive oil, olive oil, olive-pomace oil (*Article 3*).
(iii) Designation of origin (*Article 4*), special preservation conditions in terms of light, heat, etc. (*Article 4a*), mandatory particulars that have to be described in the principal field of vision of label (*Article 4b*).
(iv) Optional indications which may appear as referred to: (a, b) kind of oil extraction system applied (e.g., "first cold pressing" or "cold extraction), (c) organoleptic properties referring to taste and/or smell only if based on the results of an assessment carried out performing the official method, (d) indications related to specific chemical parameters (e.g., free acidity plus peroxide value, wax content, and ultraviolet absorptions), and (e) harvesting year in case of 100% of the contents of the container comes from that harvest (Article 5).
(v) Indication related to blends of olive oil and other vegetable oils (Article 6).
(vi) Address of manufacturer, packer, or seller (Article 7).

With regard to Article 7 in the EU Reg. 29/2012, it is reported that "*...at the request of the Member State in which the address of the manufacturer, packer or seller appearing on the labelling is located, the party concerned shall supply documentation in support of the indications referred to in Articles 4, 5 and 6, based on one or more of the following elements (a) factual elements or scientifically established facts (b) results of analyses or automatic recordings taken on representative samples (c) administrative or accounting information kept in accordance with*

Union and/or national rules...." Despite the fact that results of analyses can be requested, the documental traceability represents still, in many cases, the unique way to verify the conformity to some declarations on labels. In particular, at present, no designated analytical methods are available to verify the compliance with obligations defined in *Article 4* and *Article 6*. As it is possible to understand, this kind of verification of conformity plays a key role in protecting consumers from misleading and in detecting and preventing fraudulent practices in the olive oil sector. For this reason, in this paragraph, a brief examination of these two key aspects is presented.

According to *Article 4*, EVOOs and VOOs must bear the geographical designation of origin on the label, where "designation of origin" means reference to a geographical area on the label. The Regulation specifies references for the following cases:

(a) A reference to the Member State, to the Union, or to the third country in the case of olive oils originating from one Member State or third country;
(b) In the case of blends of olive oils originating from more than one Member State or third country, one of the following mentions is appropriate:
    (i) "Blend of olive oils of European Union origin" or a reference to the Union
    (ii) "Blend of olive oils not of European Union origin" or a reference to origin outside the Union
    (iii) "Blend of olive oils of European Union origin and not of European Union origin" or a reference to origin within the Union and outside the Union.
(c) A reference to a protected designation of origin (PDO) or a protected geographical indication (PGI) in accordance with the provisions of the product specification concerned.

[...*the designation of origin mentioning a Member State or the Union shall correspond to the geographical area in which the olives concerned were harvested or in which the mill where the oil was extracted from the olives is situated...*] [...*if the olives have been harvested in a Member State or third country other than that in which the mill where the oil was extracted from the olives is situated, the designation of origin shall contain the following wording: '(extra) virgin olive oil obtained in (the Union or the name of the Member State concerned) from olives harvested in (the Union or the name of the Member State or third country concerned)*...].

Although the large number of studies performed on this topic of geographical origin traceability (Bontempo et al., 2019; Conte et al., 2019; Peršurić et al., 2018; Sayago et al., 2018), a shared and validated analytical method is not still prompt for a possible introduction in the official legislation. In the course of the OLEUM project ("*Advanced solutions for assuring authenticity and quality of olive oil at global scale*", funded by the European Commission within the Horizon 2020 Program, grant agreement no. 635690), a specific large set of samples was collected taking into consideration oils that came from a single European Union Member State or third country or blends of European Union or non-European Union origin. Many different analytical approaches, based on spectroscopic (e.g., FTIR, NIR, NMR), isotopic, and chromatographic techniques able to realize profiling or fingerprinting of specific major or minor fractions of olive oil components, were tested. Some interesting results are coming from this very articulated analytical work, and among these, two were recently published: In Quintanilla-Casas et al. (2020), the considered markers were the semi-volatile sesquiterpene compounds analyzed by SPME-GC-MS coupled with a specific chemometric elaboration of data. Under the fingerprinting approach, peak identification and quantitation steps are not necessary, permitting to save time and to use all information on specific sesquiterpene ions contained in a specific region of the trace. Classification models by partial least square discriminant analysis (PLS-DA) evidenced a higher percentage of correct classification for untargeted (fingerprinting) than for targeted (profiling) approach. In Palagano et al. (2020), an approach solely untargeted by a flash gas chromatography (FGC) was suggested by the authors. FGC is a technique that permits to obtain the GC trace of the headspace of a sample in only 2 min. Two chemometric approaches for data elaboration were performed: one linear by PLS-DA and one nonlinear by artificial neural network (ANN). The results are satisfactory for both approaches in terms of percentage of correct classified samples but with a higher performance in the case of application of ANN strategy.

According to *Article 6*, it is possible to produce and commercialize in EU countries legal blends between OO or EVOO and vegetable olive oils. At this regard, it has to be underlined that some national exceptions are possible as in the case of Spain and Italy. In fact, in the Regulation it is specified that "...*Member states may prohibit the production in their territory of blends for internal consumption, however they may not prohibit the marketing on their territory of such blends coming from other countries and they may not prohibit the production on their territory of the blends for marketing in another Member state for exportation....*" Concerning the legal blends, the Regulation highlights that if the producer aims to advertise the presence of OO or EVOO on the front label, it has to account for more than 50% of the blend "...*where the presence of oils in a blend of olive oil and other vegetable oils is highlighted on the labelling elsewhere than in the list of ingredients, using words, images or graphics, the blend concerned must bear the following trade description: 'Blend of vegetable oils (or the specific names of the vegetable oils concerned) and olive oil', directly followed by the percentage of olive oil in the blend....*"

To fight against the possible fraud linked to the fact that, currently, no designated analytical method is available to verify the compliance with the above cited obligation (Conte et al., 2019), in the framework of a specific task of the OLEUM project, researchers set up a target method based on the elaboration of a decision tree that fixes minimum contents of specific typical major and minor compounds of OO and EVOO. One of the seed oils more frequently used in this type of declared blends is sunflower oil, of the normal type or high in oleic, and therefore a focused sampling was realized considering the need to consider the different percentages of OO or EVOO in the blends (e.g., 60:40, 50:50, and 40:60 v/v proportions) and also commercial products present on the market (e.g., sampling realized from the Swedish market). Specific limits (palmitoyl diolein above 11–15%, ß/γ-tocopherol ratio below 2.4, alkane sum C21–C25 higher than 3.5–6%, total sterol content lower than

2400 mg/kg) were proposed to guarantee the presence of olive oil or extra virgin olive oil at 50% in these kind of blends, and results were just published (Gómez-Coca et al., 2020).

Over the last decades the demand for organic food products has been growing, and the development of organic farming in Europe is one of the factors that inspired the 2020 Action Plan of the European Commission ("*Farm to Fork Strategy. For a fair, healthy and environmentally-friendly food system*"). In fact, it is possible to catch in this document the intention for a further promotion of the organic farming, reaching the objective of at least 25% of the EU's agricultural land under organic farming by 2030. Among the key factors that influence the consumer to purchase organic food in place of conventional products are the awareness to protect the environment, the need for a healthier food, and a higher perception of food quality and food safety. On the other hand, the consumer diffidence about the authenticity of labeling or certification of organic food has been recognized as one of the factors that limits the consumption of organic food (Suciu et al., 2019; Vega-Zamora et al., 2018). In a review published in 2006 (Györéné et al., 2006), the authors provided an overview on the beneficial versus harmful components found in organic and conventional crops on the basis of data obtained in international comparative surveys. A higher content of antioxidant molecules belonging to different chemical classes as well as a lower contamination due to pesticide residues, nitrates and some heavy metals, generally resulted in the characterization of organic crops compared with conventional ones.

The EU Regulation 834/2007 is the legislative reference for the production and labelling of organic foods. Despite organic products for which the EU-logo is shown on the label (Fig. 56.1) are certified by recognized inspection bodies, cases of fraud are sometimes discovered. The illicit activities in this sector are driven by the higher price of organic products than conventional ones. In a recent work, researchers (van Ruth & Pagter-de Witte, 2020) investigated fraud vulnerability associated to four different organic product chains (banana, egg, olive oil, pork) evidencing (e.g., level of fraud detectability) the perception of high vulnerability for actors in olive oil supply chain in terms of economic drivers and technical opportunities.

The case of EVOO or VOO labelled as organic can be considered easier than other processed food products: in fact, according to the description of specific categories reported in the EU Reg. 29/2012, they are "*…obtained directly from olives and solely by mechanical means….*"; (Council Regulation (EC) N. 834/2007) therefore differences between organic and conventional oils are limited to those of olives as prime material (the addition of other ingredients or additives is not allowed as well as the application of technological processes/conditions that can alter the natural composition of the oil) and are related to the activities at the agronomical level applied for olive farming.

Recently, some researchers (Rodríguez Pleguezuelo et al., 2018) discussed the current situation of organic olive farming in Andalusia, which is well known as the more productive Spanish region, analyzing benefits (e.g., prevention of soil erosion, higher soil nutrient content, protection of ecosystem services) and consequences (e.g., long period for conversion, higher costs, difficulties in finding affordable effective inputs with relation to limitation/prohibition in use of pesticides and herbicides) of the shift from conventional to organic management of olive orchards, as well as the factors that drive the decision of consumer for acquiring organic olive oil.

Studies related to the compositional aspects of organic VOOs have, in some cases, highlighted significant differences in minor components having nutritional and healthy impact (e.g., higher concentrations of tocopherols and polar phenols) with respect to conventional ones. At this regard, in a work of Gutierrez et al. (1999), olives of Picual variety cultivated using both organic and conventional systems and harvested at various stages of ripeness were processed and the respective VOOs analyzed. Researchers evidenced a superior quality of organic oil when compared with conventional one, at least for the sample produced at one specific ripeness index, in terms of tocopherols and phenolic compounds as well as stability to oxidation and sensory scores. More recently, Jiménez et al. (2014) studied the characteristics of virgin olive oils from Picual and Hojiblanca olives (among the most common varieties in Andalusia), also in this case harvested at different ripening stages and grown using organic or conventional methods. Authors pointed the attention on the fact that the variety and the ripening stage can produce a stronger effect on the compositional parameters than the kind of farming system (organic or conventional). According to other works (Dolgun et al., 2010; Ninfali et al., 2008), and as reported in paragraph 56.2.1.2, the content of phenols and tocopherols did not result as significantly different in organic and conventional EVOOs.

As already discussed in the previous paragraph, few works compared pesticide and herbicide residuals in organic and conventional oils, but the data cannot be considered sufficient to standardize a discrimination method.

Summarizing, at present, in order to verify the compliance with rules for organic farming, the main instruments are represented by documental traceability and analysis of pesticide and herbicide residuals.

Acknowledgments All authors acknowledge funding from the SUSTAINOLIVE project, PRIMA programme supported by the European Union, contract n. 1811. Authors NM, CA, and EM acknowledge the FOODINTEGRITY project, Seventh Framework Programme n. 613688. Authors AB and TGT acknowledge the OLEUM project, Horizon 2020 n. 635690.

Conflict of Interest The authors have declared no conflict of interest.

References

Agrimonti, C., & Marmiroli, N. (2019). Food genomics for the characterization of PDO and PGI virgin olive oils. *European Journal of Lipid Science and Technology, 121*, 1800132.

Agrimonti, C., Vietina, M., Pafundo, S., & Marmiroli, N. (2011). The use of food genomics to ensure the traceability of olive oil. *Trends in Food Science and Technology, 22*, 237–244.

Alba, V., Sabetta, W., Blanco, A., Pasqualone, A., & Montemurro, C. (2009). Microsatellite markers to identify specific alleles in DNA extracted from monovarietal virgin olive oils. *European Food Research and Technology, 229*, 375–382.

Anastasopoulos, E., Kalogeropoulos, N., Kaliora, A. C., Kountouri, A., & Andrikopoulos, N. K. (2011). The influence of ripening and crop year on quality indices, polyphenols, terpenic acids, squalene, fatty acid profile, and sterols in virgin olive oil (Koroneiki cv.) produced by organic versus non-organic cultivation method. *International Journal of Food Science & Technology, 46*, 170–178.

Aranda, F., Gomez-Alonso, S., Rivera del Alamo, R. M., Salvador, M. D., & Fregapane, G. (2004). Triglyceride, total and 2-position fatty acid composition of Cornicabra virgin olive oil: Comparison with other Spanish cultivars. *Food Chemistry, 86*, 485–492.

Ayed, R. B., Grati-Kamoun, N., Moreau, F., & Rebaï, A. (2009). Comparative study of microsatellite profiles of DNA from oil and leaves of two Tunisian olive cultivars. *European Food Research and Technology, 229*, 757–762.

Azzini, E., Maiani, G., Turrini, A., et al. (2017). The health-nutrition dimension: A methodological approach to assess the nutritional sustainability of typical agro-food products and the Mediterranean diet. *Journal of the Science of Food and Agriculture, 98*, 3684–3705.

Baiano, A., Terracone, C., Viggiani, I., & Del Nobile, A. M. (2013). Effects of cultivars and location on quality, phenolic content and antioxidant activity of extra-virgin olive oils. *Journal of the American Oil Chemists' Society, 90*, 103–111.

Bajoub, A., Bendini, A., Fernández-Gutiérrez, A., & Carrasco-Pancorbo, A. (2018). Olive oil authentication: A comparative analysis of regulatory frameworks with especial emphasis on quality and authenticity indices, and recent analytical techniques developed for their assessment. A review. *Critical Reviews in Food Science and Nutrition, 58*, 832–857.

Barbieri, S., Brkić Bubola, K., Bendini, A., Buar-Miklavĉiĉ, M., Lacoste, F., Tibet, U., Winkelmann, O., García-González, D. L., & Gallina Toschi, T. (2020). Alignment and proficiency of virgin olive oil sensory panels: The OLEUM approach. *Food, 9*, 355.

Bazakos, C., Dulger, A. O., Uncu, A. T., Spaniolas, S., Spano, T., & Kalaitzis, P. (2012). A SNP-based PCR-RFLP capillary electrophoresis analysis for the identification of the varietal origin of olive oils. *Food Chemistry, 134*, 2411–2418.

Bazakos, C., Khanfir, E., Aoun, M., Spano, T., Zein, Z. E., Chalak, L., Riachy, M. E., Abou-Sleymane, G., Ali, S. B., Grati Kammoun, N., & Kalaitzis, P. (2016). The potential of SNP-based PCR-RFLP capillary electrophoresis analysis to authenticate and detect admixtures of Mediterranean olive oils. *Electrophoresis, 37*, 1881–1890.

Becerra-Herrera, M., Vélez-Martínc, A., Ramos-Merchante, A., Richter, P., Beltrána, R., & Sayago, A. (2018). Characterization and evaluation of phenolic profiles and color as potential discriminating features among Spanish extra virgin olive oils with protected designation of origin. *Food Chemistry, 241*, 328–337.

Beltrán, G., Jiménez, A., Del Rio, C., Sánchez, S., Martínez, L., Uceda, M., & Aguilera, M. P. (2010). Variability of vitamin E in virgin olive oil by agronomical and genetic factors. *Journal of Food Composition and Analysis, 23*, 633–639.

Bontempo, L., Paolini, M., Franceschi, P., Ziller, L., García-González, D. L., & Camin, F. (2019). Characterisation and attempted differentiation of European and extra-European olive oils using stable isotope ratio analysis. *Food Chemistry, 276*, 782–789.

Bronzini de Caraffa, V., Gambotti, C., Giannettini, J., Maury, J., Berti, L., & Gandemer, G. (2008). Using lipid profiles and genotypes for the characterization of Corsican olive oils. *European Journal of Lipid Science and Technology, 110*, 40–47.

Busconi, M., Foroni, C., Corradi, M., Bongiorni, C., Cattapan, F., & Fogher, C. (2003). DNA extraction from olive oil and its use in the identification of the production cultivar. *Food Chemistry, 83*, 127–134.

Busconi, M., Sebastiani, L., & Fogher, C. (2006). Development of SCAR markers for germplasm characterisation in olive tree (Olea europea L.). *Molecular Breeding, 1*, 59–68.

Carmona, R., Zafra, A., Seoane, P., Castro, A. J., Guerrero-Fernández, D., Castillo-Castillo, T., Medina-García, A., Cánovas, F. M., Aldana-Montes, J. F., Navas-Delgado, I., Alché, J. D., & Claros, M. G. (2015). ReprOlive: A database with linked data for the olive tree (Olea europaea L.) reproductive transcriptome. *Frontiers in Plant Science, 6*, 625.

Carvalho, J., Puertas, G., Gaspar, J., Azinheiro, S., Diéguez, L., Garrido-Maestu, A., Manuel Velázquez, M., Barros-Velázquez, J., Cardoso, S., & Prado, M. (2018). Highly efficient DNA extraction and purification from olive oil on a washable and reusable miniaturized device. *Analytica Chimica Acta, 1020*, 30–40.

Casadei, E., Valli, E., Panni, F., Donarski, J., Farrús Gubern, J., Lucci, P., Conte, L., Lacoste, F., Maquet, A., Brereton, P., Bendini, A., & Gallina Toschi, T. (2020). Emerging trends in olive oil fraud and possible countermeasures. *Food Control, 124*, 107902.

Cevallos-Cevallos, J. M., Reyes-De-Corcuera, J. I., Etxeberria, E., Danyluk, M. D., & Rodrick, G. E. (2009). Metabolomic analysis in food science: A review. *Trends in Food Science & Technology, 20*, 557–566.

Chiappetta, A., Muto, A., Muzzalupo, R., & Muzzalupo, I. (2017). New rapid procedure for genetic characterization of Italian wild olive (Olea europaea) and traceability of virgin olive oils by means of SSR markers. *Scientia Horticulturae, 226*, 42–49.

Commission Implementing Regulation (EU). (2012). No 29/2012 of 13 January 2012 on marketing standards for olive oil. Official Journal of the European Union, 55, 14 January 2012, L12, 1–21 (current consolidated version on 6 February 2019).

Consolandi, C., Palmieri, L., Doveri, S., Maestri, E., Marmiroli, N., Reale, S., Lee, D., Baldoni, L., Tosti, N., Severgnini, M., De Bellis, G., & Castiglioni, B. (2007). Olive variety identification by ligation detection reaction in a universal array format. *Journal of Biotechnology, 129*, 565–574.

Consolandi, C., Palmieri, L., Severgnini, M., Maestri, E., Marmiroli, N., Agrimonti, C., Baldoni, L., Donini, P., De Bellis, G., & Castiglioni, B. (2008). A procedure for olive oil traceability and authenticity: DNA extraction, multiplex PCR and LDR–universal array analysis. *European Food Research and Technology, 227*, 1429–1438.

Conte, L., Bendini, A., Valli, E., Lucci, P., Moret, S., Maquet, A., Lacoste, F., Brereton, P., García-González, D. L., Moreda, W., & Gallina Toschi, T. (2019). Olive oil quality and authenticity: A review of current EU legislation, standards, relevant methods of analyses, their drawbacks and recommendations for the future. *Trends in Food Science and Technology, 105*, 483–493.

Corrado, G., Imperato, A., La Mura, M., Perri, E., & Rao, R. (2011). Genetic diversity among olive varieties of southern Italy and the traceability of olive oil using SSR markers. *The Journal of Horticultural Science and Biotechnology, 86*, 461–466.

Costa, J., Mafra, I., Beatriz, M., & Oliveira, P. P. (2012). Advances in vegetable oil authentication by DNA-based markers. *Trends in Food Science and Technology, 26*, 43–55.

Council Regulation (EC) No 834/2007 of 28 June 2007 on organic production and labelling of organic products and repealing Regulation (EEC) No 2092/91. *Official Journal of the European Union*, L 189/1–23 (current consolidated version on 1 July 2013).

Cruz, F., Julca, I., Gómez-Garrido, J., Loska, D., Marcet-Houben, M., Cano, E., Galán, B., Frias, L., Ribeca, P., Derdak, S., Gut, M., Sánchez-Fernández, M., García, J. L., Gut, I. G., Vargas, P., Alioto, T. S., & Gabaldón, T. (2016). Genome sequence of the olive tree, *Olea europaea*. *GigaScience, 5*. https://doi.org/10.1186/s13742-016-0134-5

De la Torre, F., Bautista, R., Cánovas, F. M., & Claro, M. G. (2004). Isolation of DNA from olive oil and oil sediments / application in oil fingerprinting. *Food Agriculture and Environment, 2*, 84–89.

Diamantakos, P., Giannara, T., Skarkou, M., Melliou, E., & Magiatis, P. (2020). Influence of harvest time and Malaxation conditions on the concentration of individual phenols in extra virgin olive oil related to its healthy properties. *Molecules, 25*, 2449. https://doi.org/10.3390/molecules25102449

Dolgun, O., Ozkan, G., & Erbay, B. (2010). Comparison of olive oils derived from certified organic and conventional methods. *Asian Journal of Chemistry, 22*, 2339–2384.

Doveri, S., Sullivan, D., & Lee, J. (2006). Non-concordance between genetic profiles of olive oil and fruit: A cautionary note to the use of DNA markers for provenance testing. *Journal of Agricultural and Food Chemistry, 54*, 9221–9226.

Doyle, J. J., & Doyle, J. L. (1987). A rapid isolation procedure for small quantities of fresh leaf tissues. *Phytochemical Bulletin, 19*, 11–15.

European Commission Areté Research & Consulting in Economics. (2020). *Study on the implementation of conformity checks in the olive oil sector throughout the European Union.*
Farmaki, E. G., Thomaidis, N. S., Minioti, K. S., Ioannou, E., Georgiou, C. A., & Efstathiou, C. E. (2012). Geographical characterization of greek olive oils using rare earth elements content and supervised chemometric techniques. *Analytical Letters, 458*, 920–932.
Galeano Diaz, T., Duran Meras, I., Sanchez Casas, J., & Alexandre Franco, M. F. (2005). Characterization of virgin olive oils according to its triglycerides and sterols composition by chemometric methods. *Food Control, 16*, 339–347.
Gallina Toschi, T., Bendini, A., Lozano-Sánchez, J., Segura-Carretero, A., & Conte, L. (2013). Misdescription of edible oils: Flowcharts of analytical choices in a forensic view. *European Journal of Lipid Science and Technology, 115*, 1205–1223.
Ganopoulos, I., Bazakos, C., Madesis, P., Kalaitzis, P., & Tsaftaris, A. (2013). Barcode DNA high-resolution melting (Bar-HRM) analysis as a novel close-tubed and accurate tool for olive oil forensic use. *Journal of the Science of Food and Agriculture, 93*, 2281–2286.
García-González, D. L., & Aparicio, R. (2010). Research in olive oil: Challenges for the near future. *Journal of Agricultural and Food Chemistry, 58*, 12569–12577.
García-González, D. L., Luna, G., Morales, M. T., & Aparicio, R. (2009). Stepwise geographical traceability of virgin olive oils by chemical profiles using artificial neural network models. *European Journal of Lipid Science and Technology, 111*, 1003–1013.
Ghisoni, S., Lucini, L., Angilletta, F., Rocchettia, G., Farinelli, D., Tombesi, S., & Trevisan, M. (2019). Discrimination of extra-virgin-olive oils from different cultivars and geographical origins by untargeted metabolomics. *Food Research International, 121*, 746–753.
Gomes, S., Breia, R., Carvalho, T., Carnide, V., & Martins-Lopes, P. (2018). Microsatellite high-resolution melting (SSR-HRM) to track olive genotypes: From field to olive oil. *Journal of Food Science, 83*, 2415–2423.
Gómez-Coca, R. B., Pérez-Camino, M. C., Martínez-Rivas, J. M., Bendini, A., Gallina Toschi, T., & Moreda, W. (2020). Olive oil mixtures. Part one: Decisional trees or how to verify the olive oil percentage in declared blends compounds. *Food Chemistry, 315*, 126235.
Gourdeniz, G., Ozen, B., & Tokatli, F. (2008). Classification of Turkish olive oils with respect to cultivar, geographic origin and harvest year, using fatty acid profile and mid-IR spectroscopy. *European Food Research and Technology, 227*, 1275–1281.
Granato, D., Sousa Santos, J., Galvão Maciel, L., & Sávio Nunes, D. (2016). Chemical perspective and criticism on selected analytical methods used to estimate the total content of phenolic compounds in food matrices. *Trends in Analytical Chemistry, 80*, 266–279.
Gumus, Z. P., Ertas, H., Yasar, E., & Gumus, O. (2018). Classification of olive oils using chromatography, principal component analysis and artificial neural network modelling. *Journal of Food Measurement and Characterization, 12*, 1325–1333.
Gutierrez, F., Arnaud, T., & Albi, M. A. (1999). Influence of ecological cultivation on virgin olive oil quality. *Journal of the American Oil Chemists' Society, 76*, 617–621.
Györéné, K. G., Varga, A., & Lugasi, A. (2006). A comparison of chemical composition and nutritional value of organically and conventionally grown plant derived foods. *Orvosi Hetilap, 147*, 2081–2090.
Hebert, P. D., Cywinska, A., Ball, S. L., & de Waard, J. R. (2003). Biological identifications through DNA barcodes. *Proceedings of the Biological Sciences, 270*, 313–321.
Hindson, B. J., Ness, K. D., Masquelier, D. A., Belgrader, P., Heredia, N. J., Makarewicz, A. J., Bright, I. J., Lucero, M. Y., Hiddessen, A. L., Legler, T. C., Kitano, T. K., Hodel, M. R., Petersen, J. F., Wyatt, P. W., Steenblock, E. R., Shah, P. H., Bousse, L. J., Troup, C. B., Mellen, J. C., Wittmann, D. K., Erndt, N. G., Cauley, T. H., Koehler, R. T., So, A. P., Dube, S., Rose, K. A., Montesclaros, L., Wang, S., Stumbo, D. P., Hodges, S. P., Romine, S., Milanovich, F. P., White, H. E., Regan, J. F., Karlin-Neumann, G. A., Hindson, C. M., Saxonov, S., & Colston, B. W. (2011). High-throughput droplet digital PCR system for absolute quantitation of DNA copy number. *Analytical Chemistry, 83*, 8604–8610.

Holland, P. M., Abramson, R. D., Watson, R., & Gelfand, D. H. (1991). Detection of specific polymerase chain reaction product by utilizing the 5′----3′ exonuclease activity of Thermus aquaticus DNA polymerase. *Proceedings of the National Academy of Sciences of the United States of America, 88*, 7276–7280.

İpek, A., İpek, M., Ercişli, S., & Tangu, N. A. (2017). Transcriptome-based SNP discovery by GBS and the construction of a genetic map for olive. *Functional & Integrative Genomics, 17*, 493–501.

Issaoui, M., Flamini, G., Brahmi, F., Dabbou, S., Hassine, K. B., Taamali, A., Chehab, H., Ellouz, M., Zarrouk, M., & Hammamia, M. (2010). Effect of the growing area conditions on differentiation between Chemlali and Che'toui olive oils. *Food Chemistry, 119*, 220–225.

Jimenez, B., Sánchez-Ortiz, A., Lorenzo, M. L., & Rivas, A. (2014). Effect of organic cultivation of Picual and Hojiblanca olive varieties on the quality of virgin olive oil at four ripening stages. *European Journal of Lipid Science and Technology, 116*, 1634–1646.

Jiménez, B., Rivas, A., Lorenzo, M. L., & Sánchez-Ortiz, A. (2017). Chemosensory characterization of virgin olive oils obtained from organic and conventional practices during fruit ripening. *Flavour and Fragrance Journal, 32*, 294–304.

Julca, I., Marcet-Houben, M., Cruz, F., et al. (2020). Genomic evidence for recurrent genetic admixture during the domestication of Mediterranean olive trees (Olea europaea L.). *BMC Biology, 18*, 148.

Kalogianni, D. P., Bazakos, C., Boutsika, L. M., Targem, M. B., Christopoulos, T. K., Kalaitzis, P., & Ioannou, C. (2015). Olive oil DNA fingerprinting by multiplex SNP genotyping on fluorescent microspheres. *Journal of Agricultural and Food Chemistry, 63*, 3121–3128.

Kalogiouri, N. P., Aalizadeh, R., & Thomaidis, N. S. (2017). Investigating the organic and conventional production type of olive oil with target and suspect screening by LC-QTOF-MS, a novel semi-quantification method using chemical similarity and advanced chemometrics. *Analytical and Bioanalytical Chemistry, 409*, 5413–5426.

Kaya, H. B., Cetin, O., Kaya, H., Sahin, M., Sefer, F., Kahraman, A., & Tanyolac, B. (2013). SNP discovery by Illumina-based transcriptome sequencing of the olive and the genetic characterization of Turkish olive genotypes revealed by AFLP, SSR and SNP markers. *PLoS One, 8*, e73674.

Kaya, H. B., Akdemir, D., Lozano, R., Cetin, O., Kaya, H. S., Sahin, M., Smith, J. L., Bahattin Tanyolac, B., & Jannink, J. L. (2019). Genome wide association study of 5 agronomic traits in olive (Olea europaea L.). *Scientific Reports, 9*, 18764.

Kemsley, E. K., Le Gall, G., Dainty, J. R., Watson, A. D., Harvey, L. J., Tapp, H. S., & Colquhoun, I. J. (2007). Multivariate techniques and their application in nutrition: A metabolomics case study. *British Journal of Nutrition, 98*, 1–14.

Kritioti, A., Menexes, G., & Drouza, C. (2018). Chemometric characterization of virgin olive oils of the two major Cypriot cultivars based on their fatty acid composition. *Food Research International, 103*, 426–437.

Kumar, S., Kahlon, T., & Chaudhary, S. A. (2011). A rapid screening for adulterants in olive oil using DNA barcodes. *Food Chemistry, 127*, 1335–1341.

Lopes, S. M., Mendonça, D., Sefc, K. M., Sabino Gil, F., & Câmara Machado, A. (2004). Genetic evidence of intra-cultivar variability within Iberian olive cultivars. *HortScience, 39*, 1562–1565.

López-Yerena, A., Lozano-Castellón, J. L., Olmo-Cunillera, A., Tresserra-Rimbau, A., Quifer-Rada, P., Jiménez, B., Maria Pérez, & Vallverdú-Queralt, A. (2019). Effects of organic and conventional growing systems on the phenolic profile of extra-virgin olive oil. *Molecules, 24*, 1986.

Lukic, M., Lukic, I., Krapac, M., Sladonja, B., & Pilizota, V. (2013). Sterols and triterpene diols in olive oil as indicators of variety and degree of ripening. *Food Chemistry, 136*, 251–258.

Mailer, R. J., Ayton, J., & Graham, K. (2010). The influence of growing region, cultivar and harvest timing on the diversity of Australian olive oil. *Journal of the American Oil Chemists' Society, 87*, 877–884.

Martins-Lopes, P., Gomes, S., Santos, E., & Guedes-Pinto, H. (2008). DNA markers for Portoguese olive oil fingerprinting. *Journal of Agricultural and Food Chemistry, 56*, 11786–11791.

Matos, L. C., Cunha, S. C., Amaral, J. S., Pereira, J. A., Andrade, P. B., Seabra, R. M., & Oliveira, B. (2007). Chemometric characterization of three varietal olive oils (cvs. Cobrancosa, Madural and Verdeal Transmontana) extracted from olives with different maturation indices. *Food Chemistry, 102*, 406–414.

Mohamed, M. B., Rocchetti, G., Montesano, D., Ali, S. B., Guasmi, F., Grati-Kamoun, N., & Lucin, L. (2018). Discrimination of Tunisian and Italian extra-virgin olive oils according to their phenolic and sterolic fingerprints. *Food Research International, 106*, 920–927.

Montealegre, C., Alegre, M. L. A., & Garcia-Ruiz, C. (2010). Traceability markers to the botanical origin in olive oils. *Journal of Agricultural and Food Chemistry, 58*, 28–38.

Montemurro, C., Pasqualone, A., Simeone, R., Sabetta, W., & Blanco, A. (2008). AFLP molecular markers to identify virgin olive oils from single Italian cultivars. *European Food Research and Technology, 226*, 1439–1444.

Morin, J.-F., & Lees, M. (Eds.), (2018). *Food integrity handbook. A guide to food authenticity issues and analytical solutions*. Eurofins Analytics France, La Chapelle s/Erdre. http://foodintegrity.eu. Accessed July 2020.

Muzzalupo, I., Stefanizzi, F., Salimonti, A., Falabella, R., & Perri, E. (2009). Microsatellite markers for identification of a group of Italian olive accessions. *Science in Agriculture, 66*, 685–690.

Muzzalupo, I., Pisani, F., Greco, F., & Chiappetta, A. (2015). Direct DNA amplification from virgin olive oil for traceability and authenticity. *European Food Research and Technology, 241*, 151–155.

Nescatelli, R., Bonanni, R. C., Bucci, R., Magrì, A. L., Magrì, A. D., & Marini, F. (2014). Geographical traceability of extra virgin olive oils from Sabina PDO by chromatographic fingerprinting of the phenolic fraction coupled to chemometrics. *Chemometrics and Intelligent Laboratory Systems, 139*, 175–180.

Ninfali, P., Bacchiocca, M., Biagiotti, E., Esposto, S., Servili, M., Rosati, A., & Montedoro, G. (2008). A 3-year study on quality, nutritional and organoleptic evaluation of organic and conventional extra-virgin olive oils. *Journal of the American Oil Chemists' Society, 85*, 151–158.

Pafundo, S., Agrimonti, C., & Marmiroli, N. (2005). Traceability of plant contribution in olive oil by amplified fragment length polymorphisms. *Journal of the Science of Food and Agriculture, 53*, 6995–7002.

Pafundo, S., Agrimonti, C., Maestri, E., & Marmiroli, N. (2007). Applicability of SCAR markers to food genomics: Olive oil traceability. *Journal of the Science of Food and Agriculture, 55*, 6052–6059.

Palagano, R., Valli, E., Cevoli, C., Bendini, A., & Gallina Toschi, T. (2020). Compliance with EU vs. extra-EU labelled geographical provenance in virgin olive oils: A rapid untargeted chromatographic approach based on volatile. *LWT, 130*, 109566.

Pasqualone, A., Montemurro, C., Caponio, F., & Blanco, A. (2004). Identification of virgin olive oil from different cultivars by analysis of DNA microsatellites. *Journal of Agricultural and Food Chemistry, 52*, 1068–1071.

Pasqualone, A., Montemurro, C., Summo, C., Sabetta, W., Caponio, F., & Blanco, A. (2007). Effectiveness of microsatellite DNA markers in checking the identity of protected designation of origin extra virgin olive oil. *Journal of Agricultural and Food Chemistry, 55*, 3857–3862.

Pasqualone, A., Montemurro, C., di Rienzo, V., Summo, C., Paradiso, V. M., & Caponio, F. (2016). Evolution and perspectives of cultivar identification and traceability from tree to oil and table olives by means of DNA markers. *Journal of the Science of Food and Agriculture, 96*, 3642–3657.

Pérez-Jiménez, M., Besnard, G., Dorado, G., & Hernandez, P. (2013). Varietal tracing of virgin olive oils based on plastid DNA variation profiling. *PLoS One, 8*, e70507.

Peršurić, Ž., Saftić, L., Mašek, T., & Kraljević Pavelić, S. (2018). Comparison of triacylglycerol analysis by MALDI-TOF/MS, fatty acid analysis by GC-MS and non-selective analysis by

NIRS in combination with chemometrics for determination of extra virgin olive oil geographical origin. A case study. *LWT, 95*, 326–332.

Pinatel, C., Ollivier, D., Ollivier, V., & Artaud, J. (2014). New approach to the determination of the origin of olive oils: Morphograms and morphotypes (part II). *Olivae, 119*, 48–62.

Ponchel, F., Toomes, C., Bransfield, K., Leong, F. T., Douglas, S. H., Field, S. L., Bell, S. M., Combaret, V., Puisieux, A., Mighell, A. J., Robinson, P. A., Inglehearn, C. F., Isaacs, J. D., & Markham, A. F. (2003). Real-time PCR based on SYBR-green I fluorescence: An alternative to the TaqMan assay for a relative quantification of gene rearrangements, gene amplifications and micro gene deletions. *BMC Biotechnology, 3*.

Portarena, S., Leonardi, L., Scartazza, A., Lauteri, M., Baldacchini, C., Farinelli, D., Famiani, F., Ciolfi, M., & Brugnoli, E. (2019). Combining analysis of fatty acid composition and δ13C in extra-virgin olive oils as affected by harvest period and cultivar: Possible use in traceability studies. *Food Control, 105*, 151–158.

Quintanilla-Casas, B., Bertin, S., Leik, K., Bustamante, J., Guardiola, F., Valli, E., Bendini, A., Gallina Toschi, T., Tres, A., & Vichi, S. (2020). Profiling versus fingerprinting analysis of sesquiterpene hydrocarbons for the geographical authentication of extra virgin olive oils. *Food Chemistry, 307*, 125556.

Raieta, K., Muccillo, L., & Colantuoni, V. (2015). A novel reliable method of DNA extraction from olive oil suitable for molecular traceability. *Food Chemistry, 172*, 596–602.

Ramírez-Tejero, J. A., Jiménez-Ruiz, J., de la O Leyva-Pérez, M., Barroso, J. B., & Luque, F. (2020). Gene expression pattern in olive tree organs (Olea europaea L.). *Genes, 11*, 544.

Reale, S., Doveri, S., Diaz, A., Angiolillo, A., Lucentini, L., Pilla, F., Martín, A., & Donini, P. (2006). SNP-based markers for discriminating olive (Olea europaea L.) cultivars. *Genome, 49*, 1193–1205.

Regulation (EU) No 1308/2013 of the European Parliament and of the Council of 17 December 2013 establishing a common organisation of the markets in agricultural products and repealing Council Regulations (EEC) No 922/72, (EEC) No 234/79, (EC) No 1037/2001 and (EC) No 1234/2007.

Rodrigues, N., Casal, S., Cruz, R., Peres, A. M., Baptista, P., & Pereira, J. A. (2020). GxE effects on tocopherol composition of oils from very old and genetically diverse olive trees. *Journal of the American Oil Chemists' Society, 97*, 497–507.

Rodríguez Pleguezuelo, C. R., Durán Zuazo, V. H., Francia Martínez, H. R., Martín Peinado, F. J., Moreno Martín, F., & García Tejero, I. F. (2018). Organic olive farming in Andalusia, Spain. A review. *Agronomy for Sustainable Development, 38*, 20.

Sayago, A., González-Domínguez, R., Beltrán, R., & Fernández-Recamales, A. (2018). Combination of complementary data mining methods for geographical characterization of extra virgin olive oils based on mineral composition. *Food Chemistry, 261*, 42–50.

Sayago, A., González-Domínguez, R., Urbano, J., & Fernández-Recamales, A. (2019). Combination of vintage and new-fashioned analytical approaches for varietal and geographical traceability of olive oils. *LWT, 111*, 99–104.

Scollo, F., Egea, L. A., Gentile, A., La Malfa, S., Dorado, G., & Hernandez, P. (2016). Absolute quantification of olive oil DNA by droplet digital-PCR (ddPCR): Comparison of isolation and amplification methodologies. *Food Chemistry, 213*, 388–394.

Sebastiani, L., & Busconi, M. (2017). Recent developments in olive (Olea europaea L.) genetics and genomics: Applications in taxonomy, varietal identification, traceability and breeding. *Plant Cell Reports, 36*, 1345–1360.

Skiada, V., Tsarouhas, P., & Varzakas, T. (2020). Comparison and discrimination of two major monocultivar extra virgin olive oils in the southern region of Peloponnese, according to specific compositional/traceability markers. *Food, 9*, 155.

Spaniolas, S., Bazakos, C., Awad, M., & Kalaitzis, P. (2008). Exploitation of the chloroplast trnL (UAA) intron polymorphisms for the authentication of plant oils by means of a lab-on-a-chip capillary electrophoresis system. *Journal of Agricultural and Food Chemistry, 16*, 6886–6891.

Suciu, N. A., Ferrari, F., & Trevisan, M. (2019). Organic and conventional food: Comparison and future research. *Trends in Food Science and Technology, 84*, 49–51.
Tautz, D. (1989). Hypervariability of simple sequences as a general source for polymorphic DNA markers. *Nucleic Acids Research, 17*, 6463–6471.
Tibola, C. S., da Silva, S. A., Dossa, A. A., & Patrício, D. I. (2018). Economically motivated food fraud and adulteration in Brazil: Incidents and alternatives to minimize occurrence. *Journal of Food Science, 83*, 2028–2038.
Tsatsakis, A. M., Tsakiris, I. N., Tzatzarakis, M. N., Agourakis, Z. B., Tutudaki, M., & Alegakis, A. K. (2003). Three-year study of fenthion and dimethoate pesticides in olive oil from organic and conventional cultivation. *Food Additives and Contaminants, 20*, 553–559.
Tsimidou, M. Z., Ordoudi, S. A., Nenadis, N., & Mourtzinos, I. (2016). Food fraud. In B. Caballero, P. Finglas, & F. Toldrá (Eds.), *The encyclopedia of food and health* (pp. 35–42). Academic Press.
Uncu, A. T., Frary, A., & Doganlar, S. (2015). Cultivar origin and admixture detection in Turkish olive oils by SNP-based CAPS assays. *Journal of Agricultural and Food Chemistry, 63*, 2284–2295.
Uncu, A. T., Uncu, A. O., Frary, A., & Doganlar, S. (2017). Barcode DNA length polymorphisms vs fatty acid profiling for adulteration detection in olive oil. *Food Chemistry, 221*, 1026–1033.
Uncu, A. O., Torlak, E., & Uncu, A. T. (2018). A cost-efficient and simple plant oil DNA extraction protocol optimized for DNA-based assessment of product authenticity. *Food Analytical Methods, 11*, 939–950.
Unver, T., et al. (2017). Genome of wild olive and the evolution of oil biosynthesis. *PNAS, 114*, E9413–E9422.
Van Ruth, S. M., & Pagter-de Witte, L. (2020). Integrity of organic foods and their suppliers: Fraud vulnerability across chains. *Food, 9*, 188.
Vega-Zamora, M., Torres-Ruiz, F. J., & Parras-Rosa, M. (2018). Key determinants of organic food consumption: The case of olive oil in Spain. *HortScience, 53*, 1172–1178.
Vietina, M., Agrimonti, C., Bonas, U., Marmiroli, M., & Marmiroli, N. (2011). Applicability of SSR markers to the traceability of monovarietal olive oils. *Journal of the Science of Food and Agriculture, 91*, 1381–1391.
Vietina, M., Agrimonti, C., & Marmiroli, N. (2013). Detection of plant oil DNA using high resolution melting (HRM) post PCR analysis: A tool for disclosure of olive oil adulteration. *Food Chemistry, 141*, 3820–3826.
Vos, P., Hogers, R., Bleeker, M., Reijans, M., van de Lee, T., Hornes, M., Frijters, A., Pot, J., Peleman, J., & Kuiper, M. (1995). AFLP: A new technique for DNA fingerprinting. *Nucleic Acids Research, 23*, 4407–4414.

Chapter 57
Olive Oil Extraction and Table Olive Industries

Sebastián Sánchez Villasclaras, Belén Fernández, Rafael Pacheco, Agustí Romero, Mª. Lourdes Martínez, Juan F. Hermoso, Mª. Dolores La Rubia, Rafaela Cáceres, Natividad Ramos, Elena Guzmán, Inmaculada Olivares, Ana Gaspar, and Ana Carrilho

Abstract Production of olive oils (OOs) is a key industrial activity mainly in the countries of the Mediterranean Basin. As it is well known, the production of olives has two destinations: obtaining olive oils and generating table olives. Nowadays, production of olives is mainly directed to the extraction of olive oils, although world production of table olives was 2.5 10^6 tons in 2022, with an increase of 44% in the last 15 years. In recent years, and from the perspective of the production of early harvest virgin olive oils, it is worth highlighting the introduction of cold technology in preliminary external and internal operations and in the preparatory stages of olive pastes, fundamentally in the malaxation operation. In order to differentiate the oils produced between olive-growing regions, other milling equipment different to hammer mill has been introduced. The change in this type of mill, fundamentally, has two objectives: reduce the formation of emulsions and increase the effect of dilaceration. Recently, new technologies have been introduced to replace or reduce the malaxation stage: thermomechanical extruder ("Protoreattore"), ultra vacuum malaxation ("Booster"), paste treatment by electric pulses, and use of high-power ultrasounds. This equipment, tested at an industrial level in Spanish and Italian oil mills, seems to decrease the malaxation time and increase the organoleptic quality

S. S. Villasclaras (✉) · R. Pacheco · Mª. L. Martínez · Mª. D. La Rubia · N. Ramos
E. Guzmán · I. Olivares
Center for Advanced Studies in Olive Grove and Olive Oils, University of Jaén, Science and Technology Park GEOLIT, Jaen, Spain
e-mail: ssanchez@ujaen.es; rpacheco@ujaen.es; mlmartinez@ujaen.es; mdlarubia@ujaen.ees; nramos@ujaen.es; eguzman@ujaen.es; iolivares@ujaen.es

B. Fernández · A. Romero · J. F. Hermoso · R. Cáceres
IRTA. Institute of Agrifood Research and Technology, Torre Marimon, Barcelona, Spain
e-mail: Belen.Fernandez@irta.cat; agusti.romero@irta.cat; juanf.hermoso@irta.cat; Rafaela.Caceres@irta.cat

A. Gaspar · A. Carrilho
ESPORÃO-Azeites., Herdade de sporão, Reguengos de Monsaraz, Portugal
e-mail: ana.gaspar@esporao.com; ana.carrilho@esporao.com

J. Muñoz-Rojas, R. García-Ruiz (eds.), *The Olive Landscapes of the Mediterranean*, Landscape Series 36,
https://doi.org/10.1007/978-3-031-57956-1_57

of virgin olive oils produced. Among the different historical milestones that have caused significant changes in the production of OOs, the incorporation of centrifugation technology allowed to replace extraction process by pressure, achieving a continuous separation and, with it, a leap towards the modernization of the sector. The replacement of the decanter with three outlets for those with two outlets resulted in the elimination of 'alpechines' (olive waste water), although a new olive pomace with high degrees of humidity is produced. From an environmental point of view, the change made can be considered positive, provided that there is an olive-pomace oil extraction industry that can manage these moist pomaces properly. Regarding the quality of the oils, the extraction systems using two-outlet decanter have led to products with a somewhat different composition, in general, a higher content of phenolic compounds, increasing the oxidative stability of the product. The future of the olive oil industry must move towards a greater interrelation between olive oil production, conservation of the environment and energy generation.

Keywords Olive oils · Table olives · Olive oil extraction · Olive stones · Olive pomaces · Olive mill wastewaters

57.1 Preliminary Operations

The growing demand for high-quality olive oil in a global market requires an improvement in the productive efficiency of the olive grove, which implies the use of appropriate strategies and techniques (new varieties, intensification and mechanization of the crop, localized irrigation facilities, integrated phytosanitary management etc.), to achieve competitive production costs and an improvement in oil quality (Hermoso et al., 2017; Rufat et al., 2018). The best-valued oils are the freshest, made from olives without completing their maturation, which have a greenish and intense fruity odour, with a taste tending to bitter and spicy, and better nutritional properties.

The cultivation operations, the olive harvest and the time and conditions elapsed until the oil is extracted, which include the transport, cleaning and/or washing and storage of the fruit, constitute a determining point of quality (Romero et al., 2016).

57.1.1 External

57.1.1.1 Cultivation Conditions

Sensory attributes related to the growing conditions, the growing area and the cultivation system can appear in virgin olive oil (Bianchi, 1999; García-Pizarro et al., 2024). The olive variety is the main responsible for the sensory characteristics of virgin oil. Each variety has positive attributes that allow it to be identified, but its sensory and chemical characteristics also depend on the ripening index, climate, soil and cultivation techniques (mainly irrigation, sanitary status and harvesting) and milling process (Romero, 2011).

Although agronomic factors will not be studied here in detail, a brief review is done in this section because of their clear influence on the oil quality. The generation of the sensory properties of olive oil involves multiple factors, such as the health status and maturity index of the olive, the harvest and postharvest conditions or the oil extraction conditions at an industrial level (Rufat et al., 2018). The period of the year and the harvesting system, as well as the conditions in which olives are transported and stored until they are milled, can significantly influence the quality of the oils (Boudebouz et al., 2023). The olive is the raw material; therefore, it is essential to start from fruits in perfect condition to obtain oil with good sensory and nutritional properties. Similarly, irrigation in olive groves is a strategy that has shown a positive response in the production and quality of the oil. The use of modern irrigation techniques is key to increase the productivity of olive groves and the efficient use of water resources and to obtain high-quality productions (Sastre et al., 2019, 2023). In fact, several studies have suggested that the main physical-chemical parameters of olive oil in different production models were not significantly affected by controlled deficit irrigation. Nevertheless, significant differences can be expected on fatty acid composition when irrigation dose increase, with an increase in oleic acid and a decrease in linoleic acid (Gomez del Campo & Garcia, 2013; Arbizu, 2015). Besides, the total content of phenolic compounds of the oil is negatively affected, decreasing with increased irrigation doses, especially in the most intensive plantation models.

57.1.1.2 Harvesting of the Fruit

The harvest of the olives is carried out manually or mechanically. The traditional system consists of hitting the branches with sticks, collecting the dropped fruits on nets placed on the ground around the trees. However, this is not a recommended method, due to its low hourly yield and because both olives and branches can be damaged, especially the young shoots that ensure the the following year's production. Manual harvesting can be improved with the use of electric or pneumatic combs, installed on telescopic poles of variable length. This method imitates manual harvesting, improving hourly yield, especially in young plantations or small farms, without damaging the olive or the tree. A variation of the combs is the hand-operated mechanical shakers.

In mature trees or larger groves, the most common mechanical harvesters are vibrators installed on tractors or self-propelled 'buggies', which use a hydraulic system to transfer the force to the vibrating head. Collecting meshes are first placed on the ground, under the tree. The vibrators are then attached to the trunks or main branches with clamps or rubber rolls that transmit pendular or rotating movements to the plants and fruits, which detach the olives on the nets extended on the ground. Whether the olive tree is properly pruned, with a tall vase formation system, a trunk shaker with built-in umbrella can be used. This system makes unnecessary the use of nets and considerably reduces time and personnel requirements for harvesting.

Harvesting methods that pick olives from the ground or that delayed harvesting greatly reduce quality. In all cases, the ground-flight fruit separation is a priority to reach a high olive oil quality.

In recent decades, world olive growing has undergone an important technological change, highlighting the increase in plant density and the harvesting improvement. It has gone from traditional systems, mostly in dry farming with less than 100 trees per hectare, very low productivity and manual harvest, to hedge systems with more than 1500 trees per hectare, very high productivity and fully mechanized operations. Furthermore, the super-high density system has a fast entry into production, high productivity and lower costs which makes it much more competitive compared to the traditional olive grove. Currently, there are more than 600,000 ha under this system worldwide.

57.1.1.3 Transport

The fruit transport system must ensure optimal conditions for the olives until they are unloaded and delivered at the mill. In general, any factor that breaks the cellular structure of the fruits poses a significant risk since it can facilitate the activity of microorganisms (bacteria, yeasts and moulds) or the enzymatic system of the fruit itself. Romero et al. (2016) studied the relationship between different quality parameters of the fruit entered in various cooperatives in the Priorat area (PDO 'Siurana' in the mountain area in Catalunya, Spain) and the qualitative physicochemical and sensory characteristics of the oils produced, maintaining the traceability between batches of incoming olives and oil produced. That relationship is so direct that it allows determining process defects in oil mills when the quality observed in the reception yard and the final quality of the oil obtained do not match.

57.1.2 Internal

57.1.2.1 Reception of the Fruit in the Oil Mill

Olives must be separated, according to their state, at the entrance of the oil mill for processing them in different ways according to their potential quality (Jiménez et al., 2001). The fruit reception and cleaning operations are both necessary and critical for olive oil quality.

57.1.2.2 Cleaning and Washing

Olives that arrive at the oil mill carry a variable percentage of foreign matter of a very diverse nature, such as dust, stones, leaves, stems or herbs. To obtain good-quality oils, and to greatly reduce the machinery wear, it is essential to take away all these foreign bodies. Cleaning units use air to separate objects less heavy than olives. Washing units use water to take away heavier bodies mixed with olives and chemical residues from the fruit skin. The fruits from the tree (flight), in principle and except for special circumstances, should not be subjected to washing as extractability is lost in the process and the oils obtained are less stable and fragrant (Uceda, 2004).

A particularly sensitive point is the quality status of the water used to wash the fruits, especially in those oil mills where this step is done before fruit storage. In fact, Vichi et al. (2015) and Romero et al. (2016) demonstrated the risk to oil quality associated with this operation since this is a washing system with recirculated water (due to the scarcity and high price of water in Mediterranean countries); the system becomes dirtier with each washing cycle. After a few cycles, the water carries a considerable microbial load (yeasts, bacteria and moulds) and poses a significant risk to the final quality of the oil from these 'washed' fruits, especially when the fruits are stored after the washing step and even more if big hoppers are used. Therefore, it is important to control the microbiological quality of the olives prior to their grinding. In this sense, some studies have shown an increase in lactic acid bacteria, enterobacteria, pseudomonas and moulds of up to three orders of magnitude (Vichi et al., 2011). A typical indicator of this problem is the appearance of a 'black' water films on the olives, identifiable on the fruit discharge belts from the hoppers to the crushers.

57.1.2.3 Fruit Weight and Quality Control

Once the twigs, leaves and stones are separated, the fruits are weighed to pay back the farmer. All these operations are hazardous for the oil quality due to frequent impacts, skin erosions and fruit crushing.

As already mentioned, maintaining the quality of the olive is crucial to obtain a quality oily juice. But as olive is a non-climacteric fruit, the colour, smell and nutritional composition of olives are altered during the postharvest period. In recent years there is a growing interest in the exogenous application of chemical elicitors and the usage of pre- and postharvest techniques to extend the shelf life of olives to improve their health-promoting properties. Some successful techniques have been studied, such as the immersion of the fruits in model solutions (solutions with acid pH, ascorbic acid or sodium metabisulfite) where the oxidation of the olives and the colour changes can be mitigated. Other techniques are aimed at the storing of fruits in the absence of oxygen, such as nitrogen-enriched atmospheres, or refrigerating and even freezing the olives. However, all these techniques can be costly and difficult to execute in the oil mills (Romero et al., 2018).

Currently, the simplest solution is to grind the olives as soon as possible, having a classification and handling system that allows keeping the fruits at an optimum temperature for the desired quality. Today, in-line control and classification equipment for the quality of the raw material at the entrance of the oil mill has been developed, based on image analysis or visible/near-infrared (VIS/NIR) spectroscopy for checking the quality and adulteration of olives as well as to find out the geographical origin of olive oil (Abu-Khalaf & Hmidat, 2020), which in the coming years can contribute significantly to improving quality of the oils.

57.1.2.4 Fruit Classification and Storage

Once the fruit is cleaned, and/or washed, it is stored in the lung hoppers or waiting to be ground. Among postharvest operations, fruit storage has been the most studied stage (Romero et al., 2016). This study analysed the storage time ranging from 3 days to 3 weeks and temperatures from 4 to 20 °C. The main conclusion was that the storage conditions are crucial for the quality of the oil and that the most advisable is to adapt the storage capacity to the daily milling capacity of the oil mill, in order to avoid storage times over 24 h (Hermoso et al., 2011). However, there is little information on the effects of this short-term storage on the quality of the oil (Boudebouz et al., 2023; Vichi et al., 2015).

On the other hand, excessively long storage of ripe fruits can lead to considerable increases in the acidity of the oil, as well as sensorial defects from microorganism activity (fusty, winey, vinegary or musty) and a decrease in stability (García et al., 1996). Angerosa et al. (1996) demonstrated that fruits hold a specific microbial load on the surface, already from the field. These microorganisms are a combination of moulds and yeasts, as well as enteric, lactic and acetic bacteria, that are responsible for fermentation reactions during the storage of olives (Angerosa et al., 1996; Vichi et al., 2009), starting an alteration chain leading to the typical fusty defect and the subsequent appearance of vinegary and mould flaws. Some studies showed that specific microbial fermentative reactions took place even during their grinding and whipping. The mentioned alterations during extended storages were verified, with 'Arbequina' variety, years later by Vichi et al. (2009) who identified the same species of microorganisms and their effects on quality loss of the oils. Similarly, Vichi et al. (2015) also identified volatile phenolic compounds at this stage which are responsible for uncommon sensorial defects (woody, slurry, medicine etc.).

57.2 Production of Table Olives

Table olives are the most important and popular fermented vegetable in the food industry, especially in Spain, Turkey, Italy, Egypt, Morocco and Greece. The world table olive production was of 2,578,000 tons in 2019 with an increase of 44% in the last 15 years. On the other hand, in the latter 25 years, the world consumption of table olives has multiplied by 2.8, increasing by 173% in this period. Although the producer countries show a greater increase in consumption, Egypt, Turkey, Algeria, the United States, Spain, Syria, Italy, Brazil, Iran, France and Russia stand out as the leading consumer countries; also the incipient consumption is noteworthy in countries like Germany or the United Kingdom.

57.2.1 Varieties of Olives Used

Any variety of olives can be used for table olives, but all of them must have good fruit size, flavour, a high pulp/stone ratio and easy pitting and machining.

In recent years, the varieties of olives used for the production of table olives have been growing, due to the current increase in the consumption of this product. 'Manzanilla', 'Gordal' and 'Morona' are the varieties most used in Spain as table olives. The 'Manzanilla' variety is one of the most valued internationally for its high quality. It has a medium size, small bone and an almost round shape. In Spain, the name 'Manzanilla' is designated for the variety from Seville, and the one coming from Badajoz is called 'Carrasqueña'. The 'Gordal' variety is one of the best known in the world for its large size. Most of the production of this variety is located in Seville (Spain). Also, the 'Manzanilla Cacereña' variety is early ripening, ovoid in shape and of medium size; it contains a very high pulp/stone ratio, which makes it a great variety for table olives. The 'Morona' variety (Seville, Spain) is similar to 'Manzanilla', somewhat longer, with a lower pulp/stone ratio and fast fermentation.

Besides, dual-aptitude varieties are used as table olives but also to obtain good-quality oils, such as 'Hojiblanca', 'Cacereña' or 'Verdial'.

Although Spain is among the top producers of table oil, there are also varieties from other countries. In Greece, the 'Kalamata', 'Koroneiki', 'Mastoidis' or 'Amfissis' olives are well known. The 'Ascori', 'Calabrian' and 'Gaeta' varieties are also used in Italy, and 'Picholine' variety is widely used in France. In Argentina, the 'Arauco' variety is used. In Morocco and Tunisia, the 'Moroccaine Picholine' and 'Chemlaly' varieties are the most important. In the United States and Australia, the varieties that are used are 'Barnea' from Israel and 'Leccino' and 'Chamomile' from Italy (Madrid, 2018).

57.2.2 Preparation Processes

The different preparation processes consider the following aspects (Garrido-Fernández et al., 1997):

(i) Olives colour. It is related, mainly, to the ripeness degree of the fresh fruit. There are four: green, turning colour, natural black and black (obtained from fruits that, not being fully ripe, are blackened by oxidation).

(ii) The procedure used to remove bitterness. The purpose of the different production processes is to remove the natural bitterness of the olive produced by oleuropein. The treated olives undergo treatment with an aqueous solution of sodium hydroxide (bleach); in this case, the fruits can lose all the bitterness quickly. On the contrary, oleuropein can be partially and slowly eliminated by dilution in a saline solution; in this case, the term used is natural olives.

(iii) The fermentation method of the product. The most used form is the use of brine (NaCl solution). There are other systems like 'in dry salt'. The main preparation processes are described below.

57.2.2.1 Green Olives

Currently, the most used preparations for green olives are described by the IOC (2004):

- *Treated green olives in brine (Spanish or Sevillian style).* In this process, the fruits, green to yellowish in colour, already harvested, are transferred to a treating plant. After the selection and classification steps, they are treated with a dilute sodium hydroxide solution (bleach), to eliminate the large part of the oleuropein glucoside. Afterwards, the fruits are washed with water several times, in different periods, to eliminate the bleach, and later they are placed in a 9–10 Bé brine, in which the fermentation takes place. Finally, the fermented fruits are selected and classified by size to be packaged in different ways such as whole fruits, pitted, stuffed with pepper or another type of product. This is the commercial preparation that covers the largest percentage of the international market.
- *Natural green olives.* The olives are prepared without previous alkaline treatment. They are placed in brine, in which they undergo complete or partial fermentation (Estradas, 2011).

57.2.2.2 Black Olives

Among the most used preparations for black olives, it is worth highlighting (IOC, 2004):

- *Black olives.*
- They are those obtained from fruits that, not being fully ripe, have been darkened by oxidation and have lost their bitterness through alkaline treatment. They are packed in brine and preserved by heat sterilization.
- *Natural black olives.*
- They are obtained from fruits collected at full maturity or shortly before it and may present, depending on the production area and time of collection, reddish-black, purple-black, violet, greenish-black or dark brown. They are prepared without previous alkaline treatment, placing them directly in brine where they undergo natural fermentation.
- *Natural black olives in dry salt.*
- It is a preparation of Greek origin. It is made with overripe fruits of the 'Megaritki' variety or others, placing them in alternate layers of dry salt. Then the olives lose their moisture, obtaining a product with little bitterness, salty and raisin appearance, whose consumption is destined for the local market (Madrid, 2018).

- *Kalamata-style olives.*
- The Kalamata style is characterized by the use of vinegar and oil in its preparation. It took its name from the Greek variety 'Kalamata' with which it started, although today many olive varieties are used.

Fresh fruit (short method) or partially fermented (long method) can be used. In the first case, to sweeten the fruits, two longitudinal incisions are made in the fruits, one on each side, to facilitate the diffusion of the oleuropein. The fruits are immersed in water or diluted brine, until they are sweetened. Then, they are placed in vinegar until they acquire acid flavour, and they are packed in diluted brine, adding a small amount of neutral oil, until the surface of the liquid is covered. Its acidic and strong flavour and the acidity of vinegar allow it to remain free of microorganisms and retain a good texture.

The long method is similar to the natural black olives preparation, and once the fermentation in brine is finished, the Kalamata-style preparation is carried out by the short method, although the cuts must be eliminated since the fruit has already lost its bitterness and these can cause the pulp to soften. The last step is the use of vinegar and oil as in the short method.

57.2.3 Preparation Types

57.2.3.1 Seasoned and Flavoured Olives

Seasoned or flavoured olives are highly appreciated, and its production is widely spread. However, its preparation for commercial purposes is fundamentally artisanal and is very restricted to small areas close to the distribution centre, due to the fast deterioration of this product. The olives used for this type of elaboration are green or turning colour olives, sweetened by periodic changes of water that, sometimes, are cracked or crushed, treated with bleach and washed or preserved in brine. The dressings used usually include spices, natural products (garlic, thyme, fennel, paprika, lemon, vinegar, etc.), essential oils, extracts or oleoresins.

57.2.3.2 Stuffed Olives

Stuffed olives are pitted olives, packed with different products: anchovies, tuna, salmon, almonds, peppers, lemon, cheese or natural pastes prepared with these ingredients. The green olives in brine are the type of olives used for this preparation. The olives, after the alkaline treatment, are pitted and stuffed. The use of pastes is very common as they have many economic and industrial advantages.

57.2.4 Packaging

The main purpose of packaging is to divide the production into adequate quantities to the consumption capacity of the consumers, preserving the product with the best organoleptic characteristics and free of alterations.

The table olives usually are pack in glass, cans and plastic containers, mainly polyethylene and recently in laminated cardboard containers. These containers are filled with brine and then undergo a heat treatment to stabilize the product microbiologically.

According to the forms of commercial presentation, the olives can be whole olives that maintain their original shape with stone, cracked olives, split olives, pitted olives, stuffed olives, in halves, quarters, segments, chopped, broken, etc.

Regarding the manner of arrangement of the olives in the containers, they are presented as thrown (random)-packed olives, when they are not placed in the container or place-packed when they are forming shapes in a glass container.

57.3 Extraction of Olive Oils

In the olive oil extraction process, it is necessary to differentiate the following stages: preparation of olive pastes, separation of solid and liquid phases, separation of liquid phases, downstream operations and conservation of olive oils (Fig. 57.1).

57.3.1 Preparation of Olive Pastes

Nowadays, the preparation of olive paste includes the following steps: stone separation, crushing, dilaceration and malaxation. However, in most oil mills only crushing and malaxation (key operation for obtaining olive oils of quality and adequate yield) are carried out.

57.3.1.1 Crushing

The oil is found in olives in micro-drops lodged in the vacuoles of the fruit's mesocarp. For this reason, the first obligatory step to separate the oil (regardless of the extraction system) is the crushing of the olives to destroy the tissues that constitute them. In this way, the 'olive pastes' are obtained.

During the crushing, the cells are broken and torn, leaving most of the oil droplets free. These micro-drops initially gather into larger droplets. At the same time, on the surface of the initial droplets, 'lipoproteic membranes' are formed as a consequence of their contact with the proteins dissolved or solubilized in the vegetation

Fig. 57.1 Table olive preparation facility. Fermentation stage in batch fermenters partially submerged in the exterior soil of the facility. (De Prado Quality Foods S.L., in the Alentejo region, Portugal)

water (and the rest of the water from previous treatments, olive washing and water of addition) (Leone et al., 2015).

Crushing is an operation of great responsibility in the extraction process. The way it is carried out and the equipment used have a direct influence on the following operations (malaxation, extraction of any type, decantation, centrifugation) and mainly on the yield obtained and quality of the oil (Di Giovacchino, 1996; Uceda et al., 2006).

Furthermore, is a matter of fact that volatile compounds are released during the grinding process into the oil through LOX pathway (Sánchez-Ortiz et al., 2018). Concerning the types of mills, there are two large groups: metallic crushers and traditional mills or stone mills (hardly used in Spain and Portugal). Metal mills can be classified according to their active parts, distinguishing three types: hammer, disc and grooved-cylinder mills. Without a doubt, the most used are hammer mills, although they are the least suitable for producing high-quality extra virgin olive oils; usually, the speed of the mill crosshead is close to 3000 rpm, which makes it very easy for the formation of emulsions and microemulsions to take place, resulting in a loss of industrial yields and quality in the oils produced.

It should be noted that, although to a small extent, disc mills and grooved cylinder mills have a greater dilaceration effect than hammer mills. Moreover, in some commercial firms, the rotation speed in toothed disc mills is close to 1500 rpm, and in fluted belt mills, this speed can be reduced to 400 rpm. This lower rate of rotation allows less emulsion formation in the olive pastes.

On the other hand, double screen and listello mills also have great advantages compared to hammer mills due to their lower rotation speed and to a certain extent a greater dilaceration effect.

57.3.1.2 Malaxation

The main objective of this operation is to release and form a continuous phase of oil from the drops of oil dispersed in the ground pastes.

Currently, and in accordance with industrial experience, the crushers used in the extraction of olive oils, especially the metallic ones, need the complement of dilacerators and malaxers. The dilacerators carry out the shearing of the parts that have not been sufficiently treated during the milling. In some oil mills, between the metallic crusher and the malaxer, there is a 'mechanical dilacerator', essentially made up of two stainless steel rollers with an oval section that rotate towards each other, forcing the paste to pass between them and between the fixed areas of the equipment (Civantos, 1999).

Malaxers are the only equipment in oil mills that are built without a scientific or technological basis. They are made only based on industrial experience. During the malaxation stage in the olive pastes, transport phenomena (momentum, energy and matter transport), chemical processes (triglyceride hydrolysis reactions and oxidation reactions) and biochemical processes (enzymatic and fermentative) take place. Therefore, the malaxation is a complex and partly unknown stage; so far it is unknown if the process is controlled by a transport phenomenon or by a chemical or biochemical reaction.

In industrial practice, the malaxation operation is carried out at temperatures in the range of 25–32 °C, with kneading times of 45–60 min if the purpose is to obtain extra virgin olive oils. The stirring speed is normally unchanged and is in the range of 15–20 rpm. In the production of early harvest extra virgin olive oils (premium oils), the temperature and malaxation time conditions are significantly lower (20–24 °C and 20–30 min, respectively). It must be highlighted that new studies are reporting the relationship between malaxation condition and bioactive compounds released into the EVOO (Olmo-Cunillera et al., 2021; Polari et al., 2018; Veneziani et al., 2017).

Within the production process, the operation of malaxation of the pasta is the one that has undergone the least development (Alba, 2008; Clodoveo, 2012; Di Giovacchino, 2013), so that current malaxers have different drawbacks (Ayr et al., 2015):

- They occupy the largest space within the extraction area in the oil mills, since it is a semi-continuous stage in the continuous production process. It requires more than one body in most installations, to ensure the continuous operation of the entire process and the continuous feeding to the decanter.

- Inefficient heating or cooling of the olive pastes. The low energy transfer from the walls to the centre of the malaxer body causes it to take between 20 and 30 minutes to achieve uniform heating of the pastes, which reduces the effectiveness of the malaxation. This fact is justified by the low thermal conductivity that has been determined in olive pastes.
- The negative effect that this operation may have on the quality, nutritional composition and sensory characteristics of virgin olive oils is in part due to a deficient system for causing the coalescence of oil drops, as well as to difficult cleaning of the equipment itself.

Use of Technological Co-adjuvants

In certain olive-growing areas, there are serious difficulties in the olive oil production processes, mainly with certain varieties of olives. In these cases, after the crushing stage, pastes are produced in which it is difficult to separate the oil (called 'strong or difficult pastes'). Macroscopically, retention and occlusion of the liquid phases in the pastes, oil emulsion and very fine solids in suspension are observed, and all this leads to a lower industrial extraction yield. In order to improve the behaviour of these difficult pastes in the production of olive oils, one of the alternatives is the addition of a processing aid to these pastes during the whipping stage. Among these adjuvants the most used, and allowed by current regulations, are natural microtalcs. Initially, and for three decades, the purpose of the addition was only to increase the extraction yields while keeping the nutritional quality of the oils obtained intact. However, it is shown that the use of a natural microtalc of great purity and small average particle size ($\phi \leq 2.1$ μm) not only increases the extraction yields but also improves the quality of the virgin olive oils produced, especially in relation to antioxidant compounds (tocopherols and phenolic compounds) and thus in the oxidative stability of the oil (Sanchez et al., 2018). As is well known, from a health point of view, the possibility of increasing the content of tocopherols and phenolic compounds in virgin olive oils is a key factor in antioxidant defence at the cellular level.

Other additives authorized for the extraction of olive oils are kaolinitic clay and pectinolytic enzymes. It should be noted that enzymes are not authorized by the European Union (CE/1513/2001); however, they can be used to obtain second extraction oils.

57.3.1.3 Other Operations

A current trend is to improve this procedure by applying some emerging technologies such as malaxation (Yorulmaz et al., 2010) under inert atmosphere and cold malaxation (Veneziani et al., 2017). Among the emerging technologies that have spread the most, in partial replacement of the whipping stage, is the introduction of a modified thermoextruder, an innovation carried out by the Pieralisi company. It consists of an endless contained in a cylinder, which has a programmable control unit that regulates the rotation speed. As the endless rotates, the pressure of the olive paste increases as does a continuous press. Throughout the cylinder this pressure build-up takes place in several sections of the cylinder. This system of increasing and decreasing pressure facilitates the grouping of the oil droplets that is the end of the malaxation. The great advantage is that it does it in a faster way compared to normal malaxation. In any case, the paste that comes out of this equipment goes through a small malaxer, where the mixing time is drastically reduced.

Other technologies that try to partially or totally replace the malaxation stage are the following:

- Rapid heating systems for olive paste.
- Treatment of pasta by electrical pulses (Clodoveo et al., 2013; Veneziani et al., 2019; Aymen et al., 2017).
- Microwave application. The main effect of this technique is thermal.

One of the technologies used recently, and which may have a promising future, is the pretreatment of olive paste using high-power ultrasound (Ayr et al., 2015).

57.3.2 Separation of Solid and Liquid Phases

Once the olive paste has been prepared, the separation of the solid and liquid phases can be carried out through three extraction processes: pressure, selective filtration and centrifugation. Currently, continuous centrifugation processes are mainly used in high-production countries.

In contrast, the selective filtration process is hardly used, although it is a quite interesting process. It should be noted that it can be used in conjunction with pressure and centrifugation processes.

57.3.2.1 Selective Filtration

Initially, it is worth noting that the selective filtration process is a Spanish innovation at the end of the nineteenth century, although it was not patented.

This selective filtration process, also sometimes called partial extraction, separates the loose oil that results after the milling and malaxation stages. This process is based on a difference in surface tension between liquid phases to be separated—in this case, the oil and vegetation water of the olive. The surface tension of oil is lower than water, so when the two liquids are put in contact with a porous filtering surface, the oil passes through the pores but not the water because the force with which their molecules bind is higher than in the oil.

This system has very interesting advantages from the point of view of the production of high-quality virgin olive oils by obtaining the loose oil found in olive paste, the one that has the lowest viscosity and the best organoleptic characteristics, fundamentally in relation to the profile of volatile oxidation compounds.

57.3.2.2 Pressure

As it is well known, the process is older. We have been extracting olive oils for 20 centuries, and still in some countries of the Mediterranean basin this discontinuous process is the most widespread.

The pressure process, using hydraulic presses, is based on the well-known Pascal's Principle, which indicates that the pressure exerted at any point in the fluid is transmitted equally to all points in the fluid. The pressures reached in the olive paste can range from 350 to 450 kg/cm^2 with operating times in each batch of 6–120 min.

Although fully automated pressure systems are currently marketed, it is very laborious and complicated to obtain high-quality virgin olive oils, due in part to the difficult cleaning of the different equipment and facilities required in the operation. On the other hand, the operation and labour costs are higher than in continuous systems.

57.3.2.3 Centrifugation

In the last 25 years, in the olive oil production processes, the separation of solid and liquid phases has undergone a great technological evolution in its two key stages:

- Solid-liquid separation of the olive paste through the use of the horizontal centrifuge or decanter.
- Separation of liquid phases (water and oil) using the vertical centrifuge or installations that involve only settling or settling plus filtration.

This mechanical separation is carried out by the action of centrifugal force on the olive paste using a high speed, where the solid and liquid phases are separated according to their different densities (Fig. 57.2).

In the separation of solid and liquid phases by means of centrifugation, using a decanter, two different systems have been used since 1991. Initially, industrial separation by centrifugation began around 1975 using a three-outlet decanter. Later, and due to the problem of wastewater from oil mills, in 1991 the decanter of two outlets began to be introduced in Spain, which little by little was extended to other olive oil-producing countries. Currently, two extraction systems coexist, although globally the use of the two-outlet decanter is more widespread in the world (Fig. 57.3).

The decanter is a device that basically consists of a rotor or bowl, with a frusto-conical shape; inside, adapted to its shape and with a small clearance, there is an auger. Usually, the movement of the rotor and auger is generated by an electric motor, but the auger rotates at a number of turns, generally less than the rotor. This is achieved by a reducer, driven by the main motor, which is driven by a system of pulleys and belts.

The rotor and auger can rotate in a speed range of 3000–3500 rpm. In the small decanters of some commercial companies, speeds of the order of 5000 rpm can be reached.

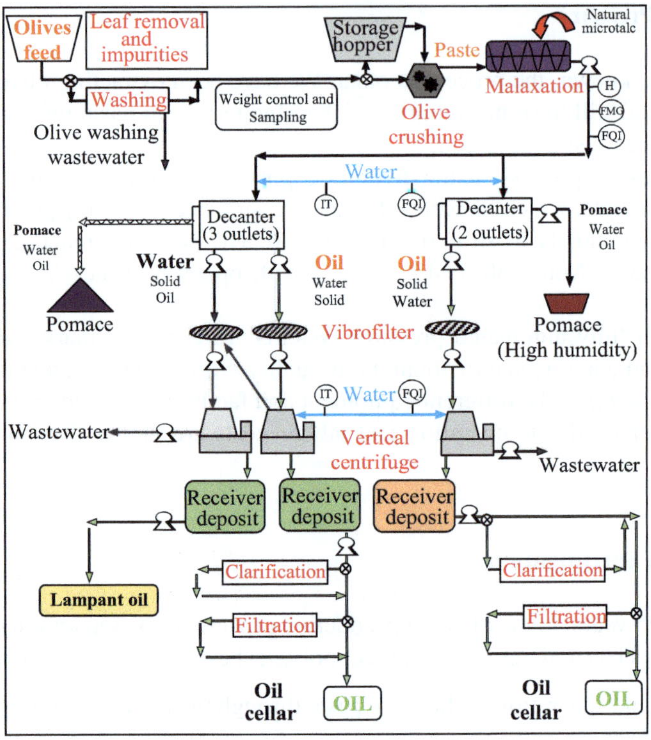

Fig. 57.2 Flowchart of olive oil production at the mill level using continuous centrifugation processes

Fig. 57.3 Formation of the centrifugation rings in the decanter

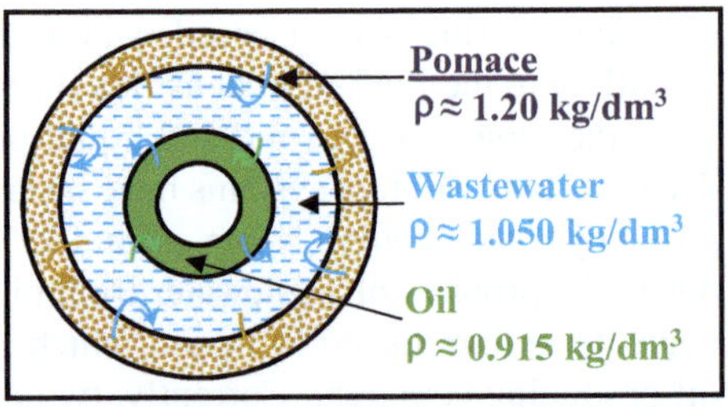

The direction of rotation of the bowl and the auger is the same, but due to the difference in angular velocity, a relative movement is originated on the surface of the auger (also, the solids adhere to the surface of the bowl due to their higher density), which makes the solids to move in the opposite direction to the advance of the propeller pitch. On the contrary, the liquid phases are pushed by the auger in the direction of the propeller's advance.

Use of Three-Outlet Decanter

This decanter is the one with the most sophisticated technology, since it allows the separation of the materials contained in the three centrifuge rings that have been formed inside (Fig. 57.2). In this system, a higher addition of water is necessary to the olive paste. The water addition is needed for several reasons: to facilitate the transport of the pastes, to maintain the operating temperature and to create liquid layers (rings) for the adequate separation of the oil, water and pomace.

Use of Two-Outlet Decanter

Historically, this was the first decanter that was used for the separation of solid and liquid phases. But it was rejected due to its bad separation, mainly concerning the oil outlet. To obtain high separation performances, the oil outlet could not be clean (Sánchez et al., 2019). This system requires less addition of water at the olive pastes that enter the decanter and has two outlets: one for the pomaces (with very high humidity, in the range 55–70%) and another for the oils.

In Spain and Portugal, the two-outlet decanter centrifugation processes are used mainly, while in Italy systems of pressure and centrifugation with three-outlet decanter are used and, to a lesser extent, the three-outlet decanter process. In general, in Greece, Morocco, Tunisia and in the rest of the producing countries, the three systems are used (pressure, centrifugation with two- and three-outlet decanter), but the most widespread is the two-outlet decanter process (Fig. 57.4).

In the three-outlet decanter process, the oil and water streams each go to a vibrating screen located on the sides of the decanter. In the two-outlet decanter process, the oil stream goes to the only vibrating screen that exists in this case, since there is no water outlet. The vegetation and addition of water are incorporated into the pomaces. The objective of the vibrating screen is to retain the solid particles that oil can carry. In the three-outlet decanter, the purpose of the vibrating screen at the water outlet is also to separate the solid particles that it carries.

57.3.3 Separation of Liquid Phases

Once the oil has been obtained by any of the most common procedures for the extraction of virgin olive oil, either the old system of presses or the centrifugal systems with two- or three-outlet decanter, it is necessary to proceed to the separation of the different phases: oil, small proportions of water and solid particles. Besides, this oil can contain microorganisms such as yeasts, fungi and bacteria (Papadimitriou et al., 2013), a fact that justifies the presence of enzymes that can contribute to the

Fig. 57.4 Continuous centrifugation process using two-outlet decanter. Olive mill of 'Picualia Oils', Bailén, Jaén (Spain)

occurrence of biochemical reactions in the aqueous phase that accompany the oil and the formation of undesirable compounds that, ultimately, by a process of mass transfer go from the aqueous phase to the oily phase, which can lead to the degradation of the olive oils that are being obtained.

This separation can be carried out using a classical decantation method or in a more efficient way through a centrifugation operation.

57.3.3.1 Decantation

Sedimentation or decantation constitutes the traditional system; the olive oil is separated from the water based on a difference in density between both phases. Currently, in oil mills, this operation can be carried out discontinuously (static decantation) or continuously (dynamic decantation) using a battery of decantation equipment.

For the separation of oil and water, so-called decantation tanks or wells have traditionally been used, consisting of small pools interconnected at the top, whereby the oil, separated, slowly passed from one well to the other and left finishing debugging. This method and structure of wells, used in traditional oil mills, has been progressively abandoned because it requires extreme cleaning and high labour costs. However, some oil mills continue to maintain and use these well rooms regardless of the existence of other liquid-liquid separation procedures.

The evolution of these decanting vessels and the need to use new inert materials for the oils led to the replacement of the wells by decanting tanks, built-in stainless steel with a cylindrical configuration and a conical bottom with an inclination of approximately 60°, to concentrate the solid waste deposited at the lowest point and facilitate its exit together with the aqueous phase during the purge.

In general, these settling deposits cause large sensory defects, the appearance of vinegar being characteristic in the oils produced (Fortini et al., 2015).

57.3.3.2 Centrifugation

Initially, it is necessary to emphasize that centrifugation systems were introduced in the oil mills to separate the liquid phases. In other words, the use of the vertical centrifuge in the oil mills predates the use of the decanter or horizontal centrifuge. The disc centrifuge or the vertical centrifuge, introduced in the virgin olive oil production process in the early years of the last century, is now crucial in industrial-scale production.

The great adaptability of vertical centrifuges to the variable contents of both moisture and impurities in the oily must makes them almost essential equipment in the olive oil clarification process. However, as a consequence of the rotational speed of the vertical centrifugation system, the oils subjected to this process show an additional increase in the content of dissolved oxygen that has a negative effect on a higher degree of oxidation thereof. Furthermore, as the addition of water is required in the process, a slight decrease in the content of phenolic compounds could occur due to the high affinity of water for this type of polar compound.

To eliminate the risk of oxidation of olive oil during vertical centrifugation, processes are being developed which involve the use of an inert nitrogen atmosphere, which would minimally affect the content of volatile compounds (in the same proportion as the use of an environment with aeration), but it would practically not alter the phenolic compound content of the oil. Another alternative proposal would be based on carrying out a stripping or washing operation with nitrogen gas after the oil has been subjected to the vertical centrifugation stage. This operating system would make it possible to reduce the oxidative effect on the oily must, preserving the content of phenolic compounds, and insignificant losses in the content of volatile compounds in the oil.

In extraction processes that use the three-outlet decanter, two vertical centrifuges are required: one for the oil outlet and another for the water outlet. In this case, the oil comes out cleaner. When using the two-outlet decanter, only a vertical centrifuge is necessary, and it is used to remove a considerable amount of water and particles that accompany the oil. Therefore, and in general, the consumption of water required in the latter case in the vertical centrifuge is higher (Sánchez et al., 2019).

Essentially, the vertical centrifuge is made up of a rolling drum and a system of dishes with a special arrangement. The rolling drum and platter system rotates at high speed in a range from 6000 to 7500 rpm, depending on the construction company. Currently, it has an automatic and programmable cleaning system. All internal parts of the centrifuge are made of special stainless steel.

In processes that use decanter with two outlets, in general, vertical centrifuges have the disadvantage of requiring high water consumption and generating a large amount of wastewater (oil washing water). Alternatives to the use of vertical centrifuges have recently been developed, based on three operating principles: decantation, filtering and a procedure called mechanized immersed separation. These alternatives to centrifugation share a common characteristic of zero or minimal water consumption and at the same time that they can lead to a reduction in energy costs.

In the production of early harvest olive oils (extra virgin olive oils, premium), the position of the oil outlet in the vertical centrifuge is a strategic place where an experienced taster performs the tasting of the oils produced and allows a previous classification of the oils destined for different deposits in the oil cellar.

57.3.4 Downstream Operations

The olive oils from the separation of liquid phases (settling equipment or receiver tank of the vertical centrifuge), in a large part of the oil mills, are taken directly to the tanks of the oil mill cellar. This way of operating is the least appropriate, but it is the most widespread in producing countries, especially in the production of olive oils where industrial yields prevail over product quality.

An alternative way to the previous procedure is to send the oil to clarification tanks, where the microparticles and microdroplets are deposited. In this equipment a very frequent purge must be carried out. This procedure allows the olive oil to enter the cellar's tanks in a cleaner way and to avoid or minimize the formation of alkyl esters (ethyl and methyl esters) in the aqueous phase, which later, through a mass transfer process, pass to the oil phase.

57.3.4.1 Use of Clarifiers

Though natural settling by gravity has been proposed as a better practice than vertical centrifugation since it is a less oxidative process, most research studies carried out, and even the real experience, suggest the opposite. In fact, Gila et al. (2016) reported the same oxidative status in virgin oils from both systems. Even more, this study pointed out a higher winey defect risk in natural settling system. The same research team (IFAPA Center, Jaén, Spain) run several experiments using transparent settling units. They concluded that oil and water flow through such decanting devices doesn't match what is expected. In some parts of the decanters, the oil remains in contact with wastewater for too long with great phenolic compounds losses and sensorial cross-contamination (vinegary, wastewater, muddy-sediment). Further, Gila et al. (2016) highlighted that vertical centrifugation is much more useful than natural settling when oily must from the decanter is dirtier.

57.3.4.2 Oil Filtration

There are different filtration systems adapted to different oil processing volumes and status of dirtiness. Vertical filtering tanks using earth are more efficient to process large volumes, while press filters are preferred for lower volumes. In both cases, several filtering media are available (diatomaceous earth, cellulose fibre or paper, cotton etc.). New filtering systems are proposed, such as inert gas flow or polypropylene bags. Furthermore, filtering operation can be carried out in more steps of different intensity. Whatever the system used, it must be inert to the oil matrix, remove moisture and suspended solids and leave minor compounds to pass through (phenolic compounds, pigments, volatiles).

Concerning the best moment to filter the oil, most published information about oil behave after filtering at different moments coincides on doing as soon as possible (Fortini et al., 2015). Even, in-line filtration has been proposed as a new step in continuous processing after vertical centrifugation. However, early filtering of ready-made oils is always challenging and expensive, mainly when green fruits are processed. Nowadays, since filtering efficiency is weak, it is accepted as a good option to choose a decanter and vertical centrifuge set-up to get cleaner oil that is stored in tanks for a few days and then proceeds with filtration.

The main sensorial defects in this step come from cross-contamination. Most usual defects are muddy-sediment, rancid, winey and musty-humid. Frangipane et al. (2023) carried out a revision about Effects of Filtration Processes on the EVOO Quality.

57.3.5 Conservation, Storage and Packaging of Olive Oils

Olive oil shelf life is the main issue, and a lot of research has been developed on this topic. Available technology makes it feasible to preserve virgin oils with a minimum quality loss. Needless to say, shelf life depends on many factors, and sometimes one particular oil batch can behave worse than expected because not all of them were considered. Currently, early filtration, room temperature control, light filtering packaging and inert gases are usual practices to enhance shelf life.

Extra premium virgin oils are more challenging because the 'premium' character is related to the volatile compounds (usually 5 and 6 carbon atom molecules, C5 and C6) of the lipoxygenase (LOX) pathway that tend to pass into the headspace during distribution step. All of this can cause the 'extra virgin' category to be maintained, but the 'premium' feature is lost (Diez-Betriu et al., 2022).

A general advice concerns washing and cleaning at any time and along the whole farm-to-fork chain.

The most common sensorial defects at this step are rancid and muddy-sediment, though winy-musty can happen sometimes. Premium oils rarely develop rancidity within a year. However, premium oils easily became 'tired-aged' which is not an official defect but depreciates oil in the 'gourmet' market.

57.4 Forthcoming Challenges Facing the Table Olive and Olive Oil Extraction Industries

The future of the table olive industry is very promising as consumption is expected to increase in the coming years. Work is being done on the dissemination of the healthy characteristics of this food. In addition, modifications are being developed in the manufacturing processes in order to reduce polluting effluents and therefore the environmental impact of this industry. On the other hand, ways to improve the shelf life of the product and its packaging forms are being studied in order to reach more consumers.

In the olive oil industry, it is very necessary to move towards an increase in the production of quality virgin olive oils and with it a progressive decrease in lampant olive oils. If this challenge is met, in parallel, the by-products generated (pomace, olive stone and olive leaf) will also be of better quality, and therefore, the value of these materials will be higher.

In both industries, these objectives of improving the quality of the virgin olive oils and table olives produced must begin in the olive grove itself and end at the consumer's own table. All stages of production will have to be impregnated with a character of sustainability.

References

Abu-Khalaf, N., & Hmidat, M. (2020). Visible/near infrared (VIS/NIR) spectroscopy as an optical sensor for evaluating olive oil quality. *Computers and Electronics in Agriculture, 173*, 105445.

Alba, J. (2008). Elaboración del Aceite de Oliva Virgen. In D. Barranco, R. Fernández-Escobar, & L. Rallo (Eds.), *El Cultivo del Olivo* (6th ed., pp. 657–698). Mundi-Prensa.

Angerosa, F., Lanza, B., & Marsilio, V. (1996). Biogenesis of 'fusty' defect in virgin olive oils. *Grasas y Aceites, 47*(3), 142–150.

Arbizu, J. (2015). *Respuesta del olivo (cv Arbequina) a riego deficitario bajo condiciones super-intensivas en La Rioja*. Doctoral Thesis, Univesity of La Rioja (Spain).

Aymen, M., Sánchez-Ortiz, A., Aguilera, M. P., Sánchez, S., Jiménez, A., & Beltrán, G. (2017). Los ultrasonidos de potencia como alternativa al batido de la pasta (OLIMERCA 2° TRIMESTRE: 36–52).

Ayr, U., Tamborrino, A., Catalano, P., Bianchi, B., & Leone, A. (2015). 3D computacional fluid dynamic simulation and experimental validation for prediction of heat transfer in a new malaxer machine. *Journal of Food Engineering, 154*, 30–38.

Bianchi, G. (1999). Extraction systems and olive oil. *OCL, 6*(1), 49–56.

Boudebouz, A., Romero, A., Hermoso, J.-F., Boqué, R., & Mestres, M. (2023). Effect of hopper loading on the formation of alkyl alcohols in olive fruits and its relationship with sensory quality losses of virgin olive oil. *Food, 12*(13), 2633. https://doi.org/10.3390/foods12132633

Civantos, L. (1999). *Obtención del Aceite de Oliva Virgen* (2nd ed., p. 112). Agrícola Española.

Clodoveo, M. L. (2012). Malaxation: Influence on virgin olive oil quality. Past, present and future – An overview. *Trends in Food Science and Technology, 25*, 13–23.

Clodoveo, M. L., Durante, V., La Notte, D., Punzi, R., & Gambacorta, G. (2013). Ultrasound-assisted extraction of virgin olive oil to improve the process efficiency. *European Journal of Lipid Science and Technology, 115*(9), 1062–1069. https://doi.org/10.1002/ejlt.201200426

Di Giovacchino, L. (1996). Olive harvesting and olive oil extraction. In D. Boskou (Ed.), *Olive oil. Chemistry and technology* (1st ed., pp. 12–51). AOCS Press.

Di Giovacchino, L. (2013). Technological aspects. In R. Aparicio & J. Harwood (Eds.), *Handbook of olive oil* (pp. 57–96). Springer.

Díez-Betriu, A., Romero, A., Ninot, A., Tres, A., Vichi, S., & Guardiola, F. (2022). Effect of freezing, fast-freezing by liquid nitrogen or refrigeration to preserve premium extra virgin olive oil during storage. *European Food Research and Technology, 248*, 2651–2663. https://doi.org/10.1007/s00217-022-04078-9

Estradas, J. M.ª (2011). *La aceituna de mesa: nociones sobre sus características, elaboración y cualidades*. 'Diputación' of Seville.

Fortini, M., Migliorini, M., Cherubini, C., Cecchi, L., Guerrini, L., Masella, P., & Parenti, A. (2015). Shelf life and quality of olive oil filtered without vertical centrifugation. *European Journal of Lipid Science and Technology, 118*(8). https://doi.org/10.1002/ejlt.201500229

Frangipane, M. T., Cecchini, M., Monarca, D., & Massantini, R. (2023). Effects of filtration processes on the quality of extra-virgin olive oil—literature update. *Food, 12*(15), 2918. https://doi.org/10.3390/foods12152918

García, J. M., Gutiérrez, F., Castellano, J. M., Perdiguero, S., Morillo, A., & Albi, M. A. (1996). Influence of storage temperature on fruit ripening and olive oil quality. *Journal of Agricultural and Food Chemistry, 44*, 264–267.

Garrido-Fernández, A., Fernandez-Diez, M. J., & Adam, M. R. (1997). Olives and table olives. In A. Garrido Fernández, M. J. Fernandez-Diaz, & M. R. Adams (Eds.), *Table olives: Production and processing* (1st ed., pp. 10–21). Chapman & Hall.

García-Pizarro, Á., Romero, A., Martí, E., Hermoso, J. F., Ninot, A., Aceña, L., & Mestres, M. (2024). Feasibility of using secondary attributes in sensory analysis to characterize protected designation of origin of olive oil. *Agronomy, 10*, 2218. https://doi.org/10.3390/agronomy14102218

Gila, A., Beltrán, G., Bejaoui, M. A., Sánchez, S., Nopens, I., & Jiménez, A. (2016). Modeling the settling behavior in virgin olive oil from a horizontal screw solid bowl. *Journal of Food Engineering, 168*, 148–153.

Gómez del Campo, M., & García, J. M. (2013). Summer deficit-irrigation strategies in a hedgerow olive cv. 'Arbequina' orchad: Effect on oil quality. *Journal of Agricultural and Food Chemistry, 61*, 8899–8905.

Hermoso, J. F., Romero, A., & Tous, J. (2011). Análisis tecno-económico de los nuevos modelos de explotación oleícola. *Vida Rural, 334*, 40–47.

Hermoso, J. F., Romero, A., Marti, E., Vichi, S., Boynuegri, P., & Caixach, J. (2017). Influencia de las operaciones de postcosecha del fruto sobre la calidad del aceite de oliva virgen. *Vida Rural, 417*, 46–51.

IOC. (2004). *Trade standard applying to table olives Resolution No. RES-2/91-IV/04.*

Jiménez, A., Uceda, M., & Beltrán, G. (2001). Efecto de la separación suelo-vuelo y la preparación del fruto en el patio sobre las características del aceite. *Oleo, 4*, 112–117.

Leone, A., Romaniello, R., Zagaria, R., Sabella, E., De Bellis, L., & Tamborrino, A. (2015). Machining effects of different mechanical crushers on pit particle size and oil drop distribution in olive paste. *European Journal of Lipid Science and Technology, 117*(8), 1271–1279. https://doi.org/10.1002/ejlt.201400485

Madrid, A. (2018). *Elaboración de aceitunas de mesa. Incluyendo la norma de calidad y el análisis sensorial.* AMV Edition.

Olmo-Cunillera, A., Lozano-Castellón, J., Pérez, M., Miliarakis, E., Tresserra-Rimbau, A., Ninot, A., Romero-Aroca, A., Lamuela-Raventós, R. M., & Vallverdú-Queralt, A. (2021). Optimizing the Malaxation conditions to produce an Arbequina EVOO with high content of bioactive compounds. *Antioxidants, 10*(11), 1819. https://doi.org/10.3390/antiox10111819

Papadimitriou, V., Dulle, M., Wachter, W., Sotiroudis, T. G., Glatter, O., & Xenakis, A. (2013). Structure and dynamics of veiled virgin olive oil: Influence of production conditions and relation to its antioxidant capacity. *Food Biophysics, 8*, 112–121.

Polari, J. J., Garcí-Aguirre, D., Olmo-García, L., Carrasco-Pancorbo, A., & Wang, S. C. (2018). Interactions between hammer mill crushing variables and Malaxation time during continuous olive oil extraction. *European Journal of Lipid Science and Technology, 120*(8), 1800097. https://doi.org/10.1002/ejlt.201800097

Romero, A. (2011). *Caracterización y diferenciación de los aceites vírgenes de oliva de la comarca del Priorat (Tarragona), dentro del mercado global de aceites de la variedad 'Arbequina.* Doctoral Thesis, Univesity of Lleida.

Romero, A., Vichi, S., Boynuegri, P., Caixach, J., Hermoso, J. F., & Marti, E. (2016). Pérdidas de calidad en aceite debidas a las operaciones de lavado y almacenaje de aceitunas. *Almazaras, 3*, 28–33.

Romero, A., Hermoso, J. F., Martí, E., Beltrán, G., Boudebouz, A., & Vichi, S. (2018). Factores de fabricación en aceites de alta calidad. *Almazaras, 12*(3), 10–22.

Rufat, J., Romero-Aroca, A. J., Arbonés, A., Villar, J. M., Hermoso, J. F., & Pascual, M. (2018). Mechanical harvesting and irrigation strategy responses on 'Arbequina' olive oil quality. *HortTechnology, 28*(5), 607–614. https://doi.org/10.21273/HORTTECH04016-18

Sánchez-Ortiz, A, Bejaoui, M. A., Quintero-Flores, A, Jiménez, A., & Beltrán, G. (2018). Biosynthesis of volatile compounds by hydroperoxide lyase enzymatic activity during virgin olive oil extraction process. *Food Research International, 111*, 220–228. https://doi.org/10.1016/j.foodres.2018.05.024

Sánchez, S., Pacheco, R., Moya, A. J., & Cuevas, M. (2019). Separación de fases sólidas y líquidas. Evolución en los últimos 25 años. *Mercacei Magazine, 100*, 104–108.

Sastre, B., Arbones, A., Rufat, J., De Lorenzo, C., Pascual, M., Benito, A., Villar, J. M., Bonet, I., Paz, S., Santos, A., Hermoso, J. F., & Pérez, M. A. (2019). Influencia del riego en la calidad del aceite en olivar superintensivo. *Vida Rural, 470*, 26–32.

Sastre, B., Amadeu Arbonés, M., Pérez-Jiménez, Á., Pascual, M., Benito, A., de Lorenzo, C., Villar, J. M., Bonet, L. J., Paz, S., Santos, Á., Hermoso, J. F., Romero, A., Farolfi, C., & Rufat, J. (2023). Influence of regulated deficit irrigation on Arbequina's crop yield and EVOOs quality and sensory profile. *Agronomy, 13*(1), 31. https://doi.org/10.3390/agronomy13010031

Uceda, M. (2004). *Variabilidad intraespecífica de los factores que determinan la calidad del aceite de oliva*', Doctoral Thesis,. Univesity of Jaen.

Uceda, M., Jiménez, A., & Beltrán, G. (2006). Trends in olive oil production. *GRASAS Y ACEITES, 57*(1), 25–31. ISSN: 0017-3495.

Veneziani, G., Esposto, S., Taticchi, A., Urbani, S., Selvaggini, R., Di Maio, I., Sordini, B., & Servili, M. (2017). Cooling treatment of olive paste during the oil processing: Impact on the yield and extra virgin olive oil quality. *Food Chemistry, 22*, 107–113. https://doi.org/10.1016/j.foodchem.2016.10.067

Veneziani, G., Esposto, S., Taticchi, A., Selvaggini, R., Sordini, B., Lorefice, A., Daidone, L., Pagano, M., Tomasone, R., & Servili, M. (2019). Extra-virgin olive oil extracted using pulsed electric field technology: Cultivar impact on oil yield and quality. *Frontiers in Nutrition, 6*, 134. https://doi.org/10.3389/fnut.2019.00134

Vichi, S., Romero, A., Gallardo-Chacón, J., Tous, J., López-Tamames, E., & Buixaderas, S. (2009). Influence of olives' storage conditions on the formation volatile phenols and their role in off-odor formation in the oil. *Journal of Agricultural and Food Chemistry, 57*(4), 1449–1455.

Vichi, S., Romero, A., Tous, J., & Caixach, J. (2011). The activity of healthy olive microbiota during virgin olive oil extraction influences oil chemical composition. *Journal of Agricultural and Food Chemistry, 59*, 4705–4714.

Vichi, S., Boynuegri, P., Caixach, J., & Romero, A. (2015). Quality losses in virgin olive oil due to washing and short-term storage before olive milling. *European Journal of Lipid Science and Technology, 117*(15), 2015–2022. https://doi.org/10.1002/ejlt.201500066

Yorulmaz, A., Tekin, A., & Turan, S. (2011). Improving olive oil quality with double protection: Destoning and malaxation in nitrogen atmosphere. *European Journal of Lipid Science and Technology, 113*(5), 637–643. https://doi.org/10.1002/ejlt.201000481

Chapter 58
Current State of Technological Applications and Novel Approaches to Value-Adding Methods for By-Products from Oil Mills

Sebastián Sánchez-Vilasclaras, Belén Fernández, Alberto J. Moya, Agustí Romero, Mª. Lourdes Martínez, Juan F. Hermoso, Gassan Hodaifa, Rafaela Cáceres, Manuel Cuevas, Juan F. García, Soledad Mateo, Juan G. Puentes, and Silvia Peinado

Abstract The olive cultivation and oil extraction processes generate wastewaters and solid wastes that have a large impact on ecosystems due to their high organic matter content and phytotoxicity. Olive mill wastewater (OMW) has a generation ratio between 0.4–0.6 and 0.8–1.2 liters of OMW per kg of olive fruits for batch pressing and continuous centrifugation system using two-outlet decanters, respectively. In addition, the semisolid waste obtained is usually dried and deoiled, resulting in a by-product known as "extracted pomace" (by-product from extraction process), which constitutes a very interesting solid biofuel with a higher heating value in the range of 13.8–15.8 MJ/kg. It is estimated that olive pomace (OP) is one of the major by-products, with a generation ratio of 0.3–0.4 kg-OP/kg of olive fruits. Olive leaves are generated at the first stage of the olive oil processing and can weigh up to 10% of the raw material that arrives at the mill. Finally, olive stones and seeds (16–25% of the total olive fruit weight) are recovered after oil separation in olive mills or from the olive table industry. Taking into account the huge amount of wastes produced during the olive oil extraction, it is not surprising the search for strategies

S. Sánchez-Vilasclaras (✉) · A. J. Moya · Mª. L. Martínez · G. Hodaifa · M. Cuevas
J. F. García · S. Mateo · J. G. Puentes · S. Peinado
Center for Advanced Studies in Olive Grove and Olive Oils, University of Jaén, Science and Technology Park GEOLIT, Mengíbar, Jaén, Spain
e-mail: ssanchez@ujaen.es; ajmoya@ujaen.es; mlmartinez@ujaen.es; ghodaifa@ujaen.es; mcuevas@ujaen.es; jfgarcia@ujaen.es; smateo@ujaen.es; jgpuentes@ujaen.es; speinado@ujaen.es

B. Fernández · A. Romero · J. F. Hermoso · R. Cáceres
IRTA, Institute of Agrifood Research and Technology, Torre Marimon, Caldes de Montbui, Barcelona, Spain
e-mail: belen.fernandez@irta.cat; agusti.romero@irta.cat; juanf.hermoso@irta.cat; rafaela.caceres@irta.cat

J. Muñoz-Rojas, R. García-Ruiz (eds.), *The Olive Landscapes of the Mediterranean*, Landscape Series 36,
https://doi.org/10.1007/978-3-031-57956-1_58

to transform these olive oil wastes into resources for olive production. In addition to this, another approach is the promotion of diversified agroecosystems with a broader biodiversity and ecosystem services, through the design and implementation of mixed farming systems (i.e., integration of livestock and olive cultivation). This chapter briefly reviews well-established and emerging valorization technologies applied mainly to by-products and OMW. Among others, there are biological-based processes such as composting or biogas production, but also thermochemical technologies such combustion, gasification, and pyrolysis (olive stones and seeds have been mainly used as an energy source), or concentration techniques such as spray-drying and the production of biosorbents. In some cases, these practices are integrated with extractive processes, using supercritical fluids or pressurized liquids and ultrasound-assisted or microwave-assisted extractions, to separate specific high-value compounds as fatty acids, antioxidants, enzymes, dietary animal supplements, and functional ingredients for food, cosmetic, and pharmaceutical industries.

Keywords Biochar · Compost of olive pomaces · Olive leaves · Olive mill wastewaters · Olive pomaces · Olive stones

58.1 Introduction

58.1.1 By-Products and Wastes from Table Olive Industries

Table olive industries generate mainly two by-products: olive stones from "pitted table olives" commercial preparation and olives that are discarded due to their low caliber. In this process, a liquid effluent residue composed of wash wastewater chiefly from lye treatments is generated. These wastewaters are the main residues and present a high pollutant impact due to their chemical composition. The organic fraction is composed of sugars, phenolic compounds (tyrosol, hydroxytyrosol, and ferulic acid among others), some nitrogenous compounds, organic acids, tannins, pectins, and oil residue compounds (Sánchez et al., 2006). On the other hand, the inorganic fraction contains high concentrations of NaCl and NaOH and trace amounts of some metals. Nowadays, the elimination of these wastes is mainly done in evaporation ponds due to their high concentration of organic and mineral compounds, with the consequent environmental and economic impact due to odors, potential leakage, and occupation of agricultural spacing (De los Santos et al., 2020). In this sense, several projects have been developed with the aim to transform these wastes into by-products, which include (i) reusing or treating the processing water (NaOH solutions or fermentation brines) by ultrafiltration and adsorption, (ii) fermentation to obtain phenolic compounds (Brenes et al., 2004; Garrido et al., 1992), (iii) NaCl removal from black olive commercial preparations (De Castro et al., 2007), (iv) application as solutions in agriculture (De los Santos et al., 2020),

(v) media for the culture of microalgae with added value (Contreras et al., 2016), and (vi) composting or biogas production (García, 2014).

Regarding the by-products already reused from this food industry, it can be mentioned the olives that are discarded due to their low caliber and the olive stones obtained from commercial pitted table olives, which contain olive oil inside. For this reason, both by-products are destined to olive oil production, where the stones are considered a by-product. It has been estimated that 12% of table olive production corresponds to these types of by-products (García, 2014), with a worldwide production of about 350,000 t/year (IOC, 2020).

Finally, olive leaves could be treated as yet another by-product, although production is usually low due to the manual harvesting method. Either way, it is used as animal feed and for electrical energy generation or direct incorporation on top of the olive grove soils.

58.1.2 By-Products and Wastes from Olive Oil Extraction

The olive grove and oil mill sectors not only produce olive oils but also generate, in parallel, a large quantity of by-products and wastes, some of them pending to be transformed and managed in an adequate and efficient way. To bring to light the importance of this issue, it should be considered that in Spain, the leading olive oil-producing country, around 3.5 million tons of olive fruits are produced annually, 75% of which are considered by-products and residues with high potential environmental impact (Sánchez, 2009). By-products and residues are generated within the olive groves, mainly when pruning, and in the olive mills during the extraction of the olive oil.

Liquid effluents generated by oil mills can produce a strong environmental impact, mainly due to their high organic load and their mineral salt content.

In all the Mediterranean Basin, and specifically in Andalusia, the olive mill subsector has caused major pollution problems. Currently, oil mills are in a very primitive state in relation to the treatment of their effluents.

58.2 Olive Pruning

58.2.1 Pruning Operation and Production

Pruning is the process by which the natural shape of the trees is modified, contributing to invigorate or restrict the branches' development. Apart from its effects on external tree appearance, pruning enables both air and light to penetrate inside the treetop and, thereby, to restore in whole or in part the tree productivity, increasing yields and fruit quality. Tree pruning frequency is highly variable among the main producing countries and highly dependent on the olive plantation density and objective of the pruning. Nevertheless, quite often it is carried out every 2 years. Biomass obtained from olive tree pruning will be composed of trunks (about 25%), branches (about 50%), and leaves (about 25%).

Since the biomass from olive tree pruning is renewable, and thus available annually, its reuse would make an essential contribution to the sustainable development. Tree pruning is enriched in lignocellulose material. Taking into account that (i) over 11.6 million ha of olive trees are cultivated worldwide (54.4% in Europe) (Urieta et al., 2018), (ii) the area dedicated to olive tree cultivation in Spain is roughly 2.7 million ha (ESYRCE, 2018), (iii) 60% of this area corresponds to Andalusian olive groves, and (iv) on average, 1 hectare of olive grove (in traditional and intensive cultivation) generates 3 tons of pruning (on wet basis) per year (Sánchez et al., 2002; Andalusian Energy Agency, 2018), it can be calculated that 34.8 million tons of olive tree pruning is generated every year, of which 8.1 million tons would be produced in Spain, out of which 4.9 million tons would be generated in Andalusia. Notwithstanding only 70% of this biomass could be accessible for its recovery.

Currently, olive tree light pruning biomass produced in about half of the area cultivated in Andalusia is burned in a controlled manner in fields (Mateo et al., 2015), emitting CO_2, although it is neutral CO_2. Another common practice is the crushing of the tree pruning in the field to cover the soils as mulch, with beneficial effects such as providing protection against soil erosion and providing habitats for organisms as a soil conditioner. However, caution should be paid in pest- or disease-infested olive groves as these could be extended if tree pruning is applied on the soil. The current inappropriate biomass use of the tree pruning promotes the search for alternatives for this lignocellulose rich material, since the most widespread applications are focused on its minority use as a raw material in power generation plants (its low density implies high transport costs) or for heating in domestic biomass boilers.

58.2.2 Physicochemical Composition of Olive Tree Pruning

The main physical characteristics include the following:

- It is a very heterogeneous material, with special dependence on the type of olive grove, growing area, climate, etc.
- It has a very low density.
- It presents dependence because of its moisture.

Table 58.1 shows the elemental analysis and both structural and nonstructural constituents for three different fractions of olive tree pruning biomass.

It is relevant to highlight the zero sulfur content of the tree pruning, which can be particularly beneficial, from an environmental point of view, for an energy recovery from olive tree pruning biomass. Furthermore, the high nitrogen content of the leaves is also noteworthy, as it can be valorized for its potential crude protein value.

Table 58.1 Chemical and biochemical composition of the olive tree pruning (Sánchez et al. 2002)

Element, %	Pruned wood	Pruned leaves	Tree pruning
C	45.5	48.7	44.6
H	6.4	7.1	6.7
N	0.3	1.4	0.8
S	0.0	0.0	0.0
Components, %			
MVM[a]	10.7	8.5	10.2
Ash	1.5	5.0	3.3
Lignin	14.7	23.1	20.8
Cellulose	31.9	5.3	26.6
Hemicellulose	27.8	16.3	19.7

[a]*MVM* moisture and volatile matter

Olive tree pruning is high in cellulose but low in lignin, although the content of the latter is in the range of other lignocellulose wastes. It should be noted that while the amount of cellulose is higher than that of the hemicellulose in the wood, the contrary is true for the leaves.

On the other hand, leaves and the whole tree pruning have a high percentage of extractives, 40–45% (w/w) and 21–29% (w/w), respectively. This nonstructural fraction contains chemical compounds such as oleuropein, hydroxytyrosol, elenolic acid, and D-mannitol, among others, with multiple applications for their antioxidant, antimicrobial, and sweetening properties.

58.2.3 *Applications*

The potential of lignocellulose biomass as a source of chemicals or fuels is currently unquestionable, although its large-scale use still presents several problems that require significant research effort.

The main applications of the use of olive tree pruning biomass are based on the energy generation by thermochemical processes, fuel production by biochemical conversion, or raw material utilization for obtaining bioproducts with high added value.

58.2.3.1 Energy Generation by Thermochemical Processes

The main thermochemical processes to obtain energy can be combustion, gasification, or pyrolysis. For this purpose, olive tree pruning biomass is often densified to generate pellets and facilitate thermal use of this raw material.

Combustion is the conversion process in which the complete oxidation of biomass occurs; it is the most elemental and oldest system of energy recovery. The energy obtained in the form of heat is used in boilers for steam production and minor heating processes (domestic applications, small-scale industry, etc.). The efficiency of these systems is limited by the thermodynamic properties of the steam cycle and by the limitations of the construction materials. In general, the efficiency is low, with energy recovery yields ranging between 20 and 22% and high CO_2 emissions. The higher heating value of the olive-pruning debris is in the range 17–20 MJ/kg, while its lower heating value is approximately 16.5 MJ/kg (García-Martín et al., 2020).

Pyrolysis is thermal degradation in the absence of oxygen. The goal is to get as much fuel as possible. It starts around 200 °C, so the material degradation always leads to a mixture of CO and CO_2, volatile compounds, short-chain hydrocarbons, furfural, and a solid carbonaceous residue. As the temperature increases, the yield in liquids decreases, and gas production rises; at temperatures above 900 °C, products with a high content of H_2, CO_2, CO, CH_4, and C_2H_4 are obtained.

The most outstanding thermochemical process of those initially mentioned is gasification. This process converts biomass into fuel gas called producer gas (syngas) composed by CO, H_2, H_2O steam, CO_2, and ash particles. This syngas can be used in many combustion systems such as boiler, furnaces, and gas engines. It is a technology based on the partial oxidation of the biomass with oxygen or steam at temperatures not below 700 °C, offering great advantages:

- It produces a lower volume of gases, so pollutants present in the gaseous stream could be more easily removed, since they would be more concentrated, so smaller and cheaper purification systems would be required for an efficient separation.
- The fuel obtained is hugely versatile and can be used in a wide range of applications with conventional equipment, being possible to transport it to distant points from the place of generation.
- Thermodynamic efficiencies are higher if the aim is to use gas from gasification and expand it in internal combustion engines and turbines by taking advantage of the heat from the exhaust gases.
- It can be applied in gas cycles that are less sensitive to economies of scale than steam cycles, in which the price of the biomass resource is extremely decisive for the project profitability, mainly in regard to the transport of the material from the point of generation to the point of consumption.
- It is much more developed than pyrolysis for an industrial operation level, and both the design of the equipment and its application are simpler.

The gasification technology is under development, with the high cost associated with the initial setup and operation of these facilities being the current main drawback.

58.2.3.2 Fuel Production by Biochemical Conversion

In these processes, the cellulose biomass is subjected to different pretreatments (liquid hot water, steam explosion, ultrasound, ozonation, dilute acid hydrolysis, biological treatments, among others) in order to facilitate the access of enzymes and/or microorganisms to the carbohydrate fibers and to enhance their hydrolysis and/or fermentation. In processes based on biochemical conversion, the following advantages should be highlighted:

- Energy consumption is reduced, and safety is increased, because these treatments generally operate under conditions of pressure and temperature not as high as thermochemical processes.
- Reduces the risk of contamination by the residual biomass valorization instead of resorting to elimination processes or alternative techniques that would cause further environmental damage.
- Effectiveness in the use of the raw material by microorganisms is higher, for its great reproductive capacity, when comparing to the growth phases that occur naturally in plants and animals. In addition, the selectivity of an enzymatic reaction, against a chemical reaction, has a positive impact on bioproduct production.

Biochemical conversion of olive tree pruning biomass could lead to the generation of liquid fuels with great applications in the transport sector: biodiesel through saponification and transesterification reactions and bioethanol through hydrolysis and fermentation processes. Limited applications by methanation of olive tree pruning material could also be considered.

58.2.3.3 Generation of Other Products and Materials

Autohydrolysis of olive tree pruning biomass can produce oligosaccharides and monomeric carbohydrates with applications in food and pharmaceutical industry from which chemicals, fuel, and biological products can be developed. Dilute acid hydrolysis under moderate reaction conditions proved to be a reliable and easy low-cost method for quantitative conversion of this raw material to monomeric sugars (Moya et al., 2008; García-Martín et al., 2013; Mateo et al., 2014).

Acid hydrolysis also leads to a valuable solid residue mainly formed by cellulose and lignin. Thus, this process can be conceived as the first stage of an integrated strategy for this biomass utilization. It is possible to obtain nanocellulose (main applications in construction, packaging, transportation, and biomedical fields), and sugars released from olive tree pruning biomass have been traditionally used to produce ethanol by fermentation, although other bioproducts such as xylitol, organic acids, 1,3-propanediol, butanol, antibiotics, or recombinant proteins can be obtained (Formenti et al., 2014). On the other hand, lignin fraction could be used in multiple applications: energy recovery (heat and electricity), manufacture of active carbon, polyurethanes and polyesters, antioxidants in polymers such as low-density polyethylene, constituents of ion exchange resins to remove metals from aqueous solutions, and reinforcing material in biocomposites as well as primary components for chemical platform (methane and phenolic compounds).

Other valuable and commercial products (furfural, hydroxymethylfurfural, etc.) have been detected in hydrolysates, so their recovery by adsorption or ion exchange would allow using them as chemical intermediates or antioxidants. In addition to these compounds, other bioproducts with antioxidant capacity (caffeic acid, ferulic acid, vanillic acid, hydroxytyrosol, oleuropein) have been found, at different concentrations, in extractives of olive tree pruning biomass.

58.3 Olive Leaves

The olive leaf fraction that comes from the harvest has been considered as a residue both at olive grove and at oil mill levels. However, during the last decade the olive leaf is being considered of some economic value, and, according to its physico-chemical composition, it could constitute in the near future the main by-product generated at the oil mill level. This by-product could reach much higher prices than the pomaces and olive stone fragments.

58.3.1 Types of Leaves and Production

Two types of olive leaves are generated: those of the oil mill from the olive fruit cleaning operation (Fig. 58.1) and those olive leaves resulting from the pruning operation. The latter are part of the branches that have been separated from the tree (Fig. 58.2). The former is generated during the fruit harvest (in the countries of the Mediterranean Basin between October and February), while the pruning leaf fraction is generated in a period after the harvest (in the countries of the Mediterranean Basin between December and March).

At the oil mill level, it is necessary to differentiate the leaf from the olive tree and that leaf that accompanies olive fruits when collected from the ground. The biochemical properties of the two can differ significantly. The first one is typically intact and maintain its quality, while the second one is usually degraded and with clear symptoms of senescence. The quality of the olive leaf by-product is also a function of the type of production. In the olive groves, the highest quality is usually that which comes from the tree of an organic or biodynamic olive grove and in the olive mill which is separated from the harvest of the trees. On the other hand, the lowest quality is that which comes from olive groves recently treated with plant protection products or that from the soil line of the oil mill.

Statistical data on annual production of olive leave in both the oil mill and tree pruning is lacking. However, according to the type of harvest and the production system for the olive grove, the annual production of the oil mill leaf and the leaf fraction of the pruning can be estimated.

Fig. 58.1 Typical aspect of the oil mill leaf. On the left a pile of leaves from the tree fruit line and on the right a pile of leaves from the ground olive line

Fig. 58.2 Pruning leaf fraction. Harvesting and crushing operation of olive pruning in the olive grove

For an approximate calculation of olive leaf production at the oil mill level, it is convenient to consider that in an adequate harvest of the fruit in the olive grove, the fraction of olive leaf that can accompany the olives ranges between 5% and 10%. In Spain, the total area dedicated to olive cultivation is roughly 2,700,000 ha. Considering an average annual production (regardless of the type of olive grove) of 3 t olives/ha (Sánchez et al., 2002), the average production of oil mill leaf would be in the range 405,000–810,000 t/year.

In relation to the production of pruning leaf, more information is available, although it is related to the total production of pruning without prior separation of the leaf fraction. In previous projects carried out by the Bioprocesses Research Group (TEP-138, "Junta de Andalucía"), the mean annual total biomass of the olive pruning of traditional and intensive olive groves during four olive campaigns was calculated in 3.0 t on wet basis (Sánchez et al., 2002). Wood (material with a diameter greater than 5 cm) is excluded from this data. On the other hand, assuming that in the pruning material about 10% is olive leaf (Sánchez et al., 2002), and taking into account the area of the olive grove in Spain, the average annual production of pruning leaf in Spain can be close to 810,000 t.

Based on the above data, the total olive leaf production in Spain, both from the olive oil mill and from pruning, can reach a value of 1,620,000 t/year. Taking into account the total cultivated area in Spain, the annual average value would be close to 1000 kg olive leaf per hectare (on wet basis), if the increase olive leaf production is included in the growing implementation of the cultures of high-density and super-intensive. It is a data of great interest, which can be extrapolated to the other olive oil-producing countries.

58.3.2 Physicochemical Characterization

Olive leaf from the pruning has different physicochemical composition compared to the olive leaf of the oil mill. In turn, the composition of both types of leaves differs according to the olive variety (Tables 58.2 and 58.3).

In general, fresh olive oil mill leaf (separated from flight olives) is valuable for its contents of fibers, proteins, mineral salts, lipids, and antioxidant compounds (fractions of phenolic compounds and tocopherols).

According to the elemental analysis, the olive leaf is high in carbon, but sulfur is hardly detected (Table 58.3). Nitrogen content is low, although it has an interesting value (Table 58.3). As is well known, the crude protein content (Pc) can be calculated from the nitrogen percentage (Becker, 1994):

$$P_c = 6.25 \cdot \%N.$$

Table 58.2 Physicochemical composition of olive leaf of different varieties

Variety	% MVM_0[a]	% HEM	% CEL	% LIG	% ASH
"Picual"	24.53 ± 0.03	8.15 ± 0.36	9.16 ± 0.55	28.09 ± 0.98	8.46 ± 0.21
"Hojiblanca"	27.73 ± 0.03	7.52 ± 0.72	11.75 ± 0.81	27.11 ± 0.67	7.26 ± 0.03
"Arbequina"	23.33 ± 0.07	8.24 ± 0.39	12.90 ± 0.65	23.77 ± 0.81	7.05 ± 0.51

MVM_0, HEM, CEL, LIG, and ASH stands for moisture and volatile matter, hemicellulose, cellulose, lignin, and ash, respectively. Data are expressed in percentage on dry weight

[a]MVM_0 is expressed on wet basis

Table 58.3 Elemental analysis of olive leaves from oil mill of different olive varieties

Variety	% N	% C	% H	% S	% O
"Picual"	1.61	48.88	6.97	0.10	42.43
"Hojiblanca"	1.65	49.58	7.0	0.10	41.59
"Arbequina"	1.54	49.19	6.90	0.09	42.27

58.3.3 Current and Potential Uses of Olive Leaves

Currently, the olive leaf is becoming more revalued due to its protein, fiber, and mineral salt contents (for animal feed) and natural antioxidants (for human consumption and even for the chemical-pharmaceutical industry) (Sánchez, 2009).

58.3.3.1 Animal Feed

The use of the olive leaf in animal feed is a very old activity in Spain and, in general, in all olive oil-producing countries. Already in the sixteenth century, its use was recommended in the feeding of a certain type of animals, a time in which the olive leaf had to be supplied to animals due to the lack of foods, because of the degree of palatability that animals demonstrated by their consumption (Herrera, 1981).

The first studies on digestibility are due to Maymone et al. (1947), but in Spain, with the exception of some works carried out during the 1960s and 1970s of the twentieth century, it was not until the 1980s of that century when researchers paid attention to this by-product.

In projects carried out by the Bioprocesses Research Group (TEP-138, "Junta de Andalucía") with different companies from 1990 to date, it has been shown that fresh oil mill leaf (separated from flight olives) is valuable for its content in fibers, proteins, mineral salts, lipids, antioxidant compounds (fractions of phenolic compounds and tocopherols), and enzymatic digestibility. In these projects, this research group has worked mainly with three varieties—"Picual," "Hojiblanca," and "Arbequina"—which are the most cultivated in Spain. Crude protein content (on a dry basis) reached values close to 12.0% in tests carried out with the "Hojiblanca" variety. In these same two projects, it was determined, in fresh oil mill leaf, total lipid contents of 23% and from 1.3% to 1.6% of total phenolic compounds, values interesting referred to dry leaf of the var. "Hojiblanca."

To date, the use of the oil mill leaf in animal feed, mainly for goats and sheep, has been carried out in three different ways: directly as a fresh mill leaf, in form of silage, and in form of granulated feed. The latter two cases constitute a complete diet for the animal. Furthermore, in the form of granulated feed, it has been tested in cattle, poultry, and pigs in different research projects.

58.3.3.2 Production of Antioxidants

Among the high-added-value compounds that can be obtained from the olive leaf, the bioactive compounds stand out, particularly the antioxidants represented by maslinic and elenoic acids phenolic compounds and tocopherols. Currently, the separation and recovery of these compounds are of great interest for the agri-food and chemical-pharmaceutical industries. Due to their high antioxidant activity, some of these compounds can be incorporated as natural additives to functional foods and to other food products and drugs.

In the phenolic fraction of the olive leaf, the compound of greatest interest is oleuropein, which in its degradation renders hydroxytyrosol as a hydrolysis product. Apart from its high antioxidant power, properties such as anti-inflammatory, anticancer, and antiviral are attributed to this latter compound (Guinda et al., 2015). On the other hand, elenolic acid is full-spectrum antimicrobial.

In relation to the production of antioxidants, the olive leaf from the pruning of the olive tree has greater advantages, as it is a cleaner, fresher, and with less degradation by-product than that from oil mills.

58.4 Olive Stones

58.4.1 Separation Processes

The industrial production of table olives and olive oils involves the total or partial separation of the stone contained in the fruit. This operation is of great importance both from a technical and economic point of view. Destoner machines for table olives are made up of a system of buckets that collects the fruits previously fermented and classified according to their size. Then, a stainless steel pusher punctures the drupe to remove the whole stone. The machine automatically detects, with advanced optical systems, the appropriate position for the fruit to be treated and prevents it to be pitted from an incorrect axis (De Jódar et al., 2020).

Improvements in the design of destoner machines have enabled to increase their rates from 220 fruits per minute in the 1970s to 2500 fruits per minute nowadays. The whole stones obtained will have sizes and shapes that depend on factors such as the variety of fruit and weather conditions. Kiliçkan and Güner (2008), for example, reported lengths and thicknesses of 16.73 mm and 7.75 mm, respectively.

On the other hand, in the industrial extraction of virgin olive oil, after the operations of grinding, malaxing, and centrifugal separation, the wet olive pomace produced can be treated in a pulp-stone separator machine to recover the fragmented endocarps. This equipment contains a cylindrical stainless-steel sieve inside which a rotor with blades rotates at high speed (around 1500 rpm), forcing the passage of the liquid and fine solid fraction through the holes of the sieve. The endocarps cannot pass through the holes due to their hardness and size and will come out of the sieve at one end. This system is not able to separate the fraction corresponding to the seed of the fruit, so it is included in the pitted olive pomace. The maximum operating capacity is between 20,000 and 25,000 kg/h of initial pomace. The pulp-stone separator must ensure that the endocarp fragments are recovered clean, without remains of other elements of the olive (such as skin or pulp), also minimizing energy consumption and the addition of water to the wet olive pomace. Most endocarp fragments (> 85% wt.) have sizes ranging from 0.7 mm to 7 mm (Gómez-de la Cruz et al., 2015; Saleh et al., 2014).

In recent years, the stoning process for wet pomace has intensified in the oil mills; this means that exhausted olive pomaces with low endocarps content are obtained in pomace oil extraction plants.

In Spain, and mainly in Andalusia, from the decade of the 1990s, different attempts were made to destoning the entire stone, prior to extracting the oils. In recent years, some authors have considered this idea and have proposed the alternative of totally or partially destoning the olives before the grinding stage in the oil mills (Restuccia et al., 2018). This possibility, which could improve the quality of virgin olive oils (Ranalli et al., 2009) and would significantly increase the available quantities of whole olive stones, is barely used at present.

58.4.2 Physicochemical Composition

The whole olive stone is mainly made up of two fractions (seed and endocarp) with clearly differentiated compositions, properties, and applications. According to Sansoucy (1985), the seed and the endocarp represent, respectively, 2.0–5.5% and 17.3–23.0% of the total weight of the fruit.

The seed is the innermost part of the stone, the embryo of the fruit, and is rich in oils (up to 27% in wet basis), dietary fibers (36–39%), and proteins (12–15%), so its use would be fundamentally for food. The oil fraction of the seed contains phenolic compounds with applications in cosmetics, food, and health (González-Hidalgo et al., 2012), compounds that could be recovered by pressurized liquid extraction (Vásquez-Villanueva et al., 2020). On the other hand, the endocarp is the lignocellulose shell that protects the seed.

The elemental analysis reveals a low percentage of sulfur (typically <1%, w/w) and nitrogen (< 0.5%, w/w) in the olive-fruit endocarps, while from a molecular point of view (considering the mean values of the data in Table 58.4), olive stones are mainly composed of lignin (34.5%), hemicellulose (28.9%), and cellulose (27.7%). The sum of extractives and mineral matter does not exceed 10% of the dry weight of this biomass. According to Font et al. (2020), the main components of the ash are potassium (32.1%) and calcium (27.8%) oxides.

Table 58.4 Chemical composition of olive-fruit endocarps

	Parameters	Amount (%, w/w)
Elemental analysis	Carbon	49.7[a], 50.4[b], 51.2[c], 51.5[d]
	Oxygen	41.8[c], 41.9 [d], 42.2[b], 43.4[a]
	Hydrogen	6.3[d], 6.49[c], 6.83[a], 6.96[b]
	Nitrogen	0.13[a], 0.2 [d], 0.4[b], 0.45[c]
	Sulfur	0.04[b], 0.06[c], 0.1[d]
Structural compounds	Lignin	29.8[e], 33.4[a], 40.2[c], 40.4[b]
	Hemicellulose	26.3[a], 28.1[e], 31.7[c], 32.2[b]
	Cellulose	26.1[a], 27.2[b], 28.8[c], 29.9 [e]
Nonstructural compounds	Extractives	6.0[e], 6.01[a]
Minerals	Ash	0.61[d], 0.69[a], 0.7[e]

[a]Cuevas et al. (2019a)
[b]Hodaifa et al. (2013)
[c]Vilardi et al. (2018)
[d]Budinova et al. (2006)
[e]Cuevas et al. (2020)

58.4.3 Energy Generation

At present, whole olive stones and fragmented endocarps are used mainly for the production of energy. For this purpose, both thermochemical and biochemical conversion routes can be used.

58.4.3.1 Thermochemical Processes

The thermochemical pathways to obtain energy from olive stones include combustion, gasification, pyrolysis, and torrefaction.

(a) Combustion.

Direct combustion is the most common method for converting olive stones to useful energy due to the excellent fuel properties of this biomass. The fragmented endocarps from olive tree fruit have relatively high HHV (higher heating value, 18.8–20.9 MJ/kg, Mata-Sánchez et al., 2013) together with low contents of sulfur, nitrogen, and mineral matter. As a result, the environmental impacts associated with the combustion process are low. The small average sizes of the endocarps, as obtained in the oil mills, lead to high bulk density values (582–821 kg/m^3), which has a positive impact on the costs of transportation and storage of this source of energy. In any case, to improve the operation of the boilers, the presence of moisture and impurities in the biomass must be controlled (Cuevas et al., 2019b; Gómez-de la Cruz et al., 2015; Mata et al., 2015).

The combustion of olive stones has been studied, with good results, both in fixed bed (Mami et al., 2018; Morán et al., 2015) and fluidized bed burners (Grammelis & Kakaras, 2005). There are also available works in the literature on the analysis of the combustion of pulverized olive stone (Rabaçal et al., 2018), as well as the co-combustion of the stone with other materials: lignite (Topal et al., 2017) and animal sludge (Vamvuka et al., 2016).

(b) Gasification.

The gasification of the olive stone makes it possible to obtain fuel gas rich in H_2 and CO. In the literature, gasification studies of olive stones carried out in downdraft (Di Blasi et al., 1999) and fluidized bed gasifiers (Skoulou et al., 2008) are available. Equivalent ratio (ER) values close to 0.2 would produce gases with a lower heating value (LHV) of about 6.5 MJ/Nm^3 in which there would be H_2 (24% v/v), CO (14.3% v/v), CO_2 (19% v/v), and around 5% v/v of CH_4 (Skoulou et al. 2008). Nevertheless, one issue found for the use of the synthesis gases is the presence of particles and tars, which could be eliminated by cyclones, scrubbers, and catalytic tar converters (Corella et al., 2004).

(c) Pyrolysis and dry torrefaction.

The pyrolysis of the olive stone generates synthetic gases of medium heating value together with solid products (biochar) and liquids (tars) that could also be used for energy purposes. The performance of the process depends on factors such as the maximum operating temperature (Sánchez & San Miguel, 2016), the heating rate (Manyà et al., 2007), or the particle size (Mancilla et al., 2013). Bartocci et al. (2015) pyrolyzed olive stones at 600 °C for 30 min to obtain syngas (44.2% wt.), tar (24.9% wt.), biochar (17.0% wt.), and water (14% wt.). The gaseous fraction, composed of H_2 (29.5% v/v), CO (31.5% v/v), and CH_4 (20.5% v/v), had an LHV of 16.0 MJ/kg.

When biochar is used for energy purposes, Sánchez and San Miguel (2016) found that the best pyrolysis conditions for olive endocarps are 278 °C during 15 min, which in practice would be equivalent to a dry torrefaction process, resulting in a biochar with HHV of 23.4 MJ/kg. The biochar produced from olive stones has been used in applications such as the manufacture of activated carbon (Hazzaa & Hussein, 2015), hydroponic culture medium (Karakaş et al., 2017), and manufacture of battery electrodes (Moreno et al., 2016).

(d) Wet torrefaction.

The moderate heating of the olive stone under an inert and aqueous environment allows generating a solid material (hydrochar) with improved energy properties compared to the raw material. In this sense, Cuevas et al. (2019b) reported that operating conditions of 225 °C-10 min led to solids with the highest HHV (23.4 MJ/kg), energy efficiency (74.7%), and comprehensive combustibility index (6.19 × 10–7 min^{-2} °C^{-3}). Álvarez-Murillo et al. (2015) achieved hydrochars with HHV of 29.59 MJ/kg after wet torrefaction (230 °C-30 h) of olive stones, although under these conditions the solid recovery yield was relatively low (46.8%).

58.4.3.2 Biochemical Conversion

The transformation of olive stone to bioethanol has been extensively studied in the last decades. In general, the results obtained are not entirely satisfactory due to the high lignin content of this biomass, its low richness in cellulose, and the low potential of hemicellulose for the production of D-glucose. The enzyme accessibility to cellulose has been improved by applying different pretreatments such as liquid hot water (Cuevas et al., 2009; Miranda et al., 2019), steam explosion (Fernández-Bolaños et al., 2001), organosolv treatment (Cuevas et al., 2015), and dilute-acid hydrolysis (Cuevas et al., 2020). The use of the hemicellulose fraction, very rich in xylans, could be carried out through its hydrolysis and the subsequent fermentation of the produced D-xylose into ethanol using nontraditional yeasts.

From an economic point of view, the production of ethanol and, in general, of bioenergy from biomass could be more feasible if products with high-added value (antioxidants, oligosaccharides, xylitol, and other bioactive molecules) were obtained in the global process. Following this idea, Goldfarb et al. (2017) proposed a scheme composed of supercritical fluid extraction of olive stones to recover phenolic compounds, followed by pyrolysis of the solid to produce biofuels, and, finally, activation of the biochar to yield adsorbent materials.

58.5 Olive Pomaces

58.5.1 Pomace Types

Olive pomaces (OP) are the by-product obtained after the olive milling step. OP contain crushed olive stones, cleaning and process water, all the fruit materials, and a low percentage of olive oil, and their generation ratio depends on the extraction process (Di Giovacchino, 2010; García-Martín et al., 2020). Pomaces are produced by batch (pressing) or continuous (centrifuging) processes. Both include washing (removal of soil, leaves, and dust), crushing, and malaxation steps.

Pressing with hydraulic press leads to a solid phase of the olive paste: the ground paste is placed between pressing mats and is subjected to pressure, to expel the liquid phases (oil and water). The production of this type of pomace is in the range 0.25–0.35 t OP/t olive fruits. The mixture of oil and water is then poured into a gravity decanter to easily separate the oil due to its different density with that of vegetation and process waters.

In the separation of solid and liquid phases by means of centrifugation, using a decanter, two different systems are used: three-outlet decanter and two-outlet decanter. In the three-outlet decanter process, water is added (0.5–1.0 dm^3 water/kg paste) to facilitate an initial separation of the solid waste (0.45–0.52 t OP/t olives, Table 58.5) from the oil must in a horizontal centrifuge. Then, the oil must go through a vertical centrifuge, where the oil is separated from the vegetable water. In

Table 58.5 Range of physicochemical characteristics of the pomaces based on the process used (Di Giovacchino, 2010; García-Martín et al., 2020)

Process	MVM[a] %	Fat matter % (in wet basis)	Fat matter % (in dry basis)	Production kg/t olives
Pressure	22–35	5.0–8.0	7.0–11.0	250–350
Three outlets	45–55	3.0–4.5	6.0–8.0	450–520
Two outlets	65–75	2.0–3.5	6.0–7.5	800–850

[a]HVM, moisture and volatile matter

the two-outlet decanter process, less water addition is required at the decanter inlet, and the vegetation and process water are incorporated into the pomace as indicated in Chap. 10; hence, the main by-product is a mixture of OP and wastewater called total pomace olive waste mill (TPOWM) (0.80–0.85 t TPOWM/t olives, Table 58.5) (García-Martín et al., 2020; Araujo et al., 2015; Dermeche et al., 2013; Di Giovacchino, 2010).

58.5.2 *Pomace Oil Extraction*

Pomaces with water content up to 48–54% can be sent to pomace units to extract residual oil (accounting for 5–8, 3–4.5, and 2–3.5% of the initial wet material for pressing-OP, 3-outlet-OP, and TPOMW, respectively; IOC, 2004) with solvents, whereas pomaces with 65–75% water content cause problems for extracting oil. The waste is called exhausted olive pomace (OP) (MORE Project, 2008).

Olive pomace oil is the oil obtained by treating OP with solvents or other physical treatments, excluding oils obtained by re-esterification processes and of any mixture with oils of other kinds. There are three types of OP oils, defined by the International Olive Council (IOC). Crude olive pomace oil is olive pomace oil with the characteristics defined by the IOC standard, intended for refining for use for human consumption or for technical use. Refined olive pomace oil is the oil obtained from crude olive pomace oil by refining methods which do not lead to alterations in the initial glyceride structure. This product may only be sold directly to the consumer if permitted in the country. Olive pomace oil is a blend of refined olive pomace oil and virgin olive oils, which can be consumed by humans. The country of retail sale may require a more specific designation.

The pomace oil extraction industry is high-energy consumer, particularly at the pomace drying stage and during extraction with solvents. By exploiting the calorific potential of EOP, it is possible to generate electrical energy and exploit the residual thermal energy for the stages of drying and extraction. Similarly, ash produced in combustion is used to manufacture fertilizers, given its high soluble potassium content.

58.5.3 Physicochemical Characterization

Olive pomaces are, in general, a solid organic waste, containing pieces of skin, pulp, stone, and oil (2–8% wet weight, depending on extraction technology). OP typically have a moisture content of 22–25% and 40–45% for traditionally press systems and 3-outlet decanter systems, respectively (Albuquerque et al., 2004; Alba-Mendoza et al., 1990). It is relatively uniform and looks like a sludge due to the high moisture and its small particle size (Alburquerque et al. 2004). The organic matter content is high, rich in organic macromolecules such as cellulose, hemicellulose, lignin, or fats; however, these compounds could be broken down during the valorization process (Estaún et al., 1985). The content of macro- and microelements is relatively high, although N is relatively low in respect to other organic by-products (Table 58.6) (Alburquerque et al., 2004). With a low content of heavy metals, OP will not pollute agricultural soils.

Table 58.6 Characterizations of fresh olive pomace and compost of this product with olive leaves and urea (Alburquerque et al., 2006)

	Units	Fresh olive pomace	Compost of olive pomace with olive leaves and urea
MVM[a]	%, fb[a]	63.9	
pH	–	5.20	8.49
EC[a]	$dS{\cdot}m^{-1}$	4.12	3.17
OM[b]	$g{\cdot}kg^{-1}$, db[a]	948.0	896.0
WSPH[b]	$g{\cdot}kg^{-1}$, db	16.1	3.3
Fat content	$g{\cdot}kg^{-1}$, db	77.5	6.6
TOC[b]	$g{\cdot}kg^{-1}$, db	515.0	470.6
TN	$g{\cdot}kg^{-1}$, db	8.9	22.7
C/N	$g{\cdot}g^{-1}$	57.9	20.7
P	$g{\cdot}kg^{-1}$, db	1.1	1.6
K	$g{\cdot}kg^{-1}$, db	25.3	37.0
Ca	$g{\cdot}kg^{-1}$, db	1.7	11.3
Mg	$g{\cdot}kg^{-1}$, db	0.8	2.3
Fe	$mg{\cdot}kg^{-1}$, db	201	624
Cu	$mg{\cdot}kg^{-1}$, db	16	24
Mn	$mg{\cdot}kg^{-1}$, db	5	38
Zn	$mg{\cdot}kg^{-1}$, db	13	40

[a]*MVM* moisture and volatile matter, *fb* fresh basis, *db* dry basis, *EC* electrical conductivity
[b]*OM* organic matter, *WSPH* water-soluble phenols, *TOC* total organic carbon

TPOMW is a thick sludge that contains olive stone, pulp, and vegetation water. Its moisture content is 65–75%. Besides, the presence of vegetation water explains that its contents of total phenolic compounds and soluble sugars are higher than those in the pomaces of press and three-outlet decanter (MORE Project, 2008). The ash content (1.4–4.0% in TPOMW) is less significant than in olive pomaces of press and three-outlet decanter (1.7–4.0%). The major mineral elements in the three types of pomaces are potassium, followed by calcium and sodium, but their contents are lower than in olive mill wastewaters. Phenolic compounds are quantitatively and qualitatively abundant in olive oil by-products (OOBP), with phenolic acids, seco-iridoids, and flavonoids being the main identified families. These are variably detected and concentrated in OOBP according to their polarity. The typical level of total phenolic compounds is 23 g/kg in TPOMW and 10 g/kg in OP of press and three-outlet decanter (IOC, 2004). Some authors found hydroxytyrosol (1.6–2.9 g/kg-OP) as the major phenolic compound (Rubio-Senent et al., 2012) and other phenolic compounds in lower concentration, such as oleuropein (Cioffi et al., 2010), tyrosol, verbascoside, caffeic acid, vanillin, vanillic acid, apigenin, luteolin, and rutin (Suárez et al., 2010).

Calorific value of pomaces and exhausted olive pomaces depends on many factors, such as the residual oil amount, the pulp-to-stone ratio in the olive pomace, or the moisture content. In general, fresh pomace, dried TPOWM, and EOP have an average higher heating value (HHV) of 7.5, 14.0, and 16.0 MJ/kg, respectively (MORE Project, 2008) (Table 58.7), very close to the HHV of major pomace components (olive pulp and stone): 22.2 MJ/kg pulp and 19.6 MJ/kg stone (Al-Ketan, 2012).

Table 58.7 Higher heating value (HHV) and energy potential of olive pomaces from different countries (ARE Liguria-Italy, UC Liguria-Italy, AGENER-Spain, IPTPO-Croatia, UP ZRS Slovenia, 2008)

Material	Dry pomace with stones and low moisture	Virgin pomace (with pomace oil, stones, and high moisture)	Stones
Average HHV	(MJ/kg)		
Spain	15.90	7.53	17.15
Italy	n.a.	7.53	19.87
Greece	14.64–16.74	n.a.	18.83
Slovenia	17.64	19.26–20.81	20.10
Croatia	n.a.	4219	18.83
Energy potential	(MWh/year)		
Spain	3,183,064	4,307,906	5,005,814
Italy	n.a.	25,000	17,898.57
Greece	231,700–267,400	n.a.	26,780–80,330
Slovenia	n.a.	4587	1570
Croatia	n.a.	17,206	4930

58.5.4 Current and Potential Uses

The wider OP use is as agricultural soil amendment, once composted, or as a component of fertilizers (López-Piñeiro et al., 2011; Roig et al., 2006), both described in the following sections. Also, OOBP have been used as raw material for energy generation in specially designed stoves or gasifiers. This latter process, gasification, has the advantage of using fresh OP and transforming it into syngas for thermal energy production or for cogeneration, but it needs larger OP quantities to be profitable (MORE Project, 2008). Pomace can also undergo anaerobic digestion where organic material is transformed into biogas which can be used to generate electricity or thermal energy; the main drawback is the production of small quantities of mud.

The recovery of phenolic compounds has a considerable interest for food matrices and/or nutraceutical industries due to their great potential as antioxidants or food preservation agents (Bouaziz et al., 2010; Molina-Alcaide & Yáñez-Ruiz, 2008). For example, olive pomace phenolic compounds included in commercial sunflower oil (without addition of other antioxidants) were compared with other food antioxidants by means of determining oxidative stability at 100 °C (Lafka et al., 2011). Results showed that phenolic extracts, combined with ascorbyl palmitate, contributed to a longer antioxidant protection than that observed with the addition of ascorbyl palmitate, BHT, and vitamin E. In another study, OP phenolic extracts were dissolved in ethanol/water and included in olive oil (Suárez et al., 2010), showing that phenol-enriched olive oil had an antioxidant capacity 73% higher than that of the control group.

OOBP have been tested as a natural supplement in feeds for small ruminants and fishes (Yanez-Ruiz & Molina-Alcaide, 2007; Sicuro et al., 2010), although it is reported that polyphenols are responsible for decreasing the nutrient digestibility (Vasta et al., 2008). In this sense, Yanez-Ruiz and Molina-Alcaide (2007) included OP in sheep and goat feeds, concluding that phenolic compounds did not cause toxicity in these animals if the inclusion levels of OP were 122 and 128 g OP/kg feed dry matter, respectively.

In any of these processes, transportation issues sometimes need to be solved due to remote distances between raw material producers and the location where it is used. Storage of fresh OP is usually an unstable process due to weather conditions, geographical diversities, and unwillingness or unawareness of growers. The legal framework is often a key issue, as it can simplify the use of OOBP (MORE Project, 2008).

58.5.5 Energy Generation

The main thermochemical methods by which this renewable energy source can be utilized are combustion (direct firing) or co-combustion (co-firing), gasification, and pyrolysis. Drying and briquetting of the raw biomass usually are compulsory previous steps to the mentioned thermochemical processes. Another type of process involves the usage of biogas (methane) generated by the anaerobic degradation of olive oil mill wastes.

These technologies aiming at energy recovery may represent an interesting alternative for a sustainable disposal of OOBP, able to reduce the environmental impact and to generate energy for sale or satisfy, for instance, the needs of oil mills, by replacing other more expensive, nonrenewable, and polluting fuels (Overend, 2019). From an energy point of view, the by-products EOP and TPOWM have favorable aspects such as ensured annual production, relative concentration in a place, proper humidity conditions (once dried with a humidity <8–10%), low sulfur content, and, finally, high heating value.

58.5.5.1 Thermochemical Processing of Pomaces

Briquetting is a low-cost technique used to agglomerate a wide range of materials into fuel blocks to be transported and utilized as a solid fuel. Considering that the need for alternative fuels will increase soon, briquettes offer a substantially better alternative to coal. Factors as shattering index, compressive strength, or water resistance of the raw material must be considered. Solid olive residues have low compression strength and shattering index, but they can be improved by adding other by-products or agro-industrial wastewaters (i.e., paper waste, which contains fibrous material, thus increasing the shattering index). Olive residues have reasonable water resistance when compared to other biomass products (MORE Project, 2008).

The *combustion process* is an easy way to exploit the energetic potential of dried OP by direct combustion into a boiler to generate heat or electricity. For example, the direct combustion in a boiler of 81.8 t dried OP/year, supplied by 3 olive mills 4 km from the facility (a 100 kW pomace biomass installation for obtaining heating and cooling for a building and to support a thermal solar energy installation for obtaining sanitary heating water), allowed the generation of 295.7 MWh/year, saving 38.9 m^3-diesel-oil/year, for covering the energy demand (thermal and cooling) of a prototype building located in Cazorla (Jaén, Spain) (MORE Project, 2008). A more complex development is the fluidized bed combustor, studied by Abu-Qudais (1996), who studied the influence of the bed height, the fluidization velocity, and the biomass flow rate. Allae Bennin et al. (2019) studied the efficiency of three different biomasses from oil mills (raw OP, EOP, and deoiled olive pomace) in a fluidized bed boiler; the main results showed that the burnt volatile matter was 65–75% at temperatures of 150–400 °C, corresponding to a very low NO_x and SO_2 emission in the resulting gas (3–49 mg kg^{-1} and 0 mg kg^{-1}, respectively, depending on the particle size and temperatures reached).

Experiments based on *gasification process* of pomace olive have been carried out by different authors. In this sense, Dogru and Erdem (2020) tested a new system configuration at pilot scale (Fig. 58.3). The experiment consisted of an updraft gasifier loaded with 500 kg/h of dried and pelleted OP, gaining an energy of

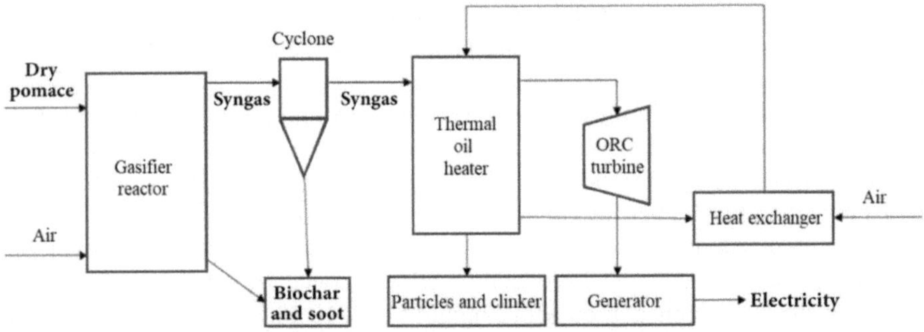

Fig. 58.3 Process flow diagram of the gasification system. (Adapted from Dogru & Erdem, 2020)

7881 MJ. The gasifier burnt the biomass with limited O_2 at temperatures of 900–1200 °C.

The result was 1251 Nm^3 syngas/h at 250–350 °C which was combusted in a thermal oil boiler after the removal of fine particles and dust in a hot gas cyclone. Also, 83 kg biochar/h was produced in the gasifier, equivalent to 1043 MJ. Thermal oil was pumped through a turbine (ORC turbine) to generate 240 kW of gross electricity (210 kW net), while the heat excess (1.3 MJ) was used for olive oil wastewater evaporation plant.

Another example of olive pomace gasification is the Spanish first biomass plant from the olive oil producer Bioland Energy's company. The main objective was to produce energy for self-consumption, which was reached in 2018. The syngas composition produced in the gasifier is CO (35–40%), CO_2 (25–35%), H_2 (20–40%), CH_4 (0–15%), hydrocarbons, and H_2O. The nominal capacity is 1 MW of electrical power and almost 3 MW of thermal power, avoiding 1.4 t CO_2 emissions annually. Finally, the output supplies the 100% of the electric energy for a wet olive pomace drying plant (Bioland Energy, 2020).

Also, the Rossano Calabro power plant (Calabria, Italy) includes a pomace gasifier and a CHP plant (4.2 MWe; 23 MWth), running with 8571 t/MWe year. The pomace is supplied by the surroundings (Puglia and Calabria). The aim of the plant is to reach a conversion ratio of 1.05 kWh/kg biomass. This can be translated into a global electric efficiency of 24% (currently efficiency is 17%) (MORE Project, 2008).

The *pyrolysis process* is a thermal decomposition without an oxidizing agent where biochar, oil, and gas are the products generated. In general, the product of pyrolysis is affected by the biomass type used as feedstock, and regarding olive pomace, the amounts of char, oil, and gas produced depend on the temperature and heating rate of the process. Ounas et al. (2011) treated dried OP (moisture 8%) and demonstrated how mass loss and mass loss rate are strongly influenced by heating rate; in this experiment with dried OP, the volatile percentage reached was 70–75%, and the conversion was 10–40% total mass (biochar accounted for 19.26% total mass) at temperatures between 265 and 270 °C. In this stage, the mass loss was due to the hemicellulose. In a second stage, a conversion of 50–80% total mass at temperatures of 327–334 °C was reached, and the mass loss was associated with cellulose.

58.5.5.2 Biogas Production from Olive Pomaces

Biogas is a gaseous mixture of methane (50–70%), carbon dioxide (20–50%), and, in less quantity, nitrogen, hydrogen, oxygen, and hydrogen sulfide, which can be used to generate energy (electricity, heat, fuel for vehicles) after an upgrading process that consists mainly in the removal of minority components and methane content concentration.

Anaerobic biogas production is an effective process for converting a broad variety of biomass to methane to substitute natural gas and medium calorific gases. For instance, biogas obtained by the anaerobic treatment of 1 m^3 of olive mill wastewater contains 60–80 kWh of energy. The methane yield of OP ranges is roughly 0.260 Nm^3 CH_4/kg COD (El Achkar et al., 2018; Fountoulakis et al., 2007). Tekin and Dalgıç (2000) used OP as the only substrate for biogas production in semicontinuous anaerobic process at 37 °C and 30 days of hydraulic residence time, reaching a production of 0.768 dm^3-biogas/day, with a CH_4 content of 80–91% volume.

The process can be carried out in relatively inexpensive and simple reactor designs and operating procedures. Efficient use of olive pomace for energy production solves two problems in one acceptable disposal of olive oil mill waste: anaerobic digestion converts the carbonaceous matter into biogas, then into clean energy production, leaving a stabilized slurry in a suitable form for application to land as fertilizer.

58.5.6 Composting

Composting is the process of transformation of organic materials into stabilized material (compost) that is carried out by aerobic microorganisms under controlled conditions. As a controlled biological method, the composting process has been studied for decades, and it is a simple, robust, efficient, and cheap process for dealing with organic waste. Composting involves several processes in which organic matter is broken down and humification and the formation of new molecules occur (Cáceres et al., 2018). Nowadays the major part of governments are boosting circular agriculture, in which composting or composting combined with other biological treatments (such as biogas production) can have a key role for recycling nutrients required for vegetal production and carbon to increase the quality of agricultural soils.

Olive pomace composting has drawbacks because of some unfavorable characteristics, which have been aforementioned (Sect. 58.5.2), but the success of the composting of this product makes compulsory the mixture of this waste with other by-products for improving its physical characteristics. Therefore, complementary materials should have structure, high nitrogen content, and low moisture. A field trial was carried out by the IOC (2004) where composting of OP and application to agricultural land were performed. Mounds of 1.0–1.5 m high and 3 m wide were prepared on a concrete platform (9 m wide to allow machinery access to turn over

Fig. 58.4 Olive pomace from oil mills that use centrifugation process with a two-outlet decanter (left) and composted olive pomace (right) in "Villanueva del Arzobispo" (Jaén, Spain)

the compost) and covered with plastic sheeting to facilitate optimal composting conditions (moisture and temperature) (Fig. 58.4). Other substrates can be added such as olive and vine wood pruning, olive leaves, grape stalks, or cattle manure to achieve well-structured compost. Olive tree pruning is used after being crushed into small 1.5 cm pieces in a shredder. Besides, the addition of 2% nitrogen (urea) in relation to the total quantity of compost is recommended to stimulate microbial activity in the compost. As a result, the application of composted OP and vegetation water to the field can reduce the loss of soil fertility when organic matter content falls below 1%. This is particularly relevant in agricultural areas where the environmental conditions (hot and arid) make it particularly hard to maintain soil fertility with the ensuing serious consequences of desertification.

The composting process takes several weeks; at intermediate scale, for instance, the process can take like 50 weeks (Alburquerque et al., 2006). The main parameters to monitor in situ are temperature, oxygen, and moisture in the composting matrix. In this sense, the moisture content of the compost must be kept at about 60% by regularly moistening the compost mounds. The actions for conducting the monitoring are basically turning over and irrigation of the composting piles.

58.5.6.1 Use of Composted Pomaces

Compost from OP should be applied to soils as main use. It is essential to evaluate the maturity of the compost before use because the lack of organic matter stability leads to modifications in the physical, chemical, and plant toxicity characteristics. However, the unfavorable characteristics of these products should be considered. The relatively low N content and the long composting time can lead to an immobilization phenomenon in the soils. Apart from the soil application of the compost, other research have focused on the use of the compost as growing media (Papafotiou et al., 2004; Aviani et al., 2010); however, this application is even more demanding in terms of quality requirement of the compost than the soil application.

OOBP have been studied as organic fertilizers (Fernández-Hernández et al., 2014; Magdich et al., 2012). Several authors have shown that the high levels of phenolic compounds in OOBP may be responsible for phytotoxicity and modifications in soil microflora (Kotsou et al., 2004). However, at a maximum spreading rate of 100 m^3/ha, OOBP did not affect olive production, despite the increasing levels of phenolic content when compared to the control soil (Magdich et al., 2012). Furthermore, phenolic compounds from OOBP may act against specific plant pathogens, showing potential as natural pesticides (Cayuela et al., 2008).

58.6 Wastewaters from Olive Mills

58.6.1 Production and Physicochemical Characteristics

As indicated in Chap. 10, the elaboration of olive oils can be carried out by the traditional process (press) or by continuous centrifugation processes, using decanters of three or two outlets. Liquid effluents in extraction processes of olive oil are composed of vegetation water from olives, water added during the extraction process, and a variable percentage of solids. In general, the components forming olive mill wastewater (OMW) can be classified as follows:

- Water vegetation from olive fruit accounting for between 45 and 50% of the crude weight of the olives, which contains dissolved substances characteristic of the fruit such as mineral salts, sugars, anthocyanins, and glycosides, among others (Madueño et al., 1993).
- Process water that is added at different stages of olive oil extraction. This originally drinking water is contaminated differently, depending on the stage of the process used. Mainly three types of wastewater can be distinguished:
  1. Waters used in fruit preparation. This water is used for fruit cleaning and generally drags clays, silts, and certain fraction of fat from olive drained during storage and transport of the fruit.
  2. Water added to different process stages (olive washing machines, input in decanter, and vertical centrifuge). The volume of these waters varies depending on the extraction process used, which can be traditional or continuous (Madueño et al., 1993). These waters usually drag soft tissue of olive pulp, oil as very stable emulsion, mucilage, pectins, etc.
  3. Typical wastewater of any industrial facility (equipment cleaning, fecal waters, etc.), the impact of which is minimal in the total volume of liquid effluent.

Considering that pure vegetation water represents on average 45% of processed olive weight and taking also into account the variable amount of washing waters, the average volume of olive mill effluents ranges from 0.5 to 1.5 m^3 per ton of processed olives. A research carried out by Balice et al. (1982) on effluents sampled in 37 oil mills located in the Apulia region (Italy) found a great variability in the composition of OMW. In that study, the following streams were considered: pure vegetation water sampled just after the oil extraction centrifuge for the batch press process and for the continuous centrifuge process and effluents sampled in the flotation/settling tanks of press mills.

Summarizing, concentrations and specific volumes (per weight unit of processed olives) are quite variable, while the polluting organic load per weight unit of processed olives is practically independent of the extraction system. On average, 100–120 kg chemical oxygen demand (COD)/ton of olives, equivalent to 45–55 kg BOD_5/ton of olives, are found in these wastewaters (Balice et al., 1982). The polluting load of olive oil mills generated in Mediterranean countries depends on the milling season which lasts about 100 days. The concentration of OMW can therefore be indirectly derived from the load and the flow rate of washing waters of the oil mill to be considered (Alba-Mendoza et al., 1996). Hence, it is much more accurate to determine the polluting load of an oil mill directly on the quantity of processed olives rather than on flow rates and concentrations of wastewaters (Balice et al., 1982; Rozzi & Malpei, 1996).

The process using a decanter with three outlets is still used in many oil extraction mills. This wastewater is markedly acidic. Although BOD_5 and COD values are 145- and 21-fold greater than those found in the wastewaters produced by the process using a decanter with two outlets (2-OMW), the percentages of C and N analyzed (on a dry-weight basis) were 17.0- and 19.6-fold higher (Dermeche et al., 2013; Paredes et al., 1999). OMW has a high content of total solids, reaching 6.5–10.0%, compared with 0.59% for 2-OMW and 0.12% for untreated, highly loaded urban sewage.

Nieto et al. (2009) have determined the physical and chemical characteristics of 22 OMW samples from extraction process using a decanter of two outlets (2-OMW) in the provinces of Jaén and Cordoba (Andalusia, Spain). The wastewater samples were collected from several oil mills operating with olive-cleaning and vertical-centrifugation equipment of different trademarks. The samples were taken while the oil mill was in full operation, both from the olive-cleaning system and the outlet of vertical-centrifuge machines for oil cleaning. Although all the samples were analyzed, in the case of olives washing wastewater, the BOD_5 and COD varied in the range of 0.5–1.15 g O_2/dm^3 and 0.810–4.86 g O_2/dm^3, respectively. In the case of olive oil wastewater, the variations were from 0.465 to 0.915 g O_2/dm^3 and from 2.88 to 12.1 g O_2/dm^3, respectively.

58.6.2 Oil Mill Wastewater Treatment

Treatment and disposal of olive oil mill wastewater is currently one of the most serious environmental problems in the Mediterranean countries, such as Spain, Italy, Greece, and Portugal, where this foodstuff is produced. As a polluting source, OMW has existed for thousands of years, but their effects on the environment are, at present, more noticeable because of the following reasons: (i) olive oil production has remarkably increased over the last 30 years; (ii) the oil mills were previously small and dispersed across the country and discharged their effluents directly on the land or underground (especially in karstic areas), but they are now much larger, and some are connected to the sewers; and (iii) the public awareness of environmental problems is now much higher than in the past.

As the OMW comes from the olive vegetation water and the drinking water used in the extraction process, it can be safely assumed that is completely biodegradable. But, even if all the constituents of OMW are biodegradable by definition (they are biosynthesized), some of them, such as phenolic compounds and lipids, are decomposed at reaction rates much lower than others, such as sugars or short-chain volatile acids. Therefore, OMW represents a serious problem to environmental engineering if it intends to cope with it using a biological process, because efficient treatment of wastewaters requires fast biodegradation of the pollutants, and, at the same time, treatment plants must be small enough to be economically feasible. Hence, many other non-biological processes have been tested on oil mill effluents.

Microalgae are one of the most important bioresources that are currently receiving a lot of attention due to a multiplicity of reasons (Sánchez et al., 1996; Chisti, 2007; Hodaifa et al., 2008). The dual application of microalgae for bioremediation and biomass production for sustainable valuable sources as pharmaceuticals, health foods, carotenoids, dyes, fine chemicals, biofuels, and other applications is a feasible option and has a vast industrial and economic potential.

Pinto et al. (2003) have indicated that the treatment of OMW with two phenol-resistant algae, *Ankistrodesmus braunii* and *Scenedesmus quadricauda*, showed a limited reduction of phenol content after 5 days of treatment, regardless of algal concentration. By contrary, cultures of both algae, grown in the dark, degraded over 50% of the low molecular weight phenols contained in OMW, but they were not completely removed, but biotransformed into other non-identified, aromatic compounds.

Advanced oxidation processes (AOP) are known for their capability to mineralize a wide range of organic compounds. AOP involve the generation of highly reactive radical species, mainly hydroxyl radical (Nieto et al., 2011). AOP versatility is enhanced by the fact that there are many ways of producing hydroxyl radicals. Heterogeneous photocatalysis using titanium dioxide (TiO_2) and solar UV, potentially combined with hydrogen peroxide (H_2O_2), and homogeneous processes such as Fenton (Fe^{2+}/H_2O_2) and photo-Fenton ($Fe^{2+}/H_2O_2/UV$) reactions have been proved to be useful tools for the treatment of pesticide-containing wastewaters (Moreira et al., 2012). Among the different AOP, Fenton and solar photo-Fenton are acknowledged within the scientific community for their effectiveness to treat wastewater with high phenolic compound content (Nieto et al., 2009).

58.6.3 Use of Oil Mill Wastewaters

58.6.3.1 As Culture Medium for Microalgal Production and Biofuel Generation

Considering the high content of nutrients in oil mill wastewaters, an interesting use as culture medium for microalgal production can be found for this residue of oil industry. The microalgal biomass is generated from the growth of microorganisms in a medium achieved by mixing culture medium with oil mill wastewaters (Sanchez et al., 2001; Hodaifa et al., 2008; Murad et al., 2020). The maximum proportion of oil mill wastewaters in these mixtures is determined by the level of tolerance to toxicity of the species being worked with; this way, the highest level of utilization of this resource can be obtained. It is important to select the species of microalgae with sufficient resistance to the drastic toxicity conditions to which it will be subjected.

In addition, this microalgal biomass developed in oil mill wastewaters can be used for biofuel generation, from the extraction of fat material and the subsequent transesterification of triglycerides in alkyl esters constituting biodiesel. Different variables may affect the transesterification reaction: different short-chain alcohols, catalyst, reaction time, and operation time (Chisti, 2007). The chromatographic analysis of the fatty acid profile of the biodiesel obtained will determine whether it is richer in short- or long-chain fatty acids. Microalgae can also be used in anaerobic digestion for the production of methane (Fernández-Rodríguez et al., 2014).

58.6.3.2 Separation of Specific Biomolecules

Other specific biomolecules such as proteins, pigments such as chlorophylls and carotenoids, specific polyunsaturated fatty acids, and carbohydrates can be obtained from microalgae grown in oil mill wastewaters. On the other hand, antioxidants, antivirals, antimicrobial compounds, antitumors, agents for anti-inflammatory use, and compounds for use in cosmetics, dietary products, and functional and prebiotic foods can be produced (Malvis et al., 2019). Microalgae biomass can be used to produce bioplastic materials too (Chiellini et al., 2008). The residual biomass from these processes can potentially be used as organic fertilizer (Canter et al., 2015), for soil modification, or to feed fish and livestock (Roeselers et al., 2008).

References

Abu-Qudais, M. (1996). Fluidized bed combustion for energy production from olive cake. *Energy, 21*, 173–178.

Alba-Mendoza, J., Ruiz-Gomez, A., & Hidalgo-Casado, F. (1990). Technological evolution of the different processes for olive oil extraction. In D. R. Erickson (Ed.), *Edible fats and oils processing: Basic principles and modern practices* (pp. 341–347). American Oil Chemical Society.

Alba-Mendoza, J., Hidalgo Casado, F., Ruiz Gómez, M. A., Martínez Román, F., Moyano Pérez, M. J., Cert, A., Pérez Camino, M. C., & Ruiz Méndez, M. V. (1996). Características de los aceites de oliva de primera y segunda centrifugación. *Grasas y Aceites, 47*(3), 163–181.

Alburquerque, J. A., Gonzálvez, J., García, D., & Cegarra, J. (2004). Agrochemical characterization of 'alperujo', a solid by-product of the two-phase centrifugation method for olive oil extraction. *Bioresource Technology, 91*, 195–300.

Alburquerque, J. A., Gonzálvez, J., García, D., & Cegarra, J. (2006). Effects of bulking agent on the composting of 'alperujo' the solid by-product of the two-phase centrifugation method for olive oil extraction. *Process Biochemistry, 41*, 127–132.

Al-Ketan, O. (2012). Potential of using olive pomace as a source of renewable energy for electricity generation in the Kingdom of Jordan. *Journal of Renewable and Sustainable Energy, 4*, 063132.

Allae Bennin, M., Koukouch, A., Bakhattar, I., Asbik, M., Boushaki, T., Sarh, B., & Bonnamy, S. (2019). Characterization and combustion of olive pomace in a fixed bed boiler: Effects of particle sizes. *International Journal of Heat and Technology, 37*(1), 229–238.

Álvarez-Murillo, A., Román, S., Ledesma, B., & Sabio, E. (2015). Study of variables in energy densification of olive stone by hydrothermal carbonization. *Journal of Analytical and Applied Pyrolysis, 113*, 307–314.

Andalusian Energy Agency, Agener. (2018). *La bioenergía en Andalucía*. Consejería de Hacienda, Industria y Energía. Junta de Andalucía.

Araujo, M., Pimentel, F., Alves, R., & Oliveira, M. (2015). Phenolic compounds from olive mill wastes: Health effects, analytical approach, and application as food antioxidants. *Trends in Food Science and Technology, 45*, 200–211.

ARE Liguria-Italy, UC Liguria-Italy, AGENER-Spain, IPTPO-Croatia, UP ZRS Slovenia. (2008). *Market of olive residues for energy*. ARE Liguria-Italy, UC Liguria-Italy, AGENER-Spain, IPTPO-Croatia, UP ZRS Slovenia. Sustainable Energy Europe. Official Partner.

Aviani, I., Laor, Y., Medina, S., Krassnovsky, A., & Raviv, M. (2010). Co-composting of solid and liquid olive mill wastes: Management aspects and the horticultural value of the resulting composts. *Bioresource Technology, 101*, 6699–6706.

Balice, V., Boari, G., Cera, O., & Abbaticchio, P. (1982). *Indagine analitica sulle acque di vegetazione. Nota 1. Inquinamento* (Vol. 6–7, pp. 49–53).

Bartocci, P., D'Amico, M., Moriconi, N., Bidini, G., & Fantozzi, F. (2015). Pyrolisis of olive stone for energy purposes. *Energy Procedia, 82*, 374–380.

Becker, E. W. (1994). *Biotechnology and microbiology* (pp. 9–41). Cambridge University Press.

Bioland Energy. (2020). *Spain's first biomass plant used for self-consumption reduces emissions with 1.4 t CO2eq per year*. Creara S.L.

Bouaziz, M., Feki, I., Ayadi, M., Jemai, H., & Sayadi, S. (2010). Stability of refined olive oil and olive-pomace oil added by phenolic compounds from olive leaves. *European Journal of Lipid Science and Technology, 112*, 894–905.

Brenes, M., Romero, C., & De Castro, A. (2004). Combined fermentation and evaporation processes for treatment of wash waters from Spanish green olives. *Journal of Chemical Technology and Biotechnology, 79*, 253–259.

Budinova, T., Petrov, N., Razvigorova, M., Parra, J., & Galiatsatou, P. (2006). Removal of arsenic (III) from aqueous solution by activated carbons prepared from sol-vent extracted olive pulp and olive stones. *Industrial and Engineering Chemistry Research, 45*, 1896–1901.

Cáceres, R., Malińska, K., & Marfà, O. (2018). Nitrification within composting: A review. *Waste Management, 72*, 119–137.

Canter, C. E., Blowers, P., Handler, R. M., & Shonnard, D. R. (2015). Implications of widespread algal biofuels production on macronutrient fertilizer supplies: Nutrient demand and evaluation of potential alternate nutrient sources. *Applied Energy, 143*, 71–80.

Cayuela, M. L., Millner, P. D., Meyer, S. L., & Roig, A. (2008). Potential of olive mill waste and compost as biobased pesticides against weeds, fungi, and nematodes. *Science of the Total Environment, 399*, 11–18.

Chiellini, E., Cinelli, P., Ilieva, V. I., & Martera, M. (2008). Biodegradable thermoplastic composites based on polyvinyl alcohol and algae. *Biomacromolecules, 9*, 1007–1013.
Chisti, Y. (2007). Biodiesel from microalgae. *Biotechnology Advances, 25*, 249–306.
Cioffi, G., Pesca, M. S., De Caprariis, P., Braca, A., Severino, L., & De Tommasi, N. (2010). Phenolic compounds in olive oil and olive pomace from Cilento (Campania, Italy) and their antioxidant activity. *Food Chemistry, 121*, 105–111.
Contreras, C. G., Serrano, A., Ruiz-Filippi, G., Borja, R., & Fermoso, F. G. (2016). Culture of microalgae biomass for valorization of table olive processing water. *Grasas y Aceites, 57*(1), 86–94.
Corella, J., Toledo, J. M., & Padilla, R. (2004). Catalytic hot gas cleaning with mono-liths in biomass gasification in fluidized beds. 1. Their effectiveness for tar elimination. *Industrial and Engineering Chemistry Research, 43*, 2433–2445.
Cuevas, M., Sánchez, S., Bravo, V., Cruz, N., & García, J. F. (2009). Fermentation of en-zymatic hydrolysates from olive stones by Pachysolen tannophilus. *Journal of Chemical Technology and Biotechnology, 84*, 461–467.
Cuevas, M., Sánchez, S., García, J. F., Baeza, J., Parra, C., & Freer, J. (2015). Enhanced ethanol production by simultaneous saccharification and fermentation of pretreated olive stones. *Renewable Energy, 74*, 839–847.
Cuevas, M., Martínez-Cartas, M. L., Pérez-Villarejo, L., Hernández, L., García-Martín, J. F., & Sánchez, S. (2019a). Drying kinetics and effective water diffusivities in olive stone and olive-tree pruning. *Renewable Energy, 132*, 911–920.
Cuevas, M., Martínez-Cartas, M. L., & Sánchez, S. (2019b). Effect of short-time hydro-thermal carbonization on the properties of hydrochars prepared from olive-fruit endocarps. *Energy and Fuels, 33*, 313–322.
Cuevas, M., Saleh, M., García-Martín, J. F., & Sánchez, S. (2020). Acid and enzymatic fractionation of olive stones for ethanol production using Pachysolen tannophilus. *PRO, 8*, 195.
De Castro, A., García, P., Romero, C., Brenes, M., & Garrido, A. (2007). Industrial implementation of black ripe olives storage under acidic conditions. *Journal of Food Engineering, 80*, 1206–1212.
de Herrera, G. A. (1981). In E. Terrón (Ed.), *Agricultura General*. Ministerio de Agricultura.
De Jódar, M., Madueño, A., Lucas, A., Molina, J. M., Ruiz, A., Madueño, J. M., Justicia, M., & Baena, M. (2020). Deep learning in olive pitting machines by computer vision. *Computers and Electronics in Agriculture, 171*, 105304.
De los Santos, B., Medina, E., Brenes, M., Aguado, A., García, P., & Romero, C. (2020). Chemical composition of table olive wastewater and its relationship with the bio-fortifying capacity of tomato (Solanum lycopersicum L.) plants. *Agricultural Water Management, 227*, 105833.
Dermeche, S., Nadour, M., Larroche, C., Moulti-Mati, F., & Michaud, P. (2013). Olive mill wastes: Biochemical characterizations and valorization strategies. *Process Biochemistry, 48*, 1532–1552.
Di Blasi, C., Signorelli, G., & Portoricco, G. (1999). Countercurrent fixed-bed gasification of biomass at laboratory scale. *Industrial and Engineering Chemistry Research, 38*, 2571–2581.
Di Giovacchino, L. (2010). I sottoprodotti della lavorazione delle olive in oleificio. In L. Di Giovacchino (Ed.), *Tecnologie di lavorazione delle olive in frantoio* (pp. 243–258). Tecnica Alimentare.
Dogru, M., & Erdem, A. (2020). Autothermal fixed bed updraft gasification of olive pomace biomass and renewable energy generation via organic Rankine cicle turbine. Green energy generation from waste biomass in the Mediterranean region. *Johnson Matthey Technology Review, 64*, 119–134.
El Achkar, J. H., Rohayem, C., Salameh, D., Louka, N., Maroun, R. G., & Hobaika, Z. (2018). *Olive pomace, a source of green energy using anaerobic digestion*. In 4th International conference on renewable energies for developing countries (REDEC), Beirut, Lebanon.
Estaún, V., Calvet, C., & Grases, J. M. (1985). Chemical determination of fatty acids, organic acids and phenols, during olive marc composting process. *Acta Horticulturae, 172*, 263–270.

ESYRCE. 'Encuesta de superficies y rendimientos de cultivos'. (2018). *Ministerio de Agricultura, Pesca y Alimentación*. Government of Spain.

Fernández-Bolaños, J., Felizón, B., Heredia, A., Rodríguez, R., Guillén, R., & Jiménez, A. (2001). Steam-explosion of olive stones: Hemicellulose solubilization and enhancement of enzymatic hydrolysis of cellulose. *Bioresource Technology, 79*, 53–61.

Fernández-Hernández, A., Roig, A., Serramiá, N., Civantos, C. G., & Sánchez-Monedero, M. A. (2014). Application of compost of two-phase olive mill waste on olive grove: Effects on soil, olive fruit and olive oil quality. *Waste Management, 34*, 1139–1147.

Fernández-Rodríguez, M. J., Rincón, B., Fermoso, F. G., Jiménez, A. M., & Borja, R. (2014). Assessment of two-phase olive mill solid waste and microalgae co-digestion to improve methane production and process kinetics. *Bioresource Technology, 157*, 263–269.

Font, A., Soriano, L., de Moraes, S. M., Tashima, M. M., Monzó, J., Borrachero, M. V., & Payá, J. (2020). Design and properties of 100% waste-based ternary alkali-activated mortars: Blast furnace slag, olive-stone biomass ash and rice husk ash. *Journal of Cleaner Production, 243*, 118568.

Formenti, L. R., Norregaard, A., Bolic, A., Hernández, D. Q., Hagemann, T., Heins, A., & Larsson, H. (2014). Challenges in industrial fermentation technology research. *Biotechnology Journal, 9*, 727–738.

Fountoulakis, M., Drakopoulou, S., Terzakis, S., Georgaki, E., & Manios, T. (2007). Potential form methane production from typical Mediterranean agro-industrial by-products. *Biomass and Bioenergy, 32*, 155–161.

García, J. L. (2014). *Investigadores de la UCA transforman los residuos de la aceituna de mesa en biogás*. http://www.uca.es/es/cargarAplicacion-Noticia.do?identificador=7257. Last visit 25/08/2020.

García-Martín, J. F., Sánchez, S., & Cuevas, M. (2013). Evaluation of the effect of the dilute acid hydrolysis on sugars release from olive prunings. *Renewable Energy, 51*, 382–387.

García-Martín, J. F., Cuevas-Aranda, M., Feng, C.-H., Álvarez-Mateos, P., Torres-García, M., & Sánchez-Villasclaras, S. (2020). Energetic valorisation of olive biomass: Olive-tree pruning, olive stones and pomaces. *PRO, 8*(511), 1–38.

Garrido, A., García, P., & Brenes, M. (1992). The recycling of table olive brine using ultrafiltration and activated carbon adsorption. *Journal of Food Engineering., 17*(4), 291–305.

Goldfarb, J. L., Buessing, L., Gunn, E., Lever, M., Billias, A., Casoliba, E., Schievano, A., & Adani, F. (2017). Novel integrated biorefinery for olive mill waste management: Utilization of secondary waste for water treatment. *ACS Sustainable Chemistry and Engineering, 5*, 876–884.

Gómez-de la Cruz, F. J., Palomar-Carnicero, J. M., Casanova-Peláez, P. J., & Cruz-Peragón, F. (2015). Experimental determination of effective moisture diffusivity during the drying of clean olive stone: Dependence of temperature, moisture content and simple thickness. *Fuel Processing Technology, 137*, 320–326.

González-Hidalgo, I., Bañón, S., & Ros-García, J. M. (2012). Evaluation of table olive by-product as a source of natural antioxidants. *International Journal of Food Science and Technology, 47*(4), 674–681.

Grammelis, P., & Kakaras, E. (2005). Biomass combustion modeling in fluidized beds. *Energy and Fuels, 19*, 292–297.

Guinda, Á., Castellano, J. M., Santos-Lozano, J. M., Delgado-Hervás, T., Gutiérrrez-Adánez, P., & Rada, M. (2015). Determination of major bioactive compounds from olive leaf. *LWT-Food Science and Technology, 64*(1), 431–438.

Hazzaa, R., & Hussein, M. (2015). Adsorption of cationic dye from aqueous solution onto activated carbon prepared from olive stones. *Environmental Technology and Innovation, 4*, 36–51.

Hodaifa, G., Martínez, M. E., & Sánchez, S. (2008). Use of industrial wastewater from olive-oil extraction for biomass production of Scenedesmus obliquus. *Bioresource Technology, 99*, 1111–1117.

Hodaifa, G., Ochando-Pulido, J. M., Driss, S. B., Rodríguez-Vives, S., & Martínez-Ferez, A. (2013). Kinetic and thermodynamic parameters of iron adsorption onto olive stones. *Industrial Crops and Products, 49*, 526–534.

International Olive Council, IOC. (2004). *Recycling of vegetable water and olive pomace on agricultural land. Case of olive growing*. Project Executive Agency CFC/IOOC/04. https://www.internationaloliveoil.org/wp-content/uploads/2019/12/GOOD-PRACTICES-1.pdf

International Olive Council, IOC. (2020). *Table olives production*. https://www.internationaloliveoil.org/wp-content/uploads/2020/04/OT-W901-29-11-2019-P-1.pdf. Last visit: 27/08/2020.

Karakaş, C., Özçimen, D., & İnam, B. (2017). Potential use of olive stone biochar as a hydroponic growing medium. *Journal of Analytical and Applied Pyrolysis, 125*, 17–23.

Kiliçkan, A., & Güner, M. (2008). Physical properties and mechanical behavior of olive fruits (Olea europaea L.) under compression loading. *Journal of Food Engineering, 87*, 222–228.

Kotsou, M., Mari, I., Lasaridi, K., Chatzipavlidis, I., Balis, C., & Kyriacou, A. (2004). The effect of olive oil mill wastewater (OMW) on soil microbial communities and suppressiveness against Rhizoctonia solani. *Applied Soil Ecology, 26*, 113–121.

Lafka, T. I., Lazou, A. E., Sinanoglou, V. J., & Lazos, E. S. (2011). Phenolic and antioxidant potential of olive oil mill wastes. *Food Chemistry, 125*(1), 92–98.

López-Piñeiro, A., Albarrán, A., Rato-Nunes, J. M., Peña, D., & Cabrera, D. (2011). Long-term impacts of de-oiled two-phase olive mill waste on soil chemical properties, enzyme activities and productivity in an olive grove. *Soil and Tillage Research, 114*, 175–182.

Madueño, S. L., Rodríguez, M. J. M., & Marouez, S. M. (1993). Alpechín: un problema en vías de solución. *Tecnoambiente, 16*, 19–25.

Magdich, S., Jarboui, R., Rouina, B. B., Boukhris, M., & Ammar, E. (2012). A yearly spraying of olive mill wastewater on agricultural soil over six successive years: Impact of different application rates on olive production, phenolic compounds, phytotoxicity and microbial counts. *Science of the Total Environment, 430*, 209–216.

Malvis, A., Hodaifa, G., Halioui, M., Seyedsalehi, M., & Sánchez, S. (2019). Integrated process for olive oil mill wastewater treatment and its revalorization through the generation of high added value algal biomass. *Water Research, 151*, 332–342.

Mami, M. A., Mätzing, H., Gehrmann, H. J., Stapf, D., Bolduan, R., & Lajili, M. (2018). Investigations of the olive mill solid wastes pellets combustion in a counter-current fixed bed reactor. *Energies, 11*(1965), 1–21.

Mancilla, A., García, A. N., Pastor, M. V., León, M., Sánchez, A. J., & Gómez, D. M. (2013). Thermal decomposition of the different particles fractions of almond shells and olive stones. Thermal behaviour changes due to the milling processes. *Thermochimica Acta, 564*, 24–33.

Manyà, J. J., Ruiz, J., & Arauzo, J. (2007). Some peculiarities of conventional pyrolysis of several agricultural residues in a packed bed reactor. *Industrial and Engineering Chemistry Research, 46*, 9061–9070.

Mata, J., Pérez, J. A., Díaz, M. J., Serrano, A., Núñez, N., & López, J. (2015). New tech-niques developed to quantify the impurities of olive stone as solid biofuel. *Re-newable Energy, 78*, 566–572.

Mata-Sánchez, J., Pérez-Jiménez, J. A., Díaz-Villanueva, M. J., Serrano, A., Núñez-Sánchez, N., & López-Giménez, F. J. (2013). Statistical evaluation of quality parame-ters of olive stone to predict its heating value. *Fuel, 113*, 750–756.

Mateo, S., Puentes, J. G., Roberto, I. C., Sánchez, S., & Moya, A. J. (2014). Optimization of acid hydrolysis of olive tree pruning residue. Fermentation with Candida guilliermondii. *Biomass and Bioenergy, 69*, 39–46.

Mateo, S., Puentes, J. G., Moya, A. J., & Sánchez, S. (2015). Ethanol and xylitol production by fermentation of acid hydrolysate from olive pruning with Candida tropicalis NBRC 0618. *Bioresource Technology, 190*, 1–6.

Maymone, B., Sblendorio, A., & Ceci-Ginestrelli, D. (1947). Ricerche sulla composizione chimica, sulla digeribilita et sul valore nutritivo delle foglie di olivo (Olea europaea L.) verdi, essicate, insilate. *Annali Sper Agr, 1*, 71–88.

Miranda, I., Simões, R., Medeiros, B., Nampoothiri, K. M., Sukumaran, R. K., Rajan, D., Pereira, H., & Ferreira-Dias, S. (2019). Valorization of lignocellulosic residues from the olive oil industry by production of lignin, glucose and functional sugars. *Bioresource Technology, 292*, 121936.

Molina-Alcaide, E., & Yáñez-Ruiz, D. R. (2008). Potential use of olive by-products in ruminant feeding: A review. *Animal Feed Science and Technology, 147*(1–3), 247–264.

Morán, J. C., Míguez, J. L., Porteiro, J., Patiño, D., & Granada, E. (2015). Low-quality fuels for small-scale combustion boilers: An experimental study. *Energy and Fuels, 29*, 3064–3081.

MORE Project. (2008). *Market of Olive Residues Energy (MORE) IEE-project.* Contract no. EIE/07/273/SI2.466853. Final publishable report. https://ec.europa.eu/energy/intelligent/projects/en/projects/more#results.

Moreira, F. C., Vilar, V. J. P., Ferreira, A. C. C., Dos Santos, F. R. A., Dezotti, M., Souza, M. A., Gonçalves, C., Boaventura, R. A. R., & Alpendurada, M. F. (2012). Treatment of a pesticide-containing wastewater using combined biological and solar-driven AOPs at pilot scale. *Chemical Engineering Journal, 209*, 429–441.

Moreno, N., Caballero, A., Morales, J., & Rodríguez-Castellón, E. (2016). Improved performance of electrodes based on carbonized olive stones/S composites by impregnating with mesoporous TiO2 for advanced li–S batteries. *Journal of Power Sources, 313*, 21–29.

Moya, A. J., Bravo, V., Mateo, S., & Sánchez, S. (2008). Fermentation of acid hydrolysates from olive-tree pruning debris by Pachysolen tannophilus. *Bioprocess and Biosystems Engineering, 31*, 611–617.

Murad, M., Hodaifa, G., Malvis, A., & Sánchez, S. (2020). Kinetic growth and biochemical composition variability of Chlorella pyrenoidosa in olive oil washing wastewater cultures enriched with urban wastewater. *Journal of Water Process Engineering, 35*, 1–10.

Nieto, L. M., Hodaifa, G., Rodríguez, S. R., Casares, J. A. G., Dris, S. B., & Grueso, R. (2009). Treatment of olive-mill wastewater from a two-phase process by chemical oxidation on an industrial scale. *Water Science Technology, 59*(10), 2017–2027.

Nieto, L. M., Hodaifa, G., Vives, S. R., Casares, J. A., & Ochando, J. (2011). Degradation of organic matter in olive-oil mill wastewater through homogeneous Fenton-like reaction. *Chemical Engineering Journal, 173*, 503–510.

Ounas, A., Aboulkas, A., El Harfi, K., Bacaoui, A., & Yaacoubi, A. (2011). Pyrolysis of olive residue and sugar cane bagasse: Non-isothermal thermogravimetric kinetic analysis. *Bioresource Technology, 102*, 11234–11238.

Overend, R. P. (2019). Biomass energy heat provision for cooking and heating in developing countries. In M. Kaltschmitt (Ed.), *Energy from organic materials (biomass)* (Encyclopedia of sustainability science and technology series). Springer. https://doi.org/10.1007/978-1-4939-7813-7_315

Papafotiou, M., Phsyhalou, M., Kargas, G., Chatzipavlidis, I., & Chronopoulos, J. (2004). Olive-mill wastes compost as growing medium component for the production of poinsettia. *Scientia Horticulturae, 102*, 167–175.

Paredes, C., Cegarra, J., Roig, A., Sánchez-Monedero, M. A., & Bernal, M. P. (1999). Characterization of olive mill wastewater ('alpechín') and its sludge for agricultural purposes. *Bioresource Technology, 67*, 111–115.

Pinto, G., Pollio, A., Previtera, L., Stanzione, M., & Temussi, F. (2003). Removal of low molecular weight phenols from olive oil mill wastewater using microalgae. *Biotechnology Letters, 25*, 1657–1659.

Rabaçal, M., Pereira, S., & Costa, M. (2018). Review of pulverized combustion of non-woody residues. *Energy and Fuels, 32*, 4069–4095.

Ranalli, A., Marchegiani, D., Pardi, D., Contento, S., Pardi, D., Girardi, F., & Kotti, F. (2009). Evaluation of functional phytochemicals in destoned virgin olive oil. *Food and Bioprocess Technology, 2*, 322–327.

Restuccia, D., Clodoveo, M. L., Corbo, F., & Loizzo, M. R. (2018). De-stoning technol-ogy for improving olive oil nutritional and sensory features: The right idea at the wrong time. *Food Research International, 106*, 636–646.

Roeselers, G., van Loosdrecht, M. C. M., & Muyzer, G. (2008). Phototrophic biofilms and their potential applications. *Applied Phycology, 20*, 227–235.

Roig, A., Cayuela, M. L., & Sánchez-Monedero, M. A. (2006). An overview on olive mill wastes and their valorisation methods. *Waste Management, 26*, 960–969.

Rozzi, A., & Malpei, F. (1996). Treatment and disposal of olive mil effluents. *International Biodeterioration and Biodegradation, 38*(3–4), 135–144.

Rubio-Senent, F., Rodríguez-Gutíerrez, G., Lama-Muñoz, A., & Fernández-Bolaños, J. (2012). New phenolic compounds hydrothermally extracted from the olive oil byproduct 'alperujo' and their antioxidative activities. *Journal of Agricultural and Food Chemistry, 60*(5), 1175–1186.

Saleh, M., Cuevas, M., García, J. F., & Sánchez, S. (2014). Valorization of olive stones for xylitol and ethanol production from dilute acid pretreatment via enzymatic hydrolysis and fermentation by Pachysolen tannophilus. *Biochemical Engineering Journal, 90*, 286–293.

Sánchez, S. (2009). *Aceite de oliva, medio ambiente y energía: hacia una fuerte integración* (pp. 17–21). University of Jaen (Ed.).

Sánchez, F., & San Miguel, G. (2016). Improved fuel properties of whole table olive stones via pyrolytic processing. *Biomass and Bioenergy, 92*, 1–11.

Sánchez, S., Martínez, M. E., Espejo, M. T., & Delgado, A. (1996). Production of microalgae from olive mil wastewater. *International Biodeterioration and Biodegradation, 38*(3–4), 245–247.

Sánchez, S., Martínez, M. E., Espejo, M. T., Pacheco, R., Espínola, F., & Hodaifa, G. (2001). Mixotrophic culture of Chlorella pyrenoidosa with olive-mill wastewater as the nutrient medium. *Journal of Applied Phycology, 13*, 443–449.

Sánchez, S., Moya, A. J., Moya, M., Romero, I., Torrero, R., Bravo, V., & San Miguel, M. P. (2002). Aprovechamiento del residuo de poda del olivar mediante conversión termoquímica. *Ingeniería Química, June*, 194–202.

Sánchez, A. H., García, P., & Rejano, P. L. (2006). Trends in table olive production. Elaboration of table olives. *Grasas y Aceites, 57*, 86–94.

Sansoucy, R. (1985). *Olive by-products for animal feed.* FAO Animal Production and Health Paper-43. http://www.fao.org/3/X6545E/X6545E00.htm.

Sicuro, B., Barbera, S., Dapra, F., Gai, F., Gasco, L., & Paglialonga, G. (2010). The olive oil by-product in 'rainbow trout Onchorynchus mykyss (Walbaum)' farming: Productive results and quality of the product. *Aquaculture Research, 41*, e475–e486.

Skoulou, V., Koufodimos, G., Samaras, Z., & Zabaniotou, A. (2008). Low temperature gasification of olive kernels in a 5-kW fluidized bed reactor for H2-rich producer gas. *International Journal of Hydrogen Energy, 33*, 6515–6524.

Suárez, M., Romero, M.-P., & Motilva, M.-J. (2010). Development of a phenol-enriched olive oil with phenolic compounds from olive cake. *Journal of Agricultural of Food Chemistry, 58*, 10396–10403.

Tekin, A., & Dalgıç, A. (2000). Biogas production from olive pomace. *Resources, Conservation and Recycling, 30*, 301–313.

Topal, H., Taner, T., Hassan, S. A., Altinsoy, Y., Amirabedin, E., & Ozkaymak, M. (2017). Exergy analysis of a circulating fluidized bed power plant co-firing with olive pits: A case study of power plant in Turkey. *Energy, 140*, 40–46.

Urieta, D., Menor, A., Caño, S., Barreal, J., Velasco, M. M., & Puentes, R. (2018). *La olivicultura internacional. Difusión histórica, análisis estratégico y visión descriptiva.* Fundación Caja Rural de Jaén.

Vamvuka, D., Papas, M., Galetakis, M., & Sfakiotakis, S. (2016). Thermal valorization of an animal sludge for energy recovery via co-combustion with olive kernel in a fluidized bed unit: Optimization of emissions. *Energy and Fuels, 30*, 5825–5834.

Vásquez-Villanueva, R., Plaza, M., García, M. C., & Marina, M. L. (2020). Recovery and determination of cholesterol-lowering compounds from Olea europaea seeds employing pressurized

liquid extraction and gas chromatography-mass spectrometry. *Microchemical Journal, 156*, 104812.
Vasta, V., Nudda, A., Cannas, A., Lanza, M., & Priolo, A. (2008). Alternative feed resources and their effects on the quality of meat and milk from small ruminants. *Animal Feed Science and Technology, 147*(1–3), 223–246.
Vilardi, G., Ochando-Pulido, J. M., Verdone, N., Stoller, M., & Di Palma, L. (2018). On the removal of hexavalent chromium by olive stones coated by iron-based nanoparticles: Equilibrium study and chromium recovery. *Journal of Cleaner Production, 190*, 200–210.
Yanez-Ruiz, D. R., & Molina-Alcaide, E. (2007). A comparative study of the effect of two-stage olive cake added to alfalfa on digestion and nitrogen losses in sheep and goats. *Animal, 1*, 227–232.

Part X
Olive Landscapes in The Mediterranean Transitioning Towards Enhanced Sustainability: Key Challenges and Opportunities

Chapter 59
Weaknesses, Threats, Strengths and Opportunities of the Olive Oil Sector and Olive Groves Across the Mediterranean Countries and Regions

José Liétor Gallego, José Muñoz-Rojas, Roberto García-Ruiz, José Manuel Herrera, and Gema Parra Anguita

59.1 Why Perform a SWOT Analysis for the Mediterranean Olive Groves and Landscapes?

59.1.1 Advantages and Limitations of SWOT as an Analytical Tool for Describing the Present and Anticipating the Desired (and Undesired) Futures of the Mediterranean Olive Groves and Landscapes

It is now clear from the analysis of the olive sector and related landscapes across the six Mediterranean countries examined in this book that there exists a high level of complexity and heterogeneity in the sector. Furthermore, such complexity results in

J. L. Gallego · G. P. Anguita
Department of Plant and Animal Biology, and Ecology, Universidad de Jaén, Jaén, Spain
e-mail: jlietor@ujaen.es; gparra@ujaen.es

J. Muñoz-Rojas (✉)
MED-Mediterranean Institute for Agriculture, Environment and Development CHANGE Associated Lab for Global Change & Sustainability Department of Landscape, Environment and Planning, Universidade de Évora, Valverde - Évora, Portugal
e-mail: jmrojas@uevora.pt

R. García-Ruiz
University Institute for Research on Olive Groves and Olive Oils, University of Jaén, Jaén, Spain
e-mail: rgarcia@ujaen.es

J. M. Herrera
Mediterranean Institute for Agriculture, Environment and Development, LAB-CHANGE, Universidade de Évora (Portugal), Évora, Portugal

IVAGRO-Universidad de Cádiz, Cádiz, Spain
e-mail: josemanuel.herrera@uca.es

J. Muñoz-Rojas, R. García-Ruiz (eds.), *The Olive Landscapes of the Mediterranean*, Landscape Series 36,
https://doi.org/10.1007/978-3-031-57956-1_59

multiple social-ecological and territorial challenges that remain largely unresolved, and that in some cases can even be considered as wicked ones, and therefore unresolvable (*sensu* Duckett et al., 2016). Thus, it is imperative to obtain an overview that can synthesize the key aspects, both positive and negative, described in detail in the previous chapters of this book, so that some light can be shed on the key question posed as a starting point for the book: What are the key opportunities and limitations for the olive sector to advance towards enhanced levels of sustainability?

To gain such an overview, various methods, tools, and approaches can be applied, although some are more commonly used than others, with the SWOT analysis standing out in particular.

This has been frequently used to characterize the agricultural sector and plan for future trajectories of change, including perspectives and goals that encompass the policy (EC, 2019—https://enrd.ec.europa.eu/sites/default/files/gpw10_9_swot_ramon_i_sumoy.pdf) and business sectors (Kurmanalina et al., 2020).

SWOT analysis stands for Strengths, Weaknesses, Opportunities, and Threats that (in this case) the olive sector encounters both internally and in the macro environment under which it operates. Strengths and weaknesses are often restricted to the system´s (company, sector or institution) internal environment—resources, skills and limitations. Opportunities and threats are factors that are analysed in view of the prevalent macro driving forces including especially the market and also importantly other factors such as technological, political, economic, legal, environmental and social, including health and safety.

According to global executive survey performed by the Harvard Business Review & Brightline Initiative (https://www.brightline.org/resources/testing-organizational-boundaries-to-improve-strategy-execution/), only 20% of the strategic targets set by public and business organizations worldwide are currently realized. The remaining 80% of these strategic targets are not achieved because of aspects such as incomprehensive planning, limited resource allocation and poor execution.

When we translate this to the olive sector, it becomes apparent that successful organizations are the ones able to predict market and socio-political trends more accurately than others, providing resources to develop products and services to leverage those trends, able to counter competitors' threats and meet customers' expected value propositions. These are all goals for which a SWOT analysis can be directly useful, thus reinforcing the relevance and pertinence of performing a SWOT analysis of the olive sector.

Nonetheless, the advantages and outreach of a SWOT analysis for synthesizing the challenges faced by the olive sector should not be overstated and should by any means be complemented by considering its several limitations, which have been clearly identified from a theoretical perspective (Gurel & Tat, 2017).

One such key limitation is the relative fuzziness and openness of the method, which renders the effectiveness and soundness of the results much dependent on the robustness and representativeness of the available data, and the experience of the team in charge of the analysis implementation. This SWOT analysis was embedded in the context of a European project, SUSTAINOLIVE (www.sustainolive.eu), with a wide set of resources and data already compiled and with actor-networks of experts and stakeholders well established that could be engaged in the eliciting and refinement of results for a SWOT analysis of the sector sufficiently wide and reliable across the overall Mediterranean macro-region.

– What did we know from previous SWOT analyses, and what did we not yet know?

Although an unified and overall SWOT analysis of olive groves across the Mediterranean is not in place, there are a number of partial analyses and reports that have provided with extremely useful indication about the key challenges facing the sector.

Quite a few examples of SWOT assessments are found as part of sectoral reports in Mediterranean regions, such as Andalucía, Spain, where the olive sector acts as keystone of the regional economy and territorial structure, or in Alentejo, Portugal, where the sector is considered as strategic in the regional development plans. Examples of these include the SWOT assessments in the II Strategic Plan for the Province of Jaén (Herrador Lindes & Martín Mesa, 2020), which contains a full axis on olives and olive oil, and which are bounded by several subprograms, strategic and implementation plans approved by the regional government of Andalucía (Junta de Andalucía, 2013). Also interesting are the Strategic Assessment Reports of the olive sector in Alentejo performed by CONSULAI & Juan Vilar Consultores Estratégicos (2019) and EDIA (2020). Many other reports are found that synthesize the state-of-the-art using SWOT assessments of the olive sector, focusing on aspects such as trade and distribution (MAPA, 2003). In summary, multiple official reports and academic documents are published that address the limitations and opportunities across intermediate administrative scales from the local to the national, naturally also reaching beyond the European context (Karray, 2006)

Nonetheless, when examined in detail, such SWOT assessment reports are relatively limited in their spatial and thematic scope, with barely none of them overcoming the provincial, regional, or at most, national level of analysis, but also lacking a complex sustainability perspective. Thus, the key aspects that can be extracted from these individual reports and documents are somehow fragmented, lacking the international comparative overview that is required for the design and implementation of policies and related management strategies and actions plans that are largely dependent on both the international legislative framework (e.g. CAP 2020–2026, EU Green Deal 2020, EU Farm to Fork Strategy, 2020). In addition, they also lag in their consideration of the complex sets of interplays occurring between national and regional market and policy frameworks that are gradually becoming more delocalized and globalized, thus hampering the implementation of common rules and standards.

In spite of these limitations, there are several key aspects that can be elicited as "known-knowns" about the olive sector and landscapes in the Mediterranean at the beginning of the twenty-first century. These include confirmation about the overall robustness of the olive economy, underpinned by an ever increasing global demand and recognition of the health benefits of olive oil, and also the fitness of olive crops and landscapes to help tackle key social-ecological challenges in the Mediterranean, including climate change and water scarcity and irregularity, food security and related health isues. Nonetheless, it also becomes apparent that better knwoledge is still required regarding aspects as important for the sector as territorial cohesion, sustainability strategies and capacities and responses to globalization, all of which are impossssible without an overview such as the one hereby provided.

59.2 How Did We Perform a SWOT Analysis?

59.2.1 Sources of Information: Surveys with Farmers, Desk-Based Analysis and Informed Opinions

To progress beyond the fragmented overview of the sector currently in place, SUSTAINOLIVE consortium planned and implemented a characterization of the sector over the six key regions and countries across the Mediterranean. Characterization tasks comprised the following:

- A large survey with over 600 farmers across the Mediterranean region
- An inventory of official and lay information about the sector, considered from a social-ecological perspective
- An ontology of agroecological terms of use in olive cultivation and sustainability
- A booklet on policy recommendations to boost agroecological transition to the olive oil sector
- A booklet of good practices in sustainable olive culture
- A SWOT assessment developed and implemented in close interaction with key actors in each region and country to synthesize knowledge gained through all other tasks

Results from all of these actions were then statistically tested for their robustness and level of confidence and are fully available for those interested in the SUSTAINOLIVE website (https://sustainolive.eu/resources/?lang=en). The goal was to progress beyond the largely sectoral nature of previous characterization exercises by proving an underpinning framework linked to sustainability standards and objectives and also to gain a comparative international overview that cannot be extracted from the joint examination of national and regional results when these have been produced using different criteria and at different times.

Under this working framework, the results obtained in the survey to farmers and producers were examined in depth to help set an initial list of, both apparent and hidden, Strengths, Weaknesses, Opportunities and Threats for the wide diversity of olive grove farms and businesses identified throughout the project. Additionally, the survey results allow to assess, and contest, the diverse sets of solutions that are currently in place in the different regions.

To elicit each of the four variables of the SWOT analysis, a half-day workshop was originally projected in 2020 in each of the six countries participating in the project that was subsequently shifted to individual online interviews with selected stakeholders due to the restrictions imposed by the COVID 19 pandemic. With each of these actors, the main typologies of farms and agronomic practices and their related ecological, socio-territorial and pedo-climatic characteristics were considered, and participants were asked to confirm and enrich the list. Stakeholders were then asked to identify the main SWOT variables for the sector and how they shift with each of the diverse farming typologies identified. Furthermore, they were also asked to help identify the main SWOT variables of the diverse productive and industrial practices and models that are present in each region, and their relation to each of the farm model typologies identified.

Participants in the workshops were selected to represent farmers and their associations, the industrial and transformation sector, government and regulation, consumers and lobbyists, environmentalists and quality control and economic agents, including those actors related to finance, export and trade. A set of guidelines to conduct the SWOT analysis were discussed by all national lead partners of the SUSTAINOLIVE project. The remaining SUSTAINOLIVE partners in each country and region, representing industrial, research and associative interests, were directly involved in the interviews and also helped identify any key local and regional stakeholders missing.

It was agreed that at least two representatives per country should be targeted for each of the following social groups:

1. Producers, ideally among those that were more closely acquainted with the purposes of SUSTAINOLIVE
2. Lobbies (for diverse interests, including farming, the environment, industry and local issues)
3. Public administration, especially those related to farming and the environment, across levels ranging from the local to the national
4. Representatives of universities and research, with expertise and knowledge about the olive sector
5. Others, including, e.g. specialists in farming sustainability

Before the survey was circulated to all SUSTAINOLIVE country leaders to identify and conduct the surveys in their own countries, the meanings of strengths, weaknesses, opportunities and threats, with examples from the olive oil sector, were defined.

In total, 60 experts were identified (2 per category and per country), and a total of 44 responses (in the form of assessment reports) were received. Although farmers and researchers or academics add up to more than half of those that responded, this largely differs among countries. The overall composition of respondents per country is given in Table 59.1.

Table 59.1 Overall composition of respondents per country

Expert category	Spain	Portugal	Italy	Greece	Morocco	Tunisia	Total
Producer	1	1	6	2	1	2	13
University	2	1	1	3	0	3	10
Lobby	0	2	4	0	2	0	8
Public Administration	2	3	1	0	1	1	8
Specialist in farming sustainability	1	2	1	1	0	0	5
Total	6	9	13	6	4	6	44

59.2.2 *SWOT Indicators: Typology, Identification, Refinement and Hierarchization*

The questions from the survey to farmers were retrieved and classified by a small discussion team of SUSTAINOLIVE partners to identify those more pertinent as indicators for each of the four SWOT variables: strengths, weaknesses, opportunities and threats. Based on this preliminary assessment, a series of variables were selected that belong to the following generic categories of indicators:

A. Farmer characteristics
B. Farm and olive grove characteristics
C. Farm bio-physical conditions and land management practices
D. Olive oil production and cooperativism
E. Farm sustainability challenges
F. Markets perception and strategies of farming
G. Social benefits

For each of the indicators included under each of the abovementioned categories, two key questions were formulated to the experts selected:

1. Do you agree that, in your national/regional context, this may be considered as a relevant indicator of strength/weakness/opportunity/threat? (Yes/No)
2. In case you agree with the previous question, how much importance would you assign to this strength/weakness/opportunity/threat? (Little/Intermediate/Much)

The survey was formalized in a standardized questionnaire containing 50 questions on strengths, 76 on weaknesses, 58 on opportunities and 60 on threats that was then translated into each language in the six countries examined (Spain, Portugal, Italy, Greece, Tunisia and Morocco) and transferred by national leading partners to experts selected by themselves.

Once the experts' responses were obtained, the following protocol was defined for the responses statistical treatment and analysis (the statistical decision-making tool from which this protocol derives can be checked here: https://sustainolive.eu/download/10782/):

(a) Transforming variables to a quantitative scale

Since the consulted experts determine the importance of the variables (strengths, weaknesses, opportunities and threats) based on a qualitative scale (little, intermediate and high importance), the first step was to transform the qualitative variables into quantitative: If the respondents considered the variable has no importance, we assign it 0. For the 3 importance ranges (low, medium, high), we assigned 1, 2 and 3, respectively.

(b) Deciding on which are the priority variables within each country

Firstly, the mean importance for each variable was calculated and ranged between the lowest and the highest means. To discriminate which strengths, weaknesses, opportunities or threats obtained the highest statistically significant values, the

80th percentile was calculated for each of the variables. Those variables whose importance mean value was equal or greater than the 80th percentile value were selected as priority variables.

(c) Deciding on how to make comparisons among blocks (topics) of questions and responses.

It was clear to us that it would be extremely interesting to determine if there were significant differences in the responses obtained between different blocks of questions (e.g., farmer characteristics *versus* olive oil production *versus* markets demands) by different types of stakeholders. To address this:

- For each country, the values of relevance indicated by stakeholders for strengths, weaknesses, opportunities and threats were compared considering the response of each expert as a replicate. In case importance value fitted a normal distribution, a one-way ANOVA was applied to find significant differences. If the values did not fit into a normal distribution, a non-parametric Kruskal Wallis test was applied.
- With the intent of finding significant differences among blocks of questions (p-value=0.01), multiple comparison tests of mean ranks after non-parametric analysis or post hoc Fisher LSD tests for parametric analysis were respectively performed. Blocks of questions were thus considered as independent variables (factors).
- A p-value of 0.05 was the significance level applied for the post-hoc multiple comparisons.

(d) Deciding on how to make comparison among experts' categories

It was also clear to us that it would be interesting to determine if there are significant differences between responses by different categories of experts.

- For each country, the values of importance for strengths, weaknesses, opportunities and threats were compared by grouping each individual category of experts consulted (producers, universities, public administrations, lobbies and specialists in farming sustainability). Due to the fact that the data obtained from the interviews did not fit with a normal distribution, non-parametric Kruskal Wallis tests were then applied.
- In the case that significant differences among blocks of questions (p-value 0.01) were not encountered, multiple comparative tests between mean ranks obtained using non-parametric analysis were performed. For operational purposes, the diverse categories of experts were considered as independent variables (factors).
- A p-value of 0.05 was the significance level applied to perform the post-hoc multiple comparisons.
- For comparisons among the same types of experts in different countries, a multiple correlation matrix based on Kendall concordance for categorical variables was performed, where a p-value of <0.01 was the significance threshold level applied.

(e) Deciding on how to make comparisons among countries.

It then became important to infer whether a variable is perceived or not as strength, weakness, opportunity or threat in a country/countries compared to the remaining ones. For each SWOT variable, the average importance values attributed by different stakeholders were compared among countries considering the response of each expert in each country as a replicate. Because any certain variable combined for all countries would in no case fit a normal distribution, non-parametric Kruskal Wallis tests were applied. No significant differences were found in the vast majority of the cases, so only tables with means and standard deviations for each of the survey questions were included in this section.

(f) Dealing with new or unexpected issues raised by expert stakeholders consulted.

Since some strengths, weaknesses, opportunities and threats have not been considered originally in our survey, and were then contributed by expert stakeholders, an extra block of responses entitled "Others suggested by experts" was added to our statistical analysis. However, inconsistencies and gaps were detected in these responses that advised us, following careful analysis, handling them with care.

59.2.3 Gaps and Limitations in the Responses Obtained and the Lists of Respondents Engaged

As evident from our list of interviewees, there are some gaps in the responses that may lead to a certain bias in the results obtained in the diverse countries. An example of such bias may arise from the absence of specialists in farming sustainability in Morocco or Tunisia, which leads us to wonder whether this is due to the relative scarcity of sustainability-oriented olive grove experts in North Africa, or to the aggravated difficulty to achieve participation targets in times of a global pandemic.

Also, big differences in the total numbers of participants may be noted among countries, with Italy and Portugal jointly gathering more than half of the total responses. The reasons behind such differences remain unresolved, and thus deeper reflections are required. When we add to the existing body of knowledge the survey work performed with farmers and producers, we find an overall picture of heterogeneities in the aspirations, barriers and strategies of diverse agents of the olive oil value chain may arise that shall ultimately be of use for defining more efficient, and better regionally tailored, policy and funding schemes and incentives.

That the farmers participating in this project were not necessarily those that are more strongly involved in sustainability experiences and tests, and that none of the other respondents (e.g. academics) were our regular partners, to avoid any potential bias.

59.3 What Did We Learn from Our SWOT Analysis?

59.3.1 Strengths

As a result of the analysis and tests of statistical robustness of the responses obtained, a series of factors that are enumerated in the following table (Table 59.2) can be considered as key strengths of the sector, considering all of our case study areas and stakeholder types.

59.3.2 Weaknesses

Equally, a series of other factors that are enumerated in Table 59.3 can be considered as key weaknesses of the sector, from an overall perspective that jointly considered all of our case study areas and stakeholder types.

59.3.3 Opportunities

The topics listed in Table 59.4 were considered by the consulted stakeholders as key overall opportunities valid for the olive sector across the Mediterranean.

59.3.4 Threats

Lastly, the key threats identified are shown in Table 59.5.

59.3.5 Comparative Results of Countries and Regions

To move beyond generic results valid for the overall Mediterranean region for each of the four SWOT components, we also looked at the differences and similarities in responses across the countries and regions. To do this, we addressed the following four questions which main implications are subsequently expanded along the following epigraphs.

Q1- Which SWOT component is predominant in each country?

As shown in Fig. 59.1, despite significant differences existing in the responses in different countries, the comparison of the importance values attributed to strengths, weaknesses, opportunities and threats provide us with the conclusion that all the

Table 59.2 Key strengths identified by experts stakeholders as contributing to the sustainability of the olive sector and landscapes across the Mediterranean region

STRENGHTS		ID code
FARMER CHARACTERISTICS	The farmer has University level education	S1
	The farmer has a specific education and or training qualification in agriculture	S2
FARM AND OLIVE GROVE CHARACTERISTICS	Sufficient mean annual rainfall	S3
	Total area of land farmed ≥100 ha (latifundio) - considered in the short term (yearly timescales)	S4
	Low to medium-density intensive olive farm (150-400 trees per hectare) - considered in the short term (yearly timescales)	S5
	Traditional/extensive olive farming (<150 trees per hectare) - considered in the medium-long term (decadal timescales)	S6
	Rain-fed olive groves	S7
	≤50 % of the olive farm irrigated	S8
	Application of sensor system for monitoring irrigation	S9
	Efforts in the mechanization of cultivation practices, especially harvesting have been made in the last 5 years	S10
	Use of the olive grove for livestock or animal production	S11
	Protected Denomination of Origin (PDO)	S12
	Fully productive stage (not too young and not too old trees) is predominant - considered in the short term (yearly timescales)	S13
FARM BIO-PHYSICAL CONDITIONS AND LAND MANAGEMENT PRACTICES	Herbaceous vegetation is maintained in the interrows (between the olive trees)	S14
	Pruning wastes are applied as part of management practices	S15
	No pests and diseases affecting the olive trees over the past 5 years	S16
	No other damages affecting the olive grove over the past 5 years	S17
	Leaf analysis have been performed over the past 5 years	S18
	Number of metrics included in leaf analysis >5	S19
	Predominance of deep soils (> 1m)	S20
	No difficulties to employ tractors for working soils	S21
	No significantly high slopes; predominant flat or nearly flat soils (0-6 %)	S22
	No signs of soil erosion detected	S23
	Soil analysis have been performed over the past 3 years	S24
OLIVE OIL PRODUCTION AND ASSOCIATIONISM	Total production of olive oil >1000 kg/hectare in the last financial year	S25
	The quality of the produced oilve oil has been improved in the last 5 years	S26
	Major or total percentage of olive oil production sold	S27
	Ratio individual/collective olive oil sales >50 in the last financial year	S28
	Farmer is a member of a cooperative	S29
	More than 3 different services provided by the cooperative in term of selling production (olives or olive oil)	S30
	Farmer is a member of a producers organization	S31
	More than 3 different services provided by the producers organization in term of selling production (olives or olive oil)	S32
	Farmer is a member of a farmers' union/association	S33
FARM SUSTAINABILITY CHALLENGES	Sustainability elements are strongly considered for olive farm management strategies	S34
	The area cultivated with environmentally-friendly techniques has increased in the last 5 years	S35
	The production of organic virgin olive oil has been started or improved in the last 5 years	S36
MARKETS PERCEPTION AND STRATEGIES OF FARMING	Decisions regarding production and farming strategies strongly consider environmental factors	S37
	Implementation of a clear plan of production for the coming 5 years is planned	S38
	Implementation of a clear market plan for the coming 5 years is planned	S39
	The farm profits from, and applies Agri-Environmental CAP Funding Schemes	S40
	The farm profits from, and applies CAP Funding Schemes related to sustainability	S41
	CAP payments represent just 0-25% of the net benefits of the farm (i.e. the farm is not dependent on CAP money for survival)	S42
	The farm receives external technical support or advice	S43
	The farmer is eager to pay for specialized support or advice	S44
	The farmer considers technicians of an association or extension/technical services to discuss a novel idea or strategy for the farm	S45
	The farmer receives support through a collaborative network (other olive farms or producers, neighbours, groupings/associations, extension/technical services)	S46
	According to the farmer, improving sustainability is a priority to turn the farm or business viable for future generations	S47
SOCIAL BENEFITS	Olive grove is an important source of employment	S48
	Olive grove is a source of local wealth that directly and indirectly nurtures other industries	S49
	The ethnographic legacy of the olive grove is a key cultural factor in the producing territories	S50
OTHER SUGGESTED BY EXPERTS	Olive oil is related to healthy foods / Mediterranean diet	S51
	Traditional olive cultivation is good for the environment	S52
	Olive oil is a natural juice with excellent sensory and nutritional characteristics	S53
	Nearby examples of sucess in the implementation of sustainability actions	S54

countries, except Tunisia, showed an optimistic attitude towards the future of the sector and its sustainability. This result *per se* may be interpreted rather openly and would require much more detailed information before a conclusion can be reached regarding the ultimate causes of these differences. Nonetheless, at the same time, it also contains important implications for future strategies to be implemented or

Table 59.3 Key weaknesses identified by experts stakeholders as contributing to the sustainability of the olive sector and landscapes across the Mediterranean region

WEAKNESSES		ID code
FARMER CHARACTERISTICS	The farmer has primary level education	W1
	The farmer hasn´t any specific educational and or training qualification in agriculture	W2
	The farmer has a low or null knowledge about market mechanisms	W3
FARM AND OLIVE GROVE CHARACTERISTICS	Total area of land farmed is <10 ha (minifundio) - considered in the short term (yearly timescales)	W4
	Superintensive olive farming system (> 800 trees per hectare) - considered in the medium-long term (yearly timescales)	W5
	Use of an irrigation system for olive groves is in place	W6
	Percentage of irrigated olive grove is ≥50%	W7
	≥1500 m^3 of irrigation water is employed per hectare	W8
	No sensor system is in place for monitoring irrigation	W9
	No mechanical technology is in place for olive harvesting	W10
	Older olive trees are predominant - considered in the short term (yearly timescales)	W11
	The farm has inherent difficulties for mechanization (inadequate planting frames, several feet per tree, high slopes, etcetera)	W12
FARM BIO-PHYSICAL CONDITIONS AND LAND MANAGEMENT PRACTICES	Herbaceous vegetation is not maintained in the interrows (between the olive trees)	W13
	Three or more pests and diseases have been detected affecting the olive trees in the farm over the past 5 years	W14
	The total costs per hectare of pests and diseases are considered as too high by the farmer	W15
	Three or more other episodes of damage have been detected affecting the olive grove over the past 5 years	W16
	No leaf analysis have been performed by the farm over the past 5 years	W17
	Increasing total fertilizers cost (euros per year and hectare)	W18
	Shallow (< 0,5 m) oil crusts (limestone, gypsum or others) are found in the olive groves	W19
	Highly significant slopes with predominant steep soils (>25%) characterize the olive grove	W20
	Stoniness (an inadequate proportion of stony soils) interferes with agricultural practices in the olive grove	W21
	Signs of soil erosion have been detected by the farmer	W22
	Erosion severely affects agricultural practices at the farm	W23
OLIVE OIL PRODUCTION AND ASSOCIATIONISM	Bulk sales prevails over retail sales	W24
	Mean value of tree olive oil acidity is ≥0,8	W25
	The farmer doesn´t know the mean value of tree olive oil acidity	W26
	Ratio individual/collective olive oil sales were <0 in the last financial year	W27
	A maximum of one service is provided by the cooperative in term of selling production (olives or olive oil)	W28
	High levels of olive oil self-consumption by the cooperative members results in a lack of short-term market power	W29
	Farmer is not a member of a producers organization	W30
	Only a single service provided by the producers organization in term of selling production (olives or olive oil)	W31
	The farmer is not a member of a farmers' union/association	W32
	Only one or two services (or none) are provided by the farmers´union/association in term of selling production (olives or olive oil)	W33
FARM SUSTAINABILITY CHALLENGES	Many challenges are identified by the farmer as a matter of concern regarding the local context where the olive grove is set	W34
	The farmer doesn´t know (can´t identify) the most worrisome challenges regarding the local context where the olive grove is set	W35
	Low or null investment are in place for reducing inputs and promoting sustainable practices such as integrated and organic production	W36
MARKETS PERCEPTION AND STRATEGIES OF FARMING	The farmer is not aware of the olive sector shortcomings	W37
	The farm cannot afford to enter onto an economy of scale (produce more at lower cost as production grows)	W38
	Olive groves are considered by the farmer as a generator of complementary income (and not as the main business)	W39
	The farm business professionalization process must be highly improved	W40
	The farm businesses should get more closely considered in the design of market arrangements	W41
	The farmer uses outdated marketing strategies that don't take full advantage of new technologies and outreach channels	W42
	The olive oil brand image should be improved to differentiate itself in the market	W43
	Few and/or inefficient olive oil promotional actions are carried out	W44
	The farmer considers being located in a rural area as a competitive disadvantage	W45
	The farmer doesn´t know (can´t identify) what sustainability factors should be considered for olive farm management strategies	W46
	Consumers´behaviour « are strongly considered by the farmer when designing olive farm management strategies	W47
	The farmer doesn´t know (can´t identify) if consumers´s behaviour ought to be considered for olive farm management strategies	W48
	The farmer expects that none or barely any production related changes expected will be implemented in the farm over the coming 5 years	W49
	The farmer doesn´t know if production related changes in the olive farm business will become implemented in the farm over the coming 5 years	W50
	The farmer expects that none or almost no market strategic related changes will be implemented in the olive farm business over the coming 5 years	W51
	The farmer doesn´t know if any market related changes in the olive farm business will be implemented in the farm over the coming 5 years	W52
	The farmer doesn´t know which CAP funding schemes are in place in the farm business	W53
	CAP payments represent more than 75% of the net agricultural benefits of the farm business	W54
	The farm doesn´t receive any external technical support or advice	W55
	The technical support or advice is received only annually (too low frecuency)	W56
	The farmer is not eager to pay for specialized support or advice	W57
	The farmer doesn´t rely on anyone else´s opinion (neither people around nor technicians) to discuss a novel idea or strategy for the farm	W58
	The farmer doesn´t receive support through a collaborative network (other olive farms or producers, neighbours, groupings/associations, extension/technical services)	W59
	The farmer is not willing to take any risks to boost the business	W60
	The farmer considers that any collaboration with neighbours would result in any environmental or social benefits for the farm business	W61
	Improving sustainability is not considered by the farmer as a priority to turn the farm or business viable for future generations	W62
	The farmer has no answers about the strategies to turn the farm or business viable for future generations	W63
MARKETS DEMANDS AND TENDENCIES	Increases in costs derived from the modernization of production structures and adaptation to new marketing channels	W64
	Compliance with increasingly demanding regulations and application of quality systems hampers the options for the farmer	W65
	Heterogeneity and complexity of marketing regulations and olive oil quality standards in consuming countries	W66
	Inefficiencies are detected in the management, valorization and use of the by-products produced by the olive grove industry	W67
	Inability to translate healthy properties of olive oils onto better prices for the producer, when compared to other vegetable fats	W68
	Insufficient concentration of olive oil supply chain is in place	W69
	Large and dramatic olive oil price fluctuations are too frequently encountered	W70
	Opacity is found in the allocation of prices throughout the olive oil marketing chain	W71
	Poor commercialization of virgin and extra virgin olive oils are in place when compared to other olive oil types	W72
	Disconnection exists between producers and packaging and distributing firms	W73
	Increased use of inputs (phytosanitary and fertilizers) and resources (energy and water) in the intensive olive grove are not perceived by the farmer as a risk	W74
	Changes from traditional varieties to intensive and exogenous ones with a strict focus on the market (involving loss of ecosystems heterogeneity and biodiversity)	W75
	The farmer acknowledges a need for higher investment in R & D (quality control, traceability of productions and improvement of productivity) to become competitive	W76
OTHER SUGGESTED BY EXPERTS	Cooperatives' organization working as mob	W77
	Lack of business mind in oil producers	W78
	Lack of generational renewal in oil producers	W79
	No knowledge of the functional properties of the oil by consumers	W80

Table 59.4 Key opportunities identified by experts stakeholders as contributing to the sustainability of the olive sector and landscapes across the Mediterranean region

OPPORTUNITIES		ID code
FARMER CHARACTERISTICS	The farmer is younger than 40 years old	O1
	The farmer has University level education	O2
FARM AND OLIVE GROVE CHARACTERISTICS	The olive groves are characterised by low to medium-density intensities (150-400 trees per hectare) - considered in the short term (yearly timescales)	O3
	The farm is managed as a Traditional/extensive olive grove (<150 trees per hectare) - considered in the medium-long term (decadal timescales)	O4
	The farm contains only rain-fed olive groves	O5
	≤50 % of the olive groves in the farm require irrigation	O6
	< 1500 m^3 of irrigation water is used by the farmer per hectare	O7
	Farm olive oil/olive production is Organic Certified	O8
	The olive oil is designated by a Protected Denomination of Origin (PDO)	O9
	Olive trees predominantly are at a young stage (between 1 and 5 years of age) - considered in the medium-long term (decadal timescales)	O10
FARM BIO-PHYSICAL CONDITIONS AND LAND MANAGEMENT PRACTICES	Herbaceous vegetation is maintained in the interrows (between the olive trees)	O11
	Cover crops in the olive grove are applied as part of farming management practices - considered in the medium-long term (decadal timescales)	O12
	Pruning wastes are applied as part of olive grove management practices	O13
	The number of metrics included in leaf analysis is >5	O14
	Only a single type of soil existing in the olive groves at the farm level	O15
	The farmer knows which type of soil is found in the olive grove	O16
OLIVE OIL PRODUCTION AND ASSOCIATIONISM	Total production of olive oil was >1000 kg/hectare in the last financial year	O17
	A major or total percentage of olive oil production was sold in the last financial year	O18
	The mean value of tree olive oil acidity is <0,3	O19
	The ratio individual/collective olive oil sales were >50% in the last financial year	O20
	More than 3 different services are provided by a cooperative in relation to selling the farmers production (olives or olive oil)	O21
	More than 3 different services are provided by a producers organization in relation to selling the farmers production (olives or olive oil)	O22
FARM SUSTAINABILITY CHALLENGES	Few challenges are identified by the farmer as a matter of concern regarding the local context where the olive grove is set	O23
	Sustainability targets are considered as extremely relevant for the olive farm management strategies	O24
MARKETS PERCEPTION AND STRATEGIES OF FARMING	Decisions regarding production and farming strategies at the farm level strongly consider environmental targets and conditions	O25
	The farmer plans to expand the existing scale of operations for olive oil in the coming 5 years	O26
	Implementation of a clear plan of production is in place at the farm level for the coming 5 years	O27
	Implementation of a clear market plan is in place at the farm level for the coming 5 years	O28
	Farmer´s expectation is for a family member to take over the farm	O29
	The farm applies for and receives Agri-Environmental CAP Funding Schemes	O30
	The farm applies for and receives CAP Funding Schemes related to sustainability targets	O31
	Technical support or advice is received very frequently by the farmer (more than once a month)	O32
	The farmer engages with technicians of an association or extension/technical services to discuss a novel idea or strategy for the farm	O33
	The farmer receives support through a collaborative network (other olive farms or producers, neighbours, groupings/associations, extension/technical services)	O34
	The farmer certainly considers that collaborating with neighbours will likely result in environmental benefits for the farm	O35
	The farmer certainly considers that collaborating with neighbours will likely result in socioeconomic benefits for the farm	O36
	According to the farmer, improving sustainability is a priority to turn the farm or business viable for future generations	O37
MARKETS DEMANDS AND TENDENCIES	Trend at the farm level to increase consumption of natural, healthy and quality products	O38
	Worldwide increase in olive oil consumption is yaking place, especially in non-producing countries with high per capita income	O39
	A greater consumer loyalty to olive oils is achieved	O40
	A greater level of consumption of eco-friendly products is achieved	O41
	A greater level of consumption of PDO products is in place	O42
	A general ackowledgement is in place of the healthy and therapeutic benefits of olive oils	O43
	A wider acceptance of the healthy character of the Mediterranean diet takes place	O44
	Marketing campaigns by cooking reality shows that are currently proliferating on televisions	O45
	Olive oil is in fashion	O46
	Increase in the use of olive oils in hotel and restaurant industries occurrs broadly	O47
	Deep-rooted consumption of olive oil is secured in producing areas	O48
	Lack of market-based knowledge of virgin olive oil and its properties is in place	O49
	Political and social recognition of the olive oil industry as a strategic sector is common	O50
	Institutional support exists of the olive oil sector	O51
	Consideration of olive oil as a priority line of research is implemented in regional and national policies	O52
	Recognition of the olive grove as na asset for rural development and territorial cohesion	O53
	Novel industries exploiting by-products of the olive grove are emergent	O54
	Growing unpopularity of palm oil and its subsequent removal from many processed and ultra-processed foods	O55
	Increased awareness by consumers of food product qualities	O56
	Continuous (not seasonal) production of olive oil is secured through the supply of olives from the southern hemisphere producing countries	O57
	Emerging nature of the oleoturistic sector	O58
OTHER SUGGESTED BY EXPERTS	During the COVID pandemic, farmers have been viewed by society as essential workers and there may be more willingness to invest in research on better agricultural practices	O59
	New uses of olive by-products	O60
	Use of by-products of the olive grove for clean energy production	O61
	Obtaining functional compounds from the olive tree	O62
	Implementation of new technologies for oil extraction in oil mills	O63
	Increased global health concern caused by pandemics	O64
	Olive oil culture	O65

avoided for enhancing the sustainability standards in the sector across each of these six countries (Table 59.6).

Q2- In which countries do the SWOT surveys show the highest scores?

According to the results in Table 59.6, Greece shows the highest scores from the SWOT analysis, followed by Morocco, Portugal and Spain, and then Italy and Tunisia. What this means in practice is that in Greece, Portugal and Morocco, both positive and negative factors are identified and valued more clearly than in the other

Table 59.5 Key threats identified by experts stakeholders as hampering the sustainability of the olive sector and landscapes across the Mediterranean region

THREATS		ID code
FARMER CHARACTERISTICS	The farmer is over 65 years old	T1
	The farmer has only primary level education	T2
FARM AND OLIVE GROVE CHARACTERISTICS	The olive farm is managed as super-intensive (> 800 trees per hectare) - considered in the medium-long term (yearly timescales)	T3
	An irrigation system for olive groves is in place	T4
	Percentage of irrigated land in the olive groves in the farm is ≥50%	T5
	≥1500 m^3 of irrigation water is employed per hectare at the farm level	T6
	No certification schemes are in place in the farm business	T7
	Old olive trees are predominant - considered in the short term (yearly timescales)	T8
FARM BIO-PHYSICAL CONDITIONS AND LAND MANAGEMENT PRACTICES	Herbaceous vegetation is not maintained in the interrows (between the olive trees)	T9
	Cover crops are not applied in the farm as part of farming management practices - considered in the medium-long term (decadal timescales)	T10
	Costs per hectare of pests and diseases are considered too high by the farmer	T11
	Various types of soil exist in the olive groves of the farm	T12
	Shallow soils (< 0,5m) are predominant in the olive groves of the farm	T13
	Shallow soil crusts (limestone, gypsum or others) are found in the olive groves (< 0,5 m)	T14
	Inability to employ tractors for working soils in the olive grove prevails	T15
	Highly significant slopes exist in the farm, with predominant steep soils (>25%)	T16
	High levels of stoniness (percentage of stony soils) interferes with agricultural practices in the olive grove	T17
	Appearance of crevasses, rills or gullies on the soil is detected following rainfall and before ploughing	T18
	Erosion severely impacts agricultural practices at the farm level	T19
OLIVE OIL PRODUCTION AND ASSOCIATIONISM	The mean value of olive oil acicity measured in the tree is ≥0,8	T20
	The farmer doesn´t know the mean value of tree olive oil acidity in the last production campaign	T21
	Only a single service is provided by the cooperative to the farmer in relation to selling the production (ofolives or olive oil)	T22
	The farmer is not a member of a producers organization	T23
	Only a single (or no) service is provided by the producers organization to the farmer in relation to selling the production (of olives or olive oil)	T24
	The farmer is not a member of a farmers' union/association	T25
	Only between one and two services are provided by the farmers´union/association to the farmer in relation to selling the production (of olives or olive oil)	T26
FARM SUSTAINABILITY CHALLENGES	Many challenges are identified by the farmer in relation to the local context where the olive grove is set	T27
	The farmer doesn´t know (can´t identify) the most worrisome challenges in relation to the local context where the olive grove is set	T28
	The farmer doesn´t know (can´t identify) the relevance of sustainability targets for his farming system	T29
MARKETS PERCEPTION AND STRATEGIES OF FARMING	The farmer doesn´t know (can´t identify) if sustainability targets should be considered when designing olive farm management strategies	T30
	Aspects of onsumers behaviour are strongly considered by the farmer for implementing olive farm management strategies	T31
	The farmer doesn´t know (can´t identify) if consumers behaviour are of relevance for implementing olive farm management strategies	T32
	The farmer doesn´t know (can´t identify) to what extent shifts in regulations can affect the farming strategies	T33
	The farmer is planning to abandon farming	T34
	The farmer doesn´t know if any production-related changes in the olive farm business will be implemented in the coming 5 years	T35
	None or barely any market-related changes in the olive farm business are planned at the farm level in the coming 5 years	T36
	The farmer doesn´t know if any market-related changes in the olive farm business will be implemented in the coming 5 years	T37
	Currently, the farmer has no expectations for the succession of the farm business	T38
	The farmer doesn´t know what the future of the farm will be	T39
	The farm receives Direct Payments (Pillar I) CAP Funding Schemes - considered in the medium-long term (decadal timescales)	T40
	CAP payments represent more than 75% of the net agricultural benefits of the farm	T41
	The farmer is not eager to pay for specialized technical support or advice	T42
	The farmer doesn´t regularly engage with any other actors (neither people around nor technicians) to discuss a novel idea or strategy for the farm	T43
	The farmer doesn´t receive support nor advice through a collaborative network (other olive farms or producers, neighbours, groupings/associations, extension/technical services)	T44
	The farmer considers that any collaboration with neighbours will not likely result in environmental or social benefits for the farm business	T45
	The farmer does not know of potential alternative strategies to secure the farm or business becomes viable for future generations	T46
MARKETS DEMANDS AND TENDENCIES	Na "Uncontrolled" increase in the supply of olive oil takes place	T47
	A clear dominance of large distribution chains defines the olive industry (linked to a concentration of the demand)	T48
	Cultural barriers exist that hamper the potential for international trade	T49
	Low market share of virgin olive oils is detected at National level	T50
	Lack of knowledge exists of olive oil characteristics and related benefits among the population of producing countries	T51
	Prevalence of substitute dietary oils of lower quality but at a lower price	T52
	Legal possibilities are in place for marketing oils obtained by mixing olive oils with seed oils	T53
	No public control exists of public policy intervention mechanisms and prices	T54
	Restructuring of the European economic support is in favor of crops other than olive groves	T55
	Free Trade Agreements (WTO agreements) hamper access to word olive oil by markets of emerging economies	T56
	Farmers ageing and lack of generational replacement is prevalent	T57
	Increase in olive grove cultivated areas in traditionally non-producing countries (Argentina, Chile, Australia, etcetera), increasing competition	T58
	Arising of import duties and other trade barriers in consuming countries, hampering olive oil exports	T59
	Bad reputation caused by media news about fraud events related to olive oil quality	T60
OTHER SUGGESTED BY EXPERTS	Increase of another plant fats for consumption with more competitive prices	T61
	Xillella and other new diseases of the olive grove	T62
	Export of old or poor quality oils in emerging markets	T63

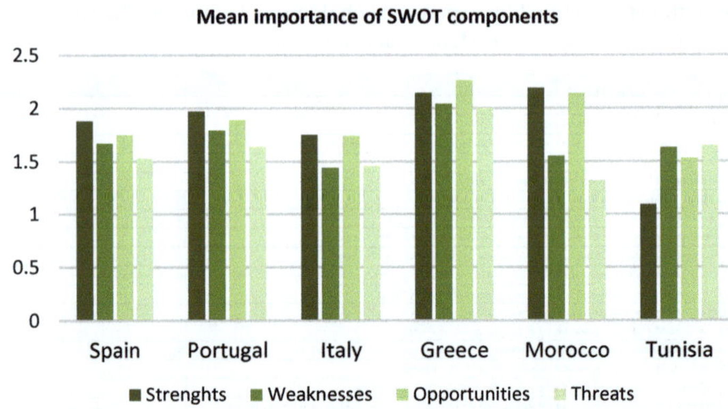

Country	p-value	Analysis	Normality	Significant Post-Hoc	Point of view
Spain	0,048	Kruskal-Wallis	No	Strenghts > Threats (P=0,047)	**Optimistic**
Portugal	0,0142	Kruskal-Wallis	No	Strenghts > Threats (P=0,010) Opportunities > Threats (P=0,049)	**Optimistic**
Italy	0,00027	One-way ANOVA	Yes	Strenghts > Weaknesses (P=0,0008) Strenghts > Threats (P=0,0029) Opportunities > Weaknesses (P=0,0013) Opportunities > Threats (P=0,0047)	**Optimistic**
Greece	0,0052	Kruskal-Wallis	No	Opportunities > Weaknesses (P=0,019) Opportunities > Threats (P=0,0068)	**Optimistic**
Morocco	<0,0001	Kruskal-Wallis	No	Strenghts > Weaknesses (P=0,000011) Strenghts > Threats (P<0,00001) Opportunities > Weaknesses (P<0,00001) Opportunities > Threats (P<0,00001)	**Optimistic**
Tunisia	<0,00001	Kruskal-Wallis	No	Threats > Strenghts (P=0,0016) Weaknesses > Strenghts (P=0,00001)	**Pesimistic**

Fig. 59.1 Mean importance value and predominance of each SWOT component in each country examined and implications for the identification of perspectives and attitudes by stakeholders

Table 59.6 Ranking of the countries based on their scores in strengths, weaknesses, opportunities and threats for enhancing olive farming sustainability

Analysis	Normality	Significant Post-Hoc
Kruskal-Wallis (p<0,01)	No	**Greece importances are significantly higher than those in the rest of countries. Portugal and Morocco importances are significantly higher than Italy & Tunisia importances** (see table below; significant differences are represented by red numbers; difference between Tunisia and Morocco is close to the significance level of P=0.01 so it has been marked in orange).
Conclusions	A ranking can be made that orders from lowest to highest the scores that national experts give to the SWOT components for the olive grove, which can be interpreted as an increasing ranking of the number of factors considered by experts that influence the olive grove business.	
	Italy/Tunisia < Spain < Portugal/Morocco < Greece	

		Spain	Italy	Greece	Tunisia	Portugal
	Mean importances	1.71	1.60	2.11	1.48	1.82
Italy	1.60	0.289496				
Greece	2.11	**<0.000001**	**<0.000001**			
Tunisia	1.48	1.000000	1.000000	**<0.000001**		
Portugal	1.82	1.000000	**0.000506**	**0.000001**	**0.004365**	
Morocco	1.80	1.000000	**0.002559**	**<0.000001**	**0.017913**	1.000000

countries, thus emphasizing the consistency and soundness of the responses obtained in these three countries, while in Spain, Italy and Tunisia, the differences are milder.

Q3- How are SWOT components scores distributed among topics in each country?

When examining Fig. 59.2, the emphasis indicated for certain countries in the responses to the previous question is reinforced. In this way, while differences in the types of factors identified among countries are evident, Greece scores consistently higher in all four types of factors (strengths, weaknesses, opportunities, and threats), farmers (socio-demographic) and bio-physical conditions, both internally and in comparison to other countries. In contrast, countries like Tunisia show significantly greater variations among the different types of factors.

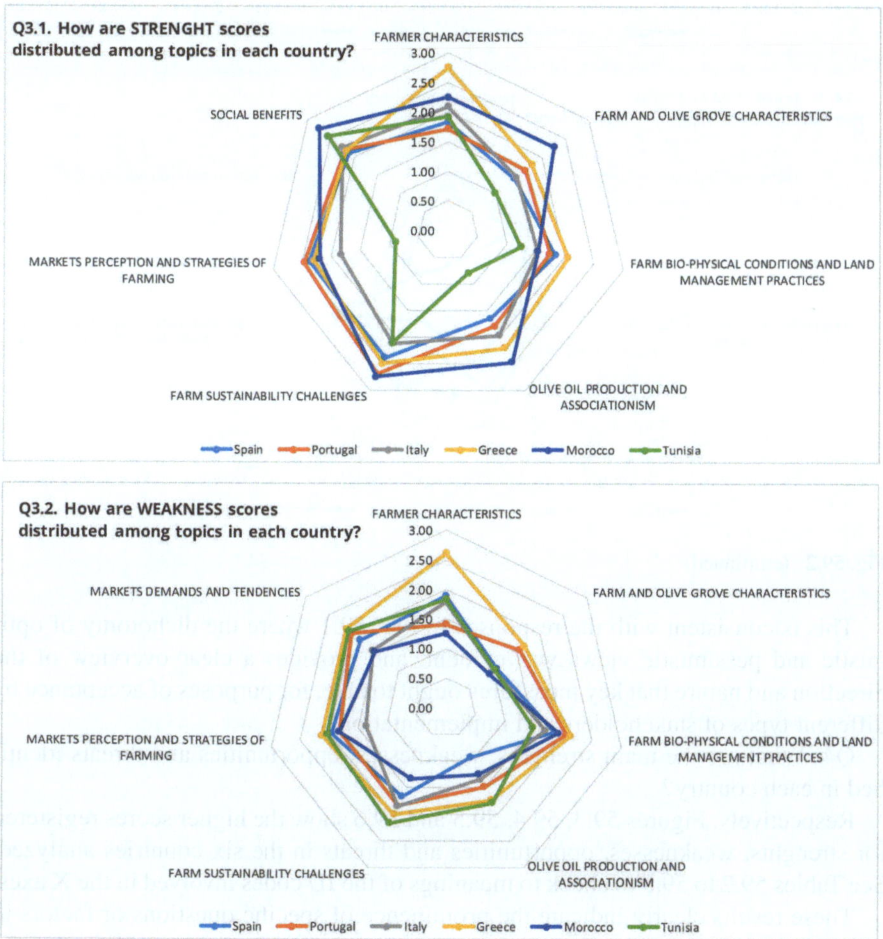

Fig. 59.2 Differences among countries in the ranks attributed to each of the four SWOT components

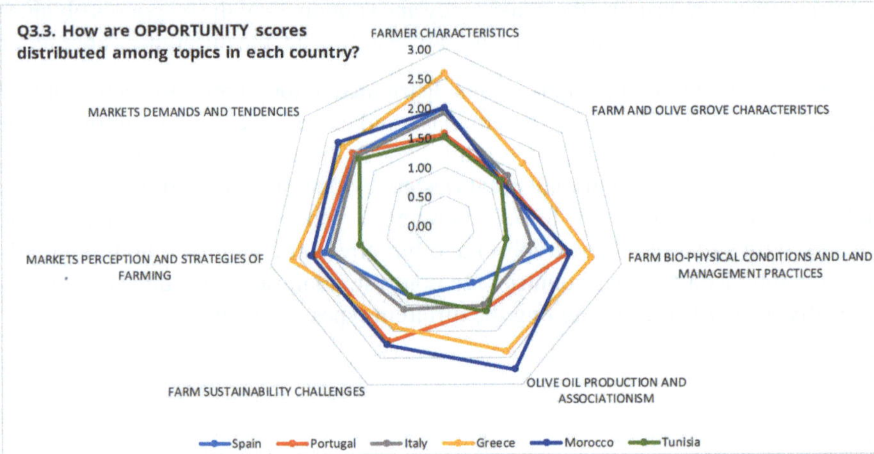

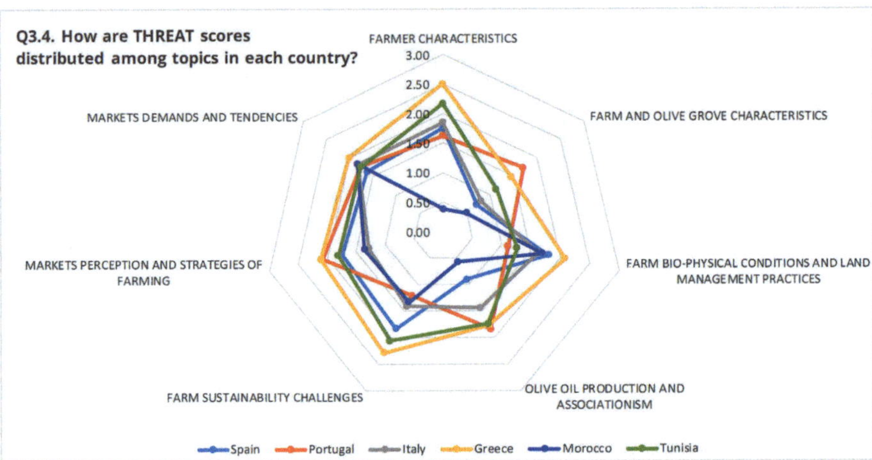

Fig. 59.2 (continued)

This is consistent with the responses in Fig. 59.1 where the dichotomy of optimistic and pessimistic views was evident, and provides a clear overview of the direction and nature that key meassures ought to take, for purposes of acceptance by different types of stakeholders and implementation.

Q4- What are the main strengths, weaknesses, opportunities and threats identified in each country?

Respectively, Figures 59.3, 59.4, 59.5 and 59.6 show the higher scores registered for strenghts, weaknesses, opportunities and threats in the six countries analyzed. See Tables 59.2 to 59.5 to check to meanings of the ID codes involved in the X axes.

These results clearly indicate the prominence of specific questions or factors in each of the countries, which can lead to better understand for each of them what the experts consulted consider should be maintained (Strengths), resolved (Weaknesses), seized (opportunities) or actively tackled (threats).

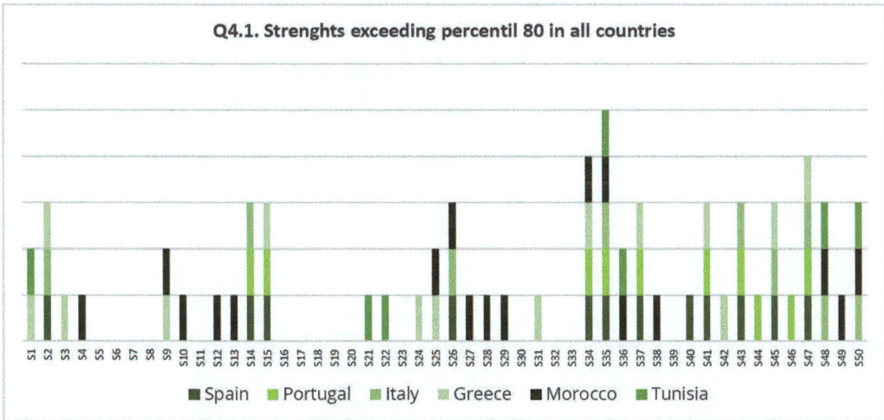

Fig. 59.3 Identified strengths surpassing the 80th percentile for Spain, Portugal, Italy, Greece, Morocco and Tunisia

59.4 And So Where to Now?

59.4.1 Key Findings Challenging Pre-existing Knowledge and Assumptions About Mediterranean Olive Groves and Landscapes and Their Current and Prospective Sustainability and Resilience

Following up from the results synthesized in the previous section, it seems that our interaction with stakeholders confirms the presumption that the olive sector across the Mediterranean is extremely diverse in the aspirations, barriers and challenges perceived by different types of stakeholders. Nonetheless, and generally speaking, a vast majority of stakeholders consulted indicate the sectors as having a bright future full of opportunities. Nonetheless, it also demands that sustainability challenges are more effectively tackled. These challenges are jointly defined by different aspects such as socio-economic, socio-political and socio-cultural characteristics, farm structure and size, bio-physical conditions, markets, cooperativism and business sustainability and strategies. In all cases, these are considered as significantly relevant, with over 120 specific factors identified across all four components or the analysis (Figures 59.3, 59.4, 59.5 and 59.6).

Key issues observed through our analysis are as follows:

- The socio-cultural constraints (lack of adequate support, knowledge and institutions) across the Mediterranean region to foster cooperation and coordination among stakeholders acting across different levels, thus hampering the overall strength of the sector.
- Marketing difficulties faced by certain types of producers (traditional, family farms and mountain farmers) to compete in the increasingly globalized market, a

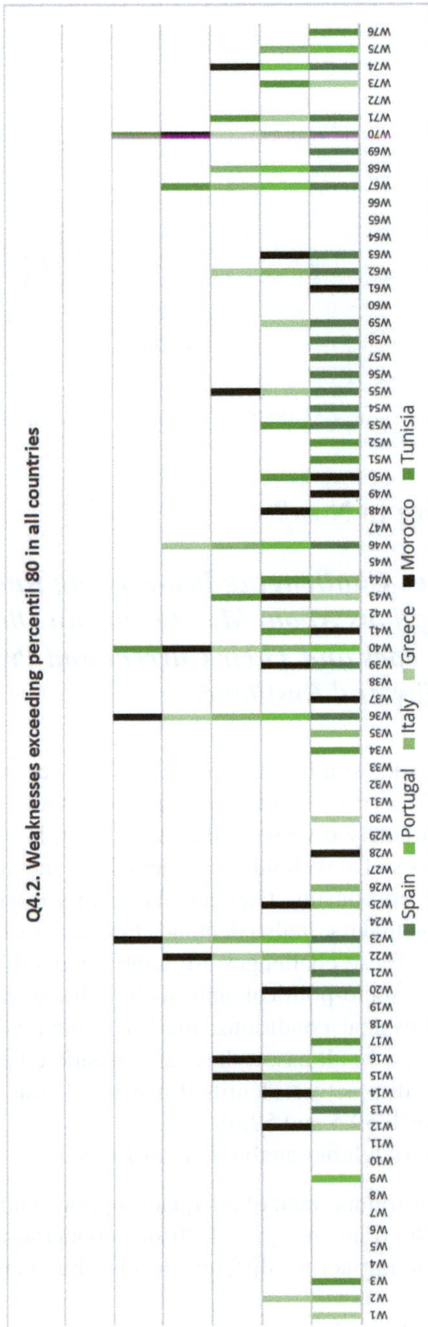

Fig. 59.4 Identified weaknesses surpassing the 80th percentile for Spain, Portugal, Italy, Greece, Morocco and Tunisia

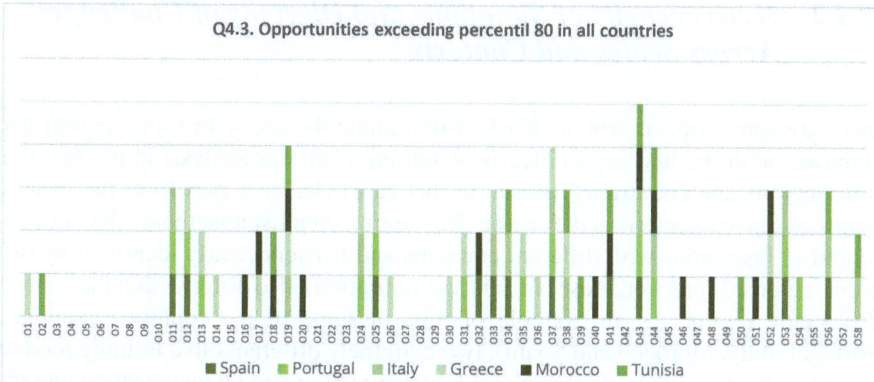

Fig. 59.5 Identified opportunities surpassing the 80th percentile for Spain, Portugal, Italy, Greece, Morocco and Tunisia

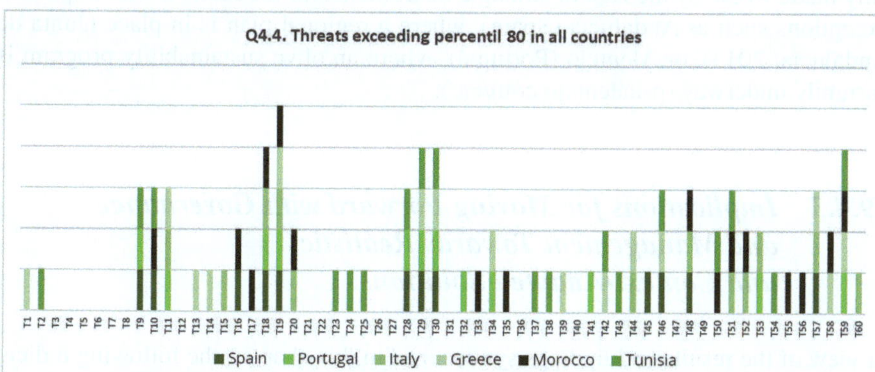

Fig. 59.6 Identified threats surpassing the 80th percentile for Spain, Portugal, Italy, Greece, Morocco and Tunisia

problem that is very much determined by access to adequate natural resources (water, productive soils), policy support, farm and proper structure and adequate institutions.

– Asymmetries in the market power of certain types of farmers and other stakeholders (lobbies, intermediaries) that predetermines the existence of winners (e.g. corporate farms and those with investment capacity to risk in markets such as organic ones) versus losers (mainly family farms and small producers that frequently are constrained to operate via local cooperatives with poor market power).

59.4.2 *Heterogeneity of Situations and Diversity of Challenges Across Scales and Contexts*

There are some aspects which clearly differentiate the sector in some regions and countries, with the first aspect clearly stemming from our analysis in the fact that some regions and countries (Tunisia, in this case) clearly depart from the generic optimism that characterizes the sector. This optimism must nonetheless be not considered as ingenuous with different concerns and barriers clearly identified by different regions. Moreover, these barriers and limitations also differ within the context of a single region. As an example of this, in some of the countries examined (Portugal, Italy, Morocco and Spain), two extremely different olive farming models coexist, with one of them favoring rapid intensification and financialization and the other pushing for alternative production models (largely organic and biodynamic), focusing on quality, shorter circuits and local varieties.

The inherent difficulties in aligining these two production models together are only made worse by the lack of strategic instruments for the sectors, with important exceptions such as Andalucía (Spain), where a regional plan is in place (Junta de Andalucía, 2013), or Alentejo (Portugal), where an olive sustainability program is currently underway (psaalentejo.com/en/).

59.4.3 *Implications for Moving Forward with Governance and Management Towards Realistic and Context-Adapted Solutions*

In view of the results and findings synthesized in this chapter, the following indications could be elicited as necessary conditions to move towards more effective and ultimately more sustainable olive sector across the Mediterranean:

- Both the farming systems and landscapes associated with olive production across the Mediterranean are extremely diverse. These range from super-intensive and corporate-dominated production systems to more traditional and marginal ones. Despite the generic preconceptions that consider the former as winner and the latter as looser, the overall perception by experts and stakeholders is that there is scope for all models to coexist and survive in a globalized and financialized economy that is nonetheless increasingly worried about food quality and sustainability, so long as opportunities identified are undertaken and threats effectively tackled.
- "The lack of enough knowledge on the environmental and economically importance of the sustainability concept and sustainable management practices" is identified as one of the most relevant threats by most of the experts in all countries, which pointed out the need in training and capacity building. In this concern, the diversity of dissemination events taken, as well as informative,

educational, and knowledge-sharing materials produced in SUSTAINOLIVE (see all of them at https://sustainolive.eu/) along with the creation of new networks among Mediterranean countries around olive grove sustainability, will boost the progress in the use of sustainable technologies across the Mediterranean basin.
- To achieve the aspirations for improvement expressed by stakeholders and actors, there are certain strategies that were identified as potentially effective by many of the stakeholders an experts consulted, including sustainable intensification, landscape approaches based on enhanced coordination and cooperation above the farm level, sustainability strategies, and strategic action planning to be implemented via public-private partnerships.
- This requires the urgent development of strategic plans and the establishment of coordinating institutions at both national and international levels, going beyond the current commercial and production objectives of the industryThe adoption of a sustainability framework as indicated in the latest EU and global policies for the food and farm sector (e.g., UN SDGs) is a key step to progress in this direction.
- In addition, there needs to be a shift in the current mindsets of key actors including farmers, market and policy agents, and lobbyists, which are still largely influenced by production goals underpinned by a bio-economy paradigm largely questioned and that should be already replaced by the eco-economy paradigm underpinning sustainable development (Silveira et al., 2018).
- Recent challenges, including the COVID-19 pandemic, potential food security issues arising from armed conflicts, and related uncertainties and fluctuations in food markets, may lead some to view this, in the words of Charles Dickens, as 'the worst of time'
- Alternatively, we propose making the most of the predominantly optimistic outlook revealed in our analysis of the sector to transition toward 'the best of times'.A task that can be best achieved if we rapidly place sustainability at the core of the sector.

Acknowledgements This work was funded by the SUSTAINOLIVE project (Grant no. 1822; PRIMA EU Programme) and by FCT-Portugal (UIDB/05183/2020—funding for all authors of the chapter who are members of MED-Universidade de Évora, and LA/P/0121/2020 for members of the CHANGE-Associated Lab).

References

CONSULAI & Juan Vilar Consultores Estratégicos. (2019). *Alentejo: a liderar a olivicultura moderna internacional. Relatório Final 2019*. CONSULAI y Juan Vilar Consultores Estratégicos, 107pp. [Available at: https://www.olivumsul.com/_files/ugd/a303d9_5993f29b65054e46a54accff8c90cf7f.pdf] Last Consulted: 23/06/2024.

Duckett, D., Feliciano, D., Martin-Ortega, J., & Munoz-Rojas, J. (2016). Tackling wicked environmental problems: The discourse and its influence on praxis in Scotland. *Landscape and Urban Planning, 154*, 44-56. https://doi.org/10.1016/j.landurbplan.2016.03.015

EDIA. (2020). Olival em Alqueva. Caraterização e perspetivas. 73 pp. [https://www.edia.pt/wp-content/uploads/2021/03/olival_digital-1.pdf]. Last Consulted: 01/04/2024.

European Commission (EC). (2019). Th2 CAP after 2020. The SWOT analysis: A key step of the future CAP Strategic Plans. DG AGRI, Unit C.1 Brussels, 21 March 2019. 1 pp. [Available at: https://ec.europa.eu/enrd/sites/default/files/gpw10_9_swot_ramon_i_sumoy.pdf] Last Consulted: 01/01/2024

Gurel, M., & TAT, M. (2017). SWOT analysis: A theoretical review. *The Journal of International Social Research, 10*(51). https://doi.org/10.17719/jisr.2017.1832

Junta de Andalucía. (2013). *Programa de desarrollo rural de andalucia 2014-2020.* Subprograma tematico del sector del oliva, 58pp.

Herrador Lindes, I., & A. Martín Mesa (2020). II Plan Estratégico de la provincia de Jaén, 2020. Ed. Fundación "Estrategias para el desarrollo económico y social de la provincia de Jaén" [Available at: https://www.planestrajaen.org/galerias/galeriaDescargas/plan-estrategico/libros/II_Plan_Estratégico_de_la_provincia_de_Jaén.pdf] Last Consulted: 01/01/2024

Karray. (2006). *Impacts of agricultural trade liberalization between the EU and Mediterranean countries*. 2004. Deliverable D18. Tunisian production and export potential for olive oil Expert Panel Analysis. 98pp. Last Consulted: 23/06/2023.

Kurmanalina, A., Bimbetova, B., Omarova, A. I., Kaiyrgaliyeva, M-G., Bekbusinova, G., Saimova, S., & Saparaliyev, D. (2020). A swot analysis of factors influencing the development of agriculture sector and agribusiness entrepreneurship. *Academy of Entrepreneurship Journal*, 26-1.

Ministerio de Agricultura, Pesca y Alimentación (MAPA). Gobierno de España. (2003). *Diagnóstico y Análisis Estratégico del Sector Agroalimentario Español.* Análisis de la cadena de producción y distribución del sector del aceite 2003. 76pp. Last Consulted: 23/06/2023.

Silveira, A., Muñoz-Rojas, J., Pinto-Correira, T., Guimarães, M. H., Ferrão, J., & Schmidt, L. (2018). The sustainability of agricultural intensification in the early 21st century: Insights from the olive oil production in Alentejo (Southern Portugal). In A. Delicado, N. Domingos, & L. de Sousa (Eds.), *Changing societies: Legacies and challenges* (The diverse worlds of sustainability) (Vol. 3, pp. 247–275). Imprensa de Ciências Sociais.

Chapter 60
Olive Groves and Landscapes in the Mediterranean: Looking into the Future – Challenges, Opportunities, and Scenarios for More Sustainable Olive Landscapes Delivering Multiple Functions and Services

José Muñoz-Rojas, Roberto Garcia-Ruiz, and José Liétor Gallego

60.1 What Do We Know Now That We Didn't Know Before?

We opened this book by arguing for the urgent need for deeper and more creative reflections about the sustainability of olive landscapes in the Mediterranean (Chap. 1). As has been thoroughly demonstrated throughout this book, olive grove landscapes are farming system that have strongly shaped the landscapes of the Mediterranean over the long and rich tract of human history in the region. Furthermore, we also anticipated the multiplicity of challenges, but also of opportunities, that are linked to this complex and rich landscape, which in itself is a cultural, economic and environmental asset for rural and urban territories across the region, and beyond. The heterogeneity of contingent situations shaping such opportunities and challenges is directly apparent in the chapter structure of this book. Thus, the reader who has followed our indications and suggestions in the initial chapter on how to approach the reading of this manual will undoubtedly have already become aware of the richness and heterogeneity of challenges and opportunities arising. Hopefully, this same reader will have learnt something new in the way, or at least challenged some preconceptions. This has certainly been our case, as book editors.

J. Muñoz-Rojas (✉)
MED-Mediterranean Institute for Agriculture, Environment and Development CHANGE-Associated Lab for Global Change & Sustainability Department of Landscape, Environment and Planning, Universidade de Évora, Valverde - Évora, Portugal
e-mail: jmrojas@uevora.pt

R. Garcia-Ruiz · J. L. Gallego
University Institute for Research on Olive Groves and Olive Oils, University of Jaén, Jaén, Spain
e-mail: rgarcia@ujaen.es; jlietor@ujaen.es

J. Muñoz-Rojas, R. García-Ruiz (eds.), *The Olive Landscapes of the Mediterranean*, Landscape Series 36,
https://doi.org/10.1007/978-3-031-57956-1_60

To close the book, the main lessons learnt and key overall findings are briefly synthesized in this chapter, which looks into the future by drafting the main issues arising in relation to the sustainability and resilience of these landscapes (Sect. 60.1), specifically discussing challenges (Sect. 60.2) and opportunities (Sect. 60.3). Finally, we close the chapter by sketching a set of plausible scenarios (Sect. 60.4) that we expect will act as a sort of window into the future that can shed light on these ancient, and yet continuously renovated, rural landscapes. These are undoubtedly landscapes without which the Mediterranean would have never come to be the cultural and environmental hot spot that we have acknowledged it to be, at least, for the past 2 millennia.

60.1.1 Heterogeneities in the Landscape Mosaic of Olive Groves Across Scales and Contexts

If one needed to indicate to only one aspect that outstands when overlooking the olive landscapes scattered across the Mediterranean, then it would need to be their heterogeneity (Fig. 60.1). This is a characteristic of olive landscapes that becomes apparent across spatial and time scales and that is reflected across all components of sustainability (environmental, economic, cultural, territorial and societal). Such cross-scalar heterogeneity is clearly reflected in the national and regional syntheses presented in this book (Chaps. 2, 3, 4, 5, 6, 7, 8, 9, 10, 11, 12, 13, 14, 15, 16, 17, 18, 19, 20, 21, 22, 23, 24, 25, 26, 27, 28, 29, 30, 31, 32, 33, 34, 35, 36, 37, 38, 39, 40, 41, 42, 43, 44, 45, 46, and 47). Nonetheless, and as we know from complex systems theory, landscape heterogeneity invariably entails a level of complexity that results in a landscape mosaic that, as all fractal systems, retains a certain order among the apparent chaos, reiterating itself across spatial-temporal scales and contexts (Forman & Wilson, 1995).

It is therefore only when a cross-scale approach is enacted that olive landscapes as diverse as those in the neighbouring regions of Andalucía (Spain) and Alentejo (Portugal), Kalamata and Crete (Greece), Puglia and Liguria (Italy), Tunisia or Morocco are perceived as different as those identified when the scale of analysis is enlarged by looking at olive landscapes within and outwith the Irrigation perimeter of Alqueva (Alentejo), or across the mountain areas of Sierra Mágina and the agricultural hills of Úbeza and Baeza, all of which are in the Andalusian province of Jaén.

Thus, only assessments that consider heterogeneities across scales from the national to the local levels will become useful to inform policies, practices and strategies that can prove effective for a sector that is as complex and diverse as the landscapes and territories of the Mediterranean where this land-use is still prevalent, and in some cases even dominant.

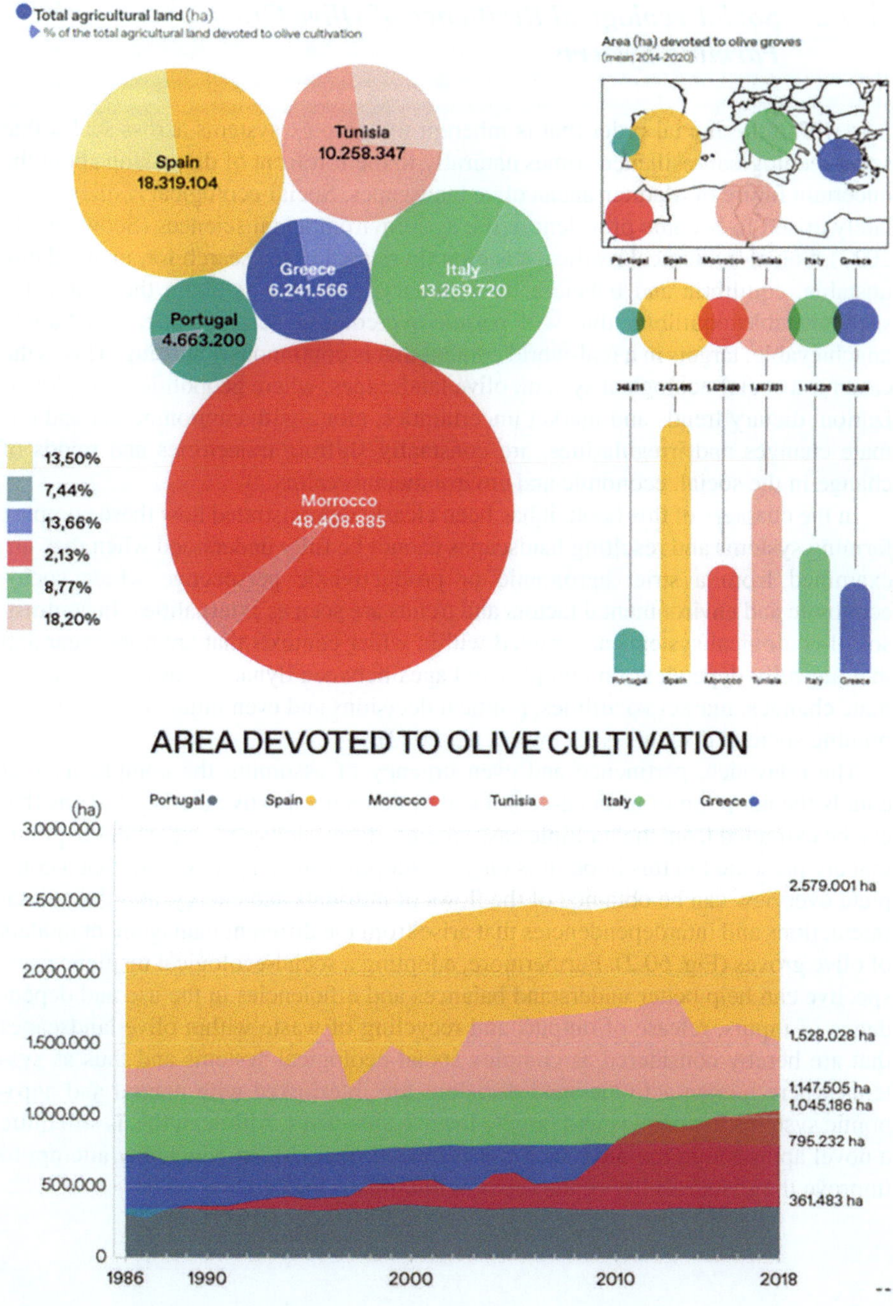

Fig. 60.1 Synthetic statistical differences among the main olive-producing Mediterranean countries, and their evolution along the past three decades

60.1.2 Social-ecological Resilience of Olive Grove Farming Systems

It is within the fractal order that is inherent to complex systems across scales that social-ecological resilience comes naturally to the forefront of discussion about the uncertain future of Mediterranean olive landscapes. Social-ecological resilience has lately arisen to become prevalent in the agri-environmental sciences (Scown et al., 2023). Social-ecological resilience is directly related to the search for, unavoidably unstable, equilibria and balances. In summary, resilience is about the search for such unstable equilibria that will permit overcoming deterministic, and largely unachievable, targets in a real-world context that is continuously shifting. This is the case for a social-ecological system, olive landscapes, where geopolitics and globalization, dietary trends and market uncertainties, along with environmental and climate changes and irregularities, are constantly shifting trajectories and trends of change in the social, economic and environmental realms.

In the chapters of this book, it has been clearly demonstrated how these complex farming systems and resulting landscapes cannot be fully understood when they are examined from a strict agronomic or productivistic perspective where socio-economic and environmental factors and trends are seen as externalities. In contrast, social-ecological systems are framed within wider contexts that are non-linear and that are fully dependent on complex linkages between dynamics as diverse as climate changes, market volatilities, political decisions and even mind-sets and underpinning societal discourses (Scown et al., 2023).

The relevance, pertinence and even urgency of assuming the complexity that entails the adoption of a social-ecological systems perspective are key lessons that can be extracted from the multiple case studies, discussions and critical descriptions that are presented in this book. It is only by adopting such a perspective that a complete overview can be obtained of the flows of nutrients and energy, and the mutual interactions and interdependencies that arise from the different management models of olive groves (Fig. 60.2). Furthermore, adopting a social-ecological resilience perspective can help better understand balances and efficiencies in the use and dependence of inputs, release of outputs and recycling of waste within olive landscapes that are hereby considered as complex social-ecological systems and thus as systems where human activities and decisions are interlinked with natural and agronomic systems in complex and largely inextricable ways. Although this is still quite a novel approach in the olive sector, it is indeed one without which any attempt to improve the sustainability of the sector will unlikely succeed.

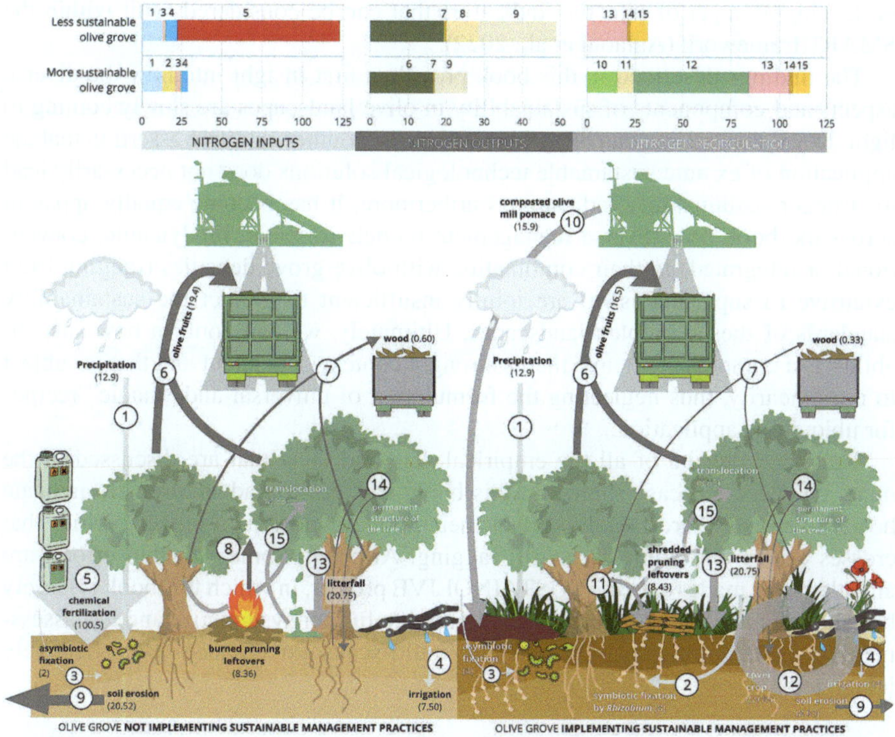

Fig. 60.2 Differences among olive groves implementing (right) and not implementing (left) sustainable practices in terms of their flows and balances of N, including emissions, capture and their recirculation. This is indicative of the differentiated social-ecological resilience of alternative olive landscape management practices

60.1.3 Sustainability of the Mediterranean Olive Economic Sector

Ultimately, it is the complex, and largely malleable, concept of sustainability that sits at the core of this book and its purposes. This is a concept that has largely risen to the height of scientific, political and societal agendas, not without leaving much futile discussions and quite a lot of greenwashing and malpractices on the way (De Vries et al., 2015). Discussions about strong versus weak sustainability, and between their financial, political, cultural, environmental and social aspects, can now be found permeating, and too often distort, legitimate concerns about the future survival of human well-being and the landscapes and ecosystems upon which human societies depend. The misuse of terms, interpretations and applications related to sustainability is a problem that is not alien to the olive sector. As a matter of fact, while initiatives have been surging to promote and increase the sustainability of the olive sector in regions across the Mediterranean (recent and notable examples include Alentejo in Portugal and Tuscany in Italy), much less effort has been placed in developing a set of effective indicators that can be considered to fit within the SMART framework (Aalaoui et al., 2022).

The findings described in this book provide a first insight into how the diverse aspects and components of sustainability in olive landscapes are slowly coming to light. Key issues arising can be synthesized in the counter-intuitive assertion that the application of ex ante sustainable technological solutions does not necessarily lead to stronger sustainability indicators. Furthermore, it has become equally apparent across the book how diverse management models (organic, biodynamic, conventional or integrated) in their combination with olive grove densities (ranging from extensive to super-intensive) are jointly insufficient to predict the sustainability standards of these complex landscapes. Ultimately, we can confirm how sustainability is a complex challenge that is strongly context-dependent and that is subject to non-linearity, thus neglecting the formulation of universal and "magic" recipes for ubiquitous application.

Thus, and despite of all the empirical data and facts that are discussed in the many chapters and case studies in this book, it is not too adventurous to indicate how these results are largely still in their infancy and that a bigger picture that crosses scales and contexts is still lagging. Although some valuable efforts are already being made (e.g., the SUSTAINOLIVE project, in which this book is largely based; see Fig. 60.3), an effort is still required to further systematize and test assessments where multi-functionality and multi-scale governance are effectively integrated.

60.1.4 Multi-functionality and Multi-level Governance

Farming system multi-functionality and multi-level governance are targets that have become the object of ardent discussions in the scientific and policy realms. In particular, it is through the framework of Ecosystem Services that the interlinkages

Fig. 60.3 Integrated sustainability index for various olive groves in southern Spain, comparing those implementing sustainable technological solutions with those that do not. Data are based on empirical results from the SUSTAINOLIVE research project. (https://sustainolive.eu/resources/?lang=en)

between these two elements have become more explicit and applicable (Brady et al., 2019). This is indeed the case of the olive sector, where the delivery of multiple services and benefits to human societies by olive landscapes through innovative governance across scales (Fig. 60.4) is not just a central topic of discussion but also one that deserves a bigger effort for theorization and implementation.

Nonetheles, from the information contained in this book, it becomes apparent that olive groves that are managed applying innovative sustainable practices can deliver more ecosystem services and thus present more opportunities to secure their future sustainability and resilience. This is true from a social, financial, political, economic and environmental standpoint. While this is now self-evident to the point that it is no longer arguable, it is much more uncertain how effective governance mechanisms to fund, promote and enact the implementation of multi-functional strategies and plans can be implemented across scales. While both policy (e.g. CAP second Pillar, Results-Based Payments) and finance (e.g. Payment for Ecosystem Services Schemes) are already leaning in that direction, they are still far from their full realisation in a context (Mediterranean) and sector (farming) that remain strongly conservative.

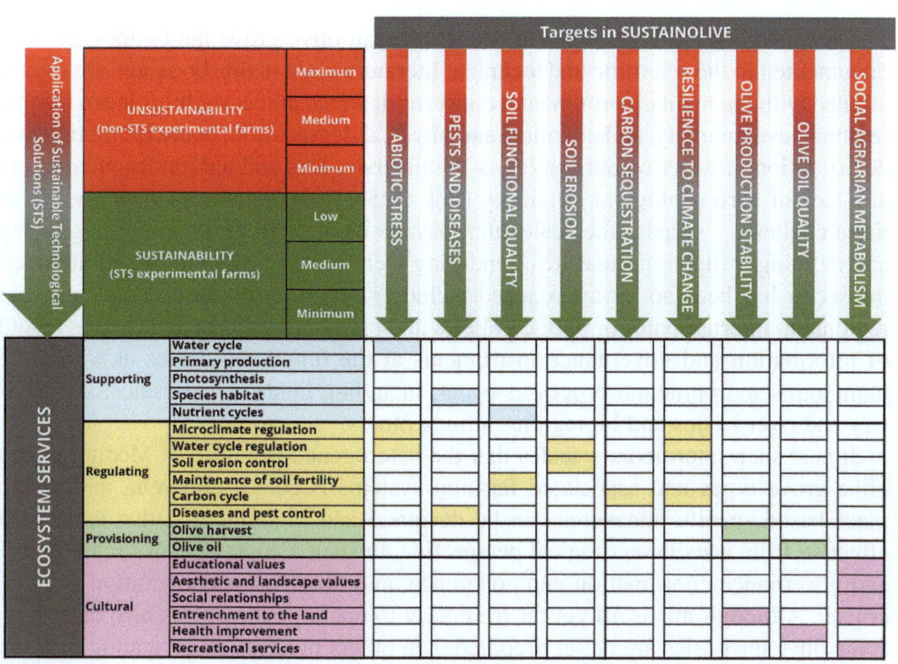

Fig. 60.4 Potential for delivery of multiple ecosystem services by olive farms and groves where sustainable technological solutions are either implemented or not across multiple governance scales and geographical contexts of the Mediterranean. (Source: SUSTAINOLIVE)

60.2 What Are the Key Challenges Arising?

In the following sections, we outline and briefly examine the key ecological (Sect. 60.2.1), agronomical (Sect. 60.2.2), socio-economic (Sect. 60.2.3) and governance (Sect. 60.2.4) challenges arising in the olive landscapes of the Mediterranean to end up by discussing (also briefly) the heterogeneity of those challenges, a point that has repeatedly been pointed out a key underpinning argument for this book.

This heterogeneity is especially relevant not only in terms of describing the challenges ahead, but especially so in our quest for prescribing potential solutions (Sect. 60.3) and forecasting plausible futures (Sect. 60.4) for a landscape system as complex as the one that is the objective of this book.

While the devil is, of course, in the detail, we expect that the outline that we hereby provide may help the reader gain some glimpses about the steps that are urgently required to make sure the ancient landscapes remain alive by adapting to a rapidly changing social-ecological context.

60.2.1 What Are the Key Ecological Challenges Arising That Remain Unresolved?

The main ecological challenges of Mediterranean olive grove landscapes are well documented in the scientific and technical literature cited in this book and are tightly aligned with the main environmental concerns detected at the Mediterranean macro-regional level, namely, soil chemical and physical degradation, overexploitation and deterioration of water resources, (agro-)biodiversity loss and habitat fragmentation, and lack of agro-ecological efficiency (Fig. 60.5). Nonetheless, we now know that these challenges surpass the scales that have been applied thus far for diagnosing and defining mitigation measures, rendering them insufficient. The scales at which the sector has been so far approached include plots to farms for management and regional to international for public policies and private strategies. This leaves a void of information and governance instruments at the functional scales at which the main agro-environmental processes occur, including landscape mosaics, sedimentary and river basins and bioregions, among others.

It is at such intermediate scales that the heterogeneous nature of Mediterranean olive grove types and associated farming systems becomes apparent, and where place-based specific measures can be designed and implemented that can prove effective from a social-ecological perspective. In such a context, cross-scale policy and governance coordination and coherence, and multi-actor cooperation become crucial. Although this is all yet far from easy to operationalize, it is also clear how some olive territories are better prepared than others in terms of mitigating existing governance gaps and limitations.

Thus, in general, it is across the European shores of the Mediterranean where public policies and private strategies are arising that have environmental concerns

Fig. 60.5 Key environmental challenges arising in Mediterranean olive groves, including landscape homogenization (upper left), groundwater overexploitation (upper right), chemical contamination (lower left) and soil erosion (lower right). (Source: SUSTAINOLIVE)

about olive grove landscapes as their main focus. Examples of these include the Olive Oil Regional Sustainability Programs (e.g. In Alentejo and Tuscany) and the latest green reform of Pillar I of the CAP (https://agriculture.ec.europa.eu/common-agricultural-policy/cap-overview/cap-2023-27_en). Nonetheless, these policies and strategic measures still suffer from a lack of scalar focus and coordination, with decisions that seem too detached from decision-makers. Without this connection, many of the environmental challenges in the sector will remain unresolved.

60.2.2 What Are the Key Agronomical Challenges Arising That Remain Unresolved?

As with other agricultural sectors, along the past 50 years, various technological revolutions have impacted the efficiency and productivity and efficiency rates of the olive sector across the Mediterranean. Each of these revolutions has been underpinned by innovations (Mairech et al., 2020). This sequence of agronomical revolutions kicked off with the (genetics-based) Green Revolution of the 1960s and is lately being culminated with Agriculture 4.0. Nonetheless, the impacts and rates of adoption of such agro-technological advances are largely heterogeneous across scales and regional contexts, resulting in a fractal picture where each of these regional contexts presents a patchwork of production models ranging from family-based, traditional and small-scale olive-culture to highly intensive, automatized and globalized farming enterprises. Naturally, multiple mixed models also co-exist with other production factors largely influencing the agronomic practices in place, resulting in distinctions between organic, biodynamic and integrated production models. This leads to multiple opportunities for combination between plantation densities, levels of technological innovation, land management practices and market strategies.

Ultimately, what is becoming more widely accepted is that the somehow bipolar discussion distinguishing between extensive and intensive agriculture in the search for a sustainable agronomic model is no longer valid. To further complicate the picture, sustainable intensification (Silveira et al., 2019) and the land-sparing vs. land-sharing debate (Herzog & Schüepp, 2013) are gradually but steadily becoming part of discussions on sustainable agricultural models for the olive sector. At the same time, it is becoming apparent that none of these seemingly contradictory production and management models can be considered as universally valid to resolve the challenges of sustainability arising in the olive sector. A combination of all of them, in the form of multiple sustainable management practices (Fig. 60.6) will undoubtedly be necessary to address the multiple targets related to the 2030 Sustainable Development Goals (https://www.springer.com/series/15486) that should be expected from olive landscapes. How to combine these in feasible, realistic and operational ways may the prime agronomic challenge for the sector (Fig. 60.6).

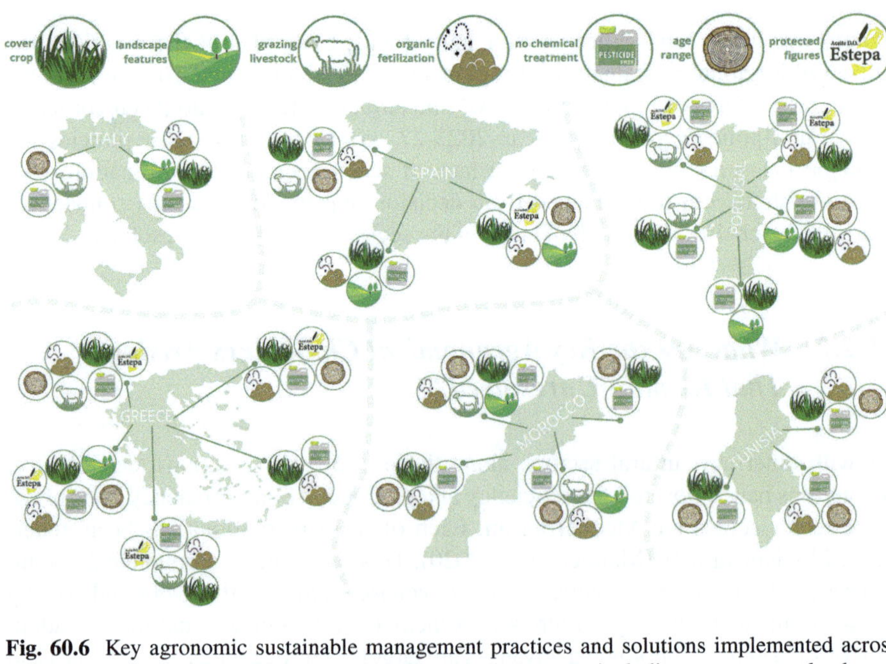

Fig. 60.6 Key agronomic sustainable management practices and solutions implemented across the main six olive oil producing countries of the Mediterranean, including cover crops, landscape features, grazing livestock, organic fertilization, non-chemical treatments, intertwining diverse tree age ranges and PDOs. (Source: SUSTAINOLIVE)

60.2.3 What Are the Key Socio-economic Challenges Arising That Remain Unresolved?

Key socio-economic challenges in the olive sector run parallel to wider discussions for the agricultural sector, specific to the Mediterranean context. Key aspects discussed in this book include the financial sustainability of small-scale agriculture (which still dominates certain marginal and mountain regions), the injustices and local externalities derived from farming globalization and financialization, the capacity and need for the public sector to intervene and regulate over markets and prices that are incrementally affected by climate and environmental changes and societal imbalances, the need for establishing efficient markets that capture multiple ecosystem services beyond agricultural production (Fig. 60.7), and the costs of opportunity to combine organic and standard farming models.

In the Mediterranean context, the presence of diverse production models and related farming systems and landscapes is, again, heterogeneous across scales and regional contexts, and thus the socio-economic challenges faced shift accordingly. This complexity hampers even further the implementation of regional cohesion targets that are at the core of the European Regional and agricultural policies (https://ec.europa.eu/regional_policy/2021-2027_en). This breach of regional cohesion standards is even more dramatic when looked across onto the South and East shores of the Mediterranean basin, where intra- and inter-regional differences in indicators related to economic development and social welfare are large, and in some cases extreme. Consequently, the main challenges arising directly relate to fairness and territorial cohesion, and to how the transformation of olive landscapes could directly contribute to achieving a much-needed sustainability transition. This is a wicked challenge (Duckett et al., 2016), whereby the economic growth model underpinned by intensification that still prevails in olive farming economies is a core key driver of the existing inequalities that current targets for sustainable development and territorial cohesion promise to tackle through alternative socio-economic and governance approaches. Whereas we already know the theory quite well, its operationalization is far from being achieved.

60.2.4 What Are the Key Policy and Governance Challenges Arising That Remain Unresolved?

Policy and governance naturally permeate all other challenges addressed, while at the same time stand at the core of any prospective solutions for enhanced sustainability standards and practices in Mediterranean olive groves. In fact, governance and policy reforms may be the more urgent and necessary ones required for improving sustainability standards in the Mediterranean context, not just in olive landscape or even in agriculture, but more widely for rural land-uses and landscapes.

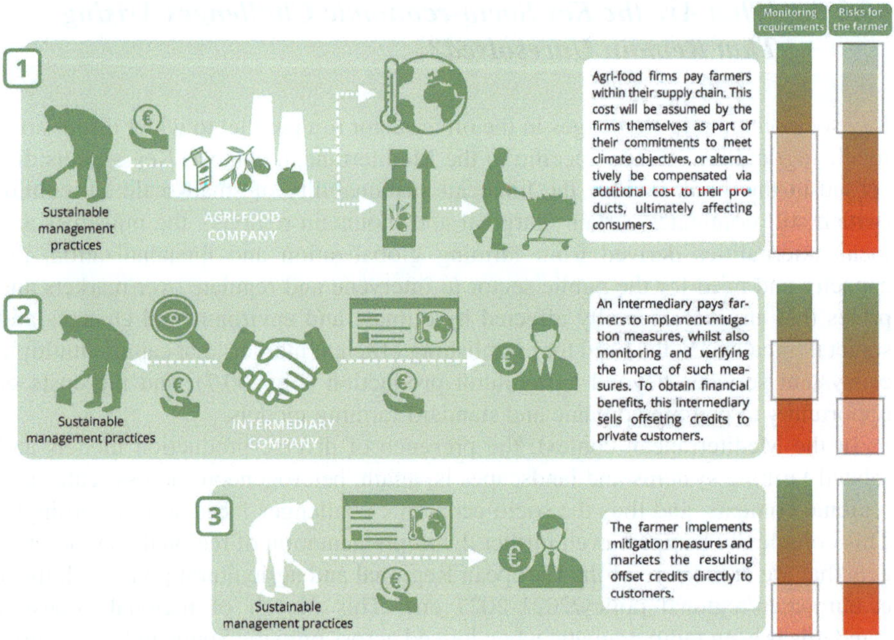

Fig. 60.7 Example of a potential alternative source of income for olive farmers via the establishment of C markets in the olive sector across the Mediterranean. (Source: SUSTAINOLIVE)

Governance and policy inefficiency are key limitations defining the Mediterranean administrative and decision-making framework (Pavić-Rogošić, 2021). Key problems detected include the lack of effective governance mechanisms for cooperation, coordination, coherence, and participation in decision-making across spatial-temporal scales and administrative levels. These problems lead to largely technocratic and top-down administrative procedures, adding up to the more commonly identified ones of corruption, endogamia, clientelism and stagnation. Although there are large differences between (and even within) the various Mediterranean countries and regions, these are problems that generally affect the macro-regional context, and ones that if unresolved, no other challenge, either bio-physical, territorial, or socio-economic, will effectively be tackled.

60.2.5 How Heterogeneous Is the Potential Impact of Such Challenges Across Territorial Contexts and Spatial-Temporal Scales?

One key aspect arising from the analyses of different contexts and scales across the book is the heterogeneity, and to a certain extent even complex and fractal nature, of the differences in agricultural practices, land management models, governance mechanisms, socio-cultural backgrounds and bio-physical conditions shaping the landscape mosaic of Mediterranean olive groves. It is now clearer that this hampers the application of agronomic practices, public policies and, ultimately, sustainable technological solutions that are valid across the whole macro region. This is certainly in conflict with the implementation of common international policies and standards (e.g. EU's CAP, or Regional Cohesion Policy) that are often criticized for imposing common standards across heterogeneous territorial contexts. Although this is far from being effectively resolved, it is certainly a target on the EU agenda. In this context, subsidiarity, decentralization and localism are gaining ground, a fact that is reflected in latest policies of relevance for the olive landscapes of Sothern Europe. On the southern and eastern shores of the Mediterranean, this is a less developed goal, still requiring significant reforms in policy, society, and governance.

Overall, such heterogeneity across levels is evident when analysed through indicators of impact of the sector in aspects as important as farmers' and consumers' health, the environment and the overall combination of all of the former (Fig. 60.8). Based on this, the following conclusions are reached:

(i) Impacts on health and the environment by farming models that apply sustainable practices are significantly reduced.
(ii) However, this effect is not ubiquitous and countries like Greece present impacts that are similar across sustainable and unsustainable farms, affirming our finding that specific farm management practices and their sustainability standards are largely context-dependent and should be explicitly designed for such specific contexts, and also that generically defined sustainable practices that are defined following a top-down approach do not necessarily lead to more sustainable outcomes and standards.
(iii) Data availability is inconsistent and should be improved across all contexts, with the EU setting a stark contrast in quality standards compared to other macro-regional contexts in the Mediterranean. Without ensuring such elevate standards of data, sustainability assessments performed across levels will remain largely of no use.

60.3 What Are the Key Opportunities for Progressing Beyond the State-of-the-Art Towards Increased Sustainability and Resilience?

To compensate for the many challenges identified for Mediterranean olive landscapes, an equal, if not even larger, number of opportunities are detected that will surely allow these landscapes to move towards enhanced standards of sustainability within a reasonable timeframe. This does not detract from the fact that the pathway forward is not an easy nor straightforward one. Surely, this will be a pathway that

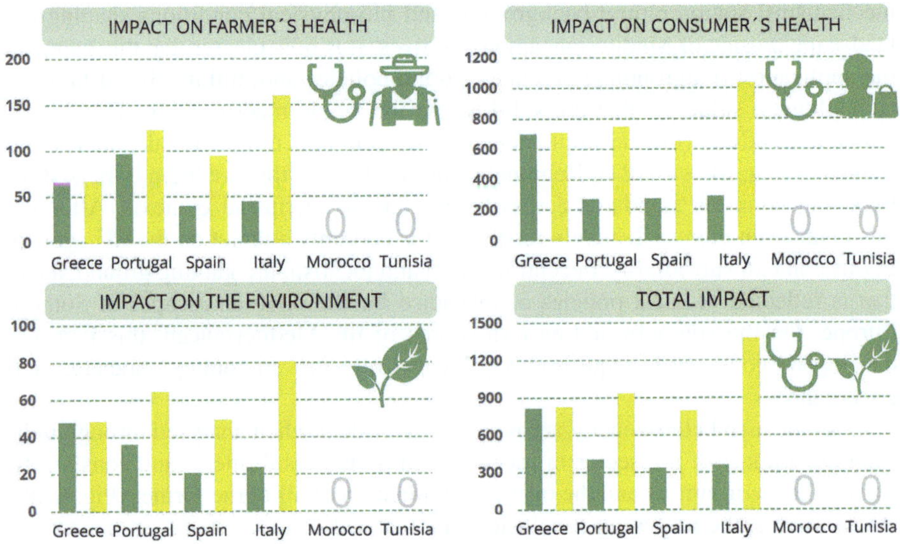

Fig. 60.8 The overall impacts on farmers' and consumers' health, as well as the environment, from the implementation—or lack thereof—of sustainable technological solutions in olive groves across six countries in the Mediterranean macro-region. (Source: SUSTAINOLIVE)

will not be achieved without drastic changes in current policy, administrative, financial and strategic planning frameworks. Furthermore, the toughest change required is the one that demands a deep revolution in the mind-sets of the many actor-networks playing significant roles in the olive sector and beyond.

In the following sections, we outline the main opportunities arising, while acknowledging that the steps and actions required to achieve them are far from linear or easy to reach. By no means do we intend to provide with strict receipts valid for universal success, which is in itself as unrealistic as absurd. Nonetheless, all our suggestions relate to opportunities that apply across different regional, local and national realities, regardless of the social-ecological conditions that turn each of these contexts unique. Ultimately, we expect that the reader will concur with us that, once these opportunities are identified, it is up to the actors, networks and institutions operating in the field to interpret, read, adapt, discuss and, more importantly, act upon them.

Following a subsidiarity approach, these opportunities are considered as guidelines for directing actions on the ground and identifying contingent pathways towards gradual and adaptive change in specific Mediterranean olive landscapes.

60.3.1 Policy and Governance Opportunities: Cohesion, Coordination and Cooperation Across Scales and Sectors Under the Umbrella of a Greener CAP

These three Cs for multi-level effective governance—coordination, coherence, and cooperation (Muñoz-Rojas et al., 2022)—are the indisputable requisites for governance frameworks that can deliver sustainability on the ground. Again, these are generic principles that can nonetheless be transformed into actions when they are adopted by regional and local champions and institutions fostering change in their territorial contexts. Naturally, the adoption of these guiding principles requires a radical shift in governance structures, but one that will undoubtedly face more difficulties in certain contexts.

Generally speaking, the more evolved the governance framework towards ascending in Arnstein's ladder of public participation (Arnstein, 1969), the easier that it will be to implement these principles. This leaves the Mediterranean North in a better position for governance innovation and effective implementation than its Southern or Eastern counterparts. Certainly, recent changes in the CAP and EU territorial policies towards greener targets and subsidiarity principles, underpinned by the EU's Green Deal (https://commission.europa.eu/strategy-and-policy/priorities-2019-2024/european-green-deal_en), are triggers in this sense. Whether these will materialize in specific actions depends much upon undersigning states and regions, and not only, with the democratic and participatory fabric of civic societies playing a key role in this transition. Main barriers towards implementation include the power grabbing that is gradually being enacted across all shores of the Mediterranean by multi-national companies and investment groups of the olive oil sector.

Whether this will remain an uncontested trend largely depends on the societies that inhabit these territories. In addition, it is important to highlight the relevance of administrative effectiveness and effectivity, an additional requisite that is often low in the Mediterranean context. An illustration of this problem is found in Andalucía (Spain), host to the largest extension of olive groves and the equally largest olive oil producing region worldwide. With an olive grove law (Junta de Andalucía, 2011) and an olive regional plan (Junta de Anadlucía, 2015), it is still a region where governance deficiencies are still acute, resulting in multiple environmental, social and economic problems, reinforcing to our call for a radical change that reaches well beyond the approval of new regulations and administrative instruments.

60.3.2 Agronomic Opportunities: How to Secure Sufficient Space and Scope for Management Models Ranging from Biodynamic and Small Scale Towards Intensified and Precision Farming?

This is an opportunity that links directly to the land sparing versus land haring debate and also one where the targets for organic agriculture lately set by the EU (https://agriculture.ec.europa.eu/farming/organic-farming/organic-action-plan_en) are relevant. In general, it has become clear from this book that no single land management model or landscape configuration will solely resolve the multi-functional targets expected from the Mediterranean olive landscapes where ecological, sociocultural, agronomic and ecological targets are simultaneously pursued. The adequacy of one or other model to specific territories will directly depend on contingent territorial conditions that are ultimately defined by the availability of water for irrigation and of funding for mechanization, soil productivity and capability, specialized knowledge and workforce in the area, market accessibility and connectivity, presence of networks and connections and ultimately of public and private support. Again, this is a fractal process whereby any single region or territory may show a capacity for hosting various of these models in simultaneous. Andalucía (Spain) and Alentejo (Portugal) are examples of this case.

The ability to establish a heterogeneous, diverse, and thus also resilient, olive sector will only be achieved where and when certain key concepts are effectively translated into practice (Fig. 60.9), including especially those that are directly linked to sustainability, namely, sustainable intensification (Silveira et al., 2019) and Agriculture 4.0 (Oleiro Araújo et al., 2021) for intensive olive-culture, and internationalisation and market expansion for traditional and extensive olive landscapes.

60.3.3 Opportunities for Territorial Integration Through Landscape Multi-functionality and Ecosystem Services

The co-existence of multiple olive production models and associated farming systems and landscape mosaics is a competitive advantage of Mediterranean olive territories and one of its greatest assets. Nonetheless, it is one that is rapidly deteriorating, either due to landscape homogenization in the most fertile and water-accessible areas, where rapid and at times uncontrolled intensification dominates, or through abandonment in marginal farming regions. The main challenge is thereby creating effective incentives to foster the resilience of marginal olive groves and at the same time to promote the sustainable intensification of intensive ones. Although multiple instrumental options exist, these remain largely underdeveloped in the Mediterranean context. Such instruments include Payment for Ecosystem Services Schemes (Morgado et al., 2022), C Markets (Galán-Martín et al., 2022) and

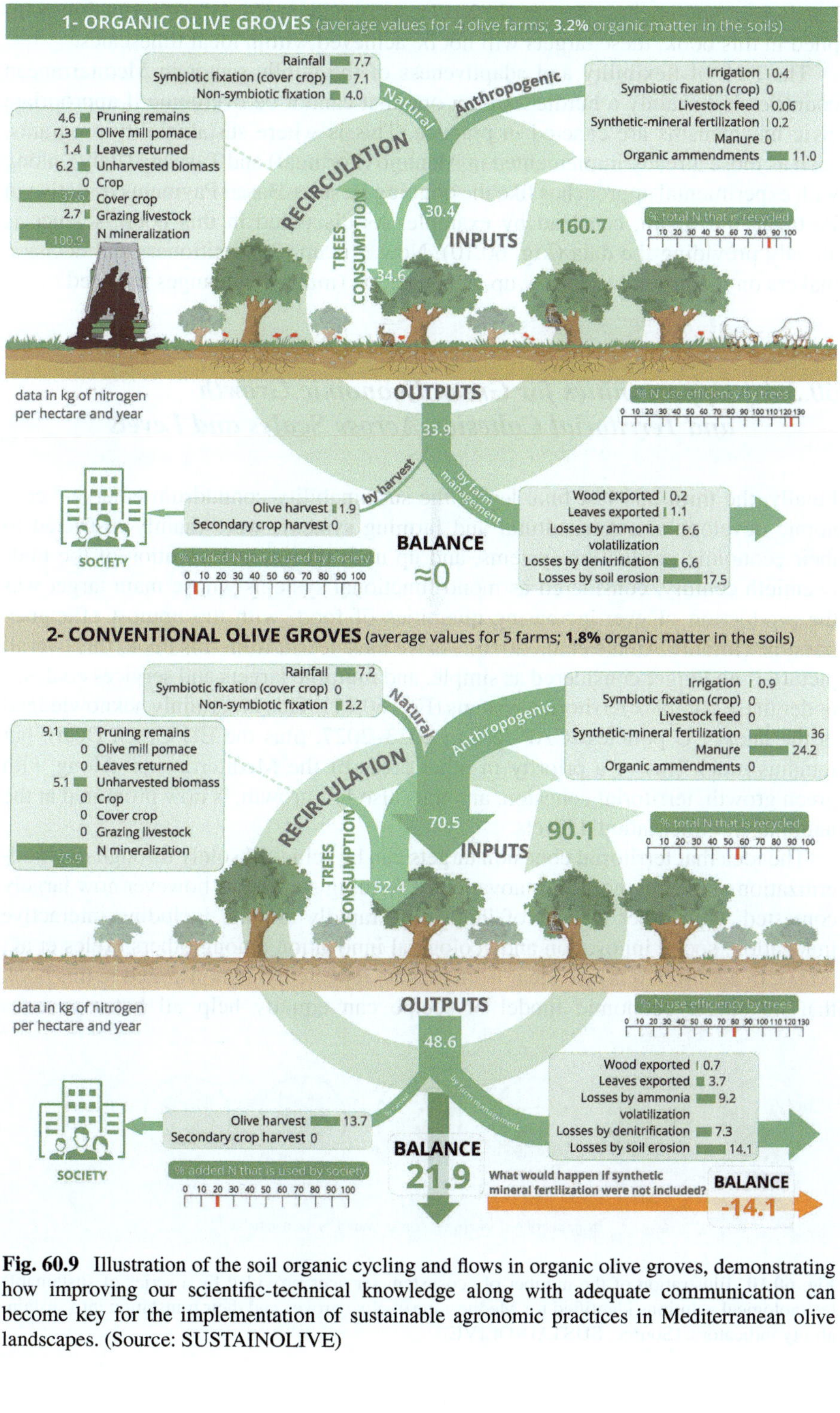

Fig. 60.9 Illustration of the soil organic cycling and flows in organic olive groves, demonstrating how improving our scientific-technical knowledge along with adequate communication can become key for the implementation of sustainable agronomic practices in Mediterranean olive landscapes. (Source: SUSTAINOLIVE)

Landscape Approaches (Arts et al., 2017). As is evident from the information compiled in this book, these targets will not be achieved within ideal timescales.

The lack of flexibility and adaptiveness of the public sector in Mediterranean countries is certainly a hurdle, but not one that cannot be overcome if appropriate civic mechanisms are enacted in practice. This is where sustainability programs, such as those already implemented in Alentejo (Portugal) and Tuscany (Italy), along with experimental approaches like the ongoing Results-Based Payments initiative in Portugal and Spain, can lead by example. As discussed in this book, science is already providing the data (Fig. 60.10). Now it is up to practitioners and decision makers on the ground to follow up and enact the (multiple) changes required.

60.3.4 Opportunities for Green Economic Growth and Territorial Cohesion Across Scales and Levels

Finally, the third and last final leg of the sustainability conundrum emerges: economic development. Agricultural and farming systems were mainly restricted to their economic production systems, and up until the green revolution of the mid-twentieth century, considered as mono-functional systems whose main target was the production of ever-increasing quantities of food, with the upmost efficiency possible (Infante-Amate et al., 2016). As we have learnt from this book, this overall picture is no longer considered as simple, and multiple targets and services coalesce under these complex territorial systems (Fig. 60.11). This is certainly acknowledged in the latest EU policies (CAP reform 2023–2027, plus the EU Green Deal), but remains much less of a priority in other areas of the Mediterranean. Along with green growth, territorial cohesion, and thus also fair growth, is now promoted at the national and international levels.

The idea that territorial cohesion targets can be achieved solely through the modernization and technological innovation of farming systems is however now largely contested, with other aspects of innovation rapidly arising, including interactive innovation, social innovation and ecological innovation among others (Teles et al., 2023). In addition, it is also clear from the fact repeatedly demonstrated in this book that no single economic model or recipe can equally help all heterogeneous

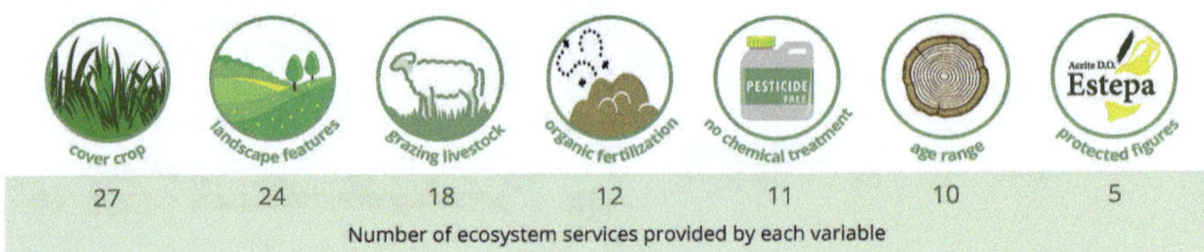

Fig. 60.10 Illustration of the number of ecosystem services provided by a series of sustainable technological solutions identified for Mediterranean olive groves and their translation into sustainability indicators. (Source: SUSTAINOLIVE)

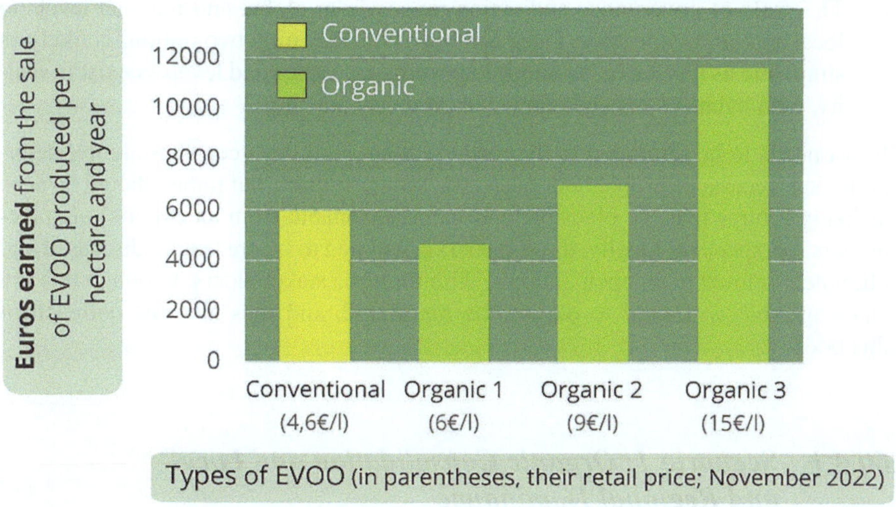

Fig. 60.11 Figures showing the profitability (measured in €) of diverse olive groves, showing how organic ones that follow the green economy principles can become more economically efficient for the farmer in the mid to longer term. (Source: SUSTAINOLIVE)

territories in the Mediterranean move towards sustainable territorial development. Thus, the ball is now in the hands of Mediterranean societies to decide how they want to make best use of the multiple opportunities that olive landscapes entail when they are considered as assets for economic sectors as diverse as tourism, nature and cultural heritage conservation, agro-industrial innovation and technological development and others. Some of these territories (e.g. Jaen's olive grove landscapes recent candidacy as UNESCO world heritage) have already set some steps in this direction. It remains to be seen whether others will follow.

60.4 And So Where to Now? Plausible Scenarios of Future Change

To conclude the book, we very briefly describe a set of scenarios that arise from the combination of two main factors:

(i) Production and innovation models are favoured, ranging from technological innovation to what has been termed by certain academic as retroinnovations (Zagata et al., 2020). Generally speaking, technological innovations are directly linked to intensive production models, whereas retroinnovation relate to extensive and family-based farming systems. As we have seen in this book, organic and even biodynamic production modes are transversal to this classification and can equally be found under extensive to super-intensive densities, despite if the former still largely dominate.

(ii) The scale of governance and action ranges from global and national levels to local and landscape ones. Once again, considering only two options is likely as simplistic as it is naïve, as multiple governance scales and levels coexist in reality, with balances between them continuously shifting.

It just needs to be reminded to the reader at this point that scenarios are not deterministic forecasts of probable or even less, secure futures, but rather should be seen as likely futures that are plausible and seem reasonable from an experts' or stakeholders' perspective. Ideally, these scenarios will ned to be presented, discussed and ultimately refined in an open dialogue with multiple stakeholders, although this is a challenge that will need to outlive the timeframe, and thus also the outreach of this book.

60.4.1 Scenario 1: Organic Consolidation and Local and Regional Dominance

This is a scenario where organic production dominates the markets, overcoming the multiple financial and regulatory barriers encountered to compete with conventional production. It is a scenario that would seem likely unachievable without a more effective development of sustainable intensification and that would be largely dependent on an increase in the standard of living of local populations and of environmental standards of local farming systems, across the Mediterranean region.

It is therefore an unlikely scenario in this regional context, although one that is likely to become more prevalent in N European regions, where a big proportion of the demand for olive oil is increasingly located. Nonetheless, it is also a scenario that presents an inherent heterogeneity at the local and regional scales, with territories where alternatives for conventional and family-based extensive farming systems are the only option becoming more likely organic. This includes mountain and marginal regions, especially where local olive varieties persist and where a strong network of local cooperatives and associations are in place.

60.4.2 Scenario 2: A Second Green Revolution Through Precision Agriculture and Technological Down-Scaling

This is a scenario that is already prevalent in those agricultural Mediterranean regions with access to ample water resources for irrigation, connectivity to market nodes and actor-networks, financial resources for implementing mechanization and technological innovations, and productive soils. In addition, favourable land prices for investment are also a positive factor for this scenario.

However, in most of these cases, the model under implementation is far too similar to the original green revolution aims and strategies, with food safety, globalized markets and deregulation as key drivers. To advance towards a second green revolution, a shift from globalization towards glocalization (Anderson, 2022) would be required, along with further advancing with the Agriculture 4.0 agenda. This would allow such a revolutionary trend to gain ecological sustainability and efficiency in the use of resources. This scenario is likely to prevail in regions where the adequate conditions required for implementation of intensive farming are maintained.

60.4.3 Scenario 3: Landscape Integration: Towards a Heterogeneous Mosaic of Place-Based Governance and Management Models

This scenario is the one that can be found in Mediterranean regions where more productive and better-connected agricultural landscapes co-exist with marginal ones. This includes Andalucía (Spain), Alentejo (Portugal) and Puglia (Italy), where olive groves have been long a key landscape element, and also others in Morocco and Tunisia where intensification and financialization of agriculture are recently raising. However, it is one scenario that seems unlikely in traditional olive-growing regions with natural limitations for intensification, including some in Central and Northern Italy and Portugal, and many of the Greek regions.

The current trajectory of the areas where this scenario prevails is one towards the bipolarization of landscapes with marginal farms too frequently being abandoned. The main obstacle is an effective governance strategy that facilitates the implementation of alternative approaches for these declining areas.

Under these future scenarios, however, Integrative Landscape Strategies defined at the regional to local levels (Reed et al., 2015) would secure that coherence, coordination and cooperation are effectively achieved, and that a healthy and sustainable rural landscape mosaic is preserved.

60.4.4 Scenario 4: Globalized Competitiveness with Winning and Losing Regions and Management Models

This last scenario basically consists of a picture on which international competition in the global markets dominates the fate of different olive regions and landscapes. To a certain extent, this could somehow be considered as a business-as-usual scenario, where government and social intervention is limited, and where olive landscapes are incrementally lost in areas where traditional agriculture dominates. Under this scenario, quantity prevails over quality, and food security, price and market competition are the main drivers shaping agricultural landscape trajectories. This is a scenario with intense and largely negative impacts on landscapes and local societies, and one that we are likely heading toward unless more creative and innovative alternatives are envisioned and enacted.

Last, this is a scenario where increasing competition from other olive-producing regions across the world (South Africa, Australia, California, and potentially also in other BRICS including Brazil and China) can outcompete the Mediterranean macro-region, which would, in a worst-case-scenario, end up loosing the majority of its olive landscapes, with those remaining becoming a niche for self-production and marginal ecological products targeting gourmet and financially competitive international demand, which will in all likelihood remain marginal at a global scale.

60.4.5 Key Lessons and Recommendations for Potential Pathways for Progress Forward

In summary, returning to original questions (see Chap. 1), the reader will by now have obtained as many new questions as resolved answers. All of these new questions can be distilled into one seemingly simple question.

If one key lesson can be drawn from the extensive range of information in this book, it is that no single direction could possibly exist for the heterogeneous, rich and diverse range of contingent situations that one can find in Mediterranean rural landscapes, and thus that no single (nor simple) recipe will resolve the many challenges encountered. However, one key positive message prevails: enhancing sustainable standards through SMART strategies and indicators, while remaining adaptive and flexible, may be the best way forward for the future survival of these valuable and widely cherished landscapes that we have inherited over centuries, and even millennia. Whether we can ultimately achieve this complex and ambitious targets depends largely on each of us.

Acknowledgements This work was funded by the SUSTAINOLIVE project (Grant n° 1822; PRIMA EU Programme) for all co-authors, and by FCT-Portugal (UIDB/05183/2020—funding for José Muñoz-Rojas, as member of MED-Universidade de Évora, and LA/P/0121/2020 as member of the CHANGE- Associated Lab-Universidade de Évora).

References

Alaoui, A., Barão, L., Ferreira, C. S. S., & Hessel, R. (2022). An overview of sustainability assessment frameworks in agriculture. *Land, 11*, 537. https://doi.org/10.3390/land11040537

Anderson, K. (2022). Agriculture in a more uncertain global trade environment. *Agricultural Economics, 53*(4), 563–579. https://doi.org/10.1111/agec.12726

Arnstein, S. R. (1969). A ladder of citizen participation. *Journal of the American Institute of Planners, 35*(4), 216–224. https://doi.org/10.1080/01944366908977225

Arts, B., Buizer, M., Horlings, L., Ingram, V., Van Oosten, C., & Opdam, P. (2017). Landscape approaches: A state-of-the-art review. *Annual Review of Environment and Resources, 42*, 439–463. https://doi.org/10.1146/annurev-environ-102016-060932

Brady, M. V., Hristov, J., Wilhelmsson, F., & Hedlund, K. (2019). Roadmap for valuing soil ecosystem services to inform multi-level decision-making in agriculture. *Sustainability, 2019*(11), 5285. https://doi.org/10.3390/su11195285

De Vries, G., Terwel, B. W., Ellemers, N., & Daamen, D. D. (2015). Sustainability or profitability? How communicated motives for environmental policy affect public perceptions of corporate greenwashing. *Corporate Social Responsibility and Environmental Management, 22*, 142–154. https://doi.org/10.1002/csr.1327

Duckett, D., Feliciano, D., Martin-Ortega, J., et al. (2016) Tackling wicked environmental problems: The discourse and its influence on praxis in Scotland. *Landscape and Urban Planning, 154*, 44–56. ISSN 0169-2046. https://doi.org/10.1016/j.landurbplan.2016.03.015.

Forman, R., & Wilson, E. O. (1995). *Land mosaics: The ecology of landscapes and regions*. Cambridge University Press. https://doi.org/10.1017/9781107050327

Galán-Martín, A., Contreras, M. M., Romero, I., Ruiz, E., Bueno-Rodríguez, S., Eliche-Quesada, D., & Castro-Galiano, E. (2022). The potential role of olive groves to deliver carbon dioxide removal in a carbon-neutral Europe: Opportunities and challenges. *Renewable and Sustainable Energy Reviews, 165*(C). https://doi.org/10.1016/j.rser.2022.112609

Herzog, F., & Schüepp, C. (2013). Are land sparing and land sharing real alternatives for European agricultural landscapes. *Aspects of Applied Biology*, 109–116. https://doi.org/10.7892/BORIS.47493

Infante-Amate, J., Aguilera, E., Villa, I., Torremocha, E., Cid, A., Guzmán, G., & González de Molina, M. (2016). The making of olive landscapes in the south of Spain. A history of expansion and intensification. In M. Agnoletti & F. Emanueli (Eds.), *Biocultural diversity in Europe* (pp. 157–159). Springer. https://doi.org/10.1007/978-3-319-26315-1_8

Junta de Andalucía. (2011). Ley 5/2011, de 6 de octubre, del olivar de Andalucía. «BOJA» núm. 205, de 19/10/2011, «BOE» núm. 268, de 07/11/2011. https://www.boe.es/eli/esan/l/2011/10/06/5/con

Junta de Andalucía. (2015). Plan Director del Olivar. Decreto 103/2015, de 10 de marzo, por el que se aprueba el Plan Director del Olivar (BOJA de 19 de marzo de 2015). https://www.juntadeandalucia.es/organismos/transparencia/planificacion-evaluacion-estadistica/planes/detalle/59239.html

Mairech, H., Lopez-Bernal, A., Moriondo, M., Dibari, C., Regni, L., Proietti, P., Villaloobos, F. J., & Testi, L. (2020). Is new olive farming sustainable? A spatial comparison of productive and environmental performances between traditional and new olive orchards with the model OliveCan. *Agricultural Systems, 181*, 102816. https://doi.org/10.1016/j.agsy.2020.102816

Morgado, R., Flores Ribeiro, P., Lima Santos, J., Rego, F., Beja, P., & Moreira, F. (2022). Drivers of irrigated olive grove expansion in Mediterranean landscapes and associated biodiversity impacts. *Landscape and Urban Planning, 225*(2022), 104429. https://doi.org/10.1016/j.landurbplan.2022.104429

Muñoz-Rojas, J., Guzmán, J. R., & Loupa-Ramos, J. (2022) 11. The complexity of public policies in Iberian Montados and Dehesas. 16 pp. Acosta Naranjo, R. Pinto-Correia, T., Guimarães, H., Moreno, G. (2019) *Governance of Mediterranean silvo-pastoral systems: lessons from the Iberian Dehesas and Montados*. Expected Publication: January 2021. Routledge. ISBN: 9780367463571 https://doi.org/10.4324/9781003028437.

Oleiro Araújo, S., Peres, R. S., Barata, J., Lidon, F., & Ramalho, J. C. (2021). Characterising the agriculture 4.0 landscape – Emerging trends, challenges and opportunities. *Agronomy, 11*, 667. https://doi.org/10.3390/agronomy11040667

Pavić-Rogošić, L. (2021). *Long-term vision for rural areas: for stronger, connected, resilient, prosperous EU rural areas*. https://ec.europa.eu/commission/presscorner/detail/en/IP_21_3162

Reed, J., Deakin, L., & Sunderland, T. (2015). What are 'integrated landscape approaches' and how effectively have they been implemented in the tropics: A systematic map protocol. *Environmental Evidence, 4*, 2. https://doi.org/10.1186/2047-2382-4-2

Scown, M. W., Craig, R. K., Allen, C., Gunderson, L., Angeler, D. G., Garcia, J. H., & Garmestani, A. S. (2023). Towards a global sustainable development agenda built on social-ecological resilience (July 4, 2023). *Global Sustainability, 6*(e8), 1–14. https://doi.org/10.1017/sus.2023.8

Silveira, A., Muñoz-Rojas, J., Pinto-Correira, T., Guimarães, M. H., Ferrão, J., & Schmidt, L. (2019). The sustainability of agricultural intensification in the early 21st century: Insights from the olive oil production in Alentejo (southern Portugal). In A. Delicado, N. Domingos, & L. de Sousa (Eds.), *Changing societies: Legacies and challenges* (The diverse worlds of sustainability) (Vol. 3, pp. 247–275). Imprensa de Ciências Sociais. ISBN: 978-972-671-505-4.

Teles, F., Rodrigues, C., Ramos, F., & Botello, A. (2023). Territorial innovation in less developed regions. Governance, technologies, and sustainability (Palgrave Studies in Sub-National Governance). Palgrave Macmillan. https://doi.org/10.1007/978-3-031-20577-4. ISBN: 9783031205767.

Zagata, L., Sutherland, L.-A., Hrabák, J., & Lostak, M. (2020). Mobilising the past: Towards conceptualisation of retro-innovation. *Sociologia Ruralis, 60*, 639–660. https://doi.org/10.1111/soru.12310

Index

J. Muñoz-Rojas, R. García-Ruiz (eds.), *The Olive Landscapes of the Mediterranean*, Landscape Series 36,
https://doi.org/10.1007/978-3-031-57956-1

MIX
Papier aus verantwortungsvollen Quellen
Paper from responsible sources
FSC® C105338

If you have any concerns about our products,
you can contact us on
ProductSafety@springernature.com

In case Publisher is established outside the EU,
the EU authorized representative is:
Springer Nature Customer Service Center GmbH
Europaplatz 3, 69115 Heidelberg, Germany

Printed by Libri Plureos GmbH
in Hamburg, Germany